BIOTECHNOLOGY AND BIOPHARMACEUTICALS

TRANSFORMING PROTEINS AND GENES INTO DRUGS

RODNEY J.Y. HO, Ph.D.

University of Washington School of Pharmacy
Department of Pharmaceutics
Seattle, Washington

MILO GIBALDI, Ph.D.

University of Washington School of Pharmacy
Department of Pharmaceutics
Seattle, Washington

A John Wiley & Sons, Inc., Publication

For general information on our other products and services please contact our Customer Care Department within the U.S. at 877-762-2974, outside the U.S. at 317-572-3993 or fax 317-572-4002.

Wiley also publishes its books in a variety of electronic formats. Some content that appears in print, however, may not be available in electronic format.

Library of Congress Cataloging-in-Publication Data:
Ho, Rodney J.Y.
    Biotechnology and biopharmaceuticals : transforming proteins
and genes into drugs / Rodney J.Y. Ho, Milo Gibaldi.
        p. ; cm.
    Includes bibliographical references and index.
    ISBN 0-471-20690-3 (pbk. : alk. paper)
    1. Pharmaceutical biotechnology
    [DNLM: 1. Biopharmaceutics.   2. Chemistry, Pharmaceutical.
3. Drug Design. QV 38 H678b 2003]   I. Gibaldi, Milo.   II. Title.
    RS380 .H6 2003
    615′.19—dc21   2002153462

Printed in the United States of America

10 9 8 7 6 5 4 3 2 1

This book is dedicated to Florence and Ann

Lily, Beatrice, and Martin

CONTENTS

FOREWORD

In the past 10–15 years, dramatic changes have occurred in the processes used by the pharmaceutical industry to discover and develop new drugs, in the chemical nature of new drugs emerging from this industry, and in the way that these new drugs are used in medicine. These changes require that the scientists who discover and develop these drugs and the healthcare professionals who use these drugs to treat patients need to have not only depth but also breadth in their knowledge and experience. This book will provide these scientists and healthcare professionals, as well as students who aspire to work in these fields, with a valuable new resource of information about the processes being utilized to discover and develop the drugs of the future and the ways that these drugs can be most effectively used in patients.

The discovery and development of new drugs was once largely conducted within "fully integrated pharmaceutical companies." These companies tended to conduct these activities in a fragmented way that involved little or no communication between the scientists from the different disciplines involved in the overall process. Today, because of pressures on the industry to be more cost- and time-efficient, the discovery and development of new drugs within these companies occur in a highly integrated manner using new technologies and involving multidisciplinary project teams. Individual members of these project teams obviously need to have a depth of knowledge and experience in their respective areas of expertise. However, they also need to have sufficient breadth in their knowledge to allow them to communicate effectively with team members from other scientific disciplines.

The processes used to conduct drug discovery have changed significantly because of the introduction of new chemical (*e.g.*, combinatorial chemistry), biological (*e.g.*, genomics, proteomics), computational (*e.g.*, bioinformatics), and mechanical (*e.g.*, robotics) technologies. These technologies have not only revolutionized the processes used to discover drugs, they have also changed the nature of drug candidates emerging from the pharmaceutical industry. The industry is no longer dependent on small molecules as the only source of drug candidates. It now has the tools to discover macromolecules (*e.g.*, proteins, genes) as drug candidates. Other technologies (*e.g.*, recombinant DNA, fermentation) have also become available that permit the production of these proteins and genes in quantities and at costs that make their use in humans commercially viable. Appropriate regulatory standards that allow the FDA to approve the use of these new "biologicals" in humans have also been developed.

While some of the biotechnology companies that originally developed these revolutionary technologies have now become "fully integrated pharmaceutical companies," others have remained "boutique drug discovery companies," relying on strategic alliances with larger companies for the preclinical and clinical development of their drug candidates. These strategic alliances require that the chemists and biologists in these "boutique drug discovery companies" have not only depth in their area of expertise but also breadth in their knowledge about drug development and clinical practice. Without this breadth of knowledge, these scientists would most likely be producing "ligands" that bind specifically and with high affinity to a potential therapeutic target *in vitro*. However, these "high affinity ligands" would probably not have the "drug-like" characteristics needed to ultimately develop them clinically.

Drug development has also changed in the past 10–15 years because of the emergence of new drug delivery technologies. Like the "boutique drug discovery companies," these "boutique drug delivery companies" are dependent on strategic alliances with "fully integrated pharmaceutical companies" for the advancement of their technologies to clinical practice. Therefore, scientists and engineers in these "boutique drug delivery companies" need not only depth in their area of expertise but also a breadth of knowledge that includes aspects of drug discovery and clinical practice. Without this breadth of knowledge, these individuals would likely be developing drug delivery technologies that would never find utility in clinical practice.

The recent development of new biological (*e.g.*, genomics, proteomics) and analytical (*e.g.*, biosensors) techniques has also significantly impacted the ways in which proteins and genes, as well as small molecules, are being used in medicine. These new biological techniques permit the "profiling" of patients for their propensity to benefit therapeutically from treatment with a certain drug and/or their propensity to suffer undesirable side effects. Analytical techniques have also afforded healthcare professionals the opportunity to maximize the therapeutic effects and minimize the side effects of a drug by monitoring actual drug exposure or a surrogate marker. Many of the new drugs forthcoming from the pharmaceutical industry need to be used in conjunction with these biological and/or analytical techniques to produce optimal therapeutic effects. Thus, it is important that healthcare professionals also have a depth of knowledge in order to make optimal use of these new drugs to treat their patients.

It should be clear from this brief description of the changes that have occurred in drug discovery and development and in the practice of medicine in the past 10–15 years that scientists, healthcare professionals, and students need to have breadth as well as depth in their knowledge. While there have been books written or edited that describe the individual steps involved in the discovery and development of proteins and genes as drugs and their utilization in specific areas of clinical medicine, there has been no single, comprehensive source of information in these areas. The authors of this book should be congratulated for making available to scientists, healthcare professionals, and students all of this information in a single volume. The chapters are written in a manner that is understandable to non-experts in the field. By utilizing this book as a source of new information to increase their breadth of knowledge, scientists currently in the industry will become more effective contributors to the drug discovery and development efforts of their companies. Graduate students who gain the knowledge contained in this book will become more competitive for jobs in this industry, and healthcare professionals who gain the knowledge in this book will become more effective in using proteins and genes to treat their patients.

Ronald T. Borchardt
Solon E. Summerfield
Distinguished Professor of
Pharmaceutical Chemistry
University of Kansas

PREFACE

The list of biotechnology-based therapeutics has rapidly grown since 1982 when FDA approved the first recombinant biotech drug, insulin, for human use. The annual sales of several protein therapeutics have surpassed the billion dollar mark. With the completion of the primary DNA map for the human genome and the progress made in high-throughput technology for drug discovery, we are about to experience an explosive, never-before-seen growth in the development of therapeutic modalities. Biotechnology, the application of biologic molecules to mimic biologic processes, will play a central role in the discovery and development of protein- and gene-based drugs. While there are books discussing various aspects of biotechnology, there is no single, comprehensive source of information available for pharmaceutical scientists and health care professionals.

The general principles of pharmacology and pharmaceutics have helped us to understand the relationship between clinical outcomes and the physical-chemical properties of traditional drugs and dosage forms. These principles, however, often fail to accommodate the products of biotechnology. The therapeutic application of protein-based drugs over the past 10 years has provided a much fuller understanding of the intricacies and mechanisms of protein disposition and pharmacologic actions. To fully appreciate the complexities of these macromolecules and their biologic effects, one must understand the fundamental differences in drug design, dosage formulation, and time course of distribution, to target tissues between protein-based drugs and small organic molecules. New strategies have been developed to deliver protein-based drugs and more will follow.

We undertook the considerable task of creating this book because we believe established pharmaceutical scientists, as well as those in training, need to understand the principles underlying the discovery, development, and application of drugs of the future. An understanding and appreciation of these principles by health scientists, physicians, pharmacists, and other health care providers should allow informed decisions to improve the pharmaceutical care of patients. The ability to integrate this knowledge in the clinical setting is essential, especially for the clinical pharmacologist and pharmacist. We also believe that a single source of comprehensive information about biotechnology is needed to serve the interests of a large population of professionals. This book considers biotechnology products from the following perspectives: (1) the integration of pharmacology and biotechnology with medical sciences, (2) the unique aspects of the applications of biologics or macromolecules as therapeutic agents, (3) the impact of biotechnology on modern medicine, and

(4) the prospect of applying cutting-edge biotechnology and drug systems in shaping the future of medical practice.

■ ACKNOWLEDGMENTS

The effort to bring together research, development, regulation, and application of biotechnology products in an integrated book format would not be possible without the encouragement of co-author Milo Gibaldi. The long hours spent in discussion and editing of text related to complex biologic processes and methodologies used in creating biotechnology products, not only enriched this book but led to the inclusion of insights that were not available to either of us alone. Also we are indebted to the works of many dedicated scientists and clinicians who continually strive to improve medical therapies. Without their foresight and creative research, this book would not have been possible. The stimulating discussions and support of our colleagues in Seattle, Washington, and the pharmaceutical industry limited our excursions into theoretical fantasies and kept us on a realistic path.

As this book is an outgrowth of my 10 plus years of experience teaching pharmaceutical biotechnology to undergraduate, professional and graduate students, I am indebted to those who have helped develop and teach the course. These individuals include René H. Levy, Kenneth E. Thummel, Carl Scandella, Howard Mandelsohm, Andy Stergachis, Nancy Haigwood, Mark Kay, Jacqueline S. Gardner, Thomas K. Hazlet, Douglas Black, Paul Beaumier, and Gordon Duncan.

The preparation of the actual book required the dedicated effort of many individuals. I would like to acknowledge the effort of Julie Fu, who spent hours in assembling the data in appendixes; Colleen McCallum who helped to ensure timely receipt of manuscripts from cooperative, but very busy authors; and Kate Connolly and Tami Daley for editing the manuscript to improve readability. The committed efforts of the editorial and production staffs at John Wiley and Sons are also greatly appreciated.

Rodney J.Y. Ho

ORGANIZATION OF THE BOOK

Part I focuses on the process of taking a biologic macromolecule such as a protein found naturally in minute quantities from identification of structure and function to a therapeutic agent that can be delivered safely and effectively in a pharmaceutical dosage form to patients for a specific therapeutic indication. With the advancement of recombinant DNA technology and the enhancements in automation efficiency and computing power, we have more drug targets than we can exploit to produce, recombinantly or synthetically, drugs or pharmaceuticals that provide health benefits. Therefore it is increasingly important for drug industry decision makers, pharmaceutical scientists, and physicians to acquire the knowledge that has been gained from the experience of transforming biologic macromolecules into drugs. The first part of the book highlights some of the key differences between the discovery and development of small molecules and biopharmaceuticals.

For readers familiar with biotechnology, biopharmaceutics, and the drug development process, and for those that focus on the application of biopharmaceuticals, Part II provides a brief overview of each class of macromolecule with respect to physiological role and clinical application. Additional detail for each FDA approved, recombinantly derived biopharmaceutical, and several other interesting therapeutic proteins, for each category of macromolecule

is provided in monographs. These monographs are organized as follows: (1) general description, (2) indications, (3) dosage form, route of administration, and dosage, (4) pharmacology and pharmaceutics (i.e., clinical pharmacology, pharmacokinetics, disposition, and drug interactions), (5) therapeutic response, (6) role in therapy, and (7) other clinical applications. Readers seeking pharmacokinetic information and additional details on molecular characteristics of biopharmaceuticals are directed to the appendices.

Part III focuses on the future, on advances that will enhance our ability to develop new and already identified macromolecules into safe and effective biopharmaceuticals. Using drug delivery strategies to optimize drug distribution profiles, including drug targeting by means of physical-chemical and physiological approaches, as well as optimization of molecular properties by sequence modification and molecular redesign are key strategies needed to improve safety and efficacy and to increase the limited bioavailability of macromolecules that often requires systemic or regional administration. This part also describes gene and cell therapies, strategies that are needed when traditional drug therapy is not suitable or effective.

For potent drugs that produce severe toxicity in a small population of patients but are otherwise safe and effective for the majority of patients, laboratory-based genetic tests are in development to identify the at-risk population. As our understanding of the relationship between phamacological responses and genetic variations grows, it is important to learn how pharmacogenetic and other factors may allow pharmacists and physicians to consider the cost and benefits of individualized drug selection and dosage regimens. With automation of analytical, robotic, and computational techniques, the role of proteomics and genomics in accelerating drug discovery and predicting pharmacophores and perhaps pharmacokinetic properties may allow scientist to reduce to a minimum the number of candidate molecules needed to be synthesized or cloned. Some of these efforts have allowed the chemical synthesis of active-site mimics that are similar to classic drugs.

The book concludes with a chapter on how these scientific advances are being integrated by large and small biotechnology-driven and traditional drug companies to accelerate the drug discovery and development processes. The pharmaceutical industry is nearly universally incorporating biotech strategies as tools to accelerate the development of drug from concepts into products.

CONTRIBUTORS

Sandra Blethen, M.D.
Genentech, Inc.
San Francisco, California

Ronald T. Borchardt, Ph.D.
Department of Pharmaceutical Chemistry
University of Kansas
Lawrence, Kansas

Ernest C. Borden, M.D.
Center for Cancer Drug Discovery and
Development
Taussig Cancer Center and Lerner
Research Institute
Cleveland, Ohio

Mamta Chawla-Sarkar, Ph.D.
Center for Cancer Drug Discovery and
Development
Taussig Cancer Center and Cleveland
Clinic Foundation
Cleveland, Ohio

Gary L. Davis, M.D.
Section of Hepatobiliary Diseases
University of Florida College of Medicine
Gainesville, Florida

Shiu-Lok Hu, Ph.D.
Departments of Pharmaceutics and
Microbiology
Regional Primate Research Center
University of Washington
Seattle, Washington

Paul Masci, D.O.
Center for Cancer Drug Discovery and
Development
Taussig Cancer Center and Cleveland
Clinic Foundation
Cleveland, Ohio

Sean M. Sullivan, Ph.D.
Department of Pharmaceutics
College of Pharmacy
University of Florida
Gainesville, Florida

Part I

TRANSFORMING PROTEINS AND GENES INTO THERAPEUTICS

1

INTRODUCTION TO BIOPHARMACEUTICALS

Since the discovery in 1800 that the human body is composed of cells and proteins that are susceptible to but can also fight off pathogenic microbes, the perceived battlefield has challenged our imagination to develop biopharmaceuticals—biologically based therapeutic products. While biotechnology today is seen as the cutting edge of biological sciences, the use of biological entities as therapeutic agents has been extant since the early 1800s.

The considerable history of biopharmaceuticals notwithstanding, the refinement of our abilities to produce recombinant macro-molecules and monoclonal antibodies is a recent achievement and, as depicted in Figure 1.1, is the result of an exponential growth in the knowledge of biological processes and engineering advancements. Protein- and antibody-based therapies that require milligrams of materials for pharmaceutical use cannot be consistently, safely, and cost-effectively applied without these recent biotechnology milestones that include the basic principles of recombinant DNA and cell and fermentation technologies.

Today biotechnology-based pharmaceutical companies, while still relatively small,

Biotechnology and Biopharmaceuticals, by Rodney J. Y. Ho and Milo Gibaldi
ISBN 0-471-20690-3   Copyright © 2003 by John Wiley & Sons, Inc.

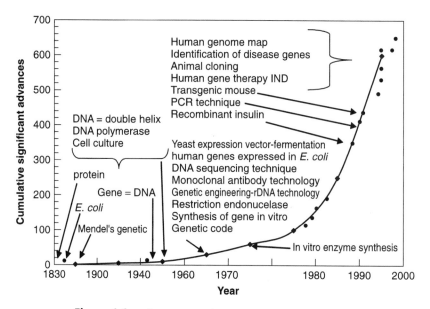

Figure 1.1. Milestones in pharmaceutical biotechnology.

realize hundred of millions of dollars or more in sales and their productivity, measured by revenue generated per employee, is comparable to that of large pharmaceutical companies (Table 1.1). A survey of top-selling biotechnology drugs identified four products (three different proteins) that have consistently achieved about one billion dollars in sales for at least four consecutive years (Table 1.2).

Biotechnology companies spend more than 20% of revenues in research and development (R&D). This is well above the 6% to 18% of revenue invested in R&D expenditure by large pharmaceutical companies (Table 1.1). The difference is due, at least in part, to the high cost of biotechnology research and perhaps to the intellectual climate at the relatively young, upstart biotechnology-based companies compared with that at the more established, large pharmaceutical companies. Given the rate at which new biotechnology-based pharmaceuticals are reaching the market and, in many cases, their therapeutic importance, it is essential

for health care professionals and pharmaceutical researchers to understand the application of biotechnology to transform biological processes and entities into pharmaceuticals and other therapeutic modalities.

As we enter the information age, the availability of vast amounts of biological and genetic data, coupled with exponential growth in computing power, mean that the rate of developing novel biopharmaceuticals is no longer limited by the ability to identify targets and clone macromolecules. We have more targets than we can develop into pharmaceuticals. Therefore, drug candidate selection must be refined with experience gained in using macromolecules as therapeutic agents. We must focus on drug candidates that will be safe and effective and also have desirable clinical pharmacokinetic profiles. Compounds that exhibit high affinity binding to receptor targets but fail to penetrate target tissue, or do not persist there long enough to produce desirable biologic responses, cannot be considered for development as biopharmaceuticals.

■TABLE 1.1. Comparison of revenue, productivity, and market share between selective list of established biotechnology-based and large pharmaceutical companies

Company	Employees	Revenue Total (in millions)	Revenue Per Employee (in thousand)	R&D Expenditure Total (in millions)	R&D Expenditure % Revenue
Biotechnology companies					
Amgen	6,400	$3,433	$536	$823	24
Genentech	3,880	$1,414	$364	$331	23
Biogen	1,350	$825	$611	$221	27
Genzyme	3,860	$777	$201	$156	20
Chiron	3,110	$684	$220	$254	37
Immunex	1,170	$559	$478	$127	23
Large pharmaceutical companies					
Merck & Co	62,300	$32,762	$526	$2,119	6
Johnson & Johnson	97,800	$27,439	$281	$2,600	9
Bristol Myers Squibb	54,500	$20,199	$371	$1,843	9
Pfizer	51,000	$16,269	$319	$2,776	17
Glaxo Wellcome	55,273	$13,566	$245	$2,049	15
Eli Lilly & Co	31,300	$9,819	$314	$1,784	18

Source: Based on year 2000 data reported by Ernst & Young LLP, extracted from annual report of the listed companies.

In this chapter, we will define biotechnology from the perspective of pharmaceuticals and follow this with a historical overview of pharmaceutical biotechnology and a discussion of how macromolecules are named and used as therapeutic agents.

■ 1.1. BIOTECHNOLOGY VERSUS PHARMACEUTICAL BIOTECHNOLOGY

Biotechnology, like beauty, is in the eye of the beholder—a last hope for a patient with Alzheimer's disease; an anathema to an environmentalist. Seeking a broad consensus, we define biotechnology as an integrated application of scientific and technical understanding of a biologic process or molecule to develop a useful product. Biologic processes of interest include cellular activities such as protein synthesis, DNA replication, transcription (to RNA), protein processing, receptor-ligand interactions at cell surfaces, fermentation of bacteria, yeast, and mammalian cells.

Our broad definition of biotechnology includes beer and wine fermentation technology to produce distinctive beverages with commercial advantages, the isolation or passaging of viruses to produce nonvirulent variants for use as vaccines, genetic manipulation to coax bacteria to express metabolic enzymes that transform petroleum products into water-soluble forms for environmental cleanup after a spill, and the development of a recombinant, disease-resistant fruit or vegetable crop with prolonged freshness. Very often biotechnology means commercialization of biological sciences by integrating discoveries from many disciplines, including

■TABLE 1.2. Sales of top 12 selling biotech drugs

Drug	Marketer	Developer	Indication	1999 ($millions)	1998 ($millions)	1997 ($millions)	1996 ($millions)
Epogen	Amgen	Amgen	Anemia	1760	1380	1161	1150
Procrit	Amgen	Ortho Biotech	Anemia	1505	1363	1169	995
Neupogen	Amgen	Amgen	Neutropenia	1260	1120	1056	1017
Humulin	Genentech	Eli Lilly	Diabetes	1088	959	936	884
Engerix-B	Genentech	SmithKline Beeham	Hepatitis B	540	887	584	568
Intron A	Biogen	Shering-Plough	Hairy cell leukemia, Kaposi's sarcoma, and Hepatitis C	650	718	598	524
Kogenate	Bayer Biological	Bayer Biological	Hemophilia A	403	429	335[a]	NA
Genotropin	Genentech	Pharmacia	Growth failure	461	395	349	391
Avonex	Biogen	Biogen	Multiple sclerosis	621	395	158[b]	NA
Betaseron	Chiron/ Berlex	Berlex/Schering AG	Multiple sclerosis	545[a]	369[a]	387	353
ReoPro	Centocor	Eli Lilly, Centocor	Cardiac ischemic complications	447	365	NA	NA
Ceredase/ Cerezyme	Genezyme	Genzyme	Gaucher's disease	479[c]	441[c]	333	265
Total				*9759*	*8821*	*7066*	*6147*

Source: Data from Ernst & Young LLP for the last four years except where indicated.
[a]Data from Schering AG annual report 1999 and expressed in Euro-dollars.
[b]Data from Biogen annual report 1998.
[c]Data from Genzyme annual report 2000, p. 52.

microbiology, biochemistry, genetics, and bioengineering.

Currently biotechnology is an integral component of many industries, in addition to pharmaceutical companies. Our treatment of biotechnology and biopharmaceuticals will focus only on the application of biotechnology with respect to biologic entities and biologic processes to develop pharmaceutical products.

■ 1.2. HISTORICAL PERSPECTIVE OF PHARMACEUTICAL BIOTECHNOLOGY

The application of biological processes to develop useful products is as old as Mendel's pea experiment, which he con-

ducted in 1866 (Figure 1.1). As a result of his avocation, Mendel developed the principles of heredity, and thereby laid the basis of modern genetics. While the listing of biotechnology in the dictionary did not occur until 1979, the fermentation technology we use today to produce recombinant proteins was first used in World War I to ferment corn starch (with the help of *Clostridium acetobutylicum*) and produce acetone for manufacturing explosives. Fermentation technology took on even greater importance with the development after World War II to produce antibiotics.

An enhanced understanding of protein structure, a detailed elucidation of cell replication and protein synthesis, and the isolation of DNA replication enzymes, including restriction enzymes and polymerases, led to the rapid development of recombinant DNA technology, permitting cloning and expression of proteins and peptides that had eluded efforts at isolation and harvest only a few years earlier. At about the same time, in 1975, scientists developed monoclonal antibody technology, which allowed large-scale preparation of purified, highly specific antibodies with mono-specific binding sites (spanning 6–10 amino acids in length) in a reproducible manner. This technology also allowed the generation and use of monoclonal antibodies as a tool to characterize and purify proteins that would selectively bind to respective antibodies with high specificity. These tools for preparation and characterization of recombinant products are essential for developing macromolecules into therapeutic products.

The biotechnology milestones are graphically presented in Figure 1.1. While each event listed in Figure 1.1 may not by itself have permitted the rapid application of biotechnology to drug development, in the aggregate they have led to the development of dozens of pharmaceutical products that could not have been realized without the availability of these technologies. The advances in technologies make the process possible or accelerate it, or just simply make the product cost-effective and much safer than the same material extracted from tissue sources. For example, the development of a yeast plasmid vector permitted mass production of hepatitis B surface antigen for vaccine development and the economical manufacture of recombinant human insulin.

Almost all of the biopharmaceuticals available today, other than vaccines, are proteins or peptides. Of considerable importance among this array of products are monoclonal antibodies. These "magic bullets" became a reality with

the introduction of Orthoclone (muromonab) in 1986. At present, monoclonal antibodies are the fastest growing category of biopharmaceuticals approved for therapeutic use. Our ability to identify novel, potentially therapeutic proteins and peptides, like monoclonal antibodies, has advanced at such a rate that we are now limited by the human effort and resources needed to develop and demonstrate the clinical efficacy and safety of these candidates.

■ 1.3. NOT ALL PROTEIN DRUGS AND VACCINES OF THE SAME NAME ARE IDENTICAL

A slight chemical modification in a small molecule can dramatically change biologic activity. For example, the addition of methyl groups at position 1, 3, and 7 of the natural substance xanthine produces the widely consumed compound caffeine; the addition of methyl groups at position 1 and 3 or 3 and 7 produces the bronchodilator theophylline or a related compound, theobromine. By the same token, the addition of a hydroxy-methyl group to the anti-herpes simplex drug acyclovir results in ganciclovir, which has anti-cytomegalovirus activity.

Xanthine

Caffeine

Theophylline

Theobromine

Acyclovir

Gancyclovir

Fexofenadine

Terfenadine

One can find many more examples where a subtle modification in a side chain leads to a new drug that produces a drastically different therapeutic or toxicological outcome. This is clearly illustrated by the nonsedating antihistamine terfenadine (Seldane), which produces cardiotoxicity when given with certain drugs that inhibits its metabolism. This product is no longer marketed. It has been replaced by its safer but no less effective carboxylic oxidative metabolite fexofenadine (Allegra).

Biopharmaceuticals based on natural proteins and peptides are often called by the same name as the biologic natural material despite differences in one or more amino-acid residues. For example, insulin, which regulates blood glucose and is used clinically to treat type 1 diabetes and some cases of type 2 diabetes, has several variants that are approved for human use. Insulin contains two polypeptides, A and B chains (Figure 1.2), that are linked together by two disulfide bridges to assume a biologically active conformation. Compared with human insulin, insulin extracted from beef tissue exhibits threonine → alanine and isoleucine → valine substitutions at posi-

tion 8 and 10 of the insulin A chain, while insulin extracted from pork tissue contains a threonine → alanine substitution at position 30 of the insulin B chain (Table 1.3). Yet both pork and beef insulins have been used successfully to treat diabetes. While trade names may differ, all the insulins, including those that are modified to produce more desirable pharmacokinetic and disposition profiles such as insulin lispro, insulin glargine, and insulin aspart, are still known as and called insulins by physician, and researchers alike. All of

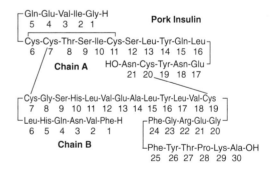

Figure 1.2. Schematic presentation of insulin A and B chains and their amino-acid sequences based on insulin extracted from pork.

■TABLE 1.3. Sequence variation among insulins available for human administration

Insulin Amino Acid, aa Position	A Chain			B Chain			
	8	10	21	20 & 21 Insertion	28	29	30
Insulin—beef	Ala	Val	Asn	None	Pro	Lys	Ala
Insulin—pork	Thr	Ile	Asn	None	Pro	Lys	Ala
Insulin—human	Thr	Ile	Asn	None	Pro	Lys	Thr
Insulin—lispro (*Humalog*)	Thr	Ile	Asn	None	Lys	Pro	Thr
Insulin—asp (*Novolog*)	Thr	Ile	Asn	None	Asp	Lys	Thr
Insulin—glargine	Thr	Ile	Gly	Arg & Arg	Pro	Lys	Thr

■TABLE 1.4. Comparison of recombinant hepatitis B vaccines dose recommendations

Group of Patients	*Recombivax HB*		*Engerix-B*	
	Dose (µg)	Volume (ml)	Dose (µg)	Volume (ml)
Infants (HB⁻), children	2.5	0.25	10	0.5
Infants (HB⁺)	5	0.5	10	0.5
Children, 11–19	5	0.5	20	1.0
Adult ≥20	10	1.0	20	1.0
Dialysis/immune-compromised patients	40	1.0ᵃ	40	2.0

ᵃSpecial formulation.

these variants of insulin are used for the same treatment indication—to control blood glucose—and are efficacious as long as the dose and dosing frequency are determined on a product-by-product basis.

The same name is also being used for some vaccines that differ in potency. As shown in Table 1.4, the two approved vaccines against hepatitis B, *Recombivax HB* and *Engerix-B*, both known as Hepatitis B Vaccine (Recombinant), when used as directed, are therapeutically equivalent in terms of their ability to induce antibodies that protect vaccinated individuals from hepatitis B virus infection. However, the dose and volume required to produce a satisfactory immune response are different for each product and for each age group.

Despite these differences, physicians use the insulins interchangeably (Table 1.4). The difference in dose between the two vaccines may be due to sequence and production variations of the recombinant proteins that constitute the preparation.

REFERENCES

Moses, V., and R.E. Cape, *Biotechnology: the science and business.* 1991, Harwood Academic Publishers, Chur Switzerland.

Ernst & Young LLP, *Convergence: the biotechnology industry report*, 2000, NY.

Kohler, G., and G. Milstein, Nature, **256**: 495–7.

A timeline of biotechnology. Accessed August 2001, http://www.bio.org/timeline/timeline.html.

2

COMPARATIVE DRUG DEVELOPMENT OF PROTEINS AND GENES VERSUS SMALL MOLECULES

The US government allocates billons of dollars every year through the Department of Health and Human Services (HHS), the Public Health Service (PHS), and the National Institutes of Health (NIH) to support research aimed at understanding disease mechanisms and finding therapeutic solutions. The drug industry also spends several billon dollars each year in pharmaceutical research to transform novel chemicals and biologics into medicinals that are safe and effective for human use. To ensure that a new drug candidate is proved effective and safe before it reaches the US market, all of the information that has been gathered about the new drug candidate is rigorously scrutinized by the Food and Drug Administration and by independent experts called upon to assist the agency in its evaluation. Whether a drug candidate is a protein, nucleic acid, or small molecule, manufactured in the United States or abroad, the same regulations apply.

In general, traditional drugs (i.e., small organic molecules) are reviewed through the new drug application (NDA) process, while macromolecules (biopharmaceuticals), including proteins, peptides, genes, and recombinant products are reviewed under the biologic license application

Biotechnology and Biopharmaceuticals, by Rodney J. Y. Ho and Milo Gibaldi
ISBN 0-471-20690-3   Copyright © 2003 by John Wiley & Sons, Inc.

(BLA) process [1]. The Center for Drug Evaluation and Research (CDER) and the Center for Biologics Evaluation and Research (CBER), within the FDA, have been responsible for the review of NDAs and BLAs, respectively. A tripartite organization, known as the International Conference on Harmonization (ICH), which includes the FDA and its counterpart agencies in the European Union and Japan, was established in 1991 with the goal of increasing the efficiency of the drug approval process for the international marketplace. The guidelines developed by the ICH, however, have yet to be fully accepted in most nations. While we recognize the importance of international licensing of drugs and biologics, we will focus on the drug development and approval process elaborated by the FDA for new molecular entities (NMEs), emphasizing the differences between traditional drugs and biopharmaceuticals.

■ 2.1. TRANSFORMING NEW MOLECULAR ENTITIES INTO DRUGS

To gain FDA approval or license for marketing, a pharmaceutical product must be shown to be safe and effective for its proposed or intended use. The drug company or sponsor must also provide evidence to show that the processes and control procedures used for synthesis, manufacture, and packaging are independently validated to ensure that the pharmaceutical product meets established standards of quality. The overall effort from the inception of a new molecular entity and the establishment of analytical, scale-up, and quality control procedures, to the collection of safety and efficacy data for consideration by the FDA as part of an NDA or BLA, is called the drug development process.

While the journey from the discovery of a drug candidate to final marketing

approval can be lengthy, the steps are well defined for both traditional drugs and biopharmaceuticals. Figure 2.1 is a schematic presentation showing that after chemical or biological synthesis and purification, the NME is rigorously and systematically evaluated in preclinical studies that include the characterization of its physical-chemical and biologic properties, the determination of its toxicity in laboratory animals and cell systems (gross toxicity, hematological and end-organ effects, carcinogenicity, mutagenicity, and teratogenicity), the establishment of its distribution and pharmacokinetic profile in laboratory animals, and the evaluation of its stability and other characteristics important to the preparation of a final dosage form.

Before animal testing, analytical and biological assay capabilities must be developed while the NME is being scaled up to produce a sufficient quantity with acceptable purity for use in subsequent studies. Drug standards and analytical methods for evaluating the bulk NME and the final product, as well as the tentative chemical, physical, and biologic specifications, are then established. In parallel, formulation studies are initiated to produce a stable dosage form that will provide a suitable platform for delivery of the NME in a reproducible manner.

Often a series of related NMEs with similar chemical structures are evaluated systematically to optimize specificity and affinity to the target molecule or cellular receptor and minimize potential drug–drug interactions by selection of molecules with a low affinity for key drug metabolizing enzymes. Those candidates with desirable *in vitro* profiles are evaluated further in laboratory animals for pharmaceutical properties that include good target bioavailability (adequate absorption after administration and an ease of distribution or transport to the site of action) and a degree of resistance to metabolism and excretion to ensure that the drug molecule

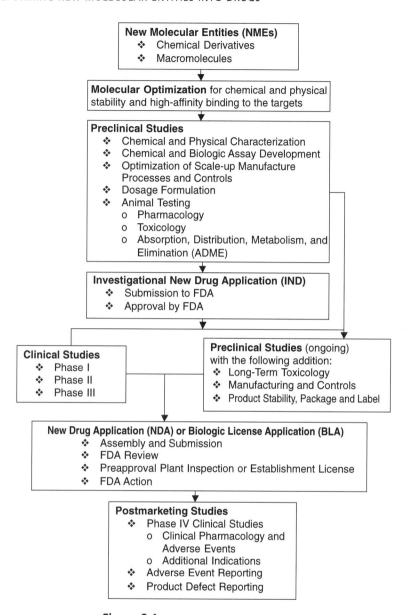

Figure 2.1. Drug development process.

persists for a sufficient time at the site of action; these are often called absorption–distribution–metabolism-excretion (ADME) studies. The studies designed to characterize pharmaceutical properties are important in drug development to predict whether a pharmacologically active compound will be useful as a therapeutic agent. Preclinical studies must also establish the range of doses needed to produce pharmacological and toxicological responses in commonly used animal models and characterize those responses at the organ, cellular, and molecular levels.

A drug candidate that completes preclinical testing and maintains promise is then considered for evaluation in human subjects. The first step in this process is the

Investigational New Drug (IND) application. The IND petition requires full disclosure of where and how the NME is manufactured and controlled for quality and stability. It also contains proposed analytical methods, pharmacology and toxicology data, and evidence of desired effects in disease models. The application lists proposed clinical investigators and contains complete human subject protocols. Under current regulations the FDA must provide a written response to the sponsor within 30 days after submission. The lack of a timely response is tacit approval for the sponsor to proceed to the clinic.

The NME can now be administered to humans. The first step in clinical evaluation is one or more phase I studies designed to assess the drug's safety and pharmacokinetic profile. Phase I studies usually involve a small number of healthy volunteers who are closely monitored after receiving escalating doses of the drug candidate. Phase I studies of drugs for cancer or HIV infection must be carried out in patients, not in healthy subjects. Ordinarily, until more information is available, the minimum dose to induce side effects is stipulated as the upper dose limit for subsequent administration to human subjects.

The efficacy of new drug candidates with acceptable safety profiles—based on phase I findings—is evaluated in phase II studies in patients with medical disorders consistent with the sponsor's proposed indication(s). Mindful of getting its drug candidate to market as quickly as possible, drug sponsors usually select patients with well-defined medical conditions and use surrogate end points to evaluate response. A sponsor may seek a very narrow indication to expedite evaluation, with the hope of expanding that indication for a larger patient population based on studies carried out after initial approval and marketing. Phase II clinical studies enroll several hundred subjects and are designed not only to assess efficacy but also to detect acute and short-term side effects and risks asso-

ciated with the investigational drug. Some phase II studies are well controlled, while others are open.

A new drug candidate that successfully completes phase II studies, safely demonstrating benefit in a well-defined patient population, is then required to undergo additional clinical testing (phase III) in designated patients, usually under rigorously controlled conditions, to collect sufficient data to evaluate the drug's effectiveness and safety with regard to an overall assessment of the benefit–risk relationship. Phase III clinical trials often involve multiple medical centers and enroll several hundred to several thousand subjects. The exact number of subjects required for phase II and phase III studies depends primarily on statistical considerations that take into account the expected differences in therapeutic end points for active drug and placebo, the patient population, and the expected variation in biotherapeutic assessment.

When phase III studies near conclusion, nonclinical and clinical data are assembled for submission to the FDA in an NDA or BLA. The staff at the FDA and members of FDA advisory committees, which are composed of experts in a therapeutic area, review the application and attempt to ensure that the benefits of the new drug outweigh its risks.

The transformation of a new molecular entity into a marketed drug product takes many years, some say 12 years on average, and costs about $200 to $350 million. Twenty years ago, following the thalidomide tragedy, a fearful FDA stalled, and the time taken by the agency to review an NDA began, in some cases, to match the time required for the drug development process and to seriously erode patent life. The pharmaceutical industry pressed Congress for relief. With the passage of the Prescription Drug User Fee Act (PDUFA) by the US Congress in 1992, which permits the FDA to charge sponsors who submit NDA and BLA documents, with the stipulation that the additional revenue be used to hire more reviewers, the approval time (from submission of the NDA or BLA to approval date) has been reduced to about one-half. The FDA's current goal is to review 90% of submitted BLA and NDA applications within 12 months. The agency will strive to reduce this target to 10 months. Surveys suggest that the agency has been meeting its goals and is on track for meeting future goals. Some industry observers, however, fear that a spate of withdrawals of marketed prescription drugs for reasons of safety may make the FDA more cautious and increase approval time.

While the efficiency of the NDA and BLA review process has greatly improved, dramatically decreasing the time for regulatory review, the same cannot be said for overall drug development time, from identification of a new drug candidate to NDA or BLA submission, which has resisted the anticipated streamlining effects of new technologies and efficiencies.

■ 2.2. DIFFERENCES BETWEEN DEVELOPMENT OF BIOTECHNOLOGY PRODUCTS OF MACROMOLECULES AND CHEMICAL PRODUCTS

Unlike traditional drugs, which can be chemically synthesized and purified to homogeneity, biological macromolecules are often derived from living sources—human and animal tissues and cells and microorganisms. Therefore most biologic macromolecules are not easily characterized and refined to a high degree of purity. In the absence of well-developed standards for these products, CBER has developed guidelines to ensure that macromolecules approved for human use are manufactured under conditions that ensure batch-to-batch uniformity and that no infectious agents are inadvertently introduced into a product.

Before 1996 the FDA approved biopharmaceuticals through a dual process that required a Product License Application (PLA) and an Establishment License Application (ELA) (Table 2.1). According to PHS rules, a "Responsible Head" had to be appointed to ensure the safety of the product manufactured by the sponsor. Because the FDA's philosophy was "the process is the product," the manufacturing facility was tightly regulated, requiring the pharmaceutical company to obtain an establishment license (through the ELA process) for manufacturing the biologic macromolecule before initiating any human clinical trials. In addition, up until 1996, any modification in manufacturing process and controls required preapproval by CBER prior to implementation.

In 1996, about 10 years after the introduction of the first recombinant DNA product for human use, the FDA modified and streamlined the approval process for biotechnology products considered to be "well characterized." These modifications, in essence, established the direction of how biologic macromolecules are researched and developed today in biotechnology-based and traditional pharmaceutical companies [2]. "Well-characterized" biotechnology products include (1) synthetic peptides consisting of fewer than 20 amino acids, (2) monoclonal antibodies and derivatives, and (3) recombinant DNA-derived products. Anticipating future developments, the FDA is also prepared to consider DNA plasmid products as well-characterized when the first medicinal in this class is submitted for approval. CBER now approves well-characterized biopharmaceuticals under the BLA process [3].

The BLA strategy has significantly improved the process for approving biopharmaceuticals through reductions in paperwork and financial burden, a particular boon for start-up biotechnology-based drug companies. Drug sponsors are no longer required to manufacture the product in-house; contract manufacturers with established resources are now permitted to make the biopharmaceutical for both human testing and marketing.

■TABLE 2.1. Regulatory requirements for development of chemicals and macromolecules into drugs

	Chemicals	Macromolecules	
		1996 and Earlier	Current[a]
Application and approval process	NDA (New Drug Application)	PLA (Product License Application) ELA (Establishment License Application)	BLA (Biologic License Application)
FDA center responsible for review	Center for Drug Evaluation and Research (CDER)	Center for Biologics Evaluation and Research (CBER)	
Compliance responsibility	No "Responsible Head" requirement	Designated "Responsible Head"	No "Responsible Head" requirement[b]
Manufacturing requirement	Any company can submit NDA without the requirement to manufacture the drug in house	PLA can only be submitted by manufacturers of significant steps in process. More than one manufacturer can be licensed for a given product	The applicant may or may not own the manufacturing facilities. No requirement for contract facility to obtain a separate license
Quality control and assurance	Final product must be made under current good manufacturing process (cGMP)— emphasis placed on the final bulk product	"The product is the manufacturing process" —cGMPs from seed stock or first step onward evaluated with equal scrutiny	Regulated under analytical procedures and method validation, chemistry, manufacturing and control (CMC) documentation
Lot release requirement	Not controlled by CDER	Every lot manufactured for marketing controlled by CBER	No longer required, but must be made available upon request by CDER
Labeling requirement	No promotional material approval required after FDA approval for product marketing	All promotional materials must be pre-approved	Promotional materials are submitted to the FDA for information
Manufacturing process modification	Document manufacturing changes in annual reports	Approval required for every manufacturing change before implementation	Submit manufacturing changes and validation documents to FDA 30 days prior to product distribution[c]

[a]Elimination of ELA and PLA requirement, 14 May 1996 and replaced by BLA Federal Register, 63, 147, 40858–40871 [9].

[b]15 October, 1996 , FDA final rule published through the President's reinventing government (REGO) initiative [3].

[c]Manufacture process modification, effective 7 October 1997 [10].

Other modifications to the regulatory requirements promulgated in 1996 and summarized in Table 2.1 are final rules on labeling, lot release, and manufacturing process modifications. With these changes in place, the development process of the well-characterized biopharmaceutical is approximately in line with that of traditional drugs [2]. In fact FDA has now committed to shift the regulatory responsibility of these recombinant products under CDER, and allowing its biologics division (CBER) to focus on vaccines, gene and cell therapies, and other blood products (Box 2.1.).

BOX 2.1. FDA PLANS TO SHIFT REVIEW AND APPROVAL OF WELL-CHARACTERIZED AND RECOMBINANT BIOTECHNOLOGY PRODUCTS FROM ITS BIOLOGICS DIVISION TO ITS DRUG DIVISION

On September 6, 2002, the Food and Drug Administration (FDA) announced its intention to consolidate review of most new biotechnology products under its drug division, the Center for Drug Evaluation and Research (CDER), rather than its biologics division, the Center for Biologics Research and Evaluation (CBER). While the plan's details are being developed by a working group, chaired by Senior Associate Commissioner Murray M. Lumpkin, it is likely that review and approval of well-characterized and recombinant biotechnology products, including therapeutic proteins and antibodies, will be the under the purview of CDER instead of CBER, possibly as early as 2003. This move has been anticipated as some recombinant proteins and therapeutic antibodies and related derivatives are now being reviewed by CDER. Recent examples of biopharmaceuticals reviewed and approved by CDER include *Intron A* (interferon- α2b); human insulins such as *Humalog, Humulin*, and *Iletin I*; and antibodies such as *Mylotarg*.

Under the revised system, CBER will continue to review blood products and vaccines. These product areas are CBER's strengths and the basis for creation of the biologics division. CBER will also be responsible for evaluating gene therapy and tissue transplantation products as the development of these novel entities comes to fruition.

While consolidation plans have been in the works for several years, many believe that the process was accelerated by Congress, particularly those members and staffers involved in the investigation of the FDA's handling of the approval application for *Erbitux*. *Erbitux* is a controversial cancer drug sponsored by ImClone Systems, a firm riddled with scandal. In a written communication to the FDA, the House Committee on Energy and Commerce stated, "There appears to be a different and better approach to the expedited review of (other) cancer drugs in CDER" than in CBER, the division that reviewed *Erbitux*. (For additional details regarding Imclone Systems and *Erbitux*, see Chapter 4, Section 4.3.2.)

It is anticipated that CDER will shorten the review time for biotechnology products, and make the review process for biologics and traditional chemical drugs more parallel. Media reports say that the biotechnology industry welcomes this move with the hope that it will expedite new drug approvals.

Currently, most recombinant proteins such as erythropoietin, colony-stimulating factors, and interferons are reviewed by CBER. Under the proposed consolidation plan, these well-characterized and recombinant products are likely to be reviewed by CDER.

Even with these regulatory changes, manufacturing and assay costs for biologic macromolecules remain much higher than for traditional drugs. Biologics not considered to be well-characterized, such as blood and tissue and their components and inactivated vaccine antigen preparations, continue to be developed, manufactured, and reviewed for approval under pre-1996 regulations.

■ 2.3. CURRENT TRENDS IN DRUG DEVELOPMENT

With the emergence of new technologies—including genomics, proteomics, molecular biology, combinatorial chemistry, computer-assisted drug design, robotics, and high-throughput screening—our ability to generate new molecular entities

that are high-affinity ligands or receptors has greatly increased. However, success in developing these new entities through preclinical and clinical testing has not improved significantly. It is estimated that less than 40% of NMEs produced by drug discovery groups in pharmaceutical companies are candidates for preclinical testing [4]. Only about 10% of them actually make it through clinical trials and are approved for marketing [5].

There is intense economic pressure on the pharmaceutical industry by those who pay for its products to make the development of drugs more efficient and thereby reduce or at least contain costs. Industry executives ignore this clamor at their own peril. A widely embraced strategy to improve productivity is to integrate drug discovery and development by bringing company researchers from basic, preclini-

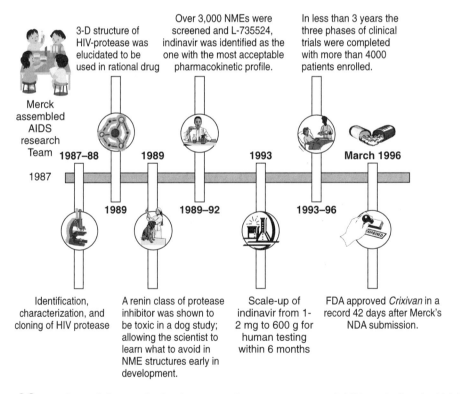

Figure 2.2. Timeline of fast-track development of an HIV protease inhibitor, indinavir (*Crixivan*) by Merck through a project research team approach. Adapted from Merck's account on *Crixivan* development.

cal, and clinical sciences together to work very early in the development of a drug candidate or even during discovery efforts as a project team. In the ideal, the project team, relying on the collective wisdom of the group, and, when needed, calling upon academic experts, is charged with anticipating problems that may be encountered in the preclinical and clinical testing of a drug candidate. The team is expected to develop a drug product with an acceptable pharmacokinetic profile, a low potential for drug–drug interactions, clear-cut efficacy, and a comfortable safety margin. The team, by identifying the target population, must strive to exclude from clinical trials patients who are unlikely to derive a benefit from the new drug and eliminate those likely to respond to its effects adversely.

Already, the project team approach has recorded successes (Figure 2.2). A project team at Merck, using advanced physical, chemical, and molecular biotechnologies coupled with rational drug design to predict the most promising lead compounds as part of the drug discovery process, significantly accelerated the development of the antiretroviral protease inhibitor indinavir (*Crixivan*), once the most widely used protease inhibitor for the treatment of HIV infection and still a very important drug. With continuous feedback from the project team, Merck scientists sequenced HIV protease, demonstrated its vital role in viral replication, cloned and purified the enzyme, and produced sufficient quantities to identify the protein's three-dimensional structure, all in about two years [6]. Similar efforts in the past had taken 8–10 years.

Based on the crystal structure of HIV protease, ideas about what an inhibitor should look like emerged. With the availability of a quantity of HIV protease sufficient for enzyme assays, the construction and evaluation of HIV-protease inhibitor candidates began. In all, about 3000 candidates were screened, in the search for those with acceptable pharmaceutical properties that also demonstrated potent antiviral activity. These efforts led to the discovery of indinavir in 1992 [7]. By anticipating problems and issues early in the development process, through the project team approach, Merck completed clinical testing of *Crixivan* in about 3 years (1993–1996) [8]. In the face of an extraordinary demand for new drugs by HIV-infected patients and their advocates, the FDA approved Crixivan in a record time of 42 days after Merck had filed the NDA.

Clearly, from this example, we can see the benefits of the multidisciplinary project-team approach. The team effort integrates pharmaceutical technology and biotechnology to identify and isolate biologic targets of sufficient purity and in sufficient quantity for *in vitro* drug screening. Experts in advanced physical and chemical sciences can help develop high-affinity ligands, and introduce strategic considerations early in the development process, and can help pharmaceutical companies to increase the efficiency of drug development in many therapeutic areas.

REFERENCES

1. CFR Regulations, *Title 21, Parts 300–314*. CFR, 1998.
2. Galluppi, G.R., M.C. Rogge, L.K. Roskos, L.J. Lesko, M.D. Green, D.W. Feigal, Jr., and C.C. Peck, *Integration of pharmacokinetic and pharmacodynamic studies in the discovery, development, and review of protein therapeutic agents: a conference report.* Clin Pharmacol Ther, 2001. **69**(6): 387–99.
3. FDA, T., FDA *final rule, October 15,1996, published through REGO (under the President's reinventing government) initiative.* 1996.
4. *"Drug discovery: filtering out failures early in the games,".* Chem Eng News, 2000. **78**: 63.
5. Borchardt, R.T., *Integrating drug discovery and development.* AAPS Newsmagazine, 2001. **4**(5): 5.
6. Navia, M.A., P.M. Fitzgerald, B.M. McKeever, C.T. Leu, J.C. Heimbach, W.K.

Herber, I.S. Sigal, P.L. Darke, and J.P. Springer, *Three-dimensional structure of aspartyl protease from human immunodeficiency virus HIV-1.* Nature, 1989. **337**(6208): 615–20.

7. Vacca, J.P., *Design of tight-binding human immunodeficiency virus type 1 protease inhibitors.* Methods Enzymol, 1994. **241**: 311–34.

8. Vacca, J.P., B.D. Dorsey, W.A. Schleif, R.B. Levin, S.L. McDaniel, P.L. Darke, J. Zugay, J.C. Quintero, O.M. Blahy, E. Roth, and et al., *L-735,524: an orally bioavailable human immunodeficiency virus type 1 protease inhibitor.* Proc Natl Acad Sci USA, 1994. **91**(9): 4096–100.

9. *Elimination of ELA and PLA requirement, May 14, 1996 and replaced by BLA.* Federal Register, 1996. **63**(147): 40858–71.

10. Devine, R., *Manufacture process modification, effective Oct 7, 1997; CBER,.* CBER, presentation, 1997.

3

BIOTECHNOLOGY INDUSTRY PERSPECTIVE ON DRUG DEVELOPMENT

■ 3.1. INTRODUCTION

The decision to select and develop a new drug candidate is complex. It is dictated by the interplay between the potential markets and the scientific advances being made in basic, translational, and applied research. The drug development process, from the inception of a therapeutic strategy to the approval and introduction of the therapeutic compound for human use, involves investments, sharing of knowledge, and collaboration among govern-ment agencies, pharmaceutical companies, universities, and other research organizations (Figure 3.1). When successful, these interactions lead to the development of drugs that provide cure, treatment, or prevention of disease.

Since the development in the 1970s of molecular tools enabling the production of proteins and peptides in sufficient quantities for therapeutic purposes, dedicated biotechnology companies have been examining, refining, and developing these technologies for commercialization. This

Biotechnology and Biopharmaceuticals, by Rodney J. Y. Ho and Milo Gibaldi
ISBN 0-471-20690-3   Copyright © 2003 by John Wiley & Sons, Inc.

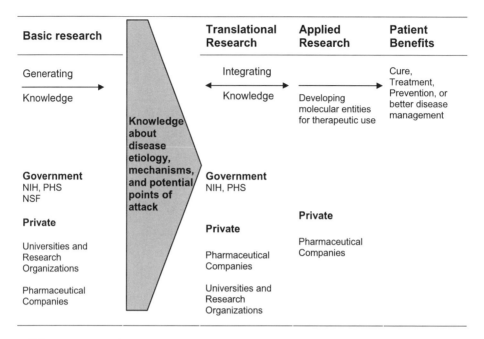

Figure 3.1. The process of pharmaceutical drug discovery and development—an integrated science that draws on the collective knowledge and resources from both government and private sectors. Abbreviations: NIH, National Institute of Health; NSF, National Science Foundation; PHS, Public Health Service.

BOX 3.1. THE BIRTH OF A DEDICATED BIOTECHNOLOGY COMPANY [1]

It has now been more than fifteen years since Robert Swanson, a young man who understood both finance and science, invited Herbert Boyer, a shy molecular biologist at the University of California, San Francisco, out for a beer. Swanson described his vision to Boyer: that the techniques and ideas that Boyer had devised for manipulating DNA could be translated into products at a private company yet to be established. As a result of that meeting, Genentech, the first well-known biotechnology corporation, was founded; Swanson and Boyer made their fortunes; and profound changes ensued in academic biomedical research.

---

pioneering work (Box 3.1) has given birth to biopharmaceuticals that are now on the market to improve human health. In more recent years, these technologies have been gradually adopted and integrated into the research and development processes of most of the major pharmaceutical companies. Today, major established pharmaceutical companies and dedicated biotechnology companies alike are using these advanced tools and resources to stay competitive in accelerating the discovery and development of pharmaceuticals.

The development process for biotechnology products is a lengthy and expensive one that carries unforeseeable risks and encounters setbacks during preclinical and clinical testing as a drug candidate is being developed. According to a study by the US Office of Technology Assessment (1991), regulatory approval and marketing of a new biopharmaceutical can cost $200 to $350 million and take from 7 to 12 years [2]. A recent study suggests that costs may run as high as $800 million. On average, the FDA approves only about one of five new molecular entities (NMEs) evaluated in human clinical trials. Furthermore as many as 5000 NMEs would have been evaluated in preclinical research to come up with the five leading candidates.

Traditionally drug discovery and development in the pharmaceutical industry relied on chemistry. Drug discovery and much of development was carried out by medicinal chemists focusing on chemical structures and corresponding functional activities of NMEs. Only limited biochemical and pharmacological information was available regarding the therapeutic target and the drug candidate. This approach was especially characteristic of drugs developed in the two or three decades following World War II. Many of these drugs and NMEs were selected based on crude biologic screening of chemical compounds. Those engaged in drug discovery were said to be playing molecular roulette. Many compounds were screened, from which one or two agents were eventually approved for human use [3].

This strategy, however, was about to change. The advent of integrated biological sciences—physiology, molecular biology, biochemistry, genetics, pharmaceutical sciences, and biotechnology—provided, for the first time, intricate details at the cellular and molecular levels and allowed insightful characterization of chemical structure–biological function relationships. Among other impacts, these advances greatly increased the efficiency of screening of chemical compounds to identify lead NMEs for preclinical testing. But bear in mind that even today the development of small molecular weight drugs begins with synthesis of a series of chemical compounds provided by medicinal and organic chemists.

In contrast with the postwar strategies of the established drug industry, dedicated biotechnology companies were created to exploit the commercial potential of technological innovation and ideas. Therefore many of the early biotechnology companies were founded with emphasis on science and technology, and were modeled after research organizations and institutions. According to one report, the US government provided over $3 billion for biotechnology research through various federal agencies in 1990 alone [2] (Table 3.1). In 2001 NIH allocated $19.7 billion to biomedical research, and the amount is expected to increase to $22.4 billions in 2002 [4].

Despite the government's largesse, the pioneering biotechnology firms survived by commercially exploiting technologic innovations and developing diagnostic tools, rather than by developing novel therapeutic entities. This was so because the transformation of recombinant proteins and peptides into biopharmaceuticals had never been charted, and venture capital to fund new therapeutic agents has not yet started to flow freely.

Today funding for a start-up biotechnology company is provided by a combination of federal grants, private parties, and venture capital. As the company matures, public stock offerings are common, and relationships may be established with large pharmaceutical companies. Frequently biotechnology firms raise cash for drug development and clinical studies by licensing key first-generation products and vital market segments to established drug companies in exchange for financial viability.

Those biotechnology companies successful in raising cash for initial start-up operation in the early 1980s still needed second and third rounds of financing to perform clinical studies to develop their basic research discoveries into marketable products. By the late 1980s, as many start-up companies were ready to initiate clinical trials, the need for capital exceeded available sources. Many companies raised the needed cash for the later phases of clinical testing through public stock offerings. The tightened funding for biotechnology companies during this time was

■TABLE 3.1. US federal funding for biotechnology
for fiscal year 1990

Agency	Amount (in millions US$)
National Institutes of Health	2900.0
National Science Foundation	167.9
Department of Agriculture	116.0
Department of Defense	98.0
Department of Energy	82.2
Agency for International Development	28.7
Food and Drug Administration	19.4
Environmental Protection Agency	8.3
Veterans Administration	7.5
National Institute of Standards and Technology	4.8
National Aeronautics and Space Administration	4.5
National Oceanic and Atmospheric Administration	2.0
Total	**3439.3**

Data source: US Office of Technology Assessment [12].

compounded by (1) the 1987 stock market crash in the United States, (2) slower than anticipated product development due to unforeseen technical problems in bringing biotechnology products to market, (3) poor investment returns, and (4) increased competition because recombinant DNA technology had become routine and the art of gene cloning was now widely available to pharmaceutical companies around the world. Nevertheless, some biotechnology companies—through creative financing, strategic alliances, and other arrangements—survived the challenges and began to launch their products by the early 1990s; many of these companies were able to post profits by the mid 1990s. Today several biotechnology-based pharmaceutical companies realize more than one billion dollars in annual profit (Table 1.1).

For the past few years, the number of biotechnology companies in the United States has hovered around 300, about 25% lower than the 403 extant 10 years ago

[2,5]. Financial viability of biotechnology compines has also remained pretty much constant over the past three years (Table 3.2).

Because the strength of most biotechnology companies is in discovery research focusing on new technologies and new molecular entities—proteins, peptides, and genes—the industry has been successful and innovative in identifying novel therapeutic agents. Often, to improve the odds of success in further developing these new molecular entities into drugs, biotechnology companies with limited financial resources must choose a target indication with the highest potential of clinical success, in terms of clinical significance and efficiency in clinical testing (refer to Box 3.2 for discussion of therapeutic indications). Sometimes the therapeutic target is obvious and indeed the point of development, but often it is not because many molecules exhibit therapeutic potential in multiple disease targets. For example, interferons, produced by lympho-

■TABLE 3.2. Financial viability of biotechnology companies

Viability[a]	1999		1998		1997	
	Companies	% Total	Companies	% Total	Companies	% Total
More than 5 years	76	25.2	75	23.7	101	30.9
3–5 years	29	9.6	23	7.3	30	9.2
2–3 years	30	10.0	38	12.0	46	14.1
1–2 years	59	19.6	76	24.0	68	20.8
Less than 1 year	107	35.5	105	33.1	82	25.1
Total public companies	**301**		**317**		**327**	

Source: Adapted from the Ernst and Young [5].

[a]Viability as measured by Ernst Young Survival Index, derived from the skew analysis. The estimates are expressed as the number of years of cash that companies have on hand based on current spending levels and on fiscal year-end numbers.

BOX 3.2. SIGNIFICANCE OF INDICATIONS AND THERAPEUTIC TARGETS AND THEIR ROLE IN DRUG DEVELOPMENT

The labeling of all drugs approved for marketing in the United States is regulated by the FDA. The label is placed on an immediate container, but also in the package insert, and in company literature, advertising, and promotion materials (CFR Title 21, part 201). The label provides summary information on (1) product description, (2) clinical pharmacology, (3) indications and usage, (4) contraindications, (5) warnings, (6) precautions, (7) adverse reactions, (8) drug abuse and dependence, (9) overdosage, (10) dosage and administration, and (11) how the product is supplied. One of the most important considerations for pharmaceutical companies is the product indications approved by FDA in the treatment, prevention, or diagnosis of disease or conditions, for which evidence of safety and effectiveness have been demonstrated by controlled clinical trials. The indications thus determine the potential patient population that will benefit from the drug from which market potential can be projected. While the final language of a product indication is negotiated at the time of FDA approval, the therapeutic target, or hypothesis, submitted within the IND application preceding a phase I clinical trial essentially sets the direction for a product. For each therapeutic target or indication, even for the same new molecular entity, the company must file a separate IND application, and evaluate the efficacy and safety of the NME independently.

---

cytes and fibroblasts in response to infectious agents, often produce wide-ranging biologic responses in animals and humans. In fact interferons are currently approved and used clinically to treat three different disease modalities—some cancers, viral infections, and the neurogenerative disorder multiple sclerosis.

■ 3.2. ROLE OF THE ORPHAN DRUG ACT

Business strategies used by biotechnology companies are (1) concentrating initially on the most obvious candidates with proven biologic activities using natural products such as human insulin, growth hormone, or interferons and (2) focusing on treating individuals with rare medical conditions so that the NME in development can be qualified for grant and tax benefits under the 1983 Orphan Drug Act. Factors influencing the choice of which indication of an NME to develop first include consideration of (1) market potential, (2) competition, (3) patents and proprietary advantages, (4) qualification as an orphan drug, and (5) reimbursement by third-party payers (i.e., Medicare, Medicaid, and insurance companies). By targeting the indications of a new drug candidate to diseases that affect fewer than 200,000 individuals in the United States, a company also receives seven years of market exclusivity. The market exclusivity discourages other companies from developing any product for the target indication. In the first seven years after enactment, more than 375 medical products received orphan designation from the FDA, and 40 orphan drugs were approved under the Orphan Drug Act. By the end of 1999, about 200 products were approved under the Orphan Drug classification (FDA data 2001) (Box 3.3).

With the Orphan Drug Act in place, it became extremely important for a pharmaceutical company to be the first to reach the market for a given orphan indication to reap the maximum financial returns on their investment. Both *Neupogen* (granulocyte colony-stimulating factor, G-CSF) and *Leukine* (granulocyte-monocyte colony-stimulating factor, GM-CSF) are capable of stimulating neutrophils and reducing neutropenia in different medical conditions (see Table 3.3). However, being approved under the Orphan Drug classification

---

BOX 3.3. THE ORPHAN DRUG ACT—AN INTERESTING INSIGHT*

The Orphan Drug Act, enacted in 1983, seeks to encourage the development of drugs for rare diseases—conditions that affect fewer than 200,000 people in the US. The act offers incentives for drug makers to invest in product development that is thought to be unlikely to offer a full return on investment. The government provides grants, tax breaks, and, most important, seven years of market exclusivity to the first manufacturer to receive approval for a product with orphan designation to treat a rare medical condition. The market exclusivity provision has proven to be controversial.

By 1991, more than 375 products had received orphan designation, and more than 40 orphan drugs were available in the US. Nine of the 15 biotechnology-derived drugs on the market have orphan drug status, as do 19 biotechnology-derived drugs under development. Some of these products, however, such as erythropoietin and human growth hormone have enjoyed remarkable commercial success, and arguments have been made that these drugs would have been developed without the Orphan Drug Act incentives because there was great opportunity for profit.

The case of erythropoietin is particularly controversial and complex. Amgen received FDA approval in June 1989 to market its Epogen to dialysis patients suffering from anemia associated with end-stage renal disease, a patient population of less than 200,000. Erythropoietin costs about $5000 per patient per year, and in the first two years on the market Amgen sold more than $300 million worth of the drug. Genetics Institute has also developed recombinant erythropoietin in parallel but has yet to receive FDA approval (as of 2002), mainly due to Amgen's orphan drug claims.

*Data source: OTA Report (1991).

■TABLE 3.3. Comparisons of indications, approval dates, and revenues of two colony-stimulating factors approved by FDA under the Orphan Drug Classification

Product Name	Company[a]	Indications	Approval Date	1998 Revenue (in $millions)
Neupogen (Filgrastim)	**Amgen** Thousand Oaks, CA	(1) To decrease the incidence of infection, as manifested by febrile neutropenia, in patients with nonmyeloid malignancies receiving myelosuppressive anticancer drugs associated with a significant incidence of severe neutropenia with fever	2–20–91	$1120.00
		(2) Treatment of neutropenia associated with bone marrow transplants	6–15–94	
		(3) Treatment of patients with severe chronic neutropenia (<500/mm^3)	12–19–94	
		(4) Mobilization of peripheral blood progenitor cells for collection in patients who will receive myeloablative or myelosuppressive chemotherapy	12–28–95	
		(5) Reduction in the duration of neutropenia, fever, antibiotic use, and hospitalization, following induction and consolidation treatment for acute myeloid leukemia	4–2–98	
Leukine (Sargramostim)	**Immunex** Seattle, WA	(1) Treatment of neutropenia associated with bone marrow transplantation	3–5–91	$64.00
		(2) To reduce neutropenia and leukopenia and decrease the incidence of death due to infection in patients with acute myelogenous leukemia	9–15–95	

[a]Immunex became a part of Amgen in 2002 and the right to market Leukine was transferred to another company as a part of the requirement by US Federal Trade Commission.

about two weeks (20 February vs. 5 March 1991) earlier than *Leukine*, Amgen's *Neupogen* was awarded market exclusivity for a wide range of conditions characterized by neutropenia, a broad indication that captures billions of dollars in annual sales. Immunex, the company that developed *Leukine*, was awarded only a very limited indication that commands less than $100 million in annual sales.

Once approved for marketing by the FDA for a specific and sometimes narrow indication, biopharmaceuticals are often tested and their makers apply for additional indications. At the same time some of these drugs, particularly those used in cancer-related areas, are prescribed off-label for nonapproved indications. According to a General Accounting Office report, oncologists prescribed drugs to 44% of patients according to the FDA approved indication (on-label). However, 56% of the patients were prescribed at least one drug off-label; 38.8% of these prescriptions were evidence-based, relying on clinical data. While the strategy to seek additional indications for an approved drug is not new, the deliberate planning to develop new and more lucrative indications outside the orphan designation after marketing of the product for an orphan indication and the promotion to prescribers of potential new indications, even before FDA approval, breaks new ground in exploiting health care needs. These business strategies expand the market and realize target populations that are many times larger than the 200,000 or fewer patients initially targeted to reap the substantial benefits of the Orphan Drug Act.

Recent development highlighted a shift in FDA position and interpretation of the market exclusivity clause. Under the orphan designation, three interferon products, *Betaseron* (IFN-β1b), *Avonex* (IFN-β1a) and *Rebif* (IFN-β1a) are currently approved for treatment of multiple sclerosis (MS). The interplay in drug safety and efficacy consideration, business strategies, and regulatory rationale permitting approval of the three drugs for MS treatment is discussed in Box 3.5.

■ 3.3. CLINICAL LEVERAGE STRATEGY IN ACCELERATING DRUG DEVELOPMENT

By the late 1990s the industry began to question the wisdom of developing a new drug candidate with multiple therapeutic targets using sequential, independent clinical trials for each indication. With the advances in basic and integrated research, a fuller picture emerged regarding the mechanisms of disease manifestation and revealed potential points of intervention for many medical conditions. Consequently new drug candidates with broad therapeutic potential are increasingly being studied in parallel.

Furthermore the ability of a biotechnology companies to raise funds increases greatly when a biopharmaceutical candidate is in clinical trials. Having a new drug in the clinic for more than one indication gives the company greater leverage to attract venture capital as well as public offerings.

The accumulated statistical data on success rate for all new drug candidates committed to clinical testing indicate that only one of four or five agents entering phase I will be approved by the FDA. To increase the chance of success of getting a biopharmaceutical to market, the alternative or newer strategy is to seek more than one indication for a given drug by initiating several clinical studies at about the same time. In theory, if a company identifies four indications for a given candidate, chances are very good that one indication will be successful and approved by the FDA. Thus the goal is to select the most promising indications, hoping to save time and increase the likelihood that one of the indications will lead to FDA approval (Table 3.4). This idea is generally known as the clinical leverage strategy. Table 3.5 lists some of the new drug candidates that are being tested in parallel human trials for multiple indications. The success rate of this strategy remains to be seen.

■ 3.4. THERAPEUTIC TARGET CONSIDERATIONS

Therapeutic target or treatment hypotheses submitted to the FDA as part of investigational new drug (IND) applications

■TABLE 3.4. Historical data on clinical trial success rates as compared to clinical leverage strategy

	Historical Data				Clinical Leverage Strategy		
			Procession Rate			Projected Procession Rate (Fraction)	
Clinical Trial	For Every 4 NDCs (⊤) Tested		Fraction (Products)	Percentage	For Each NDC— 4 Indications (⊤) Are Tested	Indications	Product
Phase I	⊤ ⊤ ⊤ ⊤		4/4	100	⊤ ⊤ ⊤ ⊤	4/4	1/1
Phase II	⊤ ⊤ ⊤		3/4	75	⊤ ⊤ ⊤	3/4	1/1
Phase III	⊤ ⊤		1.5/4	27.5	⊤ ⊤	1.5/4	1/1
FDA approval	⊤		1/4	25	⊤	1/4	1/1

■TABLE 3.5. New drug candidates in human clinical trials using clinical leverage strategy

New Drug Candidate	Indications Being Tested	Clinical status[a]	Sponsor
Pafase (platelet activation factor-AH)	ARDS[b] Asthma Post-ERCP[b] pancreatitis Severe Sepsis	Phase II Phase II Phase II Phase III	ICOS/Suncos Corp
IC351	Male erectile dysfunction Female sexual dysfunction	NDA filed Phase II	ICOS/Lily
LeukArrest (MoAb Hu23F2G)	Hemorrhagic shock Ischemic stoke Multiple sclerosis Myocardial infarction	Phase II Phase III Phase II Phase II	ICOS
Denileukin diftitox	Cutaneous T cell lymphoma Recurrent/persistent lymphoma Non-Hodgkin's lymphoma	Phase II Phase III Phase II	Ligands
Iodine-131 Anti-B1 Antibody	Non-Hodgkin's lymphoma Chronic lymphocytic Leukemia Mantle cell lymphoma	Phase II Phase I Phase II	Corixa/ GlaxoSmithKline
Bevacizumab + other drug combination	Breast cancer Colon and rectal cancer	Phase III Phase II	Genentech

[a]As of 2001.

[b]Abbreviations: ARDS, Acute respiratory distress syndrome; ERCP, Endoscopic retrograde cholangiopancreatopography.

eventually become part of the new drug application (NDA) package that supports usage and treatment indications for each new molecular entity seeking approval in the United States. For each therapeutic target or indication, the sponsoring pharmaceutical company must file separate IND applications and evaluate the efficacy and safety of the NME independently in each case. Therefore it is critical early in the process for a drug company to consider the therapeutic target with the highest impact on therapeutic outcome and return on investment.

With the introduction of molecular cloning and expression, automation, and computing power, it became even more important to choose a target disease and the right kind of molecule early in the course of development. Clearly, the selection of a disease target based on an in-depth understanding of the pathology and potential points of intervention will aid therapeutic evaluation. An additional consideration is the need for a clearly measurable therapeutic outcome or end point, based on clinical or laboratory measurements, that reflects efficacy and toxicity. Multifaceted disease states with ill-defined outcomes will be more challenging. Some of these issues will be discussed in the context of clinical testing of biopharmaceuticals (see Chapter 4, Section 4.3).

The choice of therapeutic target may vary from drug company to drug company depending on its unique resources, such as patent-protected technology and expertise in certain biological processes that lead to disease. The specific molecular characteristic of the biopharmaceutical for the therapeutic target can take the form of a protein, peptide, or antibody. These choices will have a great impact on the success of development and the time it takes to transform lead candidates into therapeutic products.

In this section we will discuss the rationale for the successful development of well-characterized biotechnology products, including (1) synthetic peptides of less than 20 amino acids, (2) recombinant DNA-derived macromolecules, and (3) monoclonal antibodies and derivatives. A drug company's strategy involves either the improvement of an already available biologic (e.g., insulin) or the development of a novel one.

3.4.1. Improving Marketed Biopharmaceuticals

The chance of success in demonstrating therapeutic efficacy is much higher for drug candidates already investigated and available. Up until 20 some years ago biologics used to ameliorate disease were largely obtained through extraction of human or animal tissue. For example, insulin extracted from beef and pork had been used to treat diabetic patients for decades before the introduction of recombinant human insulin. To achieve success, however, a recombinant protein or synthetic product must provide an improved safety profile. In addition to being competitive in production cost, the recombinant version of human insulin appears to be less immunogenic and incur fewer immune-mediated, drug-induced allergies as compared with insulin derived from animal by-products. Heightened public perception and concern regarding animal tissue-extracted contaminants such as that triggering bovine spongiform encephalopathy—better known as mad cow disease—fuels the demand for recombinant products to replace available biologic products.

Similarly, concern about minute amounts of contaminants co-purified with human growth hormone extracted from brain tissues, which could lead to the development of Jacob-Kreutzfeld syndrome, inevitably resulted in the development of recombinant human growth hormone. Inactivated hepatitis B particles isolated from hepatitis B-positive patients were originally used to prevent hepatitis B virus infection, but a small fraction of individuals became infected through vaccination. When the hepatitis B surface antigen was cloned and a recombinant form was produced, its use essentially eliminated the transmission of hepatitis B via vaccine. Today, inactivated hepatitis B vaccine is no longer marketed in the United States.

While some potential therapeutic proteins can be isolated and purified from natural sources, it may be impractical to extract a sufficient amount for therapeutic purposes. This is the case for tissue plasminogen activator (tPA). While its therapeutic potential was quickly recognized, the protein was present naturally in such small amount that it was not possible to obtain quantities sufficient for clinical evaluation. Advances in biotechnology, including the development of recombinant DNA engineering, permitted production of recombinant tPA in sufficient quantities to evaluate its effects on occluded coronary arteries and to eventually develop a product.

Once a recombinant gene is constructed and introduced in bacteria or yeast, the expressed protein may provide greater purity and quality, through tight production control, than one can obtain from an extracted protein. Added benefits of switching from a isolated natural protein to a recombinant form are reduced variation of biologic activity and less inadvertent contamination.

3.4.2. Developing a Novel Biopharmaceutical

Convention holds that one functional gene can produce one unique protein. Drugs now available on the market are known to modulate about 480 gene products of the estimated 150,000 genes spanning the entire human genome [6–9]. In theory, an extraordinary number of new therapeutic targets can be envisioned as more and more gene products are identified. Over the last decade some of these proteins have been discovered and transformed into biopharmaceuticals that are now available for improving human health and disease management.

Since the introduction of biotechnology to health sciences, many new, previously unavailable proteins and peptides for therapeutic purposes have been introduced. The agents include (1) factors that influence hematopoietic cells and blood coagulation, (2) interferons and cytokines for anti-infective and cancer therapy, (3) hormones and derivatives, (4) enzymes and derivatives, and (5) recombinant proteins for vaccines. Examples of hematopoietic growth and coagulation factors are erythropoietin, granulocyte colony-stimulating factor (G-CSF), granulocyte-macrophage colony-stimulating factor (GM-CSF), clotting factors VIIa, VIII, and IX, GPIIb/IIIa inhibitors that affect platelet function, direct thrombin inhibitors, and antihemophilic factor. Several interferons are now available for the treatment of viral infections and some cancers. Numerous hormones, including glucagon, somatotropin, calcitonin, somatostatin, thyroid-stimulating hormone, thyrotropin-releasing hormone (*Protirelin*), gonadotropin releasing hormone (*Ganirelix*, *Nafarelin*, *Leuprolide*), leuteinizing hormone-releasing hormone (*Goserelin*), and oxytocin (*Pitocin*) are now approved for use.

Some of the previously unavailable enzymes for therapeutic use include dornase (*Pulmozyme*), imiglucerase (*Cerezyme*), asparaginase, tissue-type plasminogen activator (*Activase*), and related drugs (*Retavase*). Detailed information about these products and their clinical use is provided in Part II in the form of monographs. Information relevant to pharmacokinetics and molecular characteristics can be found in Appendixes I and II.

In addition to recombinant proteins designed for treating disease, the class of biopharmaceuticals called monoclonal antibodies has been of great therapeutic interest. These antibodies have high affinity for and selective binding to unique antigens such as pharmacologically important receptors or ligands (Box 3.4). In recent years the FDA has approved more monoclonal

---

BOX 3.4. ANTIBODIES AND DERIVATIVES USED AS BIOTHERAPEUTIC AGENTS AND THEIR SPECIFIC TARGETS

- CD3 (muromonab-CD3), CD20 (Rituximab), CD25 (Basiliximab; Daclizumab), CD33- (Gemtuzumab-ozogamicin), CD52 (Alemtuzumab)
- Platelets-IIa/IIIb (Abciximab)

- TNF-α (Etanercept; Infliximab)
- HER-2 proto-oncogene product (Trastuzumab), IL-2 (*Ontak*)
- Respiratory syncytial virus (Palivizumab), (hyperimmune globulin Respigam)

BOX 3.5. HOW MARKET EXCLUSIVITY COULD BECOME NONEXCLUSIVE FOR DRUGS WITH ORPHAN DESIGNATION

Interferon Beta Use in Multiple Sclerosis

The first interferon beta for treatment of multiple sclerosis (MS)—*Betaseron* (interferon beta-1b)—received FDA approval in July 1993, based on a placebo-controlled study evaluating the incidence of exacerbations. Subsequently, a second interferon beta, *Avonex* (an interferon beta-1a), was shown to be effective for reducing the incidence of exacerbations and reducing the accumulation of physical disability. *Betaseron* received orphan drug designation prior to approval, and was still within the seven-year period of marketing exclusivity at the time *Avonex* was under review. However, Biogen, the manufacturer of *Avonex*, provided evidence that *Avonex* was not the same drug as *Betaseron* within the meaning of the orphan drug regulations, by showing a significantly better safety profile with regard to skin necrosis at injection sites. Consequently, *Avonex* and *Betaseron* were deemed to be different drugs and Biogen received marketing approval for *Avonex* in May 1996. Biogen also held an orphan drug designation for *Avonex* and had a seven-year period of marketing exclusivity that expired in May 2003.

Serono, a third manufacturer of an interferon beta product, *Rebif* (another interferon beta-1a), also conducted clinical studies in MS patients. Serono completed their studies and submitted a Biologics License Application (BLA) for *Rebif* in February 1998. The major safety and efficacy data came from a controlled study of two different doses of *Rebif* vs. placebo. Based on the findings, the FDA concluded that *Rebif* was safe and effective for the treatment of MS. However, within the framework of the orphan drug regulations, *Rebif* was regarded as the "same drug" as both *Betaseron* and *Avonex*. Serono was not able to supply evidence to establish that *Rebif* was not the "same drug," and the product was denied marketing approval until the bar of marketing exclusivity was removed, either by expiration of the exclusivity period, or by Serono providing evidence that *Rebif* was not the "same drug." Serono recognized that the period of exclusivity for *Betaseron* would expire in July 2000, leaving only the marketing exclusivity for *Avonex* as an issue after that date. Thus, in late 1999, Serono initiated another clinical study with the intent to demonstrate superior clinical efficacy of *Rebif* compared with *Avonex*.

Orphan Drug Regulations

The orphan drug regulations allow a sponsor of an orphan-designated drug a period of marketing exclusivity, free from

(*Continued on next page*)

BOX 3.5. Continued

competition from the "same drug" for the same approved indication. The regulations describe how to assess two products on a physical-chemical basis to determine if they should be regarded as the "same drug." The regulations further provide that even if the physical-chemical criteria for "same drug" are met, a demonstration of clinical superiority of the subsequent product compared to the original product will enable a determination that the two products are in fact not the "same drug." In such circumstances, the subsequent product may be given immediate marketing approval. The regulations also describe the circumstances for determining clinical superiority. A new product can be considered clinically superior if greater effectiveness has been shown or on the basis of greater safety in a substantial portion of the target population. An important aspect of this determination is that demonstration of greater effectiveness will in most cases entail direct comparative clinical trials, whereas direct comparative trials for a demonstration of superior safety are expected to be necessary in only some cases. The regulations do not state that clinical superiority must be based on overall risk–benefit being deemed superior for the subsequent product compared to the prior product. In fact, the regulations indicate that only a selected aspect may constitute a sufficient basis to reach a conclusion of clinical superiority. That is, the aspect not selected by the sponsor for focus (e.g., safety when efficacy is selected; efficacy when safety is selected) does not require a comparative assessment. Consequently, knowledge of the comparison of efficacy could be entirely lacking (and somewhat inferior efficacy a real potential), yet a clinical superiority determination, based on safety, can be reached.

Study Design of Rebif, Another Variant of Interferon-Beta 1a, That Eventually Gained FDA Approval for Marketing

The second Serono trial was a multicenter randomized study of *Rebif* at 44 mg subcutaneously three times per week compared to *Avonex* at 30 mg intramuscularly weekly. The primary efficacy outcome was the incidence of exacerbations through week 24. *Avonex* was administered according the recommended regimen in the FDA-approved labeling, and *Rebif* according to the recommended regimen in Serono's proposed labeling. Serono elected to conduct the study open label, but with a blinded clinical evaluator.

The Center for Biologics Evaluation Research (CBER) reviewed the results of the study and consulted extensively with the FDA's Office of Orphan Product Development. Regulatory officials determined that the comparative clinical study demonstrated that *Rebif* is more effective than *Avonex* in that it provides a significant therapeutic advantage over *Avonex*: 74.9% of study subjects taking *Rebif* were exacerbation-free versus 63.3% of *Avonex* subjects. According to the FDA, this is a meaningful difference because it signifies that a patient on *Rebif* is 32% less likely to experience an MS exacerbation, which can substantially lower his or her quality of life for weeks or months. MS exacerbations can be manifested by paralysis, loss of vision, loss of control of bladder and bowel function, as well as other impairments. Although *Rebif* caused injection site reactions more frequently than *Avonex*, the FDA determined that the severity and frequency of such adverse events do not render *Rebif* unlicensable, and therefore, the agency approved *Rebif* for marketing in March 2002.

*Data source: BLA document published by CBER for Rebif (2002).

antibodies than any other biotechnology product.

The first monoclonal antibody—OKT3 or *Orthoclone*—was designed to be immunosuppressive and to reduce the likelihood of transplant rejection. Now monoclonal antibodies are used to treat a wide range of diseases, including arthritis, certain kind of cancers, and microbial and viral infections.

In principal, gene-based biopharmaceuticals, such as DNA plasmids and viral vectors that carry the message for expressing specific proteins, can also be used to influence disease. Biotechnology companies are pursuing gene therapy strategies for cancer and other diseases. There is on the market a product called *Vitravene* (fomivirsen) that uses an antisense strategy to block cytomegalovirus (CMV) replication and treat CMV retinitis. Fomivirsen is an oligonucleotide that mimics gene sequences and is designed to bind directly to CMV DNA, thereby inactivating viral replication. It can be considered a form of gene therapy. The mechanism(s) by which it inhibits viral replication, however, may or may not involve direct binding of the antisense DNA to the viral DNA. Nevertheless, it provides an effective and safe means to modulate CMV retinitis.

REFERENCES

1. Bazell, R., in *The New Republic*. 1991. *Biotech bingo*, p 10–12, April 18.
2. OTA, report, *Biotechnology in a global economy*. In *OTA-BA-494, B*. 1991, US Office of Technology Assessment. 3–98.
3. U.S. Congress, H.O.R., *Office of Technology Assessment, New Developments in Biotechnology: U.S. Investment in Biotechnology, OTA-BA-360*. 1988.
4. National Institute of Health, *Summary Budget report*. 2001, NIH, USPHS.
5. Report, E.-Y., *Convergence, the biotechnology report: from thought to finish*. 2000.
6. Ward, S.J., *Impact of genomics in drug discovery*. Biotechniques, 2001. **31**(3): 626, 628, 630, passim.
7. Wickelgren, I., *Mining the genome for drugs*. Science, 1999. **285**(5430): 998–1001.
8. Ohlstein, E.H., R.R. Ruffolo, Jr., and J.D. Elliott, *Drug discovery in the next millennium*. Annu Rev Pharmacol Toxicol, 2000. **40**: 177–91.
9. Emilien, G., M. Ponchon, C. Caldas, O. Isacson, and J.M. Maloteaux, *Impact of genomics on drug discovery and clinical medicine*. Qjm, 2000. **93**(7): 391–423.

4

BIOPHARMACEUTICAL TECHNOLOGIES AND PROCESSES IN DRUG DEVELOPMENT

■ SECTION ONE ■

■ 4.1. APPLICATION OF BIOTECHNOLOGIES IN DRUG DISCOVERY AND DEVELOPMENT

In this section we will focus primarily on recombinant proteins. We will discuss strategies specific to monoclonal antibody production in the introduction to antibodies and derivatives (Chapter 10). Issues specific to peptides will be discussed in the context of optimization of molecular characteristics (Section 4.1.3). This section will conclude with a discussion of the use of recombinant DNA technology to produce proteins and genes to accelerate discovery of new molecular entities—small molecules or macromolecules—that have desirable pharmaceutical properties such as physical and chemical stability and bioavailability.

Biotechnology and Biopharmaceuticals, by Rodney J. Y. Ho and Milo Gibaldi
ISBN 0-471-20690-3   Copyright © 2003 by John Wiley & Sons, Inc.

4.1.1. Data Mining, Molecular Cloning, and Characterization

For the identification and selection of a therapeutic target, several key issues must be considered. These issues include knowledge of the molecular basis of the disease, with regard to tissues, cells, or specific molecules that modulate the disease target [1]. In other words, knowledge about the links of clinical symptoms with specific cell types and molecules will be invaluable in developing potential molecular targets. In the past this knowledge was communicated through patents and manuscripts published in scientific and medical journals. As we enter the information age, with advanced computing and high-speed communication at our disposal, knowledge critical to the selection of molecular target information is readily available to anyone with Internet access. Still more information is available through the recent completion of the Human Genome Project that cataloged the genome sequence [2,3]. Therefore we continue to add to our knowledge the genetic information that may be useful for identifying molecular targets.

What remains relatively unchanged, however, is how one uses the information available in the literature or other sources to identify the drug candidate that will have an optimal effect on the therapeutic target. The strategy to clone and express disease-related genes also remains unchanged.

The ability to manipulate DNA, to analyze the genetic code, to transcribe DNA to RNA, and subsequently to translate RNA to protein in well-defined prokaryotic and eukaryotic cells is known as genetic engineering or recombinant DNA technology. The technology developed in the late 1960s and early 1970s continued to advance rapidly by integrating the results of basic studies from chemistry, biology, biochemistry, microbiology, genetics, pharmacology, and fermentation sci-

ences [4,5]. Some of the key advances are listed below:

1. Basic understanding of DNA replication paves the way to isolate and manipulate genes of interest.
2. Elucidation of molecular processes involved in transcription and translation [6,7] leads to the development of a bacterial and mammalian protein expression system to mass-produce proteins.
3. Advances in B (antibody-producing) cell biology and tumor (e.g., melanoma) cell biology lead to the development of immortalized cells that overexpress a monospecific (monoclonal) antibody [8].
4. Advances in DNA and peptide sequencing techniques [9–14] lead to a large number of potentially useful gene and peptide sequences now widely available in the national databases.
5. Detailed understanding of gene expression (prokaryotic or eukaryotic plasmid vectors) and regulation (promoter, suppresser, feedback control, etc.) (Figure 4.1) allows large-scale manufacture of functional proteins, and forms the basis for more recent attempts at *in vivo* gene expression as part of gene therapy.

The first step in producing sufficiently pure and large enough quantities of new macromolecules or proteins is to clone the gene and express its protein in bacterial (prokaryotic) or mammalian (eukaryotic) cells. The detailed technologies and procedures to accomplish this are discussed in a number of recent publications and book chapters [4,5]. A brief description of terminology used in recombinant DNA applications is listed in Table 4.1. We will only discuss these techniques from the perspective of the strategies generally used to increase the efficiency of cloning, expression, and characterization of the recombinant products. Where appropriate, we will highlight advantages and limitations. For this purpose we will use the terms "macro-molecule" and "protein" interchangeably. The isolation of the gene encoding the protein and expressing and characterizing it will be described as a sequential process. We must be mindful that this process requires multiple iterations and simultane-ous execution of strategies. The process can be roughly divided into five categories:

1. Identify the best source for isolating the gene that encodes target protein.
2. Clone the gene.
3. Engineer an expression system for the respective protein.
4. Optimize the DNA sequences to enhance protein expression.
5. Verify the molecular and functional characteristics of the expressed re-combinant protein.

Identify the Best Source for Isolating the Gene That Encodes Target Protein

To identify an optimal source for the isola-tion of a gene, one must know the tissues, cells, bacteria, or viruses that produce the highest possible copy numbers of the target genes in DNA or RNA transcripts. Large quantities of the target gene are found in tissues or cells that express large quantities of the gene product. These tissues or cells can be used to isolate the messenger RNA that encodes for the target protein. For example, isolation of genes for bacterial and viral pro-teins will require biologic sources enriched with the respective bacteria and virus-infected cells or tissues as starting materials to isolate DNA and RNA. For most thera-peutic candidates that are endogenous pro-teins, the tissues responsible for producing these proteins will be the best sources to isolate genes of interest. Alternatively, one can use inducible elements of established tissue-culture cells to increase the mRNA that encodes for the target protein. Inducible elements increase protein expression by several-fold. Additional details on using gene-activation signals to enhance gene iso-lation are listed in Table 4.2.

From the enriched source of tissues or cells, one can isolate RNA and DNA of target genes. These tissues or cells also allow one to purify small quantities of the putative proteins, thereby permitting gen-eration of antibodies reactive to these

■TABLE 4.1. Terminology used in recombinant DNA research

Genomic DNA	All DNA Sequences of an Organism
cDNA (complementary DNA)	DNA copied from a messenger RNA molecule that encodes for respective amino acid sequences
Plasmid	A small extrachromosomal cicular DNA molecule capable of reproducing independently in a host, prokaryote, or eukaryote, cell
Vector	A plasmid containing either a cDNA or genomic DNA sequence or interest
Genomic clone	A host cell (usually bacterium) with a vector containing a fragment of genomic DNA from a different organism of interest
cDNA clone	A vector containing a cDNA molecule from another organism, which can be used to transcribe or translated into protein in selected host cells
Library	A complete set of genomic clones from an organism, or cDNA clones from one cell type
Restriction enzymes or endonucleases	Prokaryotic enzymes with exquisite sequence recognition of target DNA duplex and precise cleavage site

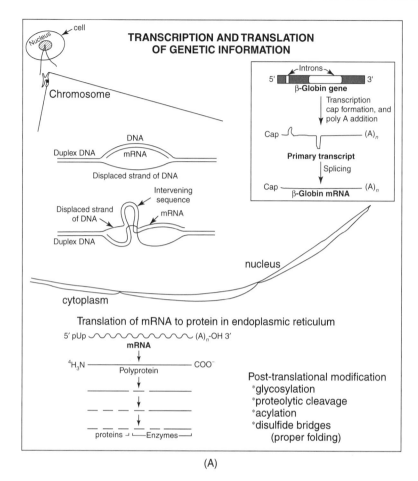

(A)

Figure 4.1. Schematic representation of protein synthesis beginning with the genetic information in DNA sequences within nuclear chromosomes (A), and concluding with post-translational protein modification (B). Some DNA gene sequences may contain introns or sequences that do not encode for the putative polypeptide and are processed within the nucleus through RNA splicing as exemplified by the well-studied β-globin gene RNA (*insert*). Other polypeptides such as those of viral gene products may express multiple proteins in a single polypeptide that is proteolytically cleaved or processed after protein translation. Additional post-translational modification may include protein glycosylation in which glycan residues are added. Terminal sugar residues, glucose or galactose, have been shown to play a significant role in the rate of protein clearance from blood circulation.

proteins for downstream verification of the protein generated by the recombinant vectors using molecular cloning strategies.

Clone the Gene

Once the enriched source of a gene that expresses a target macromolecule is identified, three general strategies are commonly used to clone the gene. They are (1) com-plementary DNA (cDNA) cloning from mRNA, (2) cloning of gene employing polymerase chain reactions (PCR) or PCR cloning, and (3) cloning the gene by first generating a genomic library or genomic cloning. In some cases, a combination of all three strategies or a variation will be used to isolate DNA fragments and reconstruct the complete DNA sequence encoding the gene product.

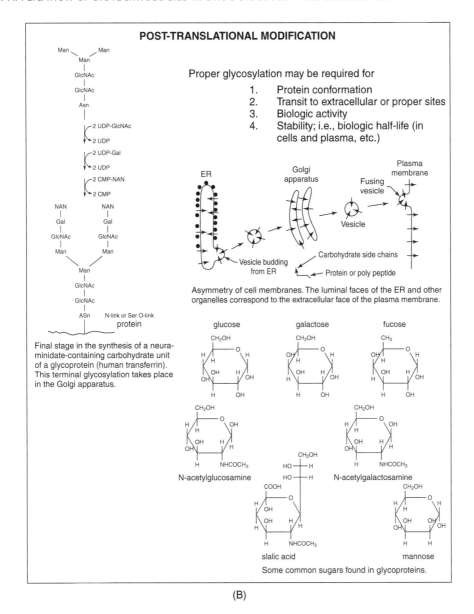

POST-TRANSLATIONAL MODIFICATION

Proper glycosylation may be required for

1. Protein conformation
2. Transit to extracellular or proper sites
3. Biologic activity
4. Stability; i.e., biologic half-life (in cells and plasma, etc.)

Asymmetry of cell membranes. The luminal faces of the ER and other organelles correspond to the extracellular face of the plasma membrane.

Final stage in the synthesis of a neura-minidate-containing carbohydrate unit of a glycoprotein (human transferrin). This terminal glycosylation takes place in the Golgi apparatus.

Some common sugars found in glycoproteins.

(B)

Figure 4.1. *Continued*

cDNA Cloning. When mRNA of the target gene, such as the insulin gene in pancreatic tissue, can be obtained, one can use the enzyme reverse transcriptase to synthesize DNA complementary (cDNA) to the messenger RNA. Reverse transcriptase is the enzyme used by retroviruses to translate their RNA to genomic DNA in completing their life cycle. Messenger RNA is only a small fraction of total RNA, but it is expressed in high concentrations in tissues that express the target protein. Therefore messenger RNA isolated from enriched sources is likely to contain RNA sequences that encode for the target protein.

Reverse transcriptase requires 10 to 15 bases complementary to target messenger

■TABLE 4.2. Gene-activation signals and tissue or cell sources that enhance isolation and identification of target genes in eukaryotes

Activation Signals

Class	Example	Target Cells, Tissues, and Genes
Hormones		
Proteins	Growth hormone	Many cells
	Prolactin	Secretory cells in breast tissue
Steroids	Estrogens	Liver, brain, reproductive organs
	Testosterone	Muscle, bone, skin, reproductive organs
Circulating or secreted factors	Nerve growth factor	Differentiating nerve cells (e.g., axons)
	Epidermal growth factor (EGF)	Many cultured cells, cells of skin and eyes
	Interleukins or lymphokines	Leukocytes or white blood cells
	Erythropoietin	Red blood cell precursors
	Interferons	Epithelial cells, white blood cells
	Platelet-derived growth factors (PDGFs)	Fibroblast cells
Environmental nutritional signals	Amino acids, nucleotides, phosphatases, glycosidases	Lower eukaryotes (cell growth, hydrolytic functions, and replications)
	Modulation of glucogenesis and protein synthesis activity	Animal cells and metabolically active tissues
	Heat shock and stress	Induction or elimination of background mRNA, allowing isolation of specific proteins and mRNAs
	Toxins, drugs, carcinogens, heavy metals	Cytochrome P-450, monooxygenase enzymes, transporter proteins in liver, kidney, gut, lung, and other tissues
	Hemorrhagic and inflammatory compounds	White blood cells

RNA to initiate cDNA synthesis. All cellular mRNA contains multiple repeats of adenine bases (poly-A tails). Therefore the complementary thymine bases (oligo-dT) can be used as a primer that binds to the mRNA template required for the reverse transcriptase to synthesize the cDNA. In the case of pancreatic mRNAs (Figure 4.2), the significantly higher mRNA for insulin compared with other proteins allowed success in isolating the insulin-specific cDNA. Subsequent insertion of cDNA into a bacterial expression vector allowed the production of functional insulin that is now marketed as a successful therapeutic product (Figure 4.2).

PCR Cloning. Unfortunately, not every gene yields measurable levels of mRNA. For these situations, amplification of specific DNA sequences is necessary. This was achieved by the invention of *in vitro* polymerase chain reaction (PCR) techniques to amplify cDNA using the DNA primers specific for the target. With this technique, a complementary oligonucleotide probe corresponding to target DNA sequences can be used to isolate the cDNA [15,16].

In the event that no partial DNA sequences are known, but functional target protein is available, we can identify the terminal amino acid sequence. The three-to-

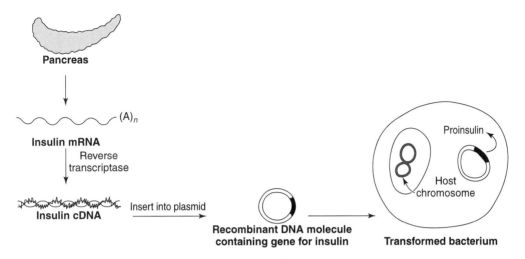

Figure 4.2. Cloning of insulin from cDNA isolated from pancreas, the main tissue responsible for synthesis of insulin, and inserted into plasmid vectors that permit expression in host cells.

five amino acid sequence can be used to predict the corresponding DNA sequences based on the degenerative genetic codes of amino acids (Appendix V). The predicted DNA sequences serve as the primer required to amplify the target gene. The highly selective amplified DNA product generated by PCR can then be cloned into expression vectors for sequence and functional characterization.

Because the efficiency of PCR performed *in vitro* is very high, we no longer require bacterial replication to amplify the quantity of DNA. Therefore, this is the method of choice for characterizing and cloning the gene for final expression of the protein products (see Figure 4.3 for brief presentation of PCR). PCR gene amplification can also be used to increase low concentrations of cDNA produced by reverse transcribing of mRNA, as described in the previous section.

A limitation of this strategy is the fidelity of enzymes—reverse transcriptase and DNA polymerase—used in PCR to produce consistently error-free DNA sequences. These enzymes typically introduce error at the rate of one base per 400

to 700 kilobases synthesized [17]. By this error rate the methodology allows error-free isolation of DNA sequences that encode up to 1333 amino acids. This translates to 180 kDa and covers most of the proteins in use as therapeutics today.

Genomic Cloning. In the event that the two strategies cited above do not provide a satisfactory outcome, a "shotgun" approach to gene cloning, otherwise known as genomic cloning, can be used to clone the entire genome of a cell (Figure 4.4). The entire genome of a cell is digested with a specific restriction nuclease to generate a very large number of fragments that are inserted into millions of bacterial plasmid vectors containing unique and sometimes overlapping genomic DNA sequences. When these plasmids are transfected into bacterial cells, each of the millions of bacteria contains some of the genomic DNA sequences, and each plasmid is a genomic DNA clone; the entire collection of plasmids is known as a genomic library.

Because nuclease digestion of genomic DNA is a random process and the higher eukaryote DNA contains introns

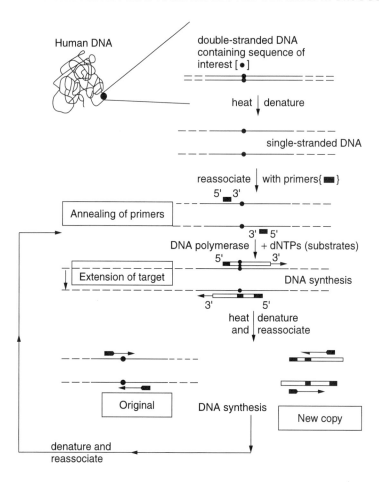

Figure 4.3. The polymerase chain reaction (PCR) to exponentially amplify a unique sequence of DNA in a test tube over 25 to 35 cycles of heating and cooling the mixture. Since the discovery of DNA polymerase by Arthur Kornberg, researchers have been dreaming about developing ways to amplify DNA without having to insert the nucleic acid into vectors and allow the bacterial host to amplify the unique sequences. A process such as this is feasible, but it is time-consuming and yields unpredictable results. While the purified DNA polymerase allows synthesis of new strands of DNA, they must be heat-denatured to produce single-stranded DNA as a starting template for the next round of DNA synthesis, a process where most DNA polymerases do not survive. Hence the reaction cannot be recycled. The search for enzymes that can survive heating and cooling has led to the discovery of heat stable, Tac-DNA polymerase that has made *in vitro* chain DNA amplification, or polymerase chain reaction, a reality. This process requires short DNA sequences to serve as primers for amplification of target sequences. Since then, many more heat-stable polymerases have been isolated and used in PCR. (Adapted from Lancet, 1988; Iss. 8599 1372–3)

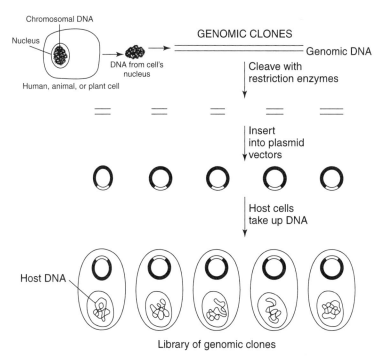

Figure 4.4. Genomic cloning of large DNA sequences. Typically chromosomal DNA is cleaved with restriction enzymes into small fragments suitable for inserting into plasmid vectors. These plasmid vectors are reintroduced into unique clones of bacterial host cells generally known as a library of genomic clones. By systematically screening for the genomic clone or clones that contain DNA sequences, the gene of interest can be identified. The expression vector containing the target gene is often required to be reconstructed from several clones, putting together DNA sequences isolated from these clones. The overall cloning process is called the genomic cloning strategy.

or noncoding sequences, an individual DNA plasmid clone generated by this strategy is unlikely to contain the entire sequence coding for a target gene product. In other words, the entire DNA sequence encoding for the polypeptide will be distributed in several clones within the genomic library. While this strategy permits, with minimum effort, generation of a genomic library with millions of DNA clones, the challenge is to find the clones that contain the DNA sequence of the target gene.

This is often done by *in situ* hybridization of the probe DNA constructed with oligonucleotides generated using the known or predicted sequences of the target gene. The bacterial cells containing these target DNA fragments will react to these probes. These probe-positive bacterial cells are expanded to collect DNA sufficient to provide for sequence analysis. Many of these clones are screened and analyzed to obtain the full genomic DNA, including sequences that are not coded for final protein synthesis. For some applications, where the target gene of interest is small and does not contain an intron, the DNA library can be cloned into vectors designed for protein expression and screened for protein function. Alternatively, if antibody to the target protein is available, it can be used to screen for bacterial clones that react to the antibody.

Regardless of the strategy one uses to identify and clone the gene of interest, these genetic clones of DNA encoding for the target protein are then analyzed initially by their physical size in relation to the size prediction based on protein data (if available). Other analyses include restriction endonuclease fragment analysis (if known), in which predicted unique fragments are generated by the enzyme, which acts on select sequences of the gene. Ultimately the gene is sequenced to deduce the entire coding sequence. It was not long ago that DNA sequencing was one of the most challenging, tedious, and time-consuming tasks in cloning of a protein. The recent advances in fluorescence dye labeling, PCR technology, and automation have expedited this process, allowing the DNA sequencing to be done with minimum effort [18–20]. Today an entire DNA sequence encoding about 4000 bases can deduced and analyzed within one week.

Engineer an Expression System for the Respective Protein

The next step in molecular cloning of the target gene product is to translate the genetic sequence into the protein product so that the functional characteristics can be verified and analyzed in detail. One can choose either prokaryotic hosts (bacteria and phage) or eukaryotic hosts (yeast, insect or mammalian cells) for this purpose (Table 4.3). The selected genes are inserted into the host cells by means of small circular DNA fragments called plasmids. Once introduced into respective host cells, the plasmid containing the gene of interest will be transcribed and translated into the protein of interest. The host-plasmid combination is known as an expression system.

The choice of prokaryotic versus eukaryotic expression systems is determined by the functional activity requirement of the protein product. When a therapeutic protein is designed, additional considerations are introduced, including post-translational modifications (i.e., glyco-

sylation, acylation, proteolytic cleavage, and protein folding). The vectors destined to be expressed in bacteria are simpler and easier to engineer. The bacterial expression system has higher replication rates and costs less to produce a unit of protein.

Most target proteins of therapeutic interest are human or mammalian in origin, but some require post-translational modification during expression for optimum biological and pharmacokinetic properties. Bacterial expression systems do not allow most post-translational modifications. In those cases, eukaryotic expression systems are chosen. These systems, however, are more complex and require more time and resources to engineer, especially in mammalian cells. Consequently, the final products produced in mammalian cells are more expensive than those produced by bacterial expression systems.

An ideal plasmid vector can be replicated and expressed in both mammalian and prokaryotic cells. Verification of gene insertion in mammalian cells is difficult, and researchers usually turn to bacterial cells for isolation of easily replicated plasmid DNA and sequence analysis. This plasmid DNA is then introduced to mammalian cells for expression.

In the past there were few plasmid vectors that could be replicated and expressed in both mammalian and prokaryotic cells. However, with a better understanding of genetic sequences essential for replication, transcription, and translation, we now have the ability to construct plasmids with nearly ideal properties. Furthermore, some plasmid constructs are now engineered to express additional tags, such as polyhistidine, as a part of protein expression. By using standard techniques to purify the tag, the efficiency of isolating target protein is greatly increased.

Regardless of which plasmid expression system one chooses, the initial goal is to verify that the target gene is inserted and cloned properly in the plasmid so that

■TABLE 4.3. Comparison of the features and requirements of gene expression systems

Expression System	Features	Selection Strategy	Some Required Elements
Prokaryotic: Bacterial plasmid	Inducible or constitutive	Drug resistance	Bacterial origin (*ori*) sequence needed for DNA replication[a]
Prokaryotic: Bacterial cells infected with bacteriophage	Transient expression	Infected cells	Viral (bacterial phage) promoter and other elements required to support viral protein synthesis
Eukaryotic: Insect cells (e.g., *Autographica californica* polyhedrosis virus)	High levels of protein expression	Infected cells	All essential elements for virus gene; target gene is usually expressed as a polyhedron fusion protein or used as a signal to express target gene
Eukaryotic: Yeast plasmid or integration into host chromosome by homologous recombination	Transient or permanent	Amino acid requirement in autotrophic strain; heavy metal induction of resistance gene	Yeast *ori* sequence; constitutive or inducible promoter; transcription terminator
Eukaryotic: Animal (recombinant) virus vectors	Transient or lytic infection	Infection in susceptible cells	All essential viral genes; strong promoter/enhancers; polyadenylation signal; intron sequences
Eukaryotic: Animal (recombinant) virus vectors	Permanent infection	Alteration of transformed host cell phenotype	All essential viral genes; strong promoter/enhancers; polyadenylation signal; intron sequences; host transforming gene sequences
Eukaryotic: Mammalian cell–transient expression	Transient expression	None: A higher copy number can be achieved using *ori* sequence that responds to factors in recipient cells	Strong constitutive or inducible promoter; polyadenylation signal; intron sequences
Eukaryotic: Mammalian cell–permanent expression	Permanent expression; achieved by integration of gene into host chromosome	Drug resistance; complementation of deleted essential gene in recipient cells	Strong constitutive or inducible promoter; polyadneylation signal; intron sequences

[a]*ori* is the unique nucleic acid sequence that serves as an origin of DNA replication.

functional characteristics of the expressed protein product can be verified. About 10 years ago, cloning of a gene and expression of a protein took up to 3 to 5 years. With the development of well-characterized plasmid constructs and standardization of molecular cloning tools, the process has been reduced to a few weeks [21–23].

Optimize the DNA Sequences to Enhance Protein Expression

The purpose of initial cloning and expression of a target gene is to demonstrate its biochemical characteristics and its functional activities, including binding affinity to the putative receptor and mediation of cellular events related to therapeutic responses. The protein produced in the initial expression system, however, may not be enough for preclinical testing, and the level of expression may be so low that even expanding the system will not allow for large-scale manufacturing. Therefore, the initial cloning of a target gene product often will be followed by optimization of the expression system.

The aim of optimization is to (1) provide high efficiency and stable protein expression, (2) include signal peptides or other DNA sequences inserted into plasmid vectors to promote excretion of target protein by the host cells, (3) allow efficient subcloning or transferring the gene into prokaryotic or eukaryotic plasmids, according to the pharmacologic and pharmacokinetic requirements for the target protein, (4) contain introns or codon modifications to enhance transcription and translation of the target gene, and (5) be adaptable to express functional proteins in limited numbers of host cells (i.e., Chinese hamster ovary cells, *E. coli*, and yeast). Sometimes, during the optimization of expression systems, a drug company must commit to one of the systems so that their scientists can begin to optimize DNA sequences for production.

Optimization of gene expression may be applied at every step of the process, from initial cloning and characterization to initiation of clinical trials. Often several rounds of optimization will be required to select the expression system that produces the highest yield with the lowest cost fermentation and purification schemes. Significant resources are allocated for optimization because the best protein expression system can lead to several hundred- to a thousandfold increased efficiency in protein produced. Under these conditions, as much as 10% of total proteins produced by the expression system are target proteins.

Verify the Molecular and Gunctional Characteristics of the Expressed Recombinant Protein

In order to verify the molecular and functional characteristics of the protein synthesized by the expression system, a number of molecular tools have been developed. These tools have been optimized over the years to allow efficient verification of the molecular characteristics of recombinant proteins. Some are also used for quality control procedures in the synthesis of recombinant proteins for pharmaceutical use.

During this process some key questions are: (1) Does the DNA sequence inserted into the expression vector contain all the sequences essential for the target protein? (2) Is the messenger RNA expressed at the predicted length and produced in high enough levels? (3) Does the recombinant protein react with antibody? (4) Is there evidence of functional activity of the recombinant protein? Methods for answering these questions are listed in Table 4.4.

When clones of cells are determined to express a putative protein of interest, a set of small-scale purification procedures is developed to further characterize the biochemical and biophysical properties of the protein. Typically, to isolate sufficiently pure

■ TABLE 4.4. Some methods used for screening and verifying molecular clones form the gene expression system

Method	Purpose	Advantages	Disadvantages
Enzyme-linked immunosorbent assay (ELISA)	Immune-based quantitation of expressed protein	Rapid, specific, quantitative, and efficient	Requires one or more high-affinity antisera or monoclonal antibodies
Spot blot	Immune-based quantitation of lysate	Quantitative and specific	Low sensitivity
Western immunoblot	Immune-based verification of the size of expressed protein	Excellent tool to detect protein size and integrity in a sensitive manner	Not a quantitative assay and not suitable for large-scale screening
RNA screening	To verify expression of transcript	Sensitive and does not require antibodies	protein detecting is indirect and problems in translation or post-translational modification may be missed
Immunoprecipitation (radiolabeled)	To determine the size and integrity of protein	Highly sensitive for use as initial characterization tool for product size and integrity	Semi-quantitative and labor-intensive—not suitable for large-scale screening
Direct colony screening using protein blot	To identify cells that express target protein	Screening of colonies in situ using a specific, high affinity serum	Not suitable for screening intracellular or membrane bound protein; must have high affinity antibody or antiserum
Immunofluorescence	To identify protein-expressing cells in the population and subcellular localization of express protein	Relatively efficient and provides intracellular details	A qualitative method requiring cell fixation, which inactivates the protein expressing cells
Fluorescence-activated cell sorter (FACS)	To predict and select cells in the population that express target protein	Sensitive and allows direct sorting and cloning of positive cells	Required proteins expressed intracellularly or on the cell surface. Not sensitive for expression system engineered to secrete the protein

protein from either cells or culture medium, column chromatography techniques are used (Box 4.1). Column chromatography techniques allow separation of proteins from contaminants based on charge, protein size, and binding affinity. In some cases the initial purification method includes isolation of a small quantity of protein using antibody immobilized on a solid support. While this immune-based purification, using immuno-affinity column chromatography, permits purification of protein for research purposes, the limited life span of an antibody bound to a column and the limited availability and reproducibility of antibody sources do not permit large-scale purification for pharmaceutical use.

Alternatively, a terminal tag such as polyhistidine can be added to the target protein to improve the efficiency of the purification procedure [24]. An affinity column designed to recognize polyhistidine could be used to isolate the tagged protein. The polyhistidine is then cleaved and the mixture dialyzed. While this approach is useful for small-scale preparations, whether it can be used for large pharmaceutical scale preparations remains to be seen.

With even a small amount of purified target protein, detailed molecular characterization can proceed. At this stage the recombinant protein is evaluated in terms of secondary and tertiary molecular structure, protein fragmentation, molecular heterogeneity, degree of glycosylation, and stability. The technologies developed to address these issues are summarized in Table 4.5. Many of these methods are discussed in texts listed in the references. During small-scale preparation of the target protein, isolation and purification tools often are refined and optimized.

A combination of these optimized tools are used for large-scale operations essential to obtaining sufficient quantities of recombinant protein for preclinical and clinical development. Additional scale-up and purification strategies are discussed in the following section. Regardless of the purifi-

cation strategy, a number of the methods used to verify the expressed protein form the basis of controlling the purity, functionality, and stability of the biopharmaceutical. Some key methods—functional (enzyme activity, cell growth, or receptor binding) assays and enzyme-linked immunosorbent assays (ELISA)—are used to study the time course of drug disposition.

4.1.2. Optimization of Cell Expression Systems to Maximize Production

While the initial goal of the discovery process is to express and purify the recombinant protein, without regard to cost and as quickly as possible, ultimately the production of a recombinant protein must be optimized to ensure that it can be produced at a reasonable cost. In most cases expression systems used to produce small quantities of recombinant protein for preliminary evaluation are not suitable to produce large quantities. Therefore, the gene inserted in the initial expression vectors must be systematically modified and recloned into vectors that can produce (1) a high yield of expressed protein, (2) stable host cell transfections, and (3) protein excretion.

Excretion of protein outside the cells allows harvesting without having to kill the cells. This process allows cells to continue to produce recombinant protein. Not all recombinant proteins can be made to be secreted from the cell. These proteins must be harvested after disruption of the cell membrane; in the process, fermentation is terminated and cells are destroyed. Isolation of proteins expressed intracellularly in cytoplasm or inclusion bodies is not only complex but also expensive. Harvesting an excreted protein from the culture medium significantly reduces the cost of making a recombinant protein. The optimization of recombinant expression systems is a continuous and important challenge in drug development and plays a key role in the profit margin of a drug company.

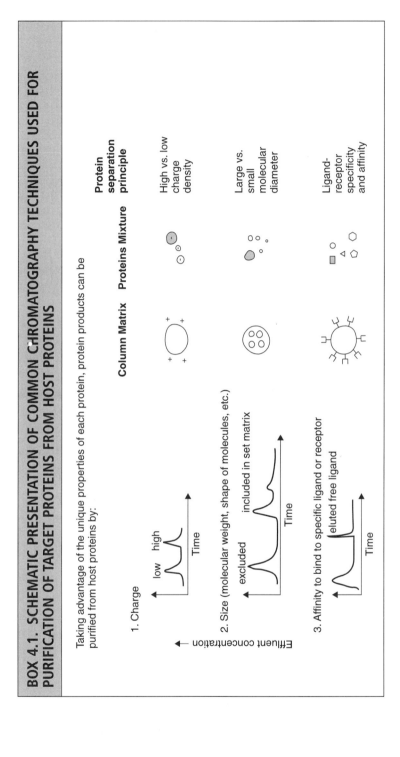

BOX 4.1. SCHEMATIC PRESENTATION OF COMMON CHROMATOGRAPHY TECHNIQUES USED FOR PURIFICATION OF TARGET PROTEINS FROM HOST PROTEINS

■**TABLE 4.5.** Some analytical techniques use in protein characterization and stability assessment

Techniques	Analysis	Protein Examples
Analytical centrifugation	Aggregation	Rop proteins [45]
Capillary electrophoresis (CE)	Degradation	Hirudin [46]
	Determination of T_m	RNase [47]
Circular dichroism (CD)	Estimation of secondary structures	α-Spectrin [48]
	Determination of T_m	muscle acylphosphatase [49]
	Conformation and interactions	IL-4 [50]
	Determination of multimers	α-Lactablumin [51]
Differential scanning calorimetry (DSC)	Determination of T_g	hGH [52]
	Determination of T_m	α-Fibroblast growth factor [53]
Electron paramagnetic resonance (EPR)	Ligand–protein interactions	rhGH, rhIFN-γ [54]
Fluorescence	Protein unfolding interaction	α/β MerP protein [55]
	Conformation	α-Antitrypsin [56]
	Degradation and aggregation	hGH [57], βFGF [58]
HPLC-ion exchange	Degradation and aggregation	hGH [52]
HPLC-reversed phase	Estimation of contamination	rh-Parathyroid hormone [59]
HPLC-size exclusion	Degradation and aggregation	hGH [52]
	Estimation of secondary structures	Chymotrysinogen [60]
Infrared spectroscopy	Determination of T_m	h-phenylalanine hydroxylase [61]
	Confirmation	IL-1α [62]
Karl Fischer	Water determination	Insulin [63]
Light scattering	Aggregation	h-Relaxin [64]
Mass spectrometry	Determination of molecular weight, degradation products, and contaminants	βFGF [58]
	Protein glycosylation	Epoietin [65]
NMR spectroscopy	Determination of 3D and secondary structures	IL-6 [66]
	Protein relaxation and softening	γ-globulin [67]
	Protein glycosyaltion	h-β-Galacto-sulfotransferase [68]
Raman spectroscopy	Determination of secondary structures	Insulin [69]
Refractometry	Ligand–protein interactions	rhGH, rhIFN-γ [54]
Two-dimensional gel electrophoresis	Protein glycosylation	Epoietin [70]
	Protein degradation	
	Product contaminant determination	
UV visible spectroscopy	Protein aggregation	αFGF [53]
	Estimation of contamination	hGH [71]
	Probing protein conformation	IL-2, insulin [72,73]

4.1.3. Optimization of Molecular Characteristics

Not only must we optimize the expression and yield of purified recombinant proteins, we must ensure that the protein is stable, has a high degree of potency and a low degree of toxicity, and has suitable pharmaceutical properties. Pharmaceutical properties encompass distribution and residence time in target tissues, and frequency of dosing, and may have an impact on efficacy and safety.

Starting with a purified recombinant protein, biochemical and biophysical tools are used to elucidate detailed structure and function at the cellular and molecular level. This process elucidates the structure-activity relationship (SAR). The mechanistic elucidation of the protein with respect to target receptor-ligand interactions and enzyme-substrate interactions has led to the construction of macromolecules with enhanced pharmaceutical properties. Some of these enhancements provide (1) increased tissue and target penetration through size reduction of the macromolecule, (2) rational design of a minimum-binding domain, optimized for receptor binding and resistance to proteolytic degradation, and (3) suitable pharmaceutical properties through domain modification. Some of the strategies in optimization of molecular characteristics are discussed below with specific examples.

Increased Tissue and Target Penetration by Reducing the Size of a Macromolecule

The simplest form of molecular optimization to enhance pharmaceutical properties is exemplified by the modification of the antibody IgG. Because IgG constitutes a major fraction of immunoglobulin and is one of the major proteins found in plasma, it was studied with regard to pharmaceutical properties including stability, distribution, and metabolism many years before recombinant DNA technology was introduced (additional details can be found in Chapter 10). Very early on, an enzyme from papaya (papain) was shown to digest the IgG molecule into three fragments: two identical (F_{ab}) fragments capable of binding to a target antigen, and a slightly larger one (F_C domain), which does not exhibit antigen binding [25]. Subjecting IgG to an enzyme abundant in the pineapple (pepsin), on the other hand, results in a single cleavage that produces one F_{ab2} fragment, consisting of an F_{ab} dimer linked via a disulfide linkage, and an F_c fragment that is similar but not identical to that generated by papain (Figure 4.5). For additional details on IgG and fragments, see Chapter 10.

With the availability of antigen binding domains—F_{ab} and F_{ab2}—that are one-third to two-thirds the size of IgG and more compact, the ability of these molecules to permeate across the endothelial cells lining blood vessels and to distribute into tissue is greatly enhanced. As a result antibody fragments, F_{ab} and F_{ab2}, can bind bacterial or tumor antigens found in cells and tissue and reverse the course of disease progression. While this strategy may provide enhanced tissue penetration, some of the functions contributed by the F_c domain, such as increased plasma residence time or prolonged half-life (up to 30 days), as well as F_c-mediated target cell lysis (via complement or cell-mediated mechanisms) will be lost.

Chemical cleavage of IgG monoclonal antibody

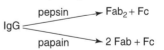

Figure 4.5. Schematic representation of enzyme specific cleavage of immunoglobulin G (IgG) by pepsin and papain. Treatment of IgG with pepsin produces two unique fragments, F_{ab2} with two antigen binding sites and F_c without binding sites. Treatment of IgG with papain generates two F_{ab} and one F_c fragments.

Alternatively, if the prolonged presence of antibody is not required for successful therapeutic application, the rapid antibody removal offered by F_{ab} fragments could be advantageous. A F_{ab} antibody fragment of the monoclonal antibody, 7E3, designed to bind to the glycoprotein receptor on human platelets and inhibit platelet aggregation, is a good example of the benefit of using small antibody fragments. Inhibition of blood clotting is needed for a limited time only, and prolonged inhibition of platelet aggregation could lead to uncontrolled bleeding.

Rational Design of Minimum Binding Domain Optimized for Receptor Binding and Resistance to Proteolytic Degradation

The overall goal of rational drug design is to draw on the knowledge of how a new molecular entity interacts with its target protein to design a therapeutic protein or peptide with superior binding affinity and pharmacokinetic profile. Molecular optimization has long depended on the chemist to design peptide analogues that are resistant to metabolism while retaining biological activity. Peptides derived in this way can be used as a starting point to further refine the molecular structure. This exercise is known as structure-based design and often leads to new molecular entities that have little resemblance to peptides and are not susceptible to peptidases or proteases. New drug candidates that are compact and not susceptible to proteases can be given orally, unlike most protein and peptide drugs, which require parenteral administration to demonstrate activity.

Somatostatin is an endogenous hormone containing 28 amino acids; it regulates a number of hormones including insulin, glucagons, and growth hormone (Box 4.2). Scientists have known for quite some time that a fully active 14 amino-acid peptide fragment contains the active motif of Phe-Trp-Lys-Thr held together by a disulfide bridge through two nearby cysteine resides.

Amino-acid abbreviations are spelled out in Appendix V. Through a series of structure-activity relationship studies, the bioactive conformation and peptide sequences that produce undesirable biologic responses were identified. Also identified were sequences susceptible to proteolysis, and a working-model compound that eliminated these sequences was proposed (Figure 4.6). This allowed the rational design of optimized somatostatin analogues with desirable biologic characteristics and activity and increased stability.

An optimized cyclic octapeptide, *Sandostatin* (octreotide acetate), is now clinically used as a more potent inhibitor than the parent somatostatin for suppressing growth hormone, glucagon, and insulin. With the reduction of amino acids from 14 to 8, the molecule became more compact and did not bind to alternate sites that could produce undesirable biologic effects (Figure 4.6). In addition, the introduction of the D isomer of tryptophan further resists proteolysis, thereby prolonging pharmacologic activity. An optimized cyclic hexapeptide in which the disulfide bridge is replaced with covalent bonds has been shown to have better pharmaceutical properties than the endogenous hormone (Box 4.3) [26–28].

A similar strategy has been used to optimize a number of peptide-based compounds with therapeutic potential, including tachykinins, enkephalins, and protease inhibitors. HIV protease is essential for producing mature, infectious virus, and two protease molecules are carried in each mature virion (Figure 4.7). With inhibitors developed specifically for HIV protease, but not human protease, one hopes to halt virus replication. Peptides with HIV protease inhibitor activity have been further refined by means of computer- and structure-based design strategies, leading to the development of new molecular entities that are stable to proteases and compact so that they can be administered orally (Table 4.6). Some protease inhibitors (e.g., ritonavir and

BOX 4.2. OPTIMIZATION OF MACROMOLECULES WITH DESIRED PHARMACEUTICAL PROPERTIES

Advantages

- Desired pharmacokinetic properties, such as, long persistence in blood, high oral bioavailability.
- Improved accessibility to tissue.
- Reduced liver uptake.
- Reduced toxicity by eliminating the domain or "sequences" that produce undesired effects.
- Reduced cost of production because optimized small peptides/fragments of the macromolecule may be chemically synthesized.

Disadvantages

- Molecular optimization of macromolecules often requires more research and development time.
- Optimization is not a well-developed area of science.
- The net benefits of rational drug design remain to be evaluated.
- Optimization sometimes yields unexpected results.

- Downsized macromolecules or peptides may show an increased renal elimination rate.

Key technology

- Cloning of genes or gene fragments and assessment of relative biologic activities
- Automated peptide synthesis allowing the construction of overlapping peptide fragments of the macromolecule.
- Advances in computer design and improved speed of numerical iteration to make three-dimensional simulation of macromolecules practical without crystallization of proteins. These simulations provide structural information about active sites, binding domains, or immunodominant sites, and confirmation. Integration of these advances allow structure-based and computed-aided drug design.
- Structure-activity relationship studies with cyclic or unnatural peptides to enhance activity and peptidase resistance.

saquinavir) are less compact and less resistant to proteases found in the GI tract than others (e.g., indinavir, amprenavir, and nelfinavir). The development of HIV protease inhibitors has been facilitated by the elucidation of the three-dimensional structure of HIV protease and knowledge gained from studies of other related aspartyl proteases. Rational iterative drug design based on structural studies of HIV protease and molecular modeling paved the way to identifying simpler compounds with higher bioavailability and less susceptibility to viral resistance. Additional refinements have included increased resistance to metabolic enzymes found in the gastrointestinal tract and liver as well as increased resistance to drug efflux transporters that may play a role in drug clearance. (Pharmaco-

kinetic concepts, including metabolism and clearance, are discussed in Chapter 5.)

Enhanced Pharmaceutical Potential through Domain Modification

Most of the proteins that are used clinically provide pharmaceutical properties that lead to safe and effective therapy for approved therapeutic indications. Some proteins, however, may influence multiple pharmacologic pathways and elicit unwanted effects. An understanding of the specific domains that mediate biologic response permits the engineering of macromolecules that are more specific and less likely to interact with receptors other than the target (see Boxes 4.4 and 4.5).

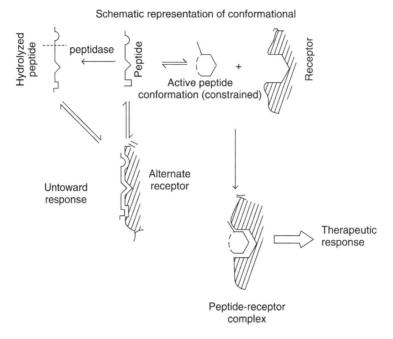

Schematic representation of conformational

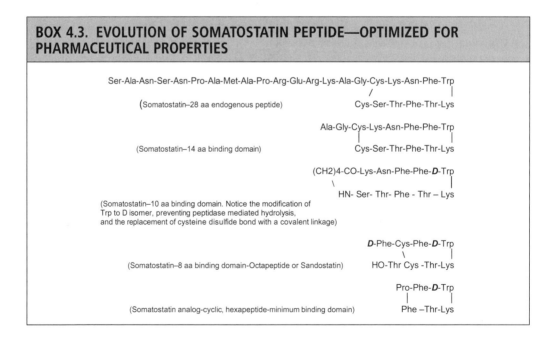

Figure 4.6. Modulation of peptide conformational equilibrium can be achieved through systematic modifications of peptide sequences that direct receptor interactions toward therapeutic response, and away from untoward effects. A systematic peptide modification may lead to reduced concentrations of peptide conformations susceptible to metabolizing enzymes such as peptidases.

BOX 4.3. EVOLUTION OF SOMATOSTATIN PEPTIDE—OPTIMIZED FOR PHARMACEUTICAL PROPERTIES

Ser-Ala-Asn-Ser-Asn-Pro-Ala-Met-Ala-Pro-Arg-Glu-Arg-Lys-Ala-Gly-Cys-Lys-Asn-Phe-Trp
/
(Somatostatin–28 aa endogenous peptide)   Cys-Ser-Thr-Phe-Thr-Lys

Ala-Gly-Cys-Lys-Asn-Phe-Phe-Trp
|
(Somatostatin–14 aa binding domain)   Cys-Ser-Thr-Phe-Thr-Lys

(CH2)4-CO-Lys-Asn-Phe-Phe-**D**-Trp
\ |
HN- Ser- Thr- Phe - Thr – Lys

(Somatostatin–10 aa binding domain. Notice the modification of
Trp to D isomer, preventing peptidase mediated hydrolysis,
and the replacement of cysteine disulfide bond with a covalent linkage)

D-Phe-Cys-Phe-**D**-Trp
\ |
(Somatostatin–8 aa binding domain-Octapeptide or Sandostatin)   HO-Thr Cys -Thr-Lys

Pro-Phe-**D**-Trp
| |
(Somatostatin analog-cyclic, hexapeptide-minimum binding domain)   Phe –Thr-Lys

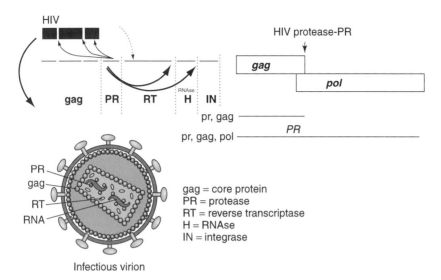

Figure 4.7. A schematic presentation of the role of protease in processing of HIV *gag* (structural protein) and polymerase *pol* essential for producing infectious virus. With inactive HIV protease, the virus generated is immature and hence not infectious. The ↓ indicates where HIV protease cleavage occurs in producing the *pol* gene product essential for viral replication.

BOX 4.4. REDUCTION OF A DOMAIN THAT CONTRIBUTES TO AN ADVERSE (INFLAMMATORY) RESPONSE OF THE CYTOKINE INTERLEUKIN-1β (IL-1β)

IL-1β is an endogenous cytokine, 269 amino acids in length, that promotes lymphocyte proliferation. However, the native protein also produces undesirable febrile responses mediated by prostaglandin E2 and IL-6. If febrile responses can be reduced, IL-1β can potentially be used therapeutically to stimulate lymphocyte growth and promote hematopoiesis. Because the nucleotide or DNA sequence of *IL-1β* is known, the protein sequence of IL-1β can be predicted and studied with overlapping peptides spanning the entire 269 amino acids to identify domains that produce proliferative and febrile responses [29]. Alternatively, the receptor binding domains can be predicted using computer simulation models to identify domains that produce

the highest proliferative response in lymphocytes with minimum stimulation of prostaglandin E2 and IL-6.

Using this strategy, researchers have been able to identify a peptide consisting of only 9 amino acids, spanning amino acids 163 through 171 in the protein [30,31]. This peptide (VQGEESNDK) has been shown to promote lymphocyte proliferation without producing significant febrile responses in animals (Figure 4.8, Table 4.7). As a result, the biologically active 9aa IL-1 peptide becomes a good therapeutic candidate with reasonable pharmaceutical profile and with far fewer side effects (i.e., febrile response), in this case febrile response that may limit its pharmacologic potential [30,31].

■TABLE 4.6. HIV protease substrates and inhibitors

Natural substrates	
Phe⌇⌇Pro, Tyr⌇⌇Pro, Leu⌇⌇Ala, Met⌇⌇Met, Phe⌇⌇Tyr, Phe⌇⌇Leu, Leu⌇⌇Phe	Substrates and cleavage sites (⌇⌇)
Inhibitors	
	Saquinavir
	Ritonavir
	Nelfinavir
	Indinavir
	Amprenavir

4.1.4. Application of Proteins and Genes as Targets to Accelerate Drug Discovery and Development

Traditionally biology has driven the identification and cloning of molecular targets in drug discovery. In recent years, however, an interesting enzyme or receptor, implicated in normal physiology or disease, has been first isolated and characterized, and the gene cloned. Following the expression of the gene using a combination of strategies as described above in a recombinant host, the desirable activity is confirmed. The well-characterized, expressed recombinant protein is then used to implement a high-throughput compound screen or to support rational drug design. This process could be time-consuming, but it delivers defined targets whose functions are understood. Many of these processes are now automated using advanced robotic technologies that provide precision, efficiency, and reproducibility.

The advent of high-throughput gene sequencing has resulted in the rapid identification of thousands of novel genes, most without known functions. Consequently pharmaceutical scientists are faced with the challenge of transforming genes of unknown function into attractive therapeutic targets, a paradigm shift sometimes called the "from-gene-to-screen" process [35].

The Human Genome Project was initiated on 1 October 1990, and the complete DNA sequence of the human genome will be realized in 2003, two years ahead of schedule. A partial blueprint of the human genome is now available [2]. Gene identification provides a basis for understanding disease at its most fundamental level. The human genome contains approximately 150,000 genes, and individual tissues express between 15,000 and 50,000 of these genes. Gene expression levels often differ in diseased tissue, with certain genes being over- or underexpressed or even entirely eliminated, or new genes being expressed. The localization of differences in gene expression is a crucial step in identifying a potential molecular target for drug discovery. Many believe that knowledge of the genetic control of cellular functions in the postgenomic era will serve as the platform of future strategies for the prevention and treatment of disease.

The identification in the 1980s of the gene thought to be responsible for cystic fibrosis took researchers about nine years to discover, whereas the gene responsible for Parkinson's disease was recently identified within a period of weeks. This extraordinary leap in the ability to associate a

BOX 4.5. MODULATION OF THE CLEARANCE OF THERAPEUTIC PROTEINS BY ADDITION OF Fc DOMAIN SEQUENCES

In the early 1970s, when recombinant proteins in large quantity first became available in pharmaceutical grade, it was recognized that most proteins less than 50 kDa are readily cleared from the circulation if they are not bound to a receptor or do not contain unique amino-acid sequences that provide prolonged residence time in blood. Studies of immunoglobulin biosynthesis and metabolism mapped the important determinants of the half-life of these molecules in plasma, regardless of their antigen specificity. These determinants were located within the F_c domain, which does not participate in antibody-antigen binding interactions. With the availability of the gene or DNA sequences corresponding to the F_c domain of IgG, researchers from a biotechnology company, Genentech, engineered a fusion protein that contains a CD4 linked to F_c domain IgG [32]. The CD4 molecule found on helper T-lymphocytes is one that human immunodeficiency virus recognizes and uses to infect these cells. The recombinant soluble CD4 (55 kDa) is cleared from blood readily, but the fusion protein CD4-F_c provides a much longer residence time in blood. Unfortunately, CD4-F_c removed the HIV gp120 envelope protein from the viral membranes, and thereby failed neutralize the virus, and it remained infectious.

The same principle was later applied by scientists at Immunex to enhance the pharmaceutical properties of the tumor necrosis factor receptor (TNFr), which, like CD4, is rapidly cleared from blood. The hybrid TNFr-F_c molecule (~150 kDa), which is now known as Enbrel, is used to reduce the inflammatory responses of rheumatoid arthritis [33,34]. The hybrid TNFr-F_c has extended the plasma half-life of TNFr from several minutes to 190 hours.

■TABLE 4.7. Elucidation of short peptide for immune-stimulatory, but not inflammatory activity of IL-1β

	Observed Effects	
Responses	IL-1β (269 aa)	IL-1 (9 aa) Peptide$_{163-171}$
Immune stimulation		
Immune restoration	+++	++
Vaccine adjuvant activity	+++	+++
Induction of IL-2 and IL-4	+	++
Inflammation		
Fever	+++	—
Corticosteroid induction	+++	—
PGE$_2$ release by fibroblast	+++	—
IL-6 induction	+++	—

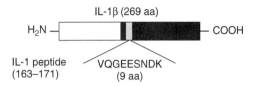

Figure 4.8. Schematic representation of amino acid 163–171 IL-1 peptide in relationship to the whole IL-1β protein.

specific gene with a disease is the result of remarkable progress in the areas of gene sequencing and information technologies [36].

Several thousand molecular targets have been cloned and are available as potential drug discovery targets. These targets include G-protein coupled receptors (GPCRs), ligand-gated ion channels, nuclear receptors and cytokines, and reuptake/transport proteins. A new potential therapeutic approach for the treatment of a known disease is published nearly every week. The sheer volume of genetic information being produced has shifted the emphasis for the generation of novel DNA sequences to the determination of which of these new targets offers biologic function with the greatest opportunity for drug discovery. Consequently target selection and validation have become the most important components of the drug discovery process [36].

An example is a chemokine called myeloid progenitor inhibitor factor-1 (MPIF-1), developed by Human Genome Sciences (HGS) for the treatment of patients with cancer to reduce the toxicity of chemotherapy. It was the first genomics-derived therapeutic product to enter clinical trials [37].

HGS was among the first biotechnology companies to mine large databases of human gene sequences, looking for previously unknown proteins that might have therapeutic value. Rather than attempting to spell out every letter of DNA's code,

the scientists at The Institute for Genomic Research (TIGR) identified and characterized messenger RNA (mRNA) copies from active genes. They copied the mRNAs back into DNAs and then spelled out a part of each gene to create expressed sequence tags (ESTs).

Researchers at HSG searched for membrane-bound proteins, including receptors for growth hormones, neurotransmitters, and cytokines that might serve as drug targets. HSG scientists also looked for protein ligands that might make good drugs themselves. By 1996 they had found about 300 genes that seemed to encode new members of known, secreted protein classes and started making proteins and testing them for activity against diseases. One of these proteins was MPIF-1. The novel chemokine reversibly stopped the proliferation of several types of bone marrow stem cells in culture, suggesting that MPIF-1 might help protect the bone marrow of cancer patients from the toxic effects of chemotherapy, which preferentially kill rapidly dividing cells. Available drugs can boost the number of blood cells after a course of chemotherapy, but there are no drugs to protect such cells from being killed in the first place [38].

Another example is a pharmaceutical agent, imatanib, developed to treat chronic myeloid leukemia (CML). The Philadelphia chromosome, discovered in 1960, was first observed as a consistent cytogenetic abnormality in the tumor cells of CML patients. This abnormal chromosome represents the reciprocal translocation of parts of the long arms of the chromosomes 9 and 22 and was the first cytogenetic evidence of the phenomenon of chromosomal translocation. The important genes in the translocation are the *Abl* gene on chromosome 9 (a tyrosine kinase) and the *Bcr* (which stands for "breakpoint cluster region") gene on chromosome 22. The BCR-ABL gene product is sufficient to cause leukemia in animal models, and its cancer-causing ability depends on tyrosine kinase activity.

CML is characterized by massive clonal expansion of cells of the myeloid lineage and progresses through three phases: chronic phase, accelerated phase, and blast phase. Current therapies include allogeneic bone marrow transplantation, which is curative but associated with substantial morbidity and mortality and limited to patients for whom a suitable donor is found, and drug regimens, including interferon-α therapy, which prolongs survival but has considerable toxicity.

In an effort to interfere with CML progression, pharmaceutical scientists at Novartis first cloned and produced recombinant BCR-ABL kinase. With the availability of this enzyme in sufficient quantity and purity, a mass *in vitro* screen of a series of enzyme inhibitors was implemented to identify drug candidates that produce an optimum pharmaceutical profile. From these, they identified STI571 (*Gleevec*) as a lead candidate to block the kinase activity of BCR-ABL. STI571 acts as a competitive inhibitor of ATP binding to the enzyme, which leads to the inhibition of tyrosine phosphorylation of proteins involved in BCR-ABL signal transduction [39,40].

Although *Gleevec* was carefully designed to inhibit the specific tyrosine kinase produced by the Philadelphia chromosome, it also produces unexpected activities. Gleevec blocks c-kit (the receptor for stem cell factor) [40], and the receptor for platelet-derived growth factor [41]. These additional activities could result in a broader array of antitumor activities, in a broader spectrum of toxicities, or both [42].

An example of drug target selection is the pairing of a G-protein coupled receptor, GPR-14, with its neuropeptide ligand, urotensin II, the most potent vasoconstrictor ever identified. GPR-14/urotensin II represents an attractive target for the treatment of disorders related to excessive vasoconstriction, such as hypertension, congestive heart failure, and coronary artery disease [43,44].

SUGGESTIONS FOR FURTHER READING

Goodman, L.S., J.G. Hardman, L.E. Limbird, and A.G. Gilman, *Goodman & Gilman's the pharmacological basis of therapeutics.* 10th ed. 2001, New York: McGraw-Hill. xxvii, 2148.

Longo, D.L., 2002 Update to Chapter 111: *Acute and chronic myeloid leukemia,* Harrison's Online (www.harrisonsonline.com), the McGraw-Hill Companies.

Alberts, B., D. Bray, J. Lewis, M. Raff, K. Roberts, and J.D. Watson, *Molecular biology of the cell.* 3rd ed. 1994, New York: Garland.

Stryer, L., *Biochemistry.* 4th ed. 1995, New York: W.H. Freeman. xxxiv.

Kornberg, A., Ten commandments: lessons from the enzymology of DNA replication. J Bacteriol, 2000. **182**(13): 3613–18.

REFERENCES

1. Goodman, L.S., J.G. Hardman, L.E. Limbird, and A.G. Gilman, *Goodman & Gilman's the pharmacological basis of therapeutics.* 10th ed. 2001, New York: McGraw-Hill. xxvii, 2148.
2. *Human genomes, public and private.* Nature, 2001. **409**(6822): 745.
3. Butler, D., *Publication of human genomes sparks fresh sequence debate.* Nature, 2001. **409**(6822): 747–8.
4. Alberts, B., D. Bray, J. Lewis, M. Raff, K. Roberts, and J.D. Watson, *Molecular biology of the cell.* 3rd ed. 1994, New York: Garland.
5. Stryer, L., *Biochemistry.* 4th ed. 1995, New York: W.H. Freeman. xxxiv, 1064.
6. Kornberg, A., *Some aspects of the enzymatic replication of DNA: the repair of partially single-stranded templates.* Proc Natl Cancer Conf, 1964. **5**: 735–41.
7. Kornberg, A., *Ten commandments: lessons from the enzymology of DNA replication.* J Bacteriol, 2000. **182**(13): 3613–18.

8. Milstein, C., and E. Lennox, *The use of monoclonal antibody techniques in the study of development cell surfaces.* Curr Top Dev Biol, 1980. **14**(Pt 2): 1–32.

9. Fairwell, T., S. Ellis, and R.E. Lovins, *Quantitative protein sequencing using mass spectrometry: thermally induced formation of thiohydantoin amino acid derivatives from N-methyl- and N-phenylthiourea amino acids and peptides in the mass spectrometer.* Anal Biochem, 1973. **53**(1): 115–23.

10. Sun, T., and R.E. Lovins, *Quantitative protein sequencing using mass spectrometry: use of low ionizing voltages in mass spectral analysis of methyl- and phenylthiohydantoin amino acid derivatives.* Anal Biochem, 1972. **45**(1): 176–91.

11. Powers, D., *Amino acid sequencing of Catostomus clarki hemoglobin. ANL-7635.* ANL Rep, 1969: 287–8.

12. Myburg, A.A., D.L. Remington, D.M. O'Malley, R.R. Sederoff, and R.W. Whetten, *High-throughput AFLP analysis using infrared dye-labeled primers and an automated DNA sequencer.* Biotechniques, 2001. **30**(2): 348–52, 354, 356–7.

13. Doolittle, L.R., G.A. Mross, L.A. Fothergill, and R.F. Doolittle, *A simple solid-phase amino acid sequencer employing a thioacetylation stepwise degradation procedure.* Anal Biochem, 1977. **78**(2): 491–505.

14. Bhown, A.S., J.E. Mole, and J.C. Bennett, *An improved procedure for high-sensitivity microsequencing: use of aminoethyl aminopropyl glass beads in the Beckman sequencer and the ultrasphere ODS column for PTH amino acid identification.* Anal Biochem, 1981. **110**(2): 355–9.

15. Mullis, K., F. Faloona, S. Scharf, R. Saiki, G. Horn, and H. Erlich, *Specific enzymatic amplification of DNA in vitro: the polymerase chain reaction.* Cold Spring Harb Symp Quant Biol, 1986. **51**(Pt 1): 263–73.

16. Mullis, K.B., and F.A. Faloona, *Specific synthesis of DNA in vitro via a polymerase-catalyzed chain reaction.* Methods Enzymol, 1987. **155**: 335–50.

17. Ling, L.L., P. Keohavong, C. Dias, and W.G. Thilly, *Optimization of the polymerase chain reaction with regard to fidelity: modified T7, Taq, and vent DNA polymerases.* PCR Methods Appl, 1991. **1**(1): 63–9.

18. Brumbaugh, J.A., L.R. Middendorf, D.L. Grone, and J.L. Ruth, *Continuous, on-line DNA sequencing using oligodeoxynucleotide primers with multiple fluorophores.* Proc Natl Acad Sci USA, 1988. **85**(15): 5610–14.

19. Prober, J.M., G.L. Trainor, R.J. Dam, F.W. Hobbs, C.W. Robertson, R.J. Zagursky, A.J. Cocuzza, M.A. Jensen, and K. Baumeister, *A system for rapid DNA sequencing with fluorescent chain-terminating dideoxynucleotides.* Science, 1987. **238**(4825): 336–41.

20. Ansorge, W., B. Sproat, J. Stegemann, C. Schwager, and M. Zenke, *Automated DNA sequencing: ultrasensitive detection of fluorescent bands during electrophoresis.* Nucleic Acids Res, 1987. **15**(11): 4593–602.

21. Parent, J.L., C. Le Gouill, M. Rola-Pleszczynski, and J. Stankova, *A highly efficient technique for cloning of refractory DNA fragments and polymerase chain reaction products.* Anal Biochem, 1994. **220**(2): 426–8.

22. Roeder, T., *Simple and efficient cloning of small polymerase chain reaction-generated DNA products.* Anal Biochem, 2000. **285**(2): 278–80.

23. Okada, T., W.J. Ramsey, J. Munir, O. Wildner, and R.M. Blaese, *Efficient directional cloning of recombinant adenovirus vectors using DNA-protein complex.* Nucleic Acids Res, 1998. **26**(8): 1947–50.

24. Strugnell, S.A., B.A. Wiefling, and H.F. Deluca, *A modified pGEX vector with a C-terminal histidine tag: recombinant double-tagged protein obtained in greater yield and purity.* Anal Biochem, 1997. **254**(1): 147–9.

25. Paul, W.E., *Fundamental immunology.* 3rd ed. 1993, New York: Raven Press. xvii, 1490.

26. Veber, D.F., R. Saperstein, R.F. Nutt, R.M. Freidinger, S.F. Brady, P. Curley, D.S. Perlow, W.J. Paleveda, C.D. Colton, A.G. Zacchei, et al., *A super active cyclic hexapeptide analog of somatostatin.* Life Sci, 1984. **34**(14): 1371–8.

27. Freidinger, R.M., D.S. Perlow, W.C. Randall, R. Saperstein, B.H. Arison, and D.F. Veber, *Conformational modifications of cyclic hexapeptide somatostatin analogs.* Int J Pept Protein Res, 1984. **23**(2): 142–50.

28. Freidinger, R.M., S.F. Brady, W.J. Paleveda, D.S. Perlow, C.D. Colton, W.L. Whitter, R. Saperstein, E.J. Brady, M.A. Cascieri, and D.F. Veber, *Synthesis of new peptides based on models of receptor-bound conformation.* Psychopharmacol Ser, 1987. **3**: 12–19.

29. March, C.J., B. Mosley, A. Larsen, D.P. Cerretti, G. Braedt, V. Price, S. Gillis, C.S. Henney, S.R. Kronheim, K. Grabstein, et al., *Cloning, sequence and expression of two distinct human interleukin-1 complementary DNAs.* Nature, 1985. **315**(6021): 641–7.

30. Tagliabue, A., P. Ghiara, and D. Boraschi, *Non-inflammatory peptide fragments of IL1 as safe new-generation adjuvants.* Res Immunol, 1992. **143**(5): 563–8; discussion 581–2.

31. Boraschi, D., P. Ghiara, G. Scapigliati, L. Villa, A. Sette, and A. Tagliabue, *Binding and internalization of the 163–171 fragment of human IL-1 beta.* Cytokine, 1992. **4**(3): 201–4.

32. Capon, D.J., S.M. Chamow, J. Mordenti, S.A. Marsters, T. Gregory, H. Mitsuya, R.A. Byrn, C. Lucas, F.M. Wurm, J.E. Groopman, et al., *Designing CD4 immunoadhesins for AIDS therapy.* Nature, 1989. **337**(6207): 525–31.

33. *Etanercept. Soluble tumour necrosis factor receptor, TNF receptor fusion protein, TNFR-Fc, TNR 001, Enbrel.* Drugs R D, 1999. **1**(1): 75–7.

34. Teng, M.N., K. Turksen, C.A. Jacobs, E. Fuchs, and H. Schreiber, *Prevention of runting and cachexia by a chimeric TNF receptor-Fc protein.* Clin Immunol Immunopathol, 1993. **69**(2): 215–22.

35. Debouck, C., and B. Metcalf, *The impact of genomics on drug discovery.* Annu Rev Pharmacol Toxicol, 2000. **40**: 193–207.

36. Ohlstein, E.H., R.R. Ruffolo, Jr., and J.D. Elliott, *Drug discovery in the next millennium.* Annu Rev Pharmacol Toxicol, 2000. **40**: 177–91.

37. Emilien, G., M. Ponchon, C. Caldas, O. Isacson, and J.M. Maloteaux, *Impact of genomics on drug discovery and clinical medicine.* Qjm, 2000. **93**(7): 391–423.

38. Wickelgren, I., *Mining the genome for drugs.* Science, 1999. **285**(5430): 998–1001.

39. Druker, B.J., M. Talpaz, D.J. Resta, B. Peng, E. Buchdunger, J.M. Ford, N.B. Lydon, H. Kantarjian, R. Capdeville, S. Ohno-Jones, and C.L. Sawyers, *Efficacy and safety of a specific inhibitor of the BCR-ABL tyrosine kinase in chronic myeloid leukemia.* N Engl J Med, 2001. **344**(14): 1031–7.

40. Shaw, T.J., E.J. Keszthelyi, A.M. Tonary, M. Cada, and B.C. Vanderhyden, *Cyclic AMP in ovarian cancer cells both inhibits proliferation and increases c-KIT expression.* Exp Cell Res, 2002. **273**(1): 95–106.

41. Carroll, M., S. Ohno-Jones, S. Tamura, E. Buchdunger, J. Zimmermann, N.B. Lydon, D.G. Gilliland, and B.J. Druker, *CGP 57148, a tyrosine kinase inhibitor, inhibits the growth of cells expressing BCR-ABL, TEL-ABL, and TEL-PDGFR fusion proteins.* Blood, 1997. **90**(12): 4947–52.

42. Tisi, M.A., Y. Xie, T.T. Yeo, and F.M. Longo, *Downregulation of LAR tyrosine phosphatase prevents apoptosis and augments NGF-induced neurite outgrowth.* J Neurobiol, 2000. **42**(4): 477–86.

43. Ames, R.S., H.M. Sarau, J.K. Chambers, et al., *Human urotensin-II is a potent vasoconstrictor and agonist for the orphan receptor GPR14.* Nature, 1999. **401**(6750): 282–6.

44. Flohr, S., M. Kurz, E. Kostenis, A. Brkovich, A. Fournier, and T. Klabunde, *Identification of nonpeptidic urotensin II receptor antagonists by virtual screening based on a pharmacophore model derived from structure-activity relationships and nuclear magnetic resonance studies on urotensin II.* J Med Chem, 2002. **45**(9): 1799–805.

45. Munson, M., S. Balasubramanian, K.G. Fleming, A.D. Nagi, R. O'Brien, J.M. Sturtevant, and L. Regan, *What makes a protein a protein? Hydrophobic core designs that specify stability and structural properties.* Protein Sci, 1996. **5**(8): 1584–93.

46. Gietz, U., R. Alder, P. Langguth, T. Arvinte, and H.P. Merkle, *Chemical degradation kinetics of recombinant hirudin (HV1) in aqueous solution: effect of pH.* Pharm Res, 1998. **15**(9): 1456–62.

47. McIntosh, K.A., S.A. Charman, L.A. Borgen, and W.N. Charman, *Analytical methods and stability assessment of liquid yeast derived sucrase.* J Pharm Biomed Anal, 1998. **17**(6–7): 1037–45.

48. Prieto, J., M. Wilmans, M.A. Jimenez, M. Rico, and L. Serrano, *Non-native local interactions in protein folding and stability: intro-*

ducing a helical tendency in the all beta-sheet alpha-spectrin SH3 domain. J Mol Biol, 1997. **268**(4): 760–78.

49. Chiti, F., N.A. van Nuland, N. Taddei, F. Magherini, M. Stefani, G. Ramponi, and C.M. Dobson, *Conformational stability of muscle acylphosphatase: the role of temperature, denaturant concentration, and pH.* Biochemistry, 1998. **37**(5): 1447–55.

50. Meyer, J.D., J.E. Matsuura, J.A. Ruth, E. Shefter, S.T. Patel, J. Bausch, E. McGonigle, and M.C. Manning, *Selective precipitation of interleukin-4 using hydrophobic ion pairing: a method for improved analysis of proteins formulated with large excesses of human serum albumin.* Pharm Res, 1994. **11**(10): 1492–5.

51. Kuhlman, B., J.A. Boice, W.J. Wu, R. Fairman, and D.P. Raleigh, *Calcium binding peptides from alpha-lactalbumin: implications for protein folding and stability.* Biochemistry, 1997. **36**(15): 4607–15.

52. Pikal, M.J., K.M. Dellerman, M.L. Roy, and R.M. Riggin, *The effects of formulation variables on the stability of freeze-dried human growth hormone.* Pharm Res, 1991. **8**(4): 427–36.

53. Tsai, P.K., D.B. Volkin, J.M. Dabora, K.C. Thompson, M.W. Bruner, J.O. Gress, B. Matuszewska, M. Keogan, J.V. Bondi, and C.R. Middaugh, *Formulation design of acidic fibroblast growth factor.* Pharm Res, 1993. **10**(5): 649–59.

54. Bam, N.B., T.W. Randolph, and J.L. Cleland, *Stability of protein formulations: investigation of surfactant effects by a novel EPR spectroscopic technique.* Pharm Res, 1995. **12**(1): 2–11.

55. Aronsson, G., A.C. Brorsson, L. Sahlman, and B.H. Jonsson, *Remarkably slow folding of a small protein.* FEBS Lett, 1997. **411**(2–3): 359–64.

56. Kwon, K.S., and M.H. Yu, *Effect of glycosylation on the stability of alpha1-antitrypsin toward urea denaturation and thermal deactivation.* Biochim Biophys Acta, 1997. **1335**(3): 265–72.

57. Zhao, F., E. Ghezzo-Schoneich, G.I. Aced, J. Hong, T. Milby, and C. Schoneich, *Metal-catalyzed oxidation of histidine in human growth hormone. Mechanism, isotope effects, and inhibition by a mild denaturing alcohol.* J Biol Chem, 1997. **272**(14): 9019–29.

58. Shahrokh, Z., G. Eberlein, D. Buckley, M.V. Paranandi, D.W. Aswad, P. Stratton, R. Mischak, and Y.J. Wang, *Major degradation products of basic fibroblast growth factor: detection of succinimide and iso-aspartate in place of aspartate.* Pharm Res, 1994. **11**(7): 936–44.

59. Nabuchi, Y., E. Fujiwara, K. Ueno, H. Kuboniwa, Y. Asoh, and H. Ushio, *Oxidation of recombinant human parathyroid hormone: effect of oxidized position on the biological activity.* Pharm Res, 1995. **12**(12): 2049–52.

60. Allison, S.D., A. Dong, and J.F. Carpenter, *Counteracting effects of thiocyanate and sucrose on chymotrypsinogen secondary structure and aggregation during freezing, drying, and rehydration.* Biophys J, 1996. **71**(4): 2022–32.

61. Chehin, R., M. Thorolfsson, P.M. Knappskog, A. Martinez, T. Flatmark, J.L. Arrondo, and A. Muga, *Domain structure and stability of human phenylalanine hydroxylase inferred from infrared spectroscopy.* FEBS Lett, 1998. **422**(2): 225–30.

62. Chang, B.S., G. Reeder, and J.F. Carpenter, *Development of a stable freeze-dried formulation of recombinant human interleukin-1 receptor antagonist.* Pharm Res, 1996. **13**(2): 243–9.

63. Strickley, R.G., and B.D. Anderson, *Solid-state stability of human insulin. I. Mechanism and the effect of water on the kinetics of degradation in lyophiles from pH 2–5 solutions.* Pharm Res, 1996. **13**(8): 1142–53.

64. Li, S., T.H. Nguyen, C. Schoneich, and R.T. Borchardt, *Aggregation and precipitation of human relaxin induced by metal-catalyzed oxidation.* Biochemistry, 1995. **34**(17): 5762–72.

65. Kawasaki, N., M. Ohta, S. Hyuga, M. Hyuga, and T. Hayakawa, *Application of liquid chromatography/mass spectrometry and liquid chromatography with tandem mass spectrometry to the analysis of the site-specific carbohydrate heterogeneity in erythropoietin.* Anal Biochem, 2000. **285**(1): 82–91.

66. Xu, G.Y., H.A. Yu, J. Hong, M. Stahl, T. McDonagh, L.E. Kay, and D.A. Cumming, *Solution structure of recombinant human interleukin-6.* J Mol Biol, 1997. **268**(2): 468–81.

67. Yoshioka, S., Y. Aso, and S. Kojima, *Softening temperature of lyophilized bovine serum albumin and gamma-globulin as measured by spin-spin relaxation time of protein protons.* J Pharm Sci, 1997. **86**(4): 470–4.

68. Honke, K., M. Tsuda, S. Koyota, Y. Wada, N. Iida-Tanaka, I. Ishizuka, J. Nakayama, and N. Taniguchi, *Molecular cloning and characterization of a human beta-Gal-3′-sulfotransferase that acts on both type 1 and type 2 (Gal beta 1-3/1-4GlcNAc-R) oligosaccharides.* J Biol Chem, 2001. **276**(1): 267–74.

69. Yeo, S.D., P.G. Debendetti, S.Y. Patro, and T.M. Przybycien, *Secondary structure characterization of microparticulate insulin powders.* J Pharm Sci, 1994. **83**(12): 1651–6.

70. Skibeli, V., G. Nissen-Lie, and P. Torjesen, *Sugar profiling proves that human serum erythropoietin differs from recombinant human erythropoietin.* Blood, 2001. **98**(13): 3626–34.

71. Hearn, M.T., M.I. Aguilar, T. Nguyen, and M. Fridman, *High-performance liquid chromatography of amino acids, peptides and proteins. LXXXIV. Application of derivative spectroscopy to the study of column residency effects in the reversed-phase and size-exclusion liquid chromatographic separation of proteins.* J Chromatogr, 1988. **435**(2): 271–84.

72. Butler, W.L., *Fourth derivative spectra.* Methods Enzymol, 1979. **56**: 501–15.

73. Brewster, M.E., M.S. Hora, J.W. Simpkins, and N. Bodor, *Use of 2-hydroxypropyl-beta-cyclodextrin as a solubilizing and stabilizing excipient for protein drugs.* Pharm Res, 1991. **8**(6): 792–5.

■ SECTION TWO ■

■ 4.2. LARGE-SCALE PRODUCTION OF RECOMBINANT PROTEINS

Protein and polypeptide-based new molecular entities that exhibit desirable pharmacologic activities are further evaluated in preclinical tests in appropriate animal models to characterize pharmacokinetics and toxicology. To do so, the drug candidate must be produced in larger scale, typically in quantities of milligrams to grams, than the amounts required for early studies in cells and *in vitro*. At this stage the yield of product per host organisms that carry recombinant genes, and the efficiency of product purification, take center stage. Both of these parameters are key determinants of the profit margins of the therapeutic protein product if marketed.

One of the key requirements in producing a given recombinant protein on a large scale is the need to identify host cells with maximum efficiency in producing proteins that are safe and effective at a reasonable cost. Theoretically, if the yield of recombinant protein recovered from each type of cell is equivalent and purification costs are similar, prokaryotic cells with the highest growth rates will be less expensive than eukaryotic cells to produce the protein of interest. In reality, choosing host cells for recombinant protein production is not always a simple or logical process (Box 4.6).

In the hope of being the first to reach the market, pharmaceutical companies often will develop plasmid vectors that can be expressed in multiple hosts. During the early phase of the scale-up process, the efficiency of recombinant protein expression in each host will be optimized, and the purification process for each host system, along with the yield and quality of the product, will be evaluated. A list of host cells and their advantages and disadvantages are presented in Table 4.8. Information regarding the pharmaceutical properties of the recombinant proteins (e.g., glycosylation, which requires post-translational modification) that may influence residence time or distribution to target tissue must also be considered in the selection of host cells.

Only eukaryotic cells are capable of

BOX 4.6. SOME ISSUES RELATED TO MANUFACTURE OF BIOTECHNOLOGY PRODUCTS

- Development of a suitable process at laboratory scale (purity adequate for intended use).
- Scale-up. Technology transfer; lack of clear boundaries dividing research, process development, and manufacturing for protein products.
- Production costs within limits defined by the market. Need to reduce production costs drives optimization for fermentation-based processes with a ripple effect downstream.
- Compliance with evolving FDA regulations and need for regulatory approvals. Need to abide by differences in regulatory environments in the United States and Europe.

- Compliance with environmental restrictions and concerns of environmentalists.
- Competitive factors. Patent protection, trade secrets, orphan drug status. Emerging competition from abroad.
- Availability of facilities and financing. Cost of construction, timing, single versus multiple-use facilities. Contract manufacturing.
- Documentation of process and validation of methods (including cleaning procedures) and facilities. Required regulatory submissions.
- Quality control and quality assurance.

■TABLE 4.8. Comparison of some key considerations in choosing host cells for recombinant protein expression in pharmaceutical scale

	Prokaryote	Eukaryote	
Consideration	*E. coli*	Yeast	Mammalian Cells (CHO, BHK)
DNA size and characteristics	4.6 Mbp, circular DNA	12.1 Mbp, chromosomal DNA	2000–3000 Mbp chromosomal DNA
Post-translational modification	None	Capable; but different from humans	Capable; similar or identical to humans
growth rate (cycles per hour)	3.33/h	0.25/h	0.02/h[a]
Cultivation method	Fermentation	Fermentation	Fermentation (suspension cells) Roller bottle (adherence cells)
Cost	Less expensive	Intermediate	>$1 million/kg

[a]Based on estimate of antibody producing hybridoma cells.

---

performing posttranslational modification. Higher eukaryotic cells produce a degree of glycosylation similar to human cells. Lower eukaryotic cells (such as yeasts) produce glycosylation characterized by branching and terminal glycosyl residues that resemble the products of human cells. While recombinant protein produced in mammalian hosts will ensure almost identical pharmaceutical properties to those of endogenous human protein, it is prudent to consider cost. Prokaryotic cells proliferate more rapidly and are more efficient and less expensive in producing recombinant proteins than eukaryotic cells, especially mammalian cells. Prokaryotic cells are particularly attractive hosts when the disposition and immunogenic properties of the recombinant protein are not critical to safety and efficacy.

4.2.1. Optimization of Product Yield through Manipulation of Genetic Construct and Recombinant Host Cells

The ultimate goal of product optimization is to increase the production of a specific recombinant protein. This is accomplished by improving the large-scale production efficiency of genetically engineered host cells. In the early stage of isolation and characterization, some information regarding the requirements of biologic and pharmacokinetic properties is gathered and used to identify the target indication(s) for which the product will be tested. This information includes consideration of factors to select the best host cells to express the recombinant proteins [1]. For example, if a protein's biologic activity is dependent on fully glycosylated product, the microbial host, *E. coli*, which is not capable of glycosylation, will not be a suitable host. On the other hand, if glycosylation of protein product need not be identical to human protein, a variant of glycosylated protein expressed in yeast is acceptable. The cost to produce the recombinant protein will be much less in yeast than that needed to produce it in mammalian cell culture.

Product yield optimization is a continuous process. It continues through preclinical and clinical development because it ultimately translates into the profit margin of the product when it is approved for marketing.

Optimization of DNA sequences in plasmid vectors by molecular biologists and production engineers is often done in collaboration with researchers in the discovery group. Some strategies include construction of elements that will enhance efficiency of protein expression and stability of the plasmids when transfected into the host cells. Once the plasmids are transfected into the host cells, these "recombinant host cells" are selected for those that are genetically stable and produce the highest yield of recombinant protein product. As many as 6000 clones of cells are screened to identify the best host cells for production. The chosen cell stock for final production is maintained by means of a master cell-bank. From each master frozen culture, a subculture stock is established for use in large-scale production. The subculture stock becomes the inoculum for every batch of product. In this way each batch of cultured cells is initiated with a common lineage of recombinant host cells. Additional details are found in Figure 4.9.

Many of these processes are not well publicized and are kept as trade secrets for most recombinant proteins. However, the optimization of host cells, such as microbial cells, to produce antibiotics or peptide derivative is well described. The screening process used to isolate and characterize cells with the highest degree of product expression is time consuming and involves multiple optimization steps. Optimization continues even after the marketing of an antibiotic. Production scientists continue to improve efficiency and lower cost per unit of product. It typically takes from 3 to 10 years to achieve a highly efficient production system (Figure 4.10). With high throughput screening, the time required for identification and development of cells that produce the highest yield may be reduced, but it may still take several years to identify a unique host-cell clone that is genetically stable and suitable for large-scale biological synthesis of recombinant products.

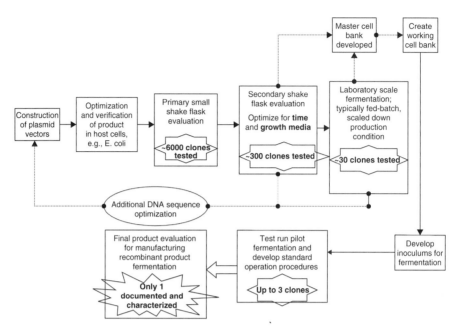

Figure 4.9. Optimization of product yield through sequential and systematic screening and selection of genetic construct and recombinant host cells that produce the highest yield and genetic stability. The schematically presented steps are required to develop a master cell-bank, and working stock for the production of biologically active recombinant proteins in pharmaceutical scale.

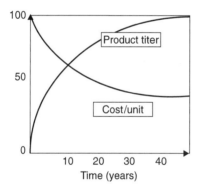

Figure 4.10. Schematic representation of the relationship between unit cost and increased product yield or titer as a function of time.

4.2.2. Large-Scale Cultivation of Host Cells for Production of Recombinant Proteins

Recombinant proteins that are genetically engineered in host cells must be cultured in large scale (up to kilograms and sometimes up to kilotons) to meet pharmaceutical needs. During the discovery or research phase, recombinant proteins are produced on a laboratory scale, to produce a few micrograms using one to two liters of cell culture. Many of the methods used for purification and isolation of proteins from these cell cultures cannot be easily scaled or proportionally increased to produce pharmaceutically useful quantities. A different set of instruments and analytical techniques are developed for large-scale cell cultures. The science of growing cells and microbes for production of pharmaceuticals and chemical compounds under well-specified conditions in large quantities is called fermentation.

Most fermentation procedures are systematically optimized in a pilot plant, which typically uses a table-top fermentor of about 30 liter capacity. These fermentors are designed to contain all the ports, valves, controls, and cleaning capabilities essential

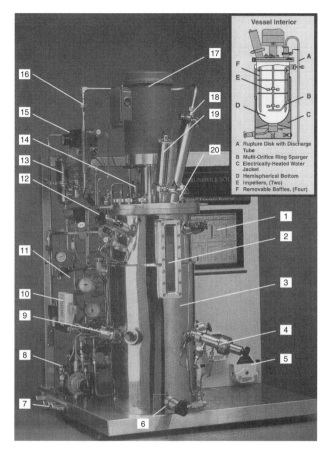

1. **Embedded Controller** with touch-screen interface
2. **Viewing Window**, allows viewing the culture above and below liquid level
3. **Sterilizable-In-Place Vessel**. Type 316L stainless steel polished to an internal finish of 15 - 20 Ra eases cleaning and prevents residue build-up
4. **Resterilizable Sample**
5. **ValvePeristaltic Pumps**
6. **Resterilizable Harvest/Drain Valve**, Easy access through removable vessel panel
7. **Services** for water, air, clean and house steam, water return and drain
8. **Steam Traps**, multiple stainless-steel 316L traps guarantee the sterilization temperature is maintained in all process lines
9. **Ports**, multiple 19 & 25 mm Ingold-type ports allow addition of RTD, pH, D.O. and other sensors
10. **Thermal Mass Flow Controller** provides precise, automatic control of air flow
11. **Open Frame Piping** facilitates access for cleaning, maintenance and servicing
12. **Re-sterilizable Inoculation/Addition Valves**
13. **Filters**, sterilizable-in-place in 316L stainless steel housings accommodate various-brand filter cartridges.
14. **Rupture Disk**, safety device of 316L stainless steel prevents over-pressurization of vessel; includes discharge tube to convey liquid to bottom of console
15. **Automatic Back-Pressure Regulator**
16. **Exhaust Line** with heat exchanger and view glass. Heat exchanger heats gas above dew point and prevents filter clogging
17. **Motor**, top-drive provides agitation for both microbial and mammalian cell culture
18. **Exhaust Condenser**
19. **Combination Light and Fill Port** provided, with sanitary Tri-Clamp fittings
20. **Headplate Ports**, (7) 28 mm ports (I.D. 19 mm) allow addition of sensors, sampling and addition devices

Figure 4.11. A 30 liter bench fermenter that can be scaled for production of recombinant proteins. The bench-top scale configuration contains all the control valves and ports necessary to monitor and control cell cultivation while maintaining sterility of the culture. The stainless steal reaction vessel allows easy cleaning and permits heat and pressure sterilization in place by connecting the vessel to a steam supply. (New Brunswick Bioflo-4500, adapted from the manufacturer's literature with permission)

for acceptable quality control and assurance during and throughout the cultivation process (Figure 4.11). They are constructed with a configuration similar to very much larger fermentors. The task of pilot plant engineers is to develop strategies that lead to the production of protein product in fermentors with a capacity of 100,000 liters (Table 4.9). A typical pilot scale, table-top fermentor is shown in Figure 4.11. Additional details on design and engineering aspects of fermentation can be found elsewhere [2,3].

Fermentation can be applied only to those cells that can grow in suspension. These cells include most prokaryotes, such as *E. coli*, and lower eukaryotes, such as yeast. However, only a small fraction of mammalian cells can be grown in suspension and adapted to a fermentation process.

Most recombinant mammalian cells are adherent cells that grow in an anchorage-dependent manner, requiring a surface support to replicate. To provide a large surface area with a minimum of cell culture medium, the adherent mammalian cells are

■**TABLE 4.9.** Batch size of cell cultures and estimated time required for fermentation

Description	Batch Size (liters)	Time (days)[a]
Laboratory shake flask	0.1	1–2
Bottles or large flask	1–2	2–4
Batch fermenter	50	4–6
Batch fermenter	2500	6–8
Batch fermenter	25,000 to 100,000	10–16

[a]Estimated based on using *E. coli* as host cells for producing recombinant proteins.

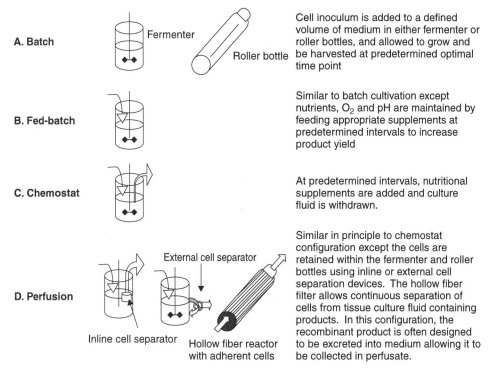

A. Batch — Fermenter, Roller bottle — Cell inoculum is added to a defined volume of medium in either fermenter or roller bottles, and allowed to grow and be harvested at predetermined optimal time point

B. Fed-batch — Similar to batch cultivation except nutrients, O_2 and pH are maintained by feeding appropriate supplements at predetermined intervals to increase product yield

C. Chemostat — At predetermined intervals, nutritional supplements are added and culture fluid is withdrawn.

D. Perfusion — Inline cell separator, External cell separator, Hollow fiber reactor with adherent cells — Similar in principle to chemostat configuration except the cells are retained within the fermenter and roller bottles using inline or external cell separation devices. The hollow fiber filter allows continuous separation of cells from tissue culture fluid containing products. In this configuration, the recombinant product is often designed to be excreted into medium allowing it to be collected in perfusate.

Figure 4.12. Configuration of large-scale, cell cultivation methods for pharmaceutical scale production of recombinant proteins.

often grown in roller bottles (Figure 4.12). Recombinant granulocyte-colony stimulating factor (G-CSF) is produced by growing recombinant Chinese hamster ovary cells using roller bottles. Alternatively, spherical support particles of 1 to 200 μm diameters (microcarriers) as well as porous microcarriers are used to increase the surface area to volume ratio.

Fermentation and alternative production techniques, such as roller bottles, can be carried out in four different ways. They are (1) batch process, (2) fed-batch process, (3) chemostat process, and (4) perfusion process. Batch and fed-batch processes require termination of cell growth while chemostat and perfusion processes allow continuous cell cultivation.

Batch Process

In a batch configuration, host cells that contain an expression vector for the recombinant product are added to a predetermined volume of growth medium (Figure 4.12A). The cells are allowed to grow until the nutrients in the medium are depleted or the excreted by-products reach inhibitory levels. At that time, the cells are harvested, and recombinant protein, found in inclusion bodies, cell-membrane fractions, or cytoplasm, are isolated after disruption of the harvested cells. Because the host cells are destroyed at the end of each run, they must be replaced every three to seven days for fermentation or every two weeks for roller-bottle or microcarrier-support production of adherent cells. To ensure uniformity and reproducibility, the FDA requires manufacturers of recombinant proteins to establish and validate a seed stock of recombinant host cells that are validated to contain the characterized expression vector and to be free of contaminants.

Fed-Batch Process

For cell culture or fermentation processes, where the growth rate of cells and thereby recombinant product formation is limited by the availability of nutrients, replenishment can be carried out in a fed-batch manner to improve product yield (Figure 4.12B). By providing a balanced mixture of nutrients, such as oxygen and amino acids required for cell growth, and chemicals to neutralize accumulating growth inhibitors, the product yield or titer and the cell density can be improved by as much as 10-fold over that of batch configuration. Fed-batch cultivation may last up to 30 days with cell densities as high as 1.4×10^7 cells/ml. At the end of the run, the cells are harvested and the recombinant products are isolated. This configuration is essential,

and often used, for mass production of recombinant proteins that are localized in cellular fractions, requiring disruption of cells to isolate the target proteins.

Chemostat Process

Continuous mass production of cells by addition of nutrients essential to promote cell growth and removal of medium and product at a predetermined time or rate is known as the chemostat process (Figure 4.12C). This strategy works best for recombinant proteins that are secreted from microbial cells and can be collected in the culture media. Typically the growth of these microbial cells is limited to a defined single substrate that can be easily replenished. Because the recombinant product can be collected, thereby preserving the cells, and because the essential substrate can be consistently replenished, production efficiency is high. However, product titer may not be as high as that obtained with fed-batch configuration but production costs overall may be lower. With the added risks of contamination due to prolonged culture, this process is not used routinely for preparation of recombinant proteins. The cells used in this cell culture configuration must be monitored to ensure their genetic and functional stability for the cultivation period, which can last up to several months.

Perfusion Configuration

The basic principle of this configuration is to retain the growth of cells in culture through removal of spent medium containing product through a filtration device and replacement with fresh medium (Figure 4.12D). In this case the cells can continue to grow under optimal conditions for long periods of time while the product is continuously harvested. Because the exchange of fresh and spent medium is facilitated through a perfusion process, this configuration is known as perfusion culture. However, the added complexity and cost of the perfusion configurations limit its use to the production of more expensive recombinant proteins such as mammalian cell products.

The medium exchange can be facilitated within the bioreactor or configured as an external attachment. A number of filter designs have been described and used for this purpose. They are the spin filter, acoustic filter, and hollow fiber culture system. The acoustic filter system is a relatively new method that uses static acoustic waves (nodes of the wave) to concentrate the cells in the effluent stream leading to sedimentation and retention in culture, while allowing the fluid to pass through. This method has been successfully tested in processing of large-scale cell cultures. The other two filtration systems rely on passage of effluent stream through porous filters that are subject to clogging. Modulating the flow rate and pattern of the effluent fluid through the system can reduce clogging.

Perfusion configurations can grow cells for up to two months, without having to reseed the cells. Consequently the process poses a risk of contamination. However, it has been shown that the preparation of some antibodies by means of hybridoma cell culture, which may last up to 30 days, can be realized with little or no contamination.

4.2.3. Downstream Processing or Purification of Recombinant Protein in Pharmaceutical Scale

Regardless of the type and configuration of cell and bioreactor combination used for recombinant product preparation, the cells are harvested at their optimal growth and viability to ensure the highest yield per unit of cell culture. The harvested biomass, which can be in the form of cell media or cell fractions, containing the putative product is subjected to a streamlined purification process tailored to the large-scale preparation of recombinant proteins. In scaling up the purification scheme (or stream) of recombinant proteins, many of the laboratory-scale preparation procedures such as centrifuging several thousand liters of cell suspension to collect a cell pellet become too expensive, if not impractical. For most proteins, the final product, designed for parenteral injection, must meet purity and sterility standards and be below maximally acceptable cellular or microbial contamination [i.e., less than 0.5 endotoxin (a component of bacterial cell wall) unit per ml of sterile solution for an injectable dosage form], as defined by the FDA. Many of these issues are considered in developing downstream or process-stream procedures for the purification of recombinant proteins.

The objective of developing a downstream purification process is to isolate the biotechnology products, employing the fewest possible steps with the simplest purification technology to achieve the required purity. The downstream process must provide reproducible purity, within a given economic framework. The lability or stability of the protein will vary with the microenvironment (i.e., bacterial, yeast, or mammalian cell lysates versus culture media or broths harvested and stored as starting materials). Minimizing the downstream process and avoiding expensive and time-consuming centrifugation, a production scientist can reduce substantially the processing time required for large-scale protein purification.

Large-scale purification methods for small molecules and short polypeptides are well established and published in the chemical engineering literature. Interested readers should refer to them for details. Although we will focus on downstream purification processes relevant to the production of recombinant products, bear in mind that many of these purification pro-

■TABLE 4.10. Estimates of cell diameters and relative density with respect to medium

Cell Type	Diameter (μm)	Relative Density (between Cell and Medium, kg/m³)
Bacterium	1–2	70
Yeast	7–10	90
Mammalian	40–50	70

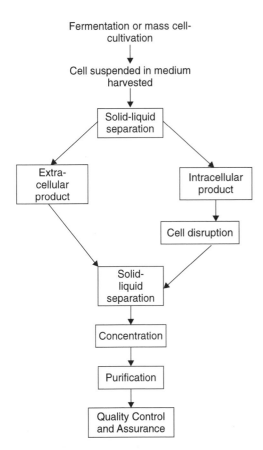

Figure 4.13. Schematic presentation of process stream to purify a recombinant protein, starting from cell suspension harvested from fermenter or cell-cultivation vessels.

cedures also apply to the isolation and characterization of monoclonal antibodies produced by hybridoma cells.

The downstream process, starting with the biomass—cells and medium har-vested from the fermentation and mass-cultivation process—can be roughly divided into four stages (Figure 4.13) [4,5]. They are (1) solid-liquid separation or clar-ification, (2) concentration, (3) purification, and (4) quality assurance and control analyses. These downstream processes and quality control procedures are essential to ensure that the purified recombinant product produced in a given batch meet or exceed the purity and quality established for a given pharmaceutical product.

Solid-Liquid Separation

Separation from culture media or broth is the primary step in collecting the product found either in cells (solid) or medium (liquid). This initial separation step is engineered based on cell size and density differences between solid and liquid (Table 4.10). In the case where the recombinant product is localized in the intracellular content such as the cytoplasm or inclusion bodies, which are highly insoluble particles found in bacteria, the cells are first isolated from the medium and then disrupted to collect the recombinant protein fraction. A number of cell disruption techniques have been developed to facilitate this step, and some are listed in Table 4.11.

Based on the differences between the solid and liquid density, batch centrifugation techniques are often used in the discovery phase as a convenient tool to separate the cell from the medium. High-speed ultracentrifugation instruments have

■**TABLE 4.11.** Some methods designed to disrupt cells

Mechanical Methods	Other Methods
Ultrasonic Homogenization Agitation with glass beads or abrasive materials	Drying Heat or osmotic shock Freeze–thaw Organic solvent Chaotropic agents Enzymes Surfactant

been developed to separate cell membrane from the medium, even when density differences are small. However, this strategy, while used in industrial scale preparations, is less attractive because of low capacity and the high cost of maintenance and energy for operating these instruments. Engineering improvements that permit centrifugation in a continuous mode include (1) the tubular bowl, (2) multi-chamber bowl, (3) disk stack bowl, and (4) decanter or scroll centrifuge (Figure 4.14). These improvements have overcome some of the capacity limitations of centrifugation, but the cost of the operation remains high.

Alternatively, separation of cells from media can be achieved with the filtration of cell suspension through membranes with defined pore size [6]. This approach takes advantage of the particle size based on size differences between cells (2–10 µm in diameter) and media (colloids of less than a few nm in diameter). Many types of filtration designs and membrane supports are available, as well as a wide range of pore sizes, to aid large-scale filtration (Figure 4.15).

Filters have been designed to reduce the fouling or clogging of solid materials at the surface of the membrane by taking into account fluid flow. Among them are the rotary drum vacuum filter (designed for

collecting yeast cells in high capacity), and the cross-flow or tangential flow filter. They have been used successfully to separate solid from liquid for harvesting fermentation products. Additional design improvements in the fabrication of membrane supports, including the hollow-fiber system, and the precision of pore size have allowed implementation of ultrafiltration strategies to separate fine particles such as membrane debris and inclusion bodies (1–5 nm in diameter) from soluble materials (less than 0.5 nm).

Other strategies for separation of solid and liquid include flocculation, a method to increase sold particle size by allowing the formation of complexes with polycations, such as cationic cellulose and polymers, inorganic salts, or mineral hydrocolloids. The resulting aggregated cells or agglomerates can be separated by gravitational sedimentation, low-speed centrifugation, or filtration with much less effort. It is interesting to note that some of these cationic agents also clear pyrogens, nucleic acids, and acidic proteins that may be problematic in purification of proteins using column chromatography (Figure 4.15).

Concentration of Putative Product before Purification

After removal of cells and debris from the culture or medium though solid-liquid separation, the fraction of recombinant protein in the liquid phase usually ranges from 2% to 15%. The large volumes associated with these low concentrations make it impractical to proceed to the next step in purification. Indeed, the volume may far exceed the capacity for chromatographic purification techniques used to isolate recombinant proteins from other soluble cellular contaminants. Therefore, a concentration step is used to reduce the volume and thereby increase the recombinant

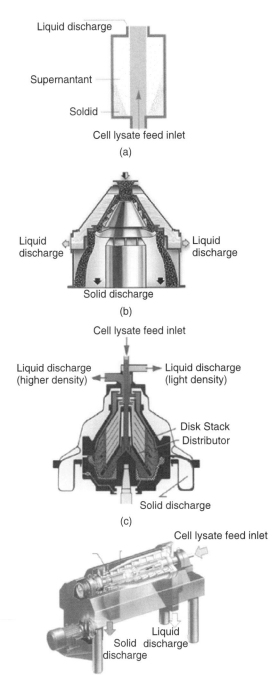

(a)

(b)

(c)

(d)

A. Tubular bowl configuration is the simplest sedimentation centrifuge and is often used in pilot pants. Due to the slender shape and small volume, a relatively high centrifugal force (*g* value) can be achieved with such centrifuges. The tubular bowl can be modified to continuously discharge the solid collected at the bottom of the bowl. The liquid can be collected efficiently through the top liquid discharge nozzle.

B. Multi-chamber centrifuge is a modification of the tubular bowl configuration. It contains a number of concentric screens that allow collection of particulate matter in the cell lysate and discharge from the instrument by gravitational force. The centrifugal force slings liquid through the effluent liquid discharge outlets.

C. Disc stack centrifuge contains a set of conical discs separated by flow channels, dividing the bowl into separate settling zones. Under the centrifugal force, the particles in the cell lysate (fed through inlet) are thrown into the space outside the disc stack at higher efficiency than the multi-chamber bowl configuration. The clarified liquid moves inward and upward to reach the annular discharge outlets where liquid of higher and lower densities can be collected. The solid accumulated at the bottom, outside of the disc stack, can be configured to discharge continuously through a nozzle or intermittently through a peripheral or axial eject mechanism.

D. Decanter or scroll centrifuge can be used for cell lysates with high solid content that can be as high as 80% of the suspension. Solids matters in the cell lysate are continuously compressed due to the rotation of the bowl and the closely fitted, tapered helical screw, operating at slightly different speed. The concentrated solids are discharged as they move toward the narrow end of the horizontal bowl.

Figure 4.14. Separation of cell lysates containing pharmaceutical products, employing large-capacity centrifuges that are configured with tubular bowl (A), multi-chamber bowl (B), disc stack bowl (C), or decanter (scroll) bowl (D).

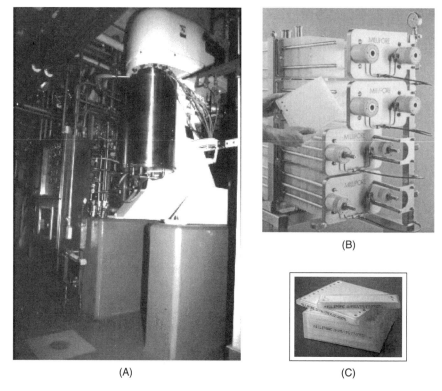

(B)

(A)   (C)

Figure 4.15. Solid-liquid separation in industrial scale using centrifugation in continuous mode (A) or passage of the suspension through manifolds such as those shown in (B) mounted with filter cartridges (C) designed for tangential or cross-flow of liquid suspension. Panel A: NIH Fredrick facility, with permission; Panels B and C: Milipore, MA, with permission.

protein concentration to make it suitable for chromatographic conditions.

Heat-assisted evaporation strategies, such as the falling-film evaporator, plate evaporator, forced-film evaporator, and centrifugal forced-film evaporator have been developed and are used to remove water from solutions of small peptides, such as antibiotics. But most recombinant proteins are heat labile and may not survive this strategy.

Other strategies have been developed to reduce the volume of recombinant protein solutions [7]. Some of the agents and strategies used to accomplish this are listed in Table 4.12 [7,8]. One approach is to precipitate the protein from solution, followed by reconstitution into a small volume of solvent.

Protein precipitation, using agents that decrease the solubility of the recombinant product, is an established technique in the pharmaceutical industry. While salts and organic solvents have traditionally been used, more specific precipitation reagents are now being tested to improve protein purification. These specific reagents are designed to provide noncovalent cross-linking of proteins, resulting in protein aggregates that precipitate from solution. They include antibodies for affinity precipitation and homo- and heterobifunctional ligands that bind to the recombinant protein, and other carriers [9]. Regardless

■TABLE 4.12. Some protein-precipitation agents and their mechanisms of action

Agents	Mechanisms of Actions
Ethanol and acetone	At low temperature, reduce dielectric constant and enhance electrostatic interactions between protein molecules
Ammonium sulfate	Removes protein-bound water molecules and promotes hydrophobic interactions between protein molecules
Charged polymers (e.g., polyacrylic acid)	Neutralize charge of protein, which decreases protein solubility in aqueous environment
Non-ionic polymers (e.g., polyethylene glycol)	Reduce water molecules available for protein solvation
pH (isoelectric point)	Lowers solubility of protein at or near isoelectric point or pI, where the protein is in net unionized form; some protein may become denatured under these conditions
Temperature	Heat-sensitive proteins expose hydrophobic domains, which exhibit decreased solubility. This process must be reversible if original conformation is required for protein activity

Figure 4.16. Precipitation and extraction of proteins in stainless steal stir-tanks in pharmaceutical scale preparations. (NIH Fredrick facility, with permission)

of the agent used for protein precipitation, the precipitant and protein solution are mixed in a stir tank (Figure 4.16). Aggregates are allowed to settle to the bottom of the tank, and the aggregate slurry containing concentrated protein is collected for further processing.

A less-destructive approach to protein concentration has been developed using membrane separation or ultrafiltration

■TABLE 4.13. Some matrices used for industrial-scale protein purification chromatography

Column Matrix or Media	Trade Name
Agarose	Sepharose, Sepharose CL, HP, FF, Suprose, Ultrogel,
Agarose and dextran composite	Superdex
Agarose and polyacrylamide composite	Ultrogel AcA
Agarose and porous kieselguhr composite	Macrosorb KA
Cellulose	Whatman TM, Cellufine, Sephacel, Cellex
Dextran, cross-linked	Sephadex
Dextran and polyacrylamide composite	Sephacryl
Ethyleneglycol-methacrylate copolymer	Fractogel TSK, Toyopearl
Hydroxyacrylic polymer	Trisacryl
Hyroxymethacrylate polymer	Spheron
Polyacrylamide	Eupergit C
Polyacrylamide, cross-linked	BioGel P
Polystyrene-divinyl benzene	Poros
Porous silica	Spherosil, Accell
Rigid organic polymer	Monobeads, TSK-PW

technologies [6,10,11] that are similar to those described for solid-liquid separation steps. The key difference is the pore size of the filter used for this purpose. In most cases, pore size is selected to retain the recombinant macromolecule while allowing the passage of water and other small molecules. In practice, however, the selection of pore size is more an art than a science, especially the choice of design and sizing configurations of the filtration system. The selection of a proper filtration system may lead to additional benefits in terms of an increase in the yield of recombinant protein and its purity.

Purification

Further purification is needed to increase the purity of the recombinant protein found in the concentrated solution. The additional steps are termed intermediate purification. In this stage, increased purity is achieved through removal of most contaminant proteins, nucleic acids, endotoxins, and viruses. This is accomplished by means of chromatography [12,13].

In the chromatography technique, proteins bind differentially to solid matix supports or media with various functional groups to provide hydrophobic, ion-exchange, and affinity interactions. Some of the matrices used for intermediate purification that provide sufficient flow rate for large-scale purification are listed in Table 4.13. Some of the functional groups attached to matrix supports and examples of proteins purified by these matrix supports are listed in Table 4.14. The chromatography technique should provide high capacity and selectivity. The matrix material must withstand multiple purification cycles with minimum loss of efficiency.

On completion of intermediate purification, most contaminants are removed. For example, endotoxin concentrations are reduced by more than two orders of magnitude, while the specific activity of recombinant protein, a measure of purity, is increased several-fold. The purified material emerging from this stage of downstream processing is composed of recombinant

■TABLE 4.14. Some functional groups attached to column matrices to purify specific proteins

Functional Group on Matrix	Example Protein
Enzyme cofactor	Enzyme
Lectins	Glycoprotein
Protein A or G	Immunoglobulin, IgG class of antibody
Hydrophobic group	Various proteins with hydrophobic domain
Metal ions	Metal ion binding proteins (metalloproteins)
Triazine dye	Dehydrogenases, kinases, and other proteins
Heparin	Coagulation factors, protein kinases
DEAE	Most proteins with domains that exhibit partial charge

protein with small amounts of contaminants and inactive protein fragments. Recombinant protein consitutes more than 90% of the purified material.

A final purification, known as the polishing step, is designed to remove trace contaminants and impurities so that a biologically active recombinant protein with a safety profile suitable for pharmaceutical application is obtained. Contaminants removed by polishing are usually fragments or inactive forms of recombinant protein and trace amounts of endotoxin. The chromatography systems used for intermediate and final purification are largely similar, but differ in capacity and separation performance. Chromatography systems for final purifications demand high performance even at the cost of a lower capacity than the column used for intermediate purification. High separation performance is needed to minimize contaminant carryover into the recombinant product destined for pharmaceutical use.

4.2.4. Quality Assurance and Quality Control

As a part of its requirements, the FDA demands documentation of manufacturing process controls, which include checks and tests that are performed at each step, from cell cultivation to purification. These checks and tests are needed to control the quality of the recombinant protein and to provide assurance that the final product meets the defined specifications to ensure safety and efficacy. The cell cultivation and purification process may introduce biological variability in the recombinant protein that may affect the safety and efficacy of the product. This variability may go undetected even with the most advanced and sensitive analytical methods. Viral contaminants, for example, that escape detection could cause harm when the product is tested clinically. Each step of the process has the potential to compromise safety unless closely monitored and controlled. Therefore, the FDA has in the past considered that the recombinant protein synthesis and purification process is the biotechnology product. More recently, however, the FDA has made exceptions for recombinant proteins that are extensively characterized by means of improved analytical methods to monitor biologic contaminants. Such products are now categorized as "well-characterized recombinant proteins."

Nevertheless, manufacturers of all recombinant products must still "guarantee" that the manufacturing site can produce a product that is free of biological contaminations (see Box 4.5). The site

---

BOX 4.5. POINTS TO CONSIDER IN QUALITY CONTROL AND ASSURANCE OF PROTEIN AND PEPTIDE PHARMACEUTICALS

- Drug versus biologic (consider which branch, CBER or CDER of FDA will review the final product)
- Quality assurance and quality control
  - Documentation of process and raw materials
  - Validation
  - cGMP compliance
- Certificate of analysis or lot release
  - Sterility
  - Endotoxin
  - Identity
  - Purity
  - Concentration
  - Activity
  - Composition (pH, salts, buffers, excipients)
  - Stability

- Purity assays (chromatographic, electrophoretic, immunochemical)
  - Reverse phase HPLC
  - Ion exchange HPLC
  - Hydrophobic interaction HPLC
  - Gel filtration HPLC
- Validation (prior to license)
  - Sterilization procedures
  - Assays
  - Cleaning (especially for multiple-use facilites)
  - Viral clearance (prior to IND)
  - Installation qualification (IQ), operation qualification (OQ), performance qualification (PQ): types of validation to show that equipment, ancillary systems, and process function as intended

---

license ensures that the manufacturing facilities meet the FDA's operation standards. The quality-assurance/quality-control (QA/QC) documentation includes standard operating procedures to verify the specification of starting materials (source, quality, amount, lot number), process flow, containments, and batch records. Other quality control procedures are checks at each stage of manufacture, processing, and purification, tracking of recombinant protein concentration, and calculation of product yield and volume.

Validation of the purification process and of the analytical tests used for characterization of the final protein product is especially important. The purification process must be validated to ensure that it is adequate to remove extraneous substances such as chemicals used in purification, column contaminants, endotoxin, antibiotics, residual cellular proteins, inactive protein, and viruses. Analytical tests are validated by using other analytical techniques to ensure that the tests can identify and establish acceptable limits for critical parameters to be used as in-process controls and to assure success of routine production. In this way, the entire process can be repeated with acceptable product quality from batch to batch.

In the case of recombinant proteins intended for use in sterile pharmaceutical products, additional process controls on microbiologic aspects of analysis must be established and validated to ensure aseptic conditions throughout the manufacturing process.

The final, purified bulk product is thoroughly characterized and compared to reference standards established by the manufacturer for new molecular entities or available from the United States Pharmacopoeia (USP) or the World Health Organization's National Institute of Biological Standards and Control. Characterization tests of proteins may include biologic potency assays, chromatographic assays, gel electrophoresis (i.e., SDS-PAGE method), immunoblot analysis, and ELISA (enzyme-linked immunosorbent assay). Data from the characterization of the protein are validated using statistical analyses, which are also independently verified. In addition, prior to release of a batch for pharmaceutical use, identity, purity, potency and lot-to-lot consistency must be recorded and compared to established standards.

QA/QC specifications are summarized in the Chemistry, Manufacturing, and Contols (CMC) section and in the Established Description sections of the Biologics License Application (BLA) for every recombinant and other biologic product. Additional details on FDA requirements for product specification and QA/QC standards are found in the "Guidance for Industry" (FDA website—www.fda.gov/cber/guidelines.htm).

SUGGESTIONS FOR FURTHER READING

FDA Points to Consider documents relating to recombinant products: FDA Web site—www.fda.gov/cber/guidelines.htm

Biotechnology principles and issues related to large scale manufacturing: Basic Biotechnology, Ratledge C. and Kristiansen B., Editors, 2001, Cambridge University Press, Cambridge, UK.

Federal law relating to biological products: Code of Federal Regulations, parts 600 to 680.

REFERENCES

1. Donahue, R.E., E.A. Wang, R.J. Kaufman, L. Foutch, A.C. Leary, J.S. Witek-Giannetti, M. Metzger, R.M. Hewick, D.R. Steinbrink, G. Shaw, and et al., *Effects of N-linked carbohydrate on the in vivo properties of human GM-CSF.* Cold Spring Harb Symp Quant Biol, 1986. **51** (Pt 1): 685–92.

2. Moo-Young, M., *Bioreactor immobilized enzymes and cells: fundamentals and applications.* 1988, New York: Elsevier Applied Science. xvi, 327.

3. Chisti, Y.A.M.-Y., M., *Fermentation technology, bioprocessing, scale-up and manufacture.*, in *Biotechnology: the science and the business*, V. Moses, R.E. Cape, and D.G. Springham, eds. 1999, New York: Hardwood Academic. 177–222.

4. Endåo, I., and Kagaku Kåogaku Kyåokai (Japan). *Bioseparation engineering: proceedings of an international conference on bioseparation engineering: "Recovery and Recycle of Resources to Protect the Global Environment," organized under the Special Research Group on Bioseparation Engineering in the Society of Chemical Engineers, Japan, Nikko, Japan, July 4–7, 1999.* 2000, New York: Elsevier Science. xi, 226.

5. Desai, M.A., *Downstream processing methods.* Methods in biotechnology; 9. 2000, Totowa, NJ: Humana Press. ix, 229.

6. Wang, W.K., *Membrane separations in biotechnology.* 2nd ed. 2001, New York: Dekker. xi, 400.

7. Walter, H., D.E. Brooks, and D. Fisher, *Partitioning in aqueous two-phase systems: theory, methods, uses, and application to biotechnology.* 1985, Orlando, FL: Academic Press. xxiv, 704.

8. Soane, D.S., *Polymer applications for biotechnology: macromolecular separation and identification.* 1992, Englewood Cliffs, NJ: Prentice Hall. xiii, 314.

9. Olson, W.P., *Separations technology: pharmaceutical and biotechnology applications.* 1995, Buffalo Grove, IL: Interpharm Press. xix, 505 ill.

10. Vieth, W.R., *Membrane systems: analysis and design* 1994, New York: Wiley. xv, 360.

11. van Reis, R., and A. Zydney, *Membrane separations in biotechnology.* Curr Opin Biotechnol, 2001. **12**(2): 208–11.

12. Janis, L.J., P.M. Kovach, R.M. Riggin, and J.K. Towns, *Protein liquid chromatographic analysis in biotechnology.* Methods Enzymol, 1996. **271**: 86–113.

13. Reif, O.W., and R. Freitag, *Comparison of membrane adsorber (MA) based purification schemes for the down-stream processing of recombinant h-AT III.* Bioseparation, 1994. **4**(6): 369–81.

■ SECTION THREE ■

■4.3. BIOLOGIC DRUG DEVELOPMENT AND APPROVAL

All pharmaceuticals marketed in the United States, including biotechnology products, must be proven safe and effective for their intended use. The FDA requires that all recombinant proteins and other biotechnology products be produced by a manufacturer holding a certified Biologic License Establishment (BLE). The pharmaceutical company is required to collect safety and efficacy data, first in animal studies and subsequently in clinical trials. This information is submitted in a new drug application package called a Biologic License Application (BLA). The data are rigorously evaluated by FDA staff, assisted by independent experts assembled as advisory panels. New drugs are approved if their benefits outweigh risks. A schematic summary of a typical clinical trial and drug approval process is presented in Figure 2.1, Chapter 2. In the following, we will expand our discussion of key issues related to clinical trials and approval process including (1) Biologic License Establishment, (2) preclinical and clinical testing requirements, (3) FDA review and approval process, and (4) globalization or harmonization of drug approval (Figure 4.17).

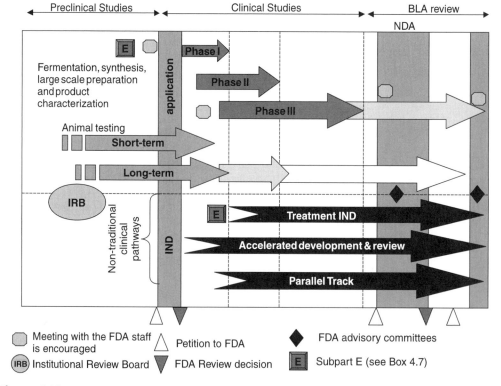

Figure 4.17. Progression of a new molecular entity through preclinical and clinical evaluation for safety and efficacy. Preclinical and clinical study results are reviewed by FDA through several pathways, including traditional and nontraditional pathways. A majority of submitted new drug applications are approved for human use. Additional details on nontraditional clinical pathways are presented in Box 4.7. (Figure adapted from an FDA report, 1999)

4.3.1. Biologic License Establishment

As a part of the FDA Modernization Act, passed in 1997, the regulatory agency has combined the Establishment License Application (ELA) and the Product License Application (PLA) into one package called the Biologic License Application (BLA) (Table 2.1 in Chapter 2). While this change has streamlined the new drug licensing process, the FDA still tightly regulates manufacturing facilities engaged in the production of biotechnology products. The FDA no longer strictly defines the biotechnology product by using the idea of "the process is the product," because a number of analytical tools that allow improved characterization of the product and of biological contaminants are now available. As a result of these changes, recombinant proteins can be manufactured in multiple facilities so long as each facility provides assurance of product quality. In the BLA process, a new manufacturing establishment may obtain a site license at the time of certification of the first batch of product manufactured in that establishment.

The manufacturing facility or establishment must file a detailed flow diagram that includes the flow of raw materials to areas allocated for storage, holding, mixing, manufacturing, purification, testing, and final release. All manufacturing must adhere to current Good Manufacturing Practice (cGMP) guidelines as briefly presented in Box 4.6. If more than one product is manufactured in the same facility, a comprehensive list of all the products manufactured in the establishment must be disclosed. Under the previous PLA/ELA dual application process, a biotechnology company was required to identify a facility to manufacture "one dedicated product."

BOX 4.6. CURRENT GOOD MANUFACTURING PRACTICE, cGMP

The FDA through the establishment of the Good Manufacturing Practice (GMP) regulation ensures that drugs produced by pharmaceutical companies consistently meet quality specification. Since it was first introduced in 1963 as a part of the Kefauver-Harris Drug Amendment of 1962, it has been updated periodically. The latest version is called current GMP (cGMP) regulation. Information is posted on the FDA's Web site (*http://www.fda.gov/cder/dmpq/cgmpregs.htm*). The cGMP regulations encompass the minimum standard expected of a manufacturing site with regard to facility, equipment, personnel, control of raw materials, production, process, packaging, labeling, holding, and distribution. In addition, the cGMP regulations also include the recording and reporting of all the manufacturing processes as well as the disposal of unused products.

The FDA inspects domestic manufacturing plants as well as foreign drug manufacturing facilities that export drugs to the United States to ensure full compliance, and provide assurance that pharmaceutical products consistently meet uniform quality standards. Failure to meet qualify standards may have impacts on a drug's efficacy and safety. The FDA can shut down a facility that fails to comply with cGMP regulations. The following lists the range of issues covered under the cGMP standard.

A. General Provisions
1. Scope.
2. Definitions.

B. Organization and Personnel
1. Responsibilities of quality control unit.
2. Personnel qualifications.

(Continued on next page)

BOX 4.6. Continued

3. Personnel responsibilities.
4. Consultants.

C. Buildings and Facilities
1. Design and construction features.
2. Lighting.
3. Ventilation, air filtration, air heating, and cooling.
4. Plumbing.
5. Sewage and refuse.
6. Washing and toilet facilities.
7. Sanitation.
8. Maintenance

D. Equipment
1. Equipment design, size, and location.
2. Equipment construction.
3. Equipment cleaning and maintenance.
4. Automatic, mechanical, and electronic equipment.
5. Filters.

E. Control of Components and Drug Product Containers and Closures
1. General requirements.
2. Receipt and storage of untested components, drug product containers, and closures.
3. Testing and approval or rejection of components, drug product containers, and closures.
4. Use of approved components, drug product containers, and closures.
5. Retesting of approved components, drug product containers, and closures.
6. Rejected components, drug product containers, and closures.
7. Drug product containers and closures.

F. Production and Process Controls
1. Written procedures; deviations.
2. Charge-in of components.
3. Calculation of yield.
4. Equipment identification.

5. Sampling and testing of in-process materials and drug products.
6. Time limitations on production.
7. Control of microbiological contamination.
8. Reprocessing.

G. Packaging and Labeling Control
1. Materials examination and usage criteria.
2. Labeling issuance.
3. Packaging and labeling operations.
4. Tamper-resistant packaging requirements for over-the-counter human drug products.
5. Drug product inspection.
6. Expiration dating.

H. Holding and Distribution
1. Warehousing procedures.
2. Distribution procedures.

I. Laboratory Controls
1. General requirements.
2. Testing and release for distribution.
3. Stability testing.
4. Special testing requirements.
5. Reserve samples.
6. Laboratory animals.
7. Penicillin contamination.

J. Records and Reports
1. General requirements.
2. Equipment cleaning and use log.
3. Component, drug product container, closure, and labeling records.
4. Master production and control records.
5. Batch production and control records.
6. Production record review.
7. Laboratory records.
8. Distribution records.
9. Complaint files.

K. Returned and Salvaged Drug Products
1. Returned drug products.
2. Drug product salvaging.

Adapted from 21 Code of Federal Regulations, Parts 210 and 211.

With the enactment of the BLA process for biotechnology-based pharmaceuticals, any facility holding a current establishment license can manufacture more than one recombinant product, thereby allowing small or start-up biotechnology companies the option of outsourcing production of recombinant products. Facilities manufacturing multiple products are required to establish procedures and documentation of measurements to prevent cross-contamination of products. The establishment must also document and validate manufacturing processes and control procedures, reference standards, analytical and specification methods, storage and shipping containers, and product stability.

Details of the regulatory requirements for establishing a site to manufacture biotechnology products can be found as a part of "Chemistry, Manufacturing, and Controls" (CMC) guidelines developed by the FDA. They are found in the following four sections of the FDA guidelines: (1) Guidance for Industry for the Submission of Chemistry, Manufacturing, and Controls Information for a Therapeutic Recombinant DNA-Derived Product or a Monoclonal Antibody Product for *In Vivo* Use (61 FR 56243, 31 October 1996); (2) Guidance for the Submission of Chemistry, Manufacturing, and Controls Information and Establishment Description for Autologous Somatic Cell Therapy Products (62 FR 1460, 10 January 1997); (3) Guidance for Industry for the Submission of Chemistry, Manufacturing, and Controls Information for Synthetic Peptide Substances; and (4) Draft Guidance for Industry for the Submission of Chemistry, Manufacturing, and Controls and Establishment Description Information for Human Plasma-Derived Biological Products or Animal Plasma or Serum-Derived Products (63 FR 3145, 21 January 1998).

In practice, the facility engaged in the manufacture of biological drug candidates usually informs the FDA early in the process to obtain an initial, often unofficial, but helpful response from the agency regarding their advice about the facility. This is almost always done much earlier than the requirements state [i.e., "... registration (of establishment) shall follow within 5 days after the submission of a biologics license application..."]" (CFR vol. 63, 147; section 607.21). Early contact with the FDA, although not a legal requirement, is essentially obligatory to avoid serious delays in product approval and marketing.

4.3.2. Preclinical and Clinical Testing

To establish safety and efficacy of a biopharmaceutical candidate for a defined therapeutic indication, the company or sponsor must provide the FDA with preclinical and clinical data assembled and summarized in the BLA. The preclinical and clinical testing process can vary depending on drug candidate and complexity of the therapeutic indication (Figure 4.18).

Preclinical Studies

The preclinical studies include chemical and physical characterization, qualitative and quantitative analytical methods development, dosage formulations, and animal testing procedures. Issues related to dosage formulations are discussed in Chapter 5. Animal testing is an integral part of preclinical studies. These data provide essential information for estimating a dosing regimen that is likely to be safe for testing in humans (Figure 2.1, Chapter 2). The FDA does not stipulate what preclinical tests the sponsor must perform; however, the agency publishes guidelines and regulations on the kinds of data expected for human testing of the new drug candidate. Two or more animal species are typically tested to ensure that drug effects not seen in one species might be detected in another. These animal tests also allow researchers to determine potential side

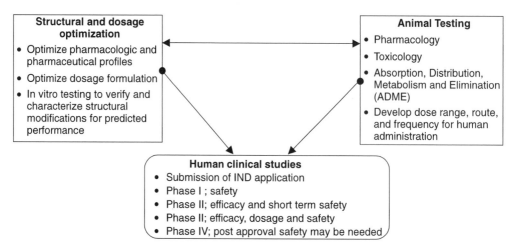

Figure 4.18. Interrelationship of preclinical and clinical studies.

effects and overall safety at much higher doses than one would ever administer to humans. Some of this information, including long-term toxicity data, guides the development of protocols for human testing and generates information for final product labeling.

In preclinical studies researchers collect data on the effects of the new drug candidate after administration in a defined dosage regimen. In other words, scientists, in a series of systematically designed controlled experiments, determine the time course of drug concentration in relationship to pharmacologic and toxicologic effects, as well as the pathological consequences of the drug. In addition, animal studies allow scientists to determine the absorption, distribution, metabolism, and excretion in the selected species. They determine how much of the drug dose is absorbed from the site of administration (intramuscular, subcutaneous, and in some cases the gastrointestinal tract), how the drug is metabolized in the body, and the rate of elimination of parent drug and metabolites from the body. The biologic activity and toxicology of the metabolites are also often defined.

Occasionally a metabolite may provide a safer or more effective pharmaceutical than the parent compound selected for development. For example, fexofenadine, a widely sold nonsedating antihistamine, is the active metabolite of terfenadine. Fexofenadine is a safer drug than terfenadine and has replaced it on the market.

Animal testing allows researchers to develop dosage forms with optimal pharmaceutical properties. If the new drug candidate is poorly absorbed into the systemic circulation, formulation strategies such as producing salts or other derivatives, modifying pH, and the addition of excipients are implemented to maximize absorption. If the new drug candidate is metabolized or inactivated rapidly, a derivative that may resist inactivation might be designed through structural modification. The interrelationship between *in vitro* optimization and *in vivo* animal testing provides information essential for administration of the drug candidate to humans in an optimized dosage regimen (Figure 4.6). Any significant change in structure of the new drug candidate requires an additional series of preclinical and clinical tests. Significant time and resources are needed to iterate *in vitro* and *in vivo* animal tests and optimize the efficacy and safety of the derivative. To increase efficiency, researchers have developed *in vitro* models that may predict *in vivo* animal test results. In this way only a few selected derivatives might be studied in laboratory animals to identify the most promising candidate for humans.

Clinical Studies

Human studies are designed to determine "Does the drug work?" To provide an answer, pharmaceutical companies, through a series of controlled clinical trials, must, according to the FDA, collect and submit "substantial evidence of effectiveness, as well as confirmation of relative safety in terms of the risk-to-benefit ratio for the disease that is to be treated." It is critical from the outset to design clinical studies that pose the right question and provide an answer to the question in the intended patient population.

The need to discuss issues related to clinical studies early is highlighted by the FDA's refuse-to-file response letter to ImClone Systems for *Erbitux*, a novel anticancer monoclonal antibody. The FDA took this drastic measure because the agency, despite all the early warnings given to the company, was not provided with the complete documentation needed for clinical evaluation. A failure to engage the FDA with the clinical trail design was aknowledged by the CEO of ImClone System, and eventually led to his resignation.

Preclinical data are compiled in the Investigational New Drug (IND) application—a request for permission to initiate human studies—and submitted to the FDA for approval. The IND application requires full disclosure of all the preclinical study results as well as where and how the drug itself is manufactured and formulated to ensure quality and stability. The IND also contains pharmacology and toxicology data, evidence of desired and unwanted

effects in disease models, and proposed analytical methods. A human subjects protocol that lists all the investigators who will participate in clinical trials must also be submitted to the FDA. In response the FDA must provide an answer within 30 days after submission. Under current FDA regulation the lack of a timely response is considered as "no objection" and permits the drug's sponsor to proceed to human testing.

When the sponsor obtains approval or tacit approval on the IND application and clearance from an independent local institutional review board (IRB), charged with the responsibility of protecting human subjects and consisting of scientists, ethicists, and nonscientist or community participants, clinical investigators will for the first time give the new drug candidate to a limited number of healthy volunteers or patients. The first human study, generally known as a phase I clinical trial, is designed to assess the most common acute side effects of the drug and determine the highest dose that the recipients can tolerate. Typically about 20 to 30 subjects are enrolled in phase I clinical trials, but the number can sometimes reach 100. From a series of blood samples collected after administration, clinical investigators can determine drug concentration in plasma and estimate the persistence of the drug in the body and determine metabolites of the parent drug. Phase I studies, however, do not reveal potential safety problems that occur infrequently. Therefore additional safety data are required from more advanced patient-based studies with larger enrollments (Table 4.15).

If no major safety problems are observed in phase I studies, the next step is to perform another clinical study enrolling patients who have the medical condition that the new drug candidate is intended to treat. The main objective is to determine whether the drug has a beneficial effect. Therefore it is critical to define the medical condition

■TABLE 4.15. Progression of drug candidates in human clinical trials.

Clinical trial	Purpose	Number of Patients	Duration	Success Rate (%)[a]
Phase I	Safety	20–100	Several months	70–75
Phase II	Efficacy, some short-term safety	Several hundred	Several months–2 years	28–33
Phase III	Overall safety and efficacy for risk and benefit analysis	Several hundred to thousands	1–4 years	~25

[a]Success rate is estimated based on average of all the investigation new drug (IND) applications submitted to FDA; according to FDA data, on the average about 25% of the original IND will clear phase III and about 20% will gain approval for marketing.

and the means to assess therapeutic outcome in a way that can be measured—clinically or by diagnostic or prognostic tests—and compared to controls.

Assessing the effectiveness of a new drug candidate can be complex and often difficult. This is because some diseases or symptoms do not follow a predictable path. For example, acute conditions such as influenza or insomnia may resolve without intervention, while chronic conditions such as multiple sclerosis or arthritis follow a varying course of progression. Depending on age, treatment, and other risk factors, heart attacks and strokes may produce variable mortality rates. Additional difficulty is introduced by subjective evaluation, which can be influenced by the expectations of patients and physicians. Some of these issues can be addressed in controlled clinical trials.

While not required by the FDA, more and more phase II trials are being conducted under well-controlled conditions. In such trials, a group of patients that are well-defined for age, weight, disease severity, general health, and other risk factors are randomly assigned (randomized) to either a control or a treatment group. It is critical to ensure that the treatment group and control group are as similar as possible so that differences will not influence the outcome of the efficacy study. The controls receive a placebo—an inactive dosage form similar to the dosage form containing the active drug—or another drug known to be effective for the disease. Sometimes, the control group receives a lower dose of the same drug candidate. When a placebo is employed, the study design is called a randomized, placebo-controlled clinical trial.

To prevent bias introduced by patients and clinical investigators during the course of a controlled study, a design feature known as "blinding" is frequently incorporated. Single-blind and double-blind studies keep patients or both patients and investigators from knowing which patients in the study receive the active drug. When possible, it is advisable to perform randomized, double-blind, placebo-controlled clinical trials. A phase II clinical trial enrolls several hundred subjects and takes several months to complete. According to data from the FDA, about one third of new drug candidates successfully complete phase II clinical trials (Table 4.15).

The new drug candidate that successfully emerges from phase I testing is then required to be tested further in at least one phase III clinical trial. This phase of human study is designed to collect data under rigorously controlled conditions that are adequate to definitively evaluate the drug's effectiveness and safety with respect to an overall analysis of benefit-risk relationship. Very often many medical centers participate in a phase III trial, enrolling several hundred to several thousand subjects. The exact number depends on the therapeutic end points and anticipated differences between the active drug group and the control group. Phase III trials enroll a much larger and more diverse patient population than do phase II trials, thereby revealing infrequent untoward events missed in earlier clinical trials.

On completion of phase III, which typically takes up to four years, the data are compiled and analyzed for overall safety and efficacy with regard to the defined indication set forward in the IND application. These data are assembled in a Biologic License Application (or New Drug Application for chemical and small peptide drug candidates) package for FDA evaluation of overall benefit and risk. About 25% of drugs with approved INDs successfully complete phase III testing and gain marketing approval.

In recent years the FDA has amended several rules regarding clinical trials, requiring that new drug candidates be studied in women, children, and the elderly, when appropriate. As the population of individuals over 65 years of age continues to grow and life expectancy continues to rise, there is concern that medications may produce different effects in the elderly than in younger patients. For example, elderly patients are more likely to take multiple drugs, suffering the consequences of drug interactions. Furthermore the elderly may not be able to metabolize or excrete drugs as well as their younger counterparts. In 1998 the FDA added the requirement of safety and efficacy data for demographic subgroups as part of the BLA submission.

The subjects enrolled in clinical studies for small molecule and biologic products must be tabulated within relevant demographic subgroups (e.g., age, gender, and race) in annual IND reports and must be included in all BLAs.

4.3.3. FDA Review and Approval Process

Before any drug is approved for marketing, the FDA reviews the data provided for the new drug candidate in BLA or NDA submissions and evaluates the benefits and risks of the new drug. The data in BLA or NDA submissions are reviewed by one of two branches within the FDA—the Center for Drug Evaluation and Research (CDER) and the Center for Biologics Evaluation and Research (CBER)—with the help of advisory panels. On completion of the review the FDA decides whether to approve the new drug, request more information, require new studies, or grant accelerated (conditional) approval (Box 4.7). On approval the FDA may request the drug's sponsor to carry out postmarketing (phase IV) studies to gather long-term data.

FDA advisory panels in specific therapeutic areas play an important role in the drug approval process. More often than not, FDA staff ask the expert members of the panel for guidance on key questions related to the safety and efficacy of the new drug candidate, and sometimes on more technical questions such as the design of clinical trials. At the conclusion of their deliberations, panel members vote to recommend or not recommend that the agency approve the new drug based separately on efficacy and safety issues; the panel also advises the regulatory agency on appropriate use of the drug, including limitations as to the indicated patient population and other safeguards. While the FDA values the advice of its advisory panels, the final decision to approve or not approve a

BOX 4.7. NONTRADITIONAL CLINICAL TESTING PATHWAYS—FAST-TRACK CLINICAL TESTING AND FDA REVIEW

Subpart E establishes procedures to expedite the development, evaluation, and marketing of new therapies intended to treat people with life-threatening and severely-debilitating illnesses (e.g., cancer and AIDS), especially where no satisfactory alternatives exist. (*Section 312, Federal Register, 21 October 1988*)

Accelerated Development and Review Process is a highly specialized mechanism for speeding the development of drugs that promise significant benefit for serious or life-threatening illnesses for which no therapy exists. This process incorporates several novel elements aimed at making sure that rapid development and review is balanced by safeguards to protect both patients and the integrity of the regulatory process.

Accelerated development/review can be used under two special circumstances: when approval is based on evidence of the product's effect on a "surrogate endpoint," and when the FDA determines that safe use of a product depends on restricting its distribution or use. A surrogate endpoint is a laboratory finding or physical sign that may not be a direct measure of how a patient feels, functions, or survives, but is still considered likely to predict therapeutic benefit for the patient.

The fundamental element of this process is that the manufacturers must continue testing after approval to demonstrate that the drug indeed provides therapeutic benefit to the patient. If not, the FDA can withdraw the product from the market more easily than usual. (*Federal Register, 15 April 1992*)

Treatment Investigational New Drugs are used to make promising new drugs available to desperately ill patients as early in the drug development process as possible. During Phase II studies, investigational new drugs are made available to patients before general marketing begins. The treatment IND track also allow the FDA to collect additional data on the drug's safety and effectiveness. To date, more than 39 Treatment INDs have been approved, mainly for AIDS and cancer indications. (*Federal Register, 22 May 1987*)

Parallel Track is another mechanism that permits dispensing investigational drugs, particularly to AIDS patients that do not qualify to participate in controlled clinical trials of promising drug candidates. The parallel track policy was developed by the US Public Health Service in response to the AIDS epidemic. (*Federal Register of 21 May 1990*)

drug rests with the agency. On occasion the FDA chooses not to follow a panel's recommendations.

The decision of whether a given IND is reviewed by CBER or CDER is not clear-cut. The designation BLA or NDA really depends on which branch of the FDA's drug division is assigned review responsibilities. CDER reviews all NDAs, while CBER reviews all BLAs. In general, most small peptides and proteins such as insulin,

peptide antibiotics, some antibodies, and antibody fragments, fall under the purview of CDER. CBER reviews new recombinant proteins of higher molecular weights such as interleukins, colony-stimulating factors, and some antibodies, as well as vaccines.

The philosophy of the FDA in reviewing BLAs and NDAs continues to evolve (see Box 4.8 for summary). The current position is reflected in the agency's statement, "No

BOX 4.8. HISTORY OF REGULATION GOVERNING DRUGS AND BIOLOGICS MARKETED IN THE UNITED STATES

- **Food and Drug Act (1906).** This first drug law required only that drugs meet standards of strength and purity. The burden of proof was on FDA to show that a drug's labeling was false and fraudulent before it could be taken off the market.
- **Federal Food, Drug and Cosmetic Act (1938).** A bill was introduced in the Senate in 1933 to completely revise the 1906 drug law—widely recognized as being obsolete. But congressional action stalled. It took a tragedy in which 107 people died from a poisonous ingredient in "Elixir Sulfanilamide" to promote passage of revised legislation that, for the first time, required a manufacturer to prove the safety of a drug before it could be marketed.
- **Durham-Humphrey Amendment (1951).** Until this law, there was no requirement that any drug be labeled for sale by prescription only. The amendment defined prescription drugs as those unsafe for self-medication and which should be used only under a doctor's supervision.
- **Kefauver-Harris Drug Amendments (1962).** News reports about the role of an FDA medical officer in keeping the drug thalidomide off the US market aroused public interest in drug regulation. Thalidomide had been associated with the birth of thousands of malformed babies in Western Europe. In October 1962 Congress passed these amendments to tighten control over drugs. Before marketing, a drug company now had to prove not only safety but also effectiveness for the product's intended use. In addition firms were required to send adverse reaction reports to the FDA, and drug advertising in medical journals was required to provide complete information to doctors—the risks, as well as

the benefits. The amendments also required that informed consent be obtained from subjects in clinical trials. (Note: In July 1998, thalidomide was approved by the FDA for an entirely different indication, with significant restrictions. Because of thalidomide's potential to cause birth defects, FDA invoked unprecedented regulatory authority to tightly control the marketing of the product in the United States.)

- **Orphan Drug Act (1983).** "Orphans drugs" are drugs and other medical products for treating rare diseases. They may offer limited potential profit to the manufacturer but may benefit people with these diseases. To foster development, this law allows drug companies to take tax deductions for about three-quarters of the cost of their clinical studies.
- **Drug Price Competition and Patient Term Restoration Act (1984).** This law expands the number of drugs suitable for an abbreviated new drug application (ANDA). ANDAs make it less costly and time-consuming for generic drugs to reach the market. Patient Term Restoration refers to the 17 years of legal protection given a firm for each drug patent. Some of that time allowance is used while the drug goes through the approval process.
- **Generic Drug Enforcement Act (1992).** This law imposes debarment and other remedies for criminal convictions based on activities relating to the approval of ANDAs.
- **Prescription Drug User Fee Act (PDUFA) (1992).** In this law, manufacturers agreed to pay user fees for certain new drug applications and supplements, an annual establishment fee, and annual product fees. Using these funds, FDA can hire

(Continued on next page)

BOX 4.8. Continued

additional staff members to increase efficiency in the NDA and BLA review process.
- **FDA Modernization Act (1997).** This act contains some of the most sweeping changes to the Food, Drug and Cosmetic Act. The act contains changes in how user fees are assessed and collected. For example, fees are waived for the

first application for small businesses, orphan products, and pediatric supplements. The act codifies FDA's accelerated approval regulations and requires guidance on fast-track policies and procedures. In addition the agency must issue guidance for NDA and BLA reviewers.

---

drug is absolutely safe . . . there is always some risk of an adverse reaction. However, when a proposed drug's benefits outweigh known risks, the FDA . . . considers it safe enough to approve." While rules and regulations governing the review and final approval by the FDA continue to evolve, what remains constant, according to the FDA position, is "Safety and effectiveness and benefit vs. risk remain the pivotal issues in drug review" (Benefit vs. Risk. How CDER approves new drugs, 2000).

NDA and BLA submissions are assigned to a team of FDA staff members that include expert chemists, pharmacologists, physicians, pharmacokineticists, statisticians, and microbiologists (see Box 4.9 for additional explanation). FDA staff critically evaluate the submitted data guided by the key question, "Does this drug work for the proposed use?" They will study how the drug is manufactured, processed, and packaged; the results of animal studies; the disposition of the drug—what happens after administration and how the body distributes and clears the drug, and what happened during the clinical studies. A number of statistical methods are used to evaluate clinical data to determine whether they demonstrate effectiveness for intended use and whether the benefit-to-risk relationship is favorable.

The pharmaceutical company must also provide a sample of the drug and proposed

professional labeling. The FDA plays a major role in shaping the final approved labeling. The label serves as a way for the FDA to control the risks that the drug product carries, by distributing it to physicians and pharmacists who then become the risk-managers for the product. The information contained in the drug's approved labeling—how the drug should be used—is part of the drug package.

As a matter of course, biotechnology companies recognize the need to discuss with FDA staff the overall plan for drug development before the IND is submitted. To avoid delays and perhaps failure, the company must understand what the FDA expects with respect to study design, the conduct of studies, and data analysis. The FDA has field reviewers available to assist drug companies with these issues as well as with issues related to regulations relevant to manufacturing and packaging facilities.

In the final analysis, approval of a BLA or NDA comes down to two key questions: Do the results of well-controlled studies provide substantial evidence of effectiveness? and Do the results show the product to be safe under the conditions of use in the proposed labeling? Safe, in this context, means that the benefits of the drug appear to outweigh its risks. On completion of the review, CDER or CBER sends a letter to the applicant sponsor stating that the drug is approved for marketing; is approvable,

BOX 4.9. TYPICAL FDA REVIEW TEAM

The members of CDER and CBER review teams who apply their special technical expertise to the review of NDAs, and BLAs:

- **Chemists** focus on how the drug is made and whether the manufacturing, controls, and packaging are adequate to ensure the identity, strength, quality, and purity of the product.
- **Pharmacologists** evaluate the effects of the drug on laboratory animals in short-term and long-term studies.
- **Physicians** evaluate the results of clinical tests—including adverse and therapeutic effects, and whether the proposed labeling accurately reflects the effects of the drug.
- **Pharmacokineticists** evaluate the rate and extent to which the active ingredient is made available to the body and

the way it is distributed, metabolized, and eliminated.
- **Statisticians** evaluate the design of each controlled study and rigorously judge the analyses and conclusions of safety and effectiveness data.
- **Microbiologists** with others evaluate the data on anti-infectives (antibiotics, antivirals, and antifungals). These drugs differ from others in that they affect the workings of microbes instead of patients. Reviewers need to know how the drug acts on these microorganisms, which ones it affects, any resistance to the drug, and clinical laboratory methods needed to evaluate the drug's effectiveness. Microbiologists also are concerned with ensuring injectable drugs are free of microorganisms.

provided that changes are made in specific areas of the application; or is not approvable because of major problems. In the case of a rejection, the applicant may amend or withdraw the application or ask for a hearing.

Once the FDA approves a drug, it can be marketed as soon as the sponsoring company gets its production, distribution, and marketing systems in place.

4.3.4. Globalization of Drug Approval

To facilitate worldwide registration and marketing of the drug product, more and more, drug research, development, and production are performed at multiple sites located in different countries. A tripartite organization, known as the International Conference on Harmonization (ICH), was established in 1991 to address the possibil-

ity of crafting drug development guidelines that would be acceptable by regulatory agencies of subscribing countries. The organization includes the FDA and its counterparts from the European Union and Japan, and advocate organizations whose members have interest and expertise in regulatory matters (Table 4.16).

The mission of the ICH is to make recommendations on ways to achieve greater consistency or harmonization in the interpretation and application of technical guidelines and requirements for product registration. The organization's goal is to develop a single set of tests to be used worldwide to answer globally uniform questions rather than comply with the patchwork of regulations now extant in different countries. This strategy may lead to increased efficiency of global drug development and swifter introductions of new medicines without compromising safety and efficacy standards. Harmonization, if

■**TABLE 4.16.** Member organizations of the International Conference on Harmonization (ICH) and their respective interests

Member Organization	Role and Relation to ICH
Regulatory agencies	
United States Food and Drug Administration (FDA)	The drug regulatory agency responsible for the approval of all drug products used in the U.S. Experts who provide technical advice for ICH are drawn from the Center for Drug Evaluation and Research and the Center for Biologics Evaluation and Research, the two branches of FDA.
European Agency for the Evaluation of Medicinal Products (EMEA)	The agency established by a commission composed of fifteen member countries of the European Union (EU). The commission is working through harmonization of technical requirements and procedures to achieve a single market for pharmaceuticals that would allow free movement of medicinal products throughout the EU.
Ministry of Health, Labor, and Welfare, Japan (MHLW)	The unit responsible for the improvement and promotion of social welfare, social security, and public health is the Pharmaceutical Affairs bureau of the ministry, It is one of the nine bureaus within the Pharmaceuticals and Cosmetics Division and is responsible for review and licensing of all medicinal products and cosmetics. In Japan, it acts as the focal point for ICH activities. Technical advice on ICH matters is obtained through MHLW's regulatory expert groups.
Advocates	
Pharmaceutical Research and Manufacturers of America (PhRMA)	The association's sixty-seven member companies are involved in the discovery, development, and manufacture of prescription drugs. Additionally, twenty-four research affiliate companies conduct research related to the development of biopharmaceuticals and vaccines. PhRMA, previously known as the Pharmaceutical Manufactures Association (PMA), provides its technical input to ICH.
European Federation of Pharmaceutical Industries and Associations (EFPIA)	With EFPIA membership from sixteen countries in Western Europe, including all of Europe's major research-based pharmaceutical companies, the Federation's work is concerned with the activities of the EMEA. A wide network of experts and country coordinators has been established to ensure that EFPIA's views are represented in ICH guidelines.
Japan Pharmaceutical Manufacturers Association (JPMA)	The key objective of JPMA is the development of a competitive pharmaceutical industry with a greater awareness and understanding of international issues. Over ninety companies are members, including all the major research-based pharmaceutical manufacturers in Japan. ICH work is coordinated through specialized committees of industry experts. JPMA promotes and encourages the adoption of international standards by its member companies.

achieved, would be an economical use of human, animal, and material resources.

The core ideas for harmonization guidelines are derived from ICH Regional Guideline Workshops, and other regional and international conferences, workshops, and symposia dealing with drug research and development and regulatory affairs. The concepts are then refined through working committees to assemble a draft harmonization proposal. Such proposals fall into the following categories:

- New types of medicinal products
- Lack of harmonization in technical requirements
- Transition to technically improved testing procedures
- Review of an existing ICH guideline
- Maintenance of an existing guideline

The expert working groups with members from the six major stakeholders (the FDA; Pharmaceutical Research and Manufacturers of America; the European Union; the European Federation of Pharmaceutical Industries Associations; Ministry of Health, Labor, and Welfare (Japan); and the Japanese Pharmaceutical Manufacturers Association) and other invited experts develop a consensus upon which the guideline draft is assembled. The proposed guideline is then systematically evaluated through a five-step scheme with final endorsement from the three regulatory agencies (Table 4.16). The progressive steps toward endorsement are described in Box 4.10.

Over the past few years, many guidelines have been developed to address (1) drug quality, (2) efficacy, (3) safety, and (4) regulatory communication. The ICH has developed more than 170 proposed guidelines; 64 of them have been adopted by the regulatory agencies in the United States, European Union, and Japan.

BOX 4.10. STEPS IN THE ICH PROCESS (ACCORDING TO THE ICH DOCUMENT)

The five-step process, which proved successful for the first phase of ICH activities, will be maintained, with appropriate modifications to accommodate the extended Expert Working Groups (EWGs):

Step 1. Consensus Building

The Regulatory Reporter prepares an initial draft of a guideline or recommendation, based on the objectives set out in the Concept Paper, and in consultation with experts designated to the EWG. The initial draft and successive revisions are circulated for comment, giving fixed deadlines for receipt of those comments.

To the extent possible, consultation will be carried out by correspondence, using fax and e-mail. Meetings of the Expert Working Group will normally only take place at the time and venue of the biannual Steering Committee meetings. Additional formal meetings of the ICH EWG need to be agreed upon in advance by the Steering Committee.

Interim reports are made at each meeting of the ICH Steering Committee. If consensus is reached within the agreed timetable, the consensus text with EWG signatures is submitted to the Steering Committee for adoption as step 2 of the ICH process.

When consensus is reached on the technical issues, all parties represented on an EWG would be invited to sign the document to indicate their agreement to the consensus text, which is then submitted to the Steering Committee. Circumstances could be envisaged, however, where not all parties are present or able to sign the consensus text. It would then be for the

(Continued on next page)

BOX 4.10. Continued

Steering Committee to decide whether to proceed to step 2.

Step 2. Start of Regulatory Action

Step 2 is reached when the Steering Committee agrees, on the basis of the report from the Expert Working Group, that there is sufficient scientific consensus on the technical issues for the draft guideline or recommendation to proceed to the next stage of regulatory consultation. This agreement is confirmed by Steering Committee Members by each of the six ICH parties signing their assent.

Step 3. Regulatory Consultation

At this stage the guideline or recommendation embodying the scientific consensus leaves the ICH process and becomes the subject of regulatory consultation in the three parts of the world. In the European Union, it is published as a draft Committee for Proprietary Medicinal Products (CPMP) Guideline, in the United States, it is published as a draft guidance in the Federal Register, and in Japan, it is translated and issued by MHLW, for internal and external consultation.

Step 4. Adoption of a Tripartite Harmonized Text

At step 4 the issue returns to the ICH forum where the Steering Committee receives a report from the Regulatory Reporter. If both regulatory and industry parties are satisfied that the consensus achieved at step 2 is not substantially altered as a result of the consultation, the text is adopted by the Steering Committee. This adoption takes place on the signatures from the three regulatory parties to ICH affirming that the guideline is recommended for adoption by the regulatory bodies in the three regions. In the event that one or more parties representing industry have strong objections to the adoption of the guideline, on the grounds that the revised draft departs substantially from the original consensus, or introduces new issues, the regulatory parties may agree that the revised text should be submitted for further consultation.

Step 5. Implementation

Having reached step 4 the tripartite harmonized text moves immediately into the final step of the process, which is regulatory implementation. This is carried out according to the same national/regional procedures that apply to other regulatory guidelines and requirements in the European Union, Japan, and the United States. Information on the regulatory action taken and implementation dates are reported back to the Steering Committee and published by the Secretariat.

Whether preventing unnecessary duplication of clinical trials and minimizing the use of animal testing, without compromising the regulatory obligations of safety and effectiveness, will be cost effective through expeditious development of pharmaceuticals with consistent quality, safety, and effectiveness remains to be seen. Nevertheless, a uniform regulatory system that permits the same information to be assembled in documents for different regulatory agencies, thereby avoiding the need to submit redundant preclinical and clinical information, manufacturing, and packaging, can translate into huge cost savings in developing pharmaceuticals to be marketed globally.

SUGGESTIONS FOR FURTHER READING

F.D.A. Points to Consider relating to recombinant products. Current version available at: http://www.fda.gov/cber/guidelines.htm

Federal law relating to biological products: Code of Federal Regulations. The most current version can be obtained from the Government Printing Office. Internet address: http://www.access.gpo.gov/cgi-bin/cfrassemble.cgi?title=200121

Vargo, S.A. *U.S. experience of regulation and inspection.* Conference on Quality Assurance in the Manufacture of Products Derived from Biotechnology 1988.

Guidance for Industry: For the submission of chemistry, manufacturing and controls and establishment description information for human plasma-derived biological products, animal plasma or serum-derived products. Current Guidance available at: http://www.fda.gov/cder/guidance/index.htm

Trijzelaar, B., *Regulatory affairs and biotechnology in Europe: III. Introduction into good regulatory practice–validation of virus removal and inactivation.* Biotherapy, 1993; 6(2):93–102.

Davis, G.C., and R.M. Riggin, *Characterization and establishment of specifications for biopharmaceuticals.* Dev Biol Stand, 1997; 91:49–54.

FDA, *An FDA center for drug evaluation and research report, Benefits vs. Risk. How CDER approves new drug* 2000.

FDA, *An FDA center for drug evaluation and research special report, from test tube to patient: improving health through human drugs* 1999.

Garnick, R.L., *Specifications from a biotechnology industry perspective.* Dev Biol Stand, 1997;91:31–6.

Hill, D.E., *Regulating biotechnology licensed products.* J Parenter Sci Technol, 1989;43(3): 139–41.

Serabian, M.A., and A.M. Pilaro, *Safety assessment of biotechnology-derived pharmaceuticals: ICH and beyond.* Toxicol Pathol, 1999; 27(1):27–31.

PHARMACOLOGY, TOXICOLOGY, THERAPEUTIC DOSAGE FORMULATIONS, AND CLINICAL RESPONSE

5.1. CLINICAL PHARMACOLOGY

5.2. DOSE AND THERAPEUTIC RESPONSE

5.3. DOSAGE FORM AND ROUTES OF ADMINISTRATION

The study of pharmacology encompasses an understanding of the characteristics of agents (i.e., drugs) that influence body structure and function. These characteristics include physical, chemical, and biological properties related to potentiating or blocking biochemical and physiological processes.

To health science practitioners and students, the term *pharmacology* relates to the knowledge of drugs useful to diagnose, prevent, and treat human diseases (i.e., clinical pharmacology). Clinical pharmacology studies are carried out in humans and are focused on collecting data needed to support the rational clinical use of drugs.

These data are required by regulatory agencies and vital for physicians. Pharmaceutical scientists also use this information as a basis for the discovery and development of new drugs. A clear understanding of basic pharmacology and the concepts unique to characterization and evaluation of therapeutic proteins and biologics, as opposed to small molecules, is essential for optimal use of these agents.

An overall appreciation of the principles of clinical pharmacology will also allow one to rationally design drugs with desirable pharmaceutical properties such as enhanced stability and activity, and a favorable disposition profile. In addition

Biotechnology and Biopharmaceuticals, by Rodney J. Y. Ho and Milo Gibaldi
ISBN 0-471-20690-3 Copyright © 2003 by John Wiley & Sons, Inc.

both scientists and clinicians need to integrate and apply the knowledge of how drug concentration affects therapeutic and toxic responses.

A fundamental assumption of drug therapy is that there is a relationship between dose or drug concentration and pharmacologic and toxicologic responses. The response may be immediate or occur some time after drug administration. While this assumption is generally valid, some protein drugs do not exhibit a simple relationship between drug concentration in blood and pharmacologic effect.

The impact of formulation on protein absorption and disposition is also an important factor in the development and use of biologic molecules. Stability of the protein drug in subcutaneous or muscle tissues and absorption rates directly influence the overall response. Various physical and chemical approaches are used to stabilize proteins and other macromolecules as a part of optimizing dosage formulations.

This section is organized in a format familiar to pharmaceutical scientists and clinical practitioners. That is, we will first discuss common features of biotechnology products and their underlying mechanisms. This essential information for therapeutic decision-making is organized in a format similar to the *Clinical Pharmacology and Toxicology* sections of product labels and monographs. The section encompasses pharmacology, pharmacokinetics (i.e., absorption, distribution, bioavailability, metabolism, and elimination), drug interactions, and immunologic responses. The format we selected should provide physicians and pharmacists, who routinely use information provided in product labels and monographs, with easy access to data relevant to optimization of therapeutic strategies for protein pharmaceuticals. The next section contains a discussion on dose-response profiles unique to protein-based pharmaceuticals. The chapter is concluded with a section on dosage form and route of administration.

■ 5.1. CLINICAL PHARMACOLOGY

As soon as a new molecular entity (NCE) is considered to be a new drug candidate (NDC) and produced in sufficient quantity, pharmacologic and pharmacokinetic data as well as short- and long-term toxicologic data are collected in animals as a part of preclinical studies. Preclinical pharmacology, pharmacokinetics, and toxicology studies are performed in rodents such as rats and mice. Promising NCEs are subsequently tested in larger animals such as dogs and nonhuman primates to validate and supplement the data collected in small animals. Preclinical data are collected from more than one species to reveal potential toxicity that may go undetected in a single species. These data also provide the basis for predicting safe and potentially effective doses for clinical studies. Additionally mechanistic elucidation of toxicology related to drug distribution, disposition, and metabolism, which cannot be done easily in humans, can be carried out in animal model systems established specifically to address these issues.

5.1.1. Pharmacology and Toxicology

Currently the FDA approves drugs based on a risk-benefit assessment; the agency considers that no prescription drug is free of risk. All prescription drugs carry some risks, and heath care professionals with prescriptive authority (physicians, nurses, and pharmacists) are the risk managers. Therefore health care professionals must have a clear understanding of the pharmacology and toxicology of biopharmaceuticals.

The pharmacology and toxicology of a drug candidate is summarized in the "Overview and General Introduction to Clinical Pharmacology" section of product labels. The FDA requires that a new drug application (NDA or BLA) must include an overall pharmacology and toxicology assessment [1,2]. While some of the pertinent issues for each medication vary, the summary information in product labels must include integrated knowledge about the pharmacology of the biopharmaceutical (i.e., mechanisms of therapeutic action, possible side effects, and interaction with other drugs).

Dose-response profiles for wanted and unwanted effects are, for the most part, derived from clinical studies. This information is also summarized in product labels. However, acute and long-term safety studies at higher than usual doses cannot be carried out in humans. Therefore such studies are performed in appropriate animal models to estimate the upper limit of the human dose. Acute and long-term toxicology data are derived from at least two animal species receiving escalating doses of drug. Side-effect monitoring during long-term repetitive dosing may include changes in behavior, carcinogenicity, mutagenicity, and untoward reproductive effects. The use of multiple species also allows identification of interspecies differences in drug responses.

Animal studies also allow early elucidation of the time course of drug concentration in blood and tissues for establishing relationships between concentration and pharmacologic and toxicologic effects. The information provided from animal studies, particularly dose ranging and long-term toxicology studies, extends our ability to extrapolate safe and effective-dose ranges for human administration.

Animal data serve as the springboard to estimating a safe and effective range of doses for human therapeutic purposes. Initial doses in phase I studies are based on preclinical pharmacokinetic and safety data. First estimates of safe and effective drug concentrations in plasma in human studies are also based on animal data. The "Clinical Studies" section in the product label includes information derived from tolerance studies of the drug (phase I), pivotal human data demonstrating efficacy at a defined dose or dose range, and a description of untoward effects observed in

healthy volunteers as well as in intended subjects (i.e., patients). Information relevant to patient population such as age, sex, and race is increasingly appreciated, but not all drug companies comply with FDA directives to provide such information. There is a growing public demand for safety and efficacy data on special populations, such as pediatric and elderly patients.

Obtaining unambiguous efficacy data in human clinical studies is almost always challenging and sometimes insurmountable. Establishing clinically relevant end points and demonstrating statistical significance between control and test groups can be difficult. For example, the difference in mortality in myocardial infarction patients treated with the thrombolytic agent alteplase and those receiving streptokinase is only 1% to 2%. Attempting to detect that difference with an acceptable level of confidence requires several thousand patients and many years. Despite these hurdles Genentech, the maker of alteplase, determined that the commercial impact of demonstrating a difference outweighed the risks of wasted resources in attempting to do so.

One of the issues unique to protein, peptide, and polymeric therapeutic products is that although they are purified, such products may contain infinitesimal amounts of cellular components and are rarely homogeneous. It is important to understand the effect of heterogeneity on pharmacology with regard to natural or engineered variations in peptide backbone and carbohydrate structures. Variations in glycosylation of protein pharmaceuticals may influence the rate of clearance of the protein (Figure 5.1). Some of the commonly used protein pharmaceuticals and the type of glycosyl group attached to each protein are listed in Table 5.1. An almost identical molecule such as interferon from two different manufacturers, even using the same recombinant host, is not likely to be identical. Often differences among the proteins produce subtle but detectable changes in biological function or disposition of the proteins that could introduce variation in therapeutic outcome.

Any significant changes in process or product formulation may also have significant impact on pharmacology, affecting therapeutic concentration range or window, bioavailability, or therapeutic

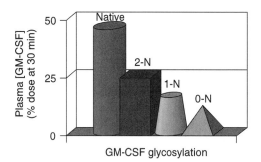

Figure 5.1. Effects of glycosylation on the plasma concentration of granulocyte-macrophage colony-stimulating factor (GM-CSF) in rats at 30 minutes after administration. The native, fully glycosylated and sialated GM-CSF is represented by the cylinder. GM-CSF with 2, 1, and 0 N-linked glycosyl groups (at amino-acid positions aspargine$_{27}$ and asparagines$_{37}$) are represented by the rectangle, tapered cylinder, and pyramid. (Data source: Donahue et al. [17])

effects. As biopharmaceuticals near the end of their term of patent protection, generic manufacturers face the challenging task of developing methods to demonstrate bioequivalence between the generic and original versions of the biopharmaceutical without resorting to very expensive clinical trials. Because there are no precedents for generic biopharmaceuticals, the FDA faces the daunting task of developing novel standards and guidelines.

5.1.2. Pharmacokinetics Principles

The methodology to predict the time course of drug concentration in plasma after administration is well described and well accepted as a pharmacokinetic principle. Today, pharmacokinetic principles are used routinely to estimate and manage dosing of medications for their safe and effective use. Such knowledge is useful not only in designing clinical trials for a new molecular entity, but also in day-to-day clinical practice (Box 5.1).

When a drug is administered by a route other than intravenous injection, the amount that reaches the circulation is frequently less that the administered dose. The

■TABLE 5.1. Types of glycosylation—N or O—linkage on some protein pharmaceuticals produced in mammalian cells

Protein Pharmaceuticals	Glycosylation Linkage to Peptide Back bone[a]
Erythropoietin (EPO)	N- and O-linked
Factors VII, VIII, IX, and X	N- and O-linked
Granulocyte colony-stimulating factor (G-CSF)	O-linked
Granulocyte-macrophage colony-stimulating factor (GM-CSF)	N-linked
Interleukin-2 (IL-2)	O-linked
Interferon-alfa (IFN-α)	N- and O-linked
Interferon-beta (IFN-β)	N-linked
Interferon-gamma (IFN-γ)	N-linked
IgG, including monoclonal antibodies	N-linked
Tissue plasminogen activator (tPA)	N-linked

[a]N-linked; addition of glycosyl group through amino linkage of sugar residues attached to asparagine amino-acid residue on protein. O-linked; addition of glycosyl group through threonine or serine amino-acid residue on protein.

percentage of the dose reaching the circulation is the bioavailability of a drug for a specified route of administration. Low bioavailability of a drug after oral administration could be the result of poor lipid solubility, poor dissolution in the gastrointestinal tract, or extensive metabolism in the gut and liver. Low bioavailablity after subcutaneous or intramuscular administration may be the result of precipitation or degradation of the product at the injection site.

When a drug reaches the circulation, it quickly distributes outside the capillary beds into well-perfused tissues but may distribute slowly or not at all to less-accessible tissues protected by barriers, such as the brain. The volume of distribution is the ratio of the amount of drug in the body divided by the drug concentration in plasma once a pseudo-equilibrium is established between blood and tissues. For small molecules, a low volume of distribution generally signifies extensive plasma protein binding that restricts distribution outside the capillary bed, while a large volume of distribution, often exceeding the total volume of body water, usually means extensive tissue binding. For large molecules such as proteins and peptides greater than 1.5 kDa, distribution is restricted to blood volume (about 5 liters). Smaller biopharmaceuticals may diffuse into extravascular fluid and reach a volume of distribution of about 20 to 30 liters. The volume of distribution of several proteins of therapeutic interest is listed in Table 5.2.

Because plasma protein and tissue binding influences the amount of drug in each body compartment, the volume of distribution of a drug may be dose dependent (nonlinear). A limited number of binding sites may result in capacity-limited binding in plasma or tissues. Transport from the blood may also be capacity limited. Examples of dose-dependent volume changes show that the volume of distribution of recombinant human tumor necrosis factor (TNF) alpha decreases sharply with a fourfold increase in dose and that the volume of distribution of recombinant human DNase,

---

BOX 5.1. PHYSICOCHEMICAL PROPERTIES OF PHARMACEUTICALS, PHYSIOLOGIC VARIABLES, AND PHARMACOKINETIC PARAMETERS

Regardless of the mathematical model—noncompartmental or compartmental—used to estimate pharmacokinetic parameters, absorption and disposition of a pharmaceutical depend on physiochemical properties such as hydrophobicity, molecular diameter, and substrate recognition by exsorption or absorption transporter molecules at absorption and distribution sites. While plasma protein binding of protein-based biopharmaceuticals may not be a significant factor, binding of therapeutic proteins to circulating receptors may have a significant impact on distribution and elimination. Gastrointestinal (GI) absorption could also be influenced by gastric emptying rate, gastrointestinal motility, and blood flow rate. Distribution of a pharmaceutical in the body may also depend on the amount of body fat, typically measured by weight-to-volume ratio. Elimination of pharmaceuticals through renal excretion may depend on activity of secretion processes such as active transport and tubular reabsorption, which is often influenced by pH, electrolytes, and urine flow rate. Many of these physiological processes may be influenced by diseases and concurrent drug therapy. The following table lists pharmacokinetic parameters, mathematical relationships, and physiologic variables that may have an impact on these parameters.

(Continued on next page)

BOX 5.1. Continued

Pharmacokinetic Parameter		Mathematical Relationship	Physiological Variables
Absorption rate	k_a	$$\frac{V_{max} \cdot [Drug]_{absorption\ site}^{a}}{K_m + [Drug]_{absorption\ site}}$$	Absorption site blood-flow rate; gastric emptying and intestinal motility rate for orally administered drugs; GI pH, precipitation at absorption site
Clearance, hepatic	CL_H	Blood flow · Fraction of drug extracted by liver	Liver blood flow; protein binding; liver metabolism
Clearance, renal	CL_R	Blood flow · Fraction of drug extracted by kidney	Renal blood flow; protein binding; active secretion, transport and reabsorption; glomerular filtration; urine pH; urine flow
Volume of distribution	V_d	$$\frac{Amount\ of\ drug\ in\ body}{[Drug]_{plasma}}$$	Binding and partition to blood, tissues, and fat; body composition and size
Half-life, elimination	$t_{1/2}$	$$0.693 \cdot \frac{Volume\ of\ distribution}{Clearance}$$	Blood flow, protein and tissue binding, metabolism, renal excretion
Elimination rate constant	k_e	$$\frac{Clearance}{Volume\ of\ distribution}$$	Blood flow, protein and tissue binding, metabolism, renal excretion
Fraction excreted unchanged		$$\frac{Renal\ clearance}{Total\ clearance}$$	Renal blood flow; protein binding; active secretion, transport and reabsorption; glomerular filtration; urine pH; urine flow; metabolism
Area under the curve, IV dosing	AUC_{iv}	$$\frac{Dose}{Clearance}$$	Protein and tissue binding, metabolism, renal excretion
Steady state plasma concentration, IV dosing	C_{ss}	$$\frac{Infusion\ rate}{Clearance}$$	Metabolism, renal excretion
Area under the curve, oral dosing	AUC_{po}	$$\frac{Dose \cdot Oral\ bioavailability}{Clearance}$$	Fraction of dose absorbed, first pass metabolism, dosage formulation
Mean residence time	MRT	$$\frac{Volume\ of\ distribution}{Clearance}$$	Route of administration; rate of absorption; IV infusion rate, metabolism and renal excretion

[a]Expression based on capacity-limited absorption (e.g., enzyme degradation) processes. Passive absorption processes can be approximated by first order kinetics. Under the usual therapeutic conditions:

$$[Drug] \ll K_m,$$

$$\therefore k_a = \frac{V_{max} \cdot [Drug]}{K_m} = k_1 \cdot [Drug],$$

Which is a first-order process.

■TABLE 5.2. Volume of distribution for selected therapeutic proteins

Therapeutic Protein	Molecular Weight (kDa)	Volume of Distribution (Liter)
Erythropoietin	30.4	2.8–3.5
Human monoclonal antibody	150	5.6
Superoxide dismutase	~32	7

■TABLE 5.3. Dose-dependent effects on volume of distribution

Therapeutic Protein	Dose	Volume of Distribution (L)
Human recombinant tumor	$25 \, mg/m^2$	66 ± 30
necrosis factor-α (rhTNF-α)	$100 \, mg/m^2$	12 ± 4
Human recombinant	$10 \, IU/kg$	$7.1 + 1.8$
erythropoietin	$50 \, IU/kg$	3.2 ± 0.3
	$1000 \, IU/kg$	5.0 ± 1.8
Human recombinant DNAse	$0.01 \, mg/kg$	5.4
	$1 \, mg/kg$	12.1

marketed for the treatment of cystic fibrosis, increases with dose (Table 5.3).

The decrease in the distribution volume of TNF with an increase in dose is thought to be due to capacity-limited extravascular transport processes (Table 5.3). The increase in volume of distribution of DNase with an increase in dose is thought to be due to capacity limited protein binding to actin in plasma (possibly an actin–vitamin D–DNase ternary complex).

Small molecules are eliminated from the body largely by means of drug metabolism enzymes in the liver and other tissues and by urinary excretion. Large molecules are also eliminated by renal and hepatic mechanisms. Proteins that are less than 40 to 50 kDa are cleared by renal filtration with little or no tubular reabsorption. Larger proteins are less likely to be filtered but may be subject to phagocytosis in hepatocytes and Kupfer cells in the liver. Protein biotransformation—denaturation, proteolysis, and oxidative metabolism—is also important.

Elimination from the body is described quantitatively in terms of drug half-life and clearance. As a first approximation, the fall in drug concentrations in plasma after absorption is complete can usually be described as a first-order process. Therefore, a plot of log drug concentration versus time is usually linear. The slope of this line is called the elimination rate constant with units of reciprocal time. The reciprocal of the elimination rate constant is called the half-life, the time required to eliminate 50% of the drug from the body. Drugs with short half-lives are rapidly eliminated from the body, while drugs with long half-lives may persist for days or weeks or even longer.

Drug clearance integrates the concepts of organ perfusion (flow rate) and the organ's ability to eliminate a drug (i.e., extraction ratio). A small-molecule drug that is completely filtered in the kidneys and totally extracted in the urine without tubular reabsorption has a clearance of 125 ml/min, the same as the glomerular filtration rate. In theory, clearance can be estimated across all eliminating organs. The sum of these organ clearances is called the total body clearance of a drug.

Clearance may vary widely across individuals, especially for drugs largely eliminated by drug metabolizing enzymes, because of genetic variability or poly-

■**TABLE 5.4.** General pharmacokinetic and pharmacodynamic features of protein versus chemical drugs

Features	Protein Drugs	Chemical
Molecular weight (MW)	>1500 Da	<1000 Da
Route of administration	Parenteral (Injection: IV, SC, IM)	All routes (PO, IV, IM, SC, TD, TOP)
Volume of distribution	Mainly limited to blood volume	Dictated by hydrophobicity and plasma and tissue protein binding
Plasma protein binding	Does not play a significant role (esp. for proteins > 50 kDa)	Important
Extracellular fluid distribution	Molecular weight and intrinsic properties of the proteins are important	Mostly depends on MW and hydrophobicity
Cell-surface receptor (ligand) interactions	May play a significant role	May be important in some cases
Phagocytosis	Aggregated proteins and macromolecules with MW > 300 kDa	Does not play a role
Elimination	Protein denaturation; proteolysis	Biotransformation (e.g., oxidation and conjugation), hydrophobicity, and protein binding
Renal	Filtration (MW < 40–50 kDa) (cleared without reabsorption); metabolism	Filtration; transport; metabolism tubular reabsorption
Hepatic	Phagocytosis; receptor-mediated endocytosis (hepatocytes); biotransformation	Hydrophobicity and molecular weight
Immunogenicity	May play a significant role for macromolecules that require repetitive administration	Usually not important

Abbreviations: MW, molecular weight; IV, intravenous, SC, subcutaneous, IM, intramuscular, PO, per oral, TD, transdermal, TOP, topical, kDa, kilo Daltons.

morphism. State of health can also determine clearance. Patients with renal or hepatic disease have decreased drug clearances compared with those having normal renal or hepatic function. Determination of the clearance of the endogenous marker creatinine is routinely used to assess renal function.

In the clinic, total body clearance of a drug is estimated by taking the ratio of the administered dose, corrected for bioavailability, to the total area under a plot of drug concentration in plasma versus time produced by that dose.

Most protein pharmaceuticals exhibit a short half-life—measured in minutes—and high clearance because they are smaller than 30 kDa and are readily cleared by glomerular filtration in the kidneys. Some pharmacokinetic features of protein phar-

■**TABLE 5.5.** Comparison of various parenteral routes on bioavailability and ease of use

Parenteral Routes	Bioavailability, Biologic Response, and Ease of Use
IV—intravenous	100% systemic availability, most rapid biologic effect, potential for significant adverse effects, most invasive of parenteral routes, half-life governed by clearance
SC—subcutaneous	Less than complete systemic availability, half-life may be limited by absorption
IM—intramuscular	Systemic availability may be less than 100%, reduced peak concentrations compared to IV administration, potentially delayed and reduced biologic effects, less invasive than IV route, half-life may depend on absorption

maceuticals compared with small-molecule drugs are listed in Table 5.4.

5.1.3. Disposition of Recombinant Protein Pharmaceuticals

Absorption and Bioavailability

Because most proteins are susceptible to protease degradation and denaturation in biologic fluids, most biopharmaceuticals must be administered by intravenous, intramuscular, or subcutaneous injection (see Table 5.5). High concentrations of proteases are found in the gastrointestinal tract, nasal mucosa, bronchioles, and alveoli, which severely limit the bioavailability of protein pharmaceuticals after oral, intranasal, and inhalation administration. Diffusional barriers to the passage of relatively large macromolecules preclude transdermal and mucosal administration of protein pharmaceuticals. Research is under way to develop methods that will protect protein drugs from proteolysis and improve transmembrane diffusion.

Tissue Distribution of Biopharmaceuticals. By design, monoclonal antibodies exhibit high-affinity binding to specific antigenic epitopes. These epitopes are present somewhere in the body; perhaps on cell surfaces in healthy or cancerous tissue. In an ideal situation, a monoclonal antibody would distribute only to the high-affinity binding sites in target tissues. Invariably, however, any protein molecule, even a monoclonal antibody, distributes into nontarget tissue sites, which can account for a major fraction of the dose. Nonspecific distribution could give rise to toxicity and loss of potency.

Distribution of proteins to tissues is controlled by the permeability (porosity) of the vasculatures and thereby influenced by the molecular size of the protein. A protein of greater than 150 kDa (~50 nm) in size will have limited distribution and may be restricted to blood volume. Infrequently a large protein has amino acid recognition sequences that allow passage across epithelial cells lining the vasculatures by transcytosis, a process that allows directional transport of protein into and out of a cell.

Another factor influencing nonspecific tissue distribution is the carbohydrate portion of the IgG molecule, which is attached to the therapeutic protein via an N-linked glycan in the constant domain (F_c). Loss of terminal sialylated residues on the carbohydrate of IgG exposes galactose and promotes receptor-mediated binding of IgG to hepatocytes. Consequently this results in an increase in nonspecific distribution to the liver. Details of desialylation of IgG and its consequences are discussed in Chapter 10. Other glycoproteins may exert similar mechanisms of nonspecific distribution.

Loss of sialylated residues may also expose terminal mannose residues, which are attractive targets for phagocytic cells. Phagocytic cells with mannose receptors are highly effective in further clearing the partially degraded IgG. While exposure of mannose residues may reduce the therapeutic effect of a monoclonal antibody by accelerating phagocytic elimination processes, phagocytosis may provide a controlled duration of action and thereby minimize toxicity. For example, thrombolytic agents designed to act at the site of a thrombus may cause bleeding if extensive nonspecific distribution occurs. Phagocytic cells in blood and liver, such as macrophages and Kupfer cells, limit nonspecific distribution of these agents.

Elimination and Metabolism

Molecular Weight. The molecular size of a protein is a critical factor in determining rate of elimination and duration of action. Small-protein molecules can be filtered in the kidneys and directly excreted, without reabsorption, into urine. The larger the molecular size of a protein, the lower its renal clearance. The limit of glomerular filtration is estimated to be 50 to 70 kDa. Because the molecular weights of peptides and proteins range from a few thousand daltons to greater than 150 kDa, many biopharmaceuticals are eliminated in part by renal excretion. The importance of renal excretion in protein elimination means that changes in renal function (e.g., due to medical disorders) may have an impact on protein therapy.

When proteins exceed a molecular weight of 200 kDa, phagocytosis plays an increasing role in elimination. Phagocytosis is also important when a partially degraded small-protein aggregate exhibits the characteristics of very large proteins. Internalization of large proteins or protein aggregates by phagocytes leads to intracellular proteolytic inactivation.

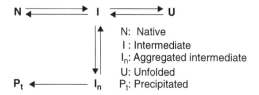

Figure 5.2. Inactivation of protein from native to various denaturation states leading to formation of protein aggregates, detectable as precipitates.

Physical and Chemical Integrity of Proteins. The primary sequence of proteins and peptides is comprised of L-amino acids linked together by covalent amide bonds. Substituent group polarity and/or charge is a critical determinant of secondary and tertiary structure and stability. Secondary structures (α-helices and β-sheets) arise from hydrophobic, ionic, and Van der Waals interactions that fold the primary amino acid chain upon itself. Most therapeutic proteins exhibit tertiary structure vital to functionality and are held together by covalent and noncovalent bonding of secondary structures (Figure 5.2).

All proteins and peptides display chemical and physical instability that affects the way they are distributed and cleared in the body and their delivery to the site of action. Physical and chemical instability is affected by primary sequences and secondary and tertiary structures and the degree of glycosylation of protein. Chemical degradation of proteins and peptides involves deamidation, racemization, hydrolysis, oxidation, beta elimination, and disulfide exchange. Physical degradation of proteins involves denaturation and aggregation.

Proteases found in blood are catalysts that accelerate hydrolysis and other chemical degradation processes. Furthermore, chemical degradation almost always leads to physical degradation.

Mechanisms of Chemical and Physical Degradation of Proteins. A folded native protein exists in a local minimum-energy conformation. Typically 5 to 15 kcal/mol of energy is needed to rapidly (in milliseconds) convert a folded protein to an unfolded conformation. In general, proteins and peptides fold in a manner that minimizes the exposure of hydrophobic regions of the molecule to water or an aqueous environment in a biological milieu. The unique structure that the molecule takes will depend on both the surrounding environment (buffered solution, biologic membrane, blood, plasma, etc.), and the primary amino acid sequence. The unfolding of a native protein leads to denaturation (Figure 5.2). The denaturation process may involve a number of intermediate conformations, leading to protein aggregates and precipitates that are readily cleared from the body. In most cases, partially denatured proteins exhibit greatly reduced or no biological activity. In some cases, partially denatured proteins are immunogenic and may play a role in eliciting an antibody response against native therapeutic proteins.

Any chemical or enzymatic process that modifies a therapeutic protein in the body is known as protein metabolism. Unfolded proteins are, in general, more susceptible to metabolism via proteolysis because of increased access to critical peptide sequences. Chemical modification of a protein can "mark" it for further degradation. This is thought to occur by altering the folding equilibrium in favor of the denatured conformation [3].

Glycosylation and Protein Stability. Many endogenous proteins and related biotechnology products exist as glycoproteins—protein chains linked to carbohydrates (glycosyl groups). Typically the carbohydrate portion of the molecule increases the stability and residence time of the protein in the blood circulation. Partially digested carbohydrates can decrease the residence time of the glycoprotein in blood.

The glycosyl groups on glycoproteins contain monosaccharides that are linked to each other by glycosidic bonds to form straight or branched chain polysaccharides. Three monosaccharide derivatives are key players in the formation and stability of glycoproteins. They are *N-acetylgalactosamine*, which links to the hydroxyl group of threonine or serine (O-glycosyl links), *N-acetylglucosamine*, which links to the terminal amine of asparagine residues (N-glycosyl links), and *N-acetylneuraminic acid* (sialic acid), which forms the terminal "cap" of a polysaccharide chain and prevents cell surface receptor's binding to other sugar residues and uptake of the entire glycoprotein by hepatocytes and macrophages.

For example, erythropoietin (epoetin) is a heavily glycosylated 165 amino-acid protein (30.4 kDa), which stimulates red blood cell production. The carbohydrate moiety comprises 40% of the total molecular size (Figure 5.3). The carbohydrate moiety is not necessary for the biologic activity of the molecule (in fact it reduces it somewhat), but the carbohydrate chain greatly reduces the clearance of the therapeutic protein. Fully glycosylated epoetin has a half-life in blood of 6 to 12 hours, whereas desialyated epoetin is cleared within minutes. Loss of the terminal sialic acid residue results in the rapid uptake of the molecule by the liver through a galactose receptor on the surface of the hepatocytes.

The carbohydrate moiety of immunoglobulin G can be an important determinant of both function and stability (Figure 5.4). It is attached to the protein via an N-linked glycosyl group in the constant domain (F_c). The polysaccharide chain influences tissue distribution (via cell surface recognition and binding) and the stability of IgG. Fully glycosylated IgG with a sialic acid cap has a half-life of 21 to 27 days. Cleavage of the terminal sialic acid residue of the polysaccharide chain promotes rapid uptake and destruction by the liver (Figure 5.5).

Chemical "Marking" Modifications. Chemical marking accelerates the

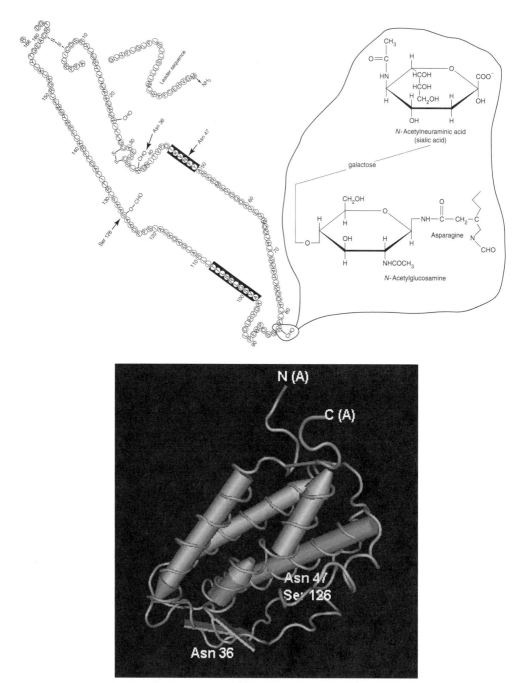

Figure 5.3. Glycosylated erythropoietin (epoetin) with an N-linked glycosyl group and terminal sialic acid cap.

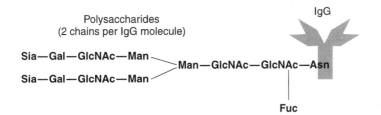

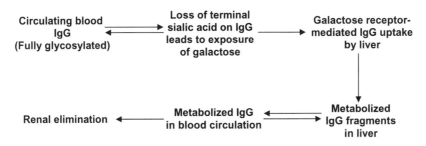

Figure 5.4. Schematic representation of glycosylated IgG. *Asn*, asparagines; *Fuc*, fucose; *GlcNAc*, N-acetyl-glucoseaimine; *Man*, Mannose; *Gal*, galactose; *Sia*, sialicacid.

Figure 5.5. Disposition of IgG.

elimination of a protein through enhanced metabolism. The process involves one of six protein modifications:

1. Deamidation of basic amino acids, glutamine and asparagine, involves spontaneous acid-catalyzed hydrolysis of the terminal amide on these amino acids. Deamidation of these basic amino acids results in formation of acidic amino-acid residues. Somatotropin, IgG, and insulin are subject to this modification. Conversion of a single basic amino acid to an acidic amino acid can have a profound effect on tertiary structure (unfolding) and the susceptibility of the amino acid backbone to proteolysis. In general, proteins rich in glutamic acid (serine, threonine, and proline) are rapidly degraded in eukaryotic cells.

2. Oxidation of methionine residues involves hydrogen peroxide (H_2O_2)-catalyzed conversion of the terminal sulfide to a sulfoxide. This marking occurs with α1-antitrypsin and parathyroid hormone and leads to the introduction of a net charge, which can affect tertiary and secondary structure.

3. Nonspecific cytochrome P450-mediated oxidation involves enzyme-catalyzed formation of reactive oxygen species (superoxide anions and hydroxyl radicals), which oxidize susceptible amino acids such as proline, arginine, lysine, and histidine.

4. Substrate-specific cytochrome P450-mediated hydroxylation involves NADPH-dependent oxidation of some cyclic peptides such as cyclosporine. Hydroxylated products of cyclosporine are biologically inactive and readily cleared from the body.

5. Disulfide exchange is a spontaneous reaction that occurs under neutral or basic conditions and involves cleavage of a disulfide bond and incorrect

reformation. As an example, disulfide bridges in insulin may unfold and be conjugated to glutathione. This process can be catalyzed by transhydrogenase cleavage of the interchain disulfide bridge and reformation with the cysteine residue of glutathione. Disulfide exchange will invariably change the tertiary structure of a protein.

6. Phosphorylation of serine or threonine residues involves an ATP-dependent addition of a phosphate group to a primary (serine) or secondary (threonine) alcohol. Phosphorylated proteins are often subject to rapid degradation.

Proteolysis. Proteolysis is the cleavage of amide bonds that comprise the backbone of proteins and peptides. The reaction can occur spontaneously in aqueous medium under acidic, neutral, or basic conditions. This process is accelerated by proteases, ubiquitous enzymes that catalyze peptide-bond hydrolysis at rates much higher than occur spontaneously. In humans, these enzymes only recognize sequences of L-amino acids but not D-amino acids. They are found in barrier tissues (nasal membranes, stomach and intestinal linings, vaginal and respiratory mucosa, ocular epithelium), blood, all internal solid organs, connective tissue, and fat. The same protease may be present in multiple sites in the body.

Several different proteases can attack a single protein at enzyme-selective amino-acid sequences. Proteases can be divided into two categories. Endopeptidases are enzymes that cleave peptide bonds between specific, nonterminal amino acids. There are endopeptidases specific for just about every amino acid. Exopeptidases are enzymes that cleave terminal peptide bonds at either the C-terminus or N-terminus.

Protein Metabolism in Eliminating Organs. Peptidases in the gastrointestinal mucosa represent the major barrier to orally administered protein or peptide drugs. Pepsins, a family of proteases specific for aspartic acid, reside in the stomach lining. The intestinal lining contains chymotrypsin, an endopeptidase specific for hydrophobic amino acids (e.g., phenylalanine, leucine, methionine, tryptophan, and tyrosine); trypsin, an endopeptidase specific for basic amino acids (e.g., arginine, lysine); elastase, an endopeptidase specific for amino acids with small, unbranched, nonaromatic functional groups (e.g., glycine, alanine, serine, valine, leucine); and carboxypeptidase A, a C-terminus exopeptidase selective for L-amino acids with aromatic or bulky functional groups (e.g., tyrosine).

The liver eliminates proteins on first pass after oral administration and on each pass of hepatic blood flow. Hepatocytes, Kupffer cells, adipocytes, and endothelial cells can all be involved in proteolysis (Figure 5.6). Proteolysis can occur in lysosomes after endocytosis of a protein and lysosomal fusion. Endocytosis of a protein may be a nonspecific or receptor-mediated process. Proteolytic products are eliminated from the liver through biliary excretion, and subsequently digested further in the intestinal tract.

The kidneys filter and metabolize proteins and peptides (Figure 5.7). The efficiency of filtration depends on the average radius of the molecule; the molecular charge of a protein may influence the apparent radius. Reabsorption, although uncommon for proteins, can take place along the luminal surface of the proximal tubule. Ordinarily proteins can bind to the brush border (luminal surface) of proximal tubules through an electrostatic interaction with lysine residues on the protein. Endocytosis and fusion with lysosomes follows binding and leads to proteolytic digestion. Resulting fragments could diffuse across the basolateral surface of tubules and be reabsorbed into the blood. Kidney metabolism is a major route of elimination for

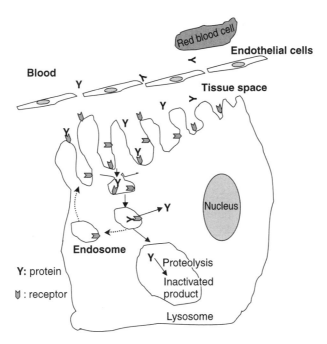

Figure 5.6. Metabolism of proteins by proteases found at the cell surface and internal cellular organelles. Intracellular uptake of proteins often involved receptor-mediated endocytosis; initial surface binding can be specific for a hormone (e.g., insulin, glucagon).

interleukins, interferons, tumor necrosis factor, and colony-stimulating factors.

5.1.4. Drug Interactions

Drug–drug interactions that influence the outcome of drug therapy are a concern for regulatory agencies, prescribers, and patients. Many drug–drug interactions are mediated by changes in metabolism of one drug, caused by a second drug, that leads to more or less rapid elimination. Examples abound of drugs becoming inactive when given with a second drug that accelerates its metabolism. There are still more examples of elevated drug levels in plasma and increased adverse effects when a drug is given with a second drug that inhibits its metabolism.

Pharmacologically or pharmacodynamically based drug–drug interactions are less well documented or understood but can be envisioned to be important [4]. A potential mechanism is the up or down regulation of pharmacologically important receptors. Such changes can alter the drug concentration-response profile and require more or less intense dosing for optimal therapeutic response.

Proteins and other macromolecules are mainly cleared by high-capacity elimination processes such as renal filtration and liver metabolism. A coadministered drug can affect these processes and lead to serious drug–drug interactions. In addition, drugs that influence receptor-mediated clearance of the therapeutic protein may also result in important drug–drug interactions.

While drug–drug interactions are well studied, and such information is essential for securing marketing approval of a new chemical drug candidate, such studies are not ordinarily required for biological drug candidates. The few reported drug–drug interactions with protein therapeutics have been attributed to modifications in protein clearance. Other mechanisms of drug–drug

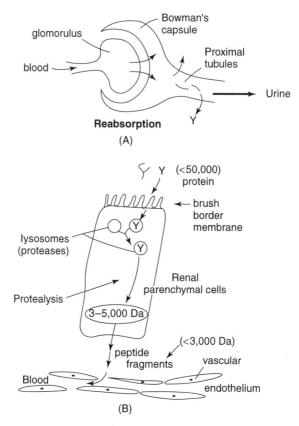

Figure 5.7. Schematic representation of therapeutic protein (A) being filtered and (B) metabolized in the kidneys. Protein reabsorption in proximal table is limited.

interactions need to be evaluated. More recent studies suggest that cytokines that modulate inflammatory processes, such as interferons and interleukins, may suppress the expression of cytochrome P450 enzymes and could result in the alteration of oxidative metabolism of protein and peptide drugs.

5.1.5. Immunologic Responses

Unlike small chemical molecules (MW < 1000), exogenous macromolecules, such as proteins and nucleic acids, when given repeatedly exhibit a high tendency to elicit immune responses in patients. It is well documented in classical immunological studies that poorly immunogenic haptens such as

dinitrophenyl derivatives must be coupled to large proteins, such as albumin or keyhole limpet hemocyanin (KLH), to elicit antibody against such haptens. Haptens are capable of binding to antibody but cannot by themselves or in the absence of carrier proteins elicit immune responses [27].

While it is generally true that recombinant proteins containing amino-acid sequences identical to those of humans are less immunogenic than mouse-based recombinant proteins, human-based recombinant proteins can, because of their size, elicit immune responses on repeated administration. The impact of immunogenicity on pharmacologic and toxic effects is only now beginning to be adequately appreciated.

Repeated administration of a therapeutic macromolecule may elicit antibodies against itself, but the degree of binding to the protein and the clinical significance of the interaction vary widely. In some cases protein immunogenicity may produce hypersensitivity or allergic reactions analogous to penicillin allergies, and alter the pharmacokinetics and pharmacodynamics of a drug. In the text that follows we will discuss (1) factors that influence immunogenicity, (2) assessment of immunogenicity, and (3) consequences of antibody formation.

Factors That Modulate Immunogenicity

Several factors ranging from the design of a recombinant protein and the way it is produced to the kind of patient to whom the protein is administered influence the immunogenicity of the protein and the likelihood of antibody being produced.

A partially manageable variable is product purity. As a recombinant protein is purified and concentrated to several thousandfold of its original concentration in host cells, impurities such as cell substrates and media components are copurified. These impurities can go undetected and can be immunogenic by themselves or act as adjuvants to elicit immune responses against the therapeutic protein. Partially denatured proteins as a result of proteolysis, aggregation, or oxidation also may lead to exposure of immunogenic domains, resulting in enhanced antibody formation. The oxidized forms of interferons are reported to be more immunogenic in mice than the native form.

Immune responses also may be mounted against therapeutic monoclonal antibodies. The class of monoclonal antibody as well as the animal species from which the distinguishing amino acid sequences of antibody are derived strongly influence immunogenicity. Suffice to say that mouse antibodies, monoclonal or polyclonal, in humans are much more immunogenic than human antibodies. However, a mouse antibody may be more immunogenic or less immunogenic than another mouse antibody, even when both are designed to bind to the same therapeutic target. The FDA assumes that all antibodies are potentially immunogenic. Therefore immunogenicity must be evaluated for each antibody product.

Alteration of amino-acid sequences also may lead to changes in the immunogenicity of a protein. Chemical conjugation of monoclonal antibodies to immunogenic bacterial toxins often result in increased immunogenicity [5]. On the other hand, sequence modifications that result from adding a highly conserved IgG heavy chain (FC domain of IgG) to receptors, such as that for tumor necrosis factor, lead to reduced antibody formation against the receptor [6,7]. Protein deglycosylation also may play a role in eliciting antibodies to otherwise unexposed antigenic determinants of the native protein. Therefore, the choice of host cell to produce native glycosyation—a posttranslational event— may have a significant impact on protein immunogenicity.

Clinical factors, such as a patient's immune status and how the biopharmaceutical is administered, also will influence immunogenicity. Results from transplant immunology studies clearly indicate that some individuals are genetically disposed to a higher degree of immune response. For example, patients exhibiting HLA type B46 and B15 tend not to produce antibody to hepatitis B antigen when presented as a recombinant vaccine [8]. Medical disorders such as autoimmune disease and kidney and liver diseases may also lead to altered immunogenicity of the pharmaceutical protein. Patients receiving chemotherapy or other immunosuppressive drugs will be less likely to produce antibody responses to protein drugs.

Finally, dose, frequency of dosing, and

route of administration also will modulate the degree and population frequency of antibody formation in patients. In general, subcutaneous injections, compared with intravenous and intramuscular injections, have a higher likelihood of antibody production. Immunogenicity tends to increase with higher doses and more frequent administration. At exceedingly high doses, however, immunological tolerance may occur.

Assessment of Immunogenicity

Immunogenicity assays are conducted early in drug development to support pharmacology and toxicology studies. The concentration of antibody in blood or serum is detected with a semiquantitative assay such as an enzyme-linked immunosorbent assay (ELISA). ELISA is an analytical tool in which the antigen, in this case a protein pharmaceutical, is coated on the surface of a test tube (solid support), allowing antigen-specific binding of a specific serum antibody. An enzyme-linked diagnostic antibody that recognizes the presence of a specific antibody then is added to the test tube along with an enzyme substrate that produces an intensity of color proportional to the amount of binding. By measuring the color intensity, an analyst can assess the antibody concentration in serum raised by the protein drug. This method allows screening of any antibody that binds to the therapeutic protein (binding antibody) in the sera of blood samples from a large number of patients.

However, not all antibody detected by ELISA corresponds to the antibody molecule that blocks the biological activity or active sites of the therapeutic protein. These neutralizing (blocking) antibodies are only a small fraction of the binding antibody population. Different cell- and receptor-based neutralizing antibody assays are developed for different protein drugs.

It is essential to evaluate potential immunogenicity of a therapeutic protein throughout clinical trials. Therefore, in planning for sample collection during clinical trials for biotechnology products, clinical investigators often consider collecting blood samples at multiple predetermined time points for immunological assays. Availability of these blood samples allows the investigators to determine the immunogenicity of the protein drug. It also reveals the time course and extent of the immune response in the population. Additional pharmacokinetic data, collected before and after induction of antibody, allow the assessment of clinical impact of immune responses generated against the administered protein. For example, a decrease in the biologic activity of a therapeutic protein after induction of an antibody response signifies the presence of drug-neutralizing antibodies.

Consequences of Antibody and Other Immune Responses

Although rare, immune response to a therapeutic protein may lead to anaphylactic or allergic reactions that require immediate medical attention. Less severe but far more frequent consequences are local injection-site reactions. Neutralizing antibodies in effect reduce the dose of drug available to exert a therapeutic response. Nonneutralizing antibodies may also reduce biological activity by increasing the size of the resulting protein-antibody complex, which may influence the tissue distribution and clearance of the pharmaceutical protein. Although both mechanisms reduce biological activity, neutralizing antibodies usually result in more severe loss of activity.

An indirect consequence of antibody formation against a therapeutic protein is inhibition of endogenous protein and its functions. For example, if antibody to interleukin-2, a T-cell growth factor, is elicited due to administered recombinant IL-2, the neutralizing antibody that blocks the activity of recombinant IL-2 may also neutralize endogenous IL-2 function, which is essential for fighting infections. Another consequence is that patients who have antibodies against a protein drug are not good candidates to receive similar protein products. Immunogenicity data and clinical consequences for some marketed protein products are summarized in Table 5.6.

The FDA advises manufacturers not only to evaluate immunogenicity of new protein products but also to reevaluate immunogenicity on any change in the manufacturing process or target patient population. Understanding of mechanisms leading to induction of immune responses against proteins and other biopharmaceuticals will allow the development of strategies appropriate for better control of unintended consequences of immunogenicity [28].

■TABLE 5.6. Immunogenicity of protein pharmaceuticals and clinical consequences

Products[a]	Frequency (in Humans)	Potential or Observed Clinical Effects
Antibodies		
Murine ($n = 6$)[b]	55–88%	Human against mouse allergic reactions
Mouse-human chimeric ($n = 3$)	1–13%	Human against mouse allergic reactions
Humanized	1–8%	Human against mouse allergic reactions
Remicade	13%	Myalgia, rash fever, polyarthralgia
Etanercept	7%	Redness at the sites of previous injections
IFN-β (beta)	15–45% in multiple sclerosis patients	Neutralizing antibodies (higher incidence with subcutaneous route of administration)
IFN-γ (gamma)	0–25%	
IFN-α (alfa)	None in chronic granulomatous	Neutralizing antibodies
Human urokinase or tissue plasminogen activator	Less than 1%	Rare antibody or allergic reactions
Streptokinase (isolated from bacteria)	1–4%	Allergic reaction; increased incidence of binding antibody with recent infection of streptococcus
Epoetin and G-CSF (human recombinant)	Rare	Binding antibody and allergic reactions
GM-CSF	2–3%	Binding and neutralizing antibodies; allergic reactions
IL-2	66–74%	Binding antibodies; but rarely are they neutralizing
IL-11	1%	Binding antibody; not neutralizing; some local subcutaneous site reactions reported

[a]Protein products were tested using therapeutic dosing schedules.

[b]Number of different molecules for the same type of protein surveyed.

■ 5.2. DOSE AND THERAPEUTIC RESPONSE

Knowledge of the dose-response or dose-effect relationship is critical for choosing an optimum regimen for patients. In the simplest case, dose is directly proportional to drug concentration in the blood and at the site of action, but biologic variability makes this assumption very tenuous. Hence knowledge of the concentration response usually provides better information. This relationship, however, is often not universally available. Biopharmaceuticals, like conventional pharmaceuticals, frequently produce a graded response in individual patients. A graded or increasing dose of a drug given to an individual may produce a corresponding increase in the magnitude of response. This relationship is often referred to as *individual dose response*. On the other hand, in the context of population studies, the fraction, typically expressed as percentage, of the population exhibiting drug effects increases as the dose is escalated. The dose-response relationship in a population exhibiting drug effects is generally known *as quantal dose response*.

Setting aside issues of compliance and administration errors, the therapeutic response experienced by each patient may be influenced by variations in pharmacokinetics (rate and extent of absorption, distribution, elimination) and pharmacodynamics (physiologic, pathologic, and genetic factors; receptor interactions; and tolerance).

The dose-response relationship for each biopharmaceutical can be analyzed, and the variation between subjects in the study population can be derived. A representative curve, presented as drug concentration versus effect intensity, is shown in Figure 5.8. It indicates that variation exists in intensity at a given drug concentration and in drug concentration at a given level of intensity. The steepness or slope of the response curve depends in part on drug characteristics. It predicts the magnitude of change in effect in relationship to a change in drug concentration. Generally speaking, in the absence of toxicity, one selects a dose that produces a response equivalent to 20% to 80% of the maximum response. Toxicity, however, may require a more conservative strategy. Although the concentration-response curve shown in Figure 5.8 is graded, in some cases the effect can be an all-or-none response. Drug potency, which is often used to compare drugs for treating similar conditions, merely indicates the concentration at which a satisfactory response occurs. A drug requiring a lower dose (concentration) to achieve a benchmark effect is considered to be more potent than another that requires a higher dose (concentration). Higher potency, however, does not necessarily mean a safer or more effective drug.

At one time scientists thought the elimination of pharmacokinetic variability

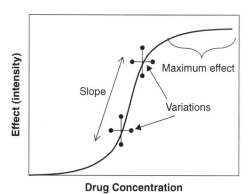

Figure 5.8. Concentration-effect relationship. A representative curve demonstrating the effects observed with increasing drug concentration, typically measured in plasma and expressed as an average and varied among individuals. Drug concentrations are plotted on a log scale. Maximum effect is achieved when no additional effect intensity can be obtained with increasing drug concentration. A similar relationship can be expressed as a function of dose administered to subjects. In this case the plot is called dose-effect or dose-response curve.

in the dose-response relationship would result in less variability in response among individual patients. Hence they proposed the exclusive use of concentration-response relationships. While this paradigm to reduce response variability often succeeds, variability among individuals in concentration-response relationships can be substantial, indicating that pharmacodynamic variability cannot be ignored.

In population studies, response variability can be analyzed with respect to the fraction of individuals exhibiting a predefined effect at a given concentration, over a range of concentrations. Such a plot is depicted in Figure 5.9A. The curve resembles the individual dose-response curve and is generally known as a *concentration-percent response or quantal concentration-effect curve*. The key difference is that the slope of this quantal concentration-response relationship reflects both intersubject variability in drug response and variations in drug concentration at which the response data are collected. Elaboration of a complete quantal concentration-effect curve is often precluded by the occurrence of adverse reactions in more people at high drug concentrations.

In order to compare the safety and efficacy of drugs and their formulations, the therapeutic index, an estimate of therapeutic effects in relation to side effects, is often used (Figure 5.9B). The dose or concentration of a drug needed to elicit a therapeutic effect in 50% of the population (median effective dose) is called the ED_{50}. Typically a median lethal dose or LD_{50} is characterized for each drug in relevant experimental animals. The relationship of median lethal dose and effective dose comparison is the therapeutic index:

$$\text{Therapeutic index, } TI = \frac{LD_{50}}{ED_{50}}$$

When there is no differentiation between LD_{50} and ED_{50} ($TI = 1$), the drug is

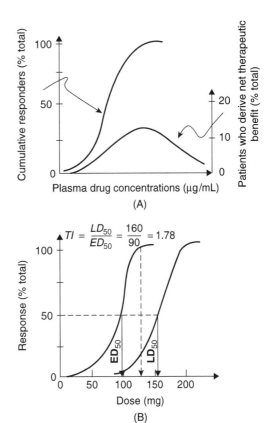

Figure 5.9. Dose-response profile in a population. (A) Relationship between responding patients, expressed as percentage of individuals, and plasma drug concentrations. With increasing drug concentration, the proportion of patients who derive therapeutic benefit, without concentration-limited side effect peaks, and then declines. (B) A schematic representation of dose-response curves. Typical therapeutic and lethal responses at indicated doses are evaluated in animal models to estimate therapeutic index, *TI*. ED_{50}, effective dose needed to produce a therapeutic response in 50% of animals, exhibiting therapeutic response; LD_{50}, effective dose needed to produce lethal effects in 50% of animals.

considered to be a toxic substance. A drug with *TI* greater than 1 is thought to be a safer or more selective drug in producing desirable effects.

The therapeutic index defined in this manner is a very crude estimate of safety and tolerance. It finds a use only in the estimation of initial doses used in phase I studies. For biologics, therapeutic index is usually derived from *in vitro* cell-culture studies. LD_{50} and ED_{50} values are replaced by the median inhibitory concentration against normal cells (IC_{50}) and median inhibitory concentration against target cells (EC_{50}):

$$\text{Therapeutic index, } TI = \frac{IC_{50}}{EC_{50}}$$

For example, acyclovir with $TI \simeq 10$ against herpes simplex virus is safer or more selective [9] than its derivative, ganciclovir ($TI = 2–3$ against cytomegalovirus) [10].

In clinical settings, choosing an optimum therapeutic dose with minimum side effects can be challenging because the dose needed to achieve a beneficial effect in a population may also produce untoward effects in many individuals. For the same drug, untoward effects may vary depending on therapeutic indication. Codeine to treat cough requires a lower drug concentration than that needed to control pain. Hence it has a higher therapeutic index for treating cough than for controlling pain.

The foregoing considered only drugs that produce a direct and instantaneous response when bound to a target receptor. When the drug leaves, the response is over. This applies to thrombolytic enzymes such as alteplase and streptokinase, as well as all other therapeutic enzymes. Some drugs, however, do not behave in this way but act indirectly so that a time lag exists between administration and evidence of a response. Among small molecules, the anticoagulant warfarin is a classic example. Many biologic response modifiers—interferons, colony-stimulating factors, and interleukins—elicit a delayed response.

Biopharmaceuticals may directly or indirectly affect multiple targets and cells. They do so by initiating a cascade of effects such as coagulation, cell growth, and tumor inhibition. Not all of the effects elicited by biopharmaceuticals, especially biologic response modifiers, are beneficial. Interleukin-2 promotes the production of lymphokine activated killer (LAK) cells with antitumor activity. But IL-2, at higher concentrations, affects endothelial cells and can cause capillary leak syndrome [11].

■ 5.3. DOSAGE FORM AND ROUTES OF ADMINISTRATION

Without exception, biotechnology-based pharmaceuticals must be formulated into dosage forms suitable for human administration in a safe and effective manner. These formulations must be stable under defined storage conditions for a defined period of time and be uniformly reconstitutable during that time. Hence biopharmaceuticals must be in compliance with stringent manufacturing guidelines and pass quality control and quality assurance standards for all bulk materials used to formulate the product. These bulk materials are formulated into appropriate dosage forms designed for specific routes of administration.

Some of the dosage formulations available for protein pharmaceuticals are listed in Table 5.7. An examination of Table 5.7 reveals that no protein drug up until this time has been formulated for oral administration. Most protein drugs are administered by means of injection (parenteral administration). Parenteral administration includes intravenous, intra-arterial, intracardiac, intraspinal or intrathecal, intramuscular, intrasynovial, intracutaneous or intradermal, subcutaneous injections, and injection directly into a dermal lesion (e.g., a wart). The parenteral route of administration requires a much higher standard of purity and sterility than oral administration. It also may require trained

■TABLE 5.7. Some dosage formulations and sites used in administration of biopharmaceuticals

Route of Administration	Dosage Formulation	Examples

Parenteral
  Intravenous,
  Intraarterial,
  Intracardiac,
  Intraspinal or Intrathecal,
  Intramuscular,
  Intrasynovial,
  Intracutaneous or Intradermal,
  Subcutaneous
	Solutions, Suspensions, Lyophilized powders to be reconstituted into solution	Blood clotting factors, colony-stimulating factors, antibodies and derivatives, interferons, interleukins, enzymes, hormones, vaccines

Local injection	Solutions	Interferon for direct injection into wart
Intrarespiratory	Aerosols	DNAse delivered to lungs to reduce mucus accumulation
Topical	Gels	Platelet-derived growth factor for wound healing
Intranasal	Solutions	Calcitonin for Paget's disease; gonadotropin-releasing hormone (GnRH) agonist for management of endometriosis
Intravitreal	Solutions	Antisense nucleotide polymer against CMV retinitis in patients with AIDS

personnel to ensure proper administration, which adds to the cost of treatment. Products intended for subcutaneous injection, however, can be self-administered or administered by a home care provider.

A small number of biopharmaceuticals are delivered by nonparenteral means. Recombinant DNase is given by inhalation aerosol to reduce the viscosity of mucus in the lungs of patients with cystic fibrosis. A platelet-derived growth factor in the form of a gel is administered topically for wound healing. Several hormones and peptides are administered in solution by the intranasal route. A solution of an antisense drug used for the treatment of cytomegalovirus retinitis is injected directly into the eye.

5.3.1. Physical and Chemical Stability

Proteins, peptides, and other polymeric macromolecules display varying degrees of chemical and physical stability. The degree of stability of these macromolecules influence the way they are manufactured, distributed, and administered. Chemical stability refers to how readily the molecule can undergo chemical reactions that modify specific amino-acid residues, the building blocks of the proteins and peptides. Chemical instability mechanisms of proteins and peptides include hydrolysis, deamidation, racemization, beta-elimination, disulfide exchange, and oxidation. Physical stability refers to how readily the molecule loses its tertiary and/or sec-

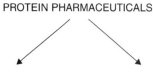

PROTEIN PHARMACEUTICALS

CHEMICAL INSTABILITY	PHYSICAL INSTABILITY
- deamidation	- denaturation
- racemization	- aggregation
- racemization	- precipitation
- oxidation	- adsorption
- beta elimination	
- disulfide exchange	

Figure 5.10. Possible mechanisms of chemical and physical instability that influence biological activities of protein pharmaceuticals. (Adapted from Manning et al. [18])

ondary structure (denaturation or unfolding), aggregates with itself, and precipitates from solution. In some cases these events lead to adsorption of proteins to storage containers and reduce the concentration of protein drug in solution, thereby the dose delivered to a patient. Both physical and chemical instability leads to loss of biological activity and, in some cases, to toxicity.

Interrelationship between Chemical and Physical Stability

A folded native protein exists in a local minimum energy conformation. Typically 5 to 15 kcal/mol free energy is sufficient to allow rapid interconversion (millisecond to second time scale) between folded and partially unfolded states. In general, proteins and peptides fold in a manner that minimizes the exposure of hydrophobic side chains of the peptide to water.

The structure that the protein molecule assumes depends on both the surrounding environment (buffer solution, storage containers, etc.), and the primary amino-acid sequence. In addition there are intrinsic properties embedded in the primary amino-acid sequence that may determine the degree of stability and hence "shelf life" of protein products. For example, immunoglobulin, a polypeptide with 16 disulfide bridges, is much more stable than interleukin-2, the unique amino-acid sequences in IgG accounting for the difference. Sequence differences also explain why IgG is a much more stable protein than IgA.

Changes in protein folding may expose functional groups susceptible to chemical modification. Partial denaturation may expose glutamine and asparagine residues prone to acid-catalyzed deamidation. Peroxide-mediated oxidation of methionine residues also may occur. Disulfide exchange may occur spontaneously under neutral and basic pH conditions. These changes may lead to further protein denaturation, resulting in aggregation, precipitation, and nonspecific adsorption to container surfaces.

5.3.2. Dosage Formulations

Proteins, peptides, antibodies and other biopharmaceuticals, like small-molecule drugs, must be characterized before and as a part of the formulation process with respect to physical-chemical stability and biological activity. While tests specific for each drug vary, some of the key analyses are the same. Common analyses, using a number of established and validated methods, concern amino-acid composition, amino-acid sequence, peptide mapping, disulfide linkages, molecular weight, endotoxin levels, residual host cell factors (proteins, DNA, and growth reagents), and inactive and degraded protein products. These methods include mass spectroscopy, high-pressure liquid chromatography (HPLC), sodium dodecyl sulfate polyacrylamide gel electrophoresis (SDS-PAGE), isoelectric focusing, and immunoassays.

With respect to endotoxin content, the FDA requires that each dose have less than 0.5 endotoxin units (EU) for each ml of drug solution with a maximum tolerated limit of 5 EU for each kg of body weight. Additional tests, such as analyses to detect the degree of posttranslational modification (e.g., glycosylation, sulfation, phosphorylation, and formylation) may be required for every batch of protein. Relevant *in vitro* and *in vivo* biological testing also is carried out to verify that a given lot of protein product exhibits potency that falls within predetermined standards.

The physiochemically and biologically characterized proteins and peptides are further formulated and subject to stability studies. The goal of these studies is to develop a unique combination of excipients, solution pH, buffer, and container that will produce an optimum dosage form. Biopharmaceutical formulations should be stable in storage and *in vivo*. Protein and peptide formulations contain excipients to stabilize protein activity and reduce inactivation or loss due to adsorption to the container, oxidation, or hydrolysis. In some cases—insulin, for example—divalent cations such as Zn^{++} are added to increase the duration of insulin effect. Additional drug delivery considerations are discussed in Chapter 13.

While formulation of proteins in solution or suspension are less costly to produce, not all therapeutic proteins can be stably stored in solution or suspension, even when refrigerated (4°C) or frozen (−20°C). In those cases freeze-dried formulations of proteins may be used as an alternative. Freeze-drying or lyophilization typically produces an amorphous form of

■TABLE 5.8. Examples of excipients used to enhance protein stability in solution and lyophilized formulations

Protein	Excipient	Formulation	Enhancement Effects
Human epidermal growth factor	Triton X-100 (0.02% w/v)	Solution	~2 fold increased stability at 60°C
	Fibronectin (0.05% w/v)		~1.5 fold stability increase at 60°C [19]
Recombinant interleukin-2 (rIL-2)	Pluronic F-127 (10% w/w)	Solution	~3 fold increased stability at 4°C [20]
Recombinant human keratinocyte growth factor	Heparin (0.5% w/v)	Solution	~50 fold increased stability at 37°C [21]
Human growth hormone	Cellobiose, trehalose, or mannitol (31:1 m/m)	Lyophilized	375–1500 fold increased stability at 50°C [22]
Recombinant human interleukin-1-receptor antagonist	Sucrose or trehalose (1% w/v)	Lyophilized	4–12 fold increased stability at 50°C [23]
Interleukin-2	Sucrose (0.5% w/v)	Lyophilized	~2 fold increased stability at 45°C [24]
Insulin	Trehalose (0.5% w/v)	Lyophilized	~2 fold increased stability at 35°C [25]
Immunoglobulin E (IgE)	Manitol (6:4 w/w)	Spray-dried	Dramatic increase in stability at 5°C and 30°C [26]

protein that can be readily rehydrated or resuspended in water just prior to use.

Lyophilization is a two-step process—freezing the protein solution and drying the frozen solid under vacuum. The drying step can be further divided into the primary event, removal of frozen water, and the secondary event, removal of nonfrozen protein-bound water, estimated to be 0.30 to 0.35 g/g protein—slightly less than surface-bound water (protein hydration shell) [12]. Protein lyophilization does not always yield increased stability compared with frozen liquid formulations. For example, the oxidation rate of lyophilized IL-2 is not slower than that of a frozen liquid formulation containing proper excipients or stabilizers [13]. Many of the factors affecting protein instability and stabilization in the freeze-drying process have been elucidated and recently reviewed [14–16]. The stability of freeze-dried protein formulations can be improved by controlling moisture content and pH, and adding cryostabilizers or protectants such as sucrose, polyols, surfactants, and polymers (Table 5.8).

REFERENCES

1. FDA, T., *FDA final rule, October 15, 1996, published through REGO (under the President's reinventing government) initiative.* 1996.
2. *Elimination of ELA and PLA requirement, May 14, 1996 and replaced by BLA.* Federal Register, 1996. **63**(147): 40858–71.
3. Daggett, V., *Validation of protein-unfolding transition states identified in molecular dynamics simulations.* Biochem Soc Symp, 2001(68): 83–93.
4. Galluppi, G.R., M.C. Rogge, L.K. Roskos, L.J. Lesko, M.D. Green, D.W. Feigal, Jr., and C.C. Peck, *Integration of pharmacokinetic and pharmacodynamic studies in the discovery, development, and review of protein therapeutic agents: a conference report.* Clin Pharmacol Ther, 2001. **69**(6): 387–99.

5. Tsutsumi, Y., M. Onda, S. Nagata, B. Lee, R.J. Kreitman, and I. Pastan, *Site-specific chemical modification with polyethylene glycol of recombinant immunotoxin anti-Tac(Fv)-PE38 (LMB-2) improves antitumor activity and reduces animal toxicity and immunogenicity.* Proc Natl Acad Sci USA, 2000. **97**(15): 8548–53.
6. Rosenberg, J.J., S.W. Martin, J.E. Seely, et al., *Development of a novel, nonimmunogenic, soluble human TNF receptor type I (sTNFR-I) construct in the baboon.* J Appl Physiol, 2001. **91**(5): 2213–23.
7. Taylor, P.C., *Anti-tumor necrosis factor therapies.* Curr Opin Rheumatol, 2001. **13**(3): 164–9.
8. Yap, I., and S.H. Chan, *A new pre-S containing recombinant hepatitis B vaccine and its effect on non-responders: a preliminary observation.* Ann Acad Med Singapore, 1996. **25**(1): 120–2.
9. Ho, R.J., B.T. Rouse, and L. Huang, *Target-sensitive immunoliposomes as an efficient drug carrier for antiviral activity.* J Biol Chem, 1987. **262**(29): 13973–8.
10. Cheng, Y.C., E.S. Huang, J.C. Lin, E.C. Mar, J.S. Pagano, G.E. Dutschman, and S.P. Grill, *Unique spectrum of activity of 9-[(1,3-dihydroxy-2-propoxy)methyl]-guanine against herpesviruses in vitro and its mode of action against herpes simplex virus type 1.* Proc Natl Acad Sci USA, 1983. **80**(9): 2767–70.
11. Lotze, M.T., M.C. Custer, and S.A. Rosenberg, *Intraperitoneal administration of interleukin-2 in patients with cancer.* Arch Surg, 1986. **121**(12): 1373–9.
12. Arakawa, T., Y. Kita, and J.F. Carpenter, *Protein–solvent interactions in pharmaceutical formulations.* Pharm Res, 1991. **8**(3): 285–91.
13. Hora, M.S., R.K. Rana, C.L. Wilcox, N.V. Katre, P. Hirtzer, S.N. Wolfe, and J.W. Thomson, *Development of a lyophilized formulation of interleukin-2.* Dev Biol Stand, 1992. **74**: 295–303; discussion 303–6.
14. Arakawa, T., Prestrelski, S.J., Kenny, W.C., Carpenter, J.F., *Factors affecting short-term and long term stabilities of proteins.* Adv Drug Deliv Rev, 1993. **10**: 1–28.
15. Cleland, J.L., M.F. Powell, and S.J. Shire, *The development of stable protein formulations:*

a close look at protein aggregation, deamidation, and oxidation. Crit Rev Ther Drug Carrier Syst, 1993. **10**(4): 307–77.

16. Wang, W., *Lyophilization and development of solid protein pharmaceuticals.* International Journal of Pharmaceutics, 2000. **203**(1–2): 1–60.

17. Donahue, R.E., E.A. Wang, R.J. Kaufman, L. Foutch, A.C. Leary, J.S. Witek-Giannetti, M. Metzger, R.M. Hewick, D.R. Steinbrink, G. Shaw, and et al., *Effects of N-linked carbohydrate on the in vivo properties of human GM-CSF.* Cold Spring Harb Symp Quant Biol, 1986. **51 Pt 1**: 685–92.

18. Manning, M.C., K. Patel, and R.T. Borchardt, *Stability of protein pharmaceuticals.* Pharm Res, 1989. **6**(11): 903–18.

19. Son, K., and C. Kwon, *Stabilization of human epidermal growth factor (hEGF) in aqueous formulation.* Pharm Res, 1995. **12**(3): 451–4.

20. Wang, P.L., and T.P. Johnston, *Enhanced stability of two model proteins in an agitated solution environment using poloxamer 407.* J Parenter Sci Technol, 1993. **47**(4): 183–9.

21. Chen, B.L., T. Arakawa, E. Hsu, L.O. Narhi, T.J. Tressel, and S.L. Chien, *Strategies to suppress aggregation of recombinant keratinocyte growth factor during liquid formulation development.* J Pharm Sci, 1994. **83**(12): 1657–61.

22. Costantino, H.R., K.G. Carrasquillo, R.A. Cordero, M. Mumenthaler, C.C. Hsu, and K. Griebenow, *Effect of excipients on the sta-*

bility and structure of lyophilized recombinant human growth hormone. J Pharm Sci, 1998. **87**(11): 1412–20.

23. Chang, B.S., R.M. Beauvais, T. Arakawa, L.O. Narhi, A. Dong, D.I. Aparisio, and J.F. Carpenter, *Formation of an active dimer during storage of interleukin-1 receptor antagonist in aqueous solution.* Biophys J, 1996. **71**(6): 3399–406.

24. Prestrelski, S.J., K.A. Pikal, and T. Arakawa, *Optimization of lyophilization conditions for recombinant human interleukin-2 by dried-state conformational analysis using Fourier-transform infrared spectroscopy.* Pharm Res, 1995. **12**(9): 1250–9.

25. Strickley, R.G., and B.D. Anderson, *Solid-state stability of human insulin. II. Effect of water on reactive intermediate partitioning in lyophiles from pH 2-5 solutions: stabilization against covalent dimer formation.* J Pharm Sci, 1997. **86**(6): 645–53.

26. Costantino, H.R., J.D. Andya, P.A. Nguyen, N. Dasovich, T.D. Sweeney, S.J. Shire, C.C. Hsu, and Y.F. Maa, *Effect of mannitol crystallization on the stability and aerosol performance of a spray-dried pharmaceutical protein, recombinant humanized anti-IgE monoclonal antibody.* J Pharm Sci, 1998. **87**(11): 1406–11.

27. Unanue, E.R., and B. Benacerrat, *Textbook of Immunology, 2nd ed.* Williams and Wilkins, Baltimore. 1986.

28. Stein, K.E., Division of Monoclonal Antibodies, CDER/FDA, presented 2122/01.

Part II

THERAPEUTICS BASED ON BIOTECHNOLOGY

This part of the book is concerned with biopharmaceuticals that have received the approval of the FDA and are currently available in the United States by prescription (except for insulin, which is available without prescription). All but one agent, a DNA-based therapy, are proteins. The chapters are organized by molecular function rather than therapeutic indication. We believe this novel approach is useful when dealing with the products of biotechnology because these drugs relate to one another in terms of how they act at the molecular level rather than their indication. Consequently a host of structurally unrelated drugs are indicated for cancer, while monoclonal antibodies are approved for treating a wide range of medical conditions.

Drug categories discussed in this part of the book include hematopoietic growth and coagulation factors (Chapter 6), interferons and cytokines (Chapter 7), hormones (Chapter 8), enzymes (Chapter 9),

antibodies and derivatives (Chapter 10), vaccines (Chapter 11), and other products (Chapter 12). Each chapter consists first of an overview of the history, molecular characteristics, and application of the respective categories of biopharmaceuticals, followed by a succinct summary or monograph of each drug in the category.

Several sources of information were vital in the preparation of individual monographs of biologics that are approved and marketed in the United States. They include online databases provided by the Food and Drug Administration, drug manufacturers literature, and drug information available online from Thomson Micromedex.

The overview material for each chapter, written by individuals with extensive knowledge and practical experience, provides invaluable insight. The properties, characteristics, and clinical application of each biopharmaceutical are discussed in

Biotechnology and Biopharmaceuticals, by Rodney J. Y. Ho and Milo Gibaldi
ISBN 0-471-20690-3   Copyright © 2003 by John Wiley & Sons, Inc.

the monographs. Each monograph informs the readers on the general description, therapeutic indication(s), administration, pharmacology and pharmaceutics, thera- peutic response, and role in therapy of a biopharmaceutical. We are not aware of any such compilation currently available to biomedical researchers and clinicians.

6

HEMATOPOIETIC GROWTH FACTORS AND COAGULATION FACTORS

■ **SECTION ONE** ■
6.1. OVERVIEW

■ **SECTION TWO** ■
6.2. MONOGRAPHS

■ SECTION ONE ■

Hematopoiesis—the production of blood cells—is a tightly regulated system, exquisitely responsive to functional demands including infection, allergic reaction, immune challenge, hemorrhage, inflammation, and hypoxia. Beyond the production of blood cells, the integrity of the circulatory system also requires platelets, growth factors, and coagulation factors. Cells in the circulatory system sustain life by delivering oxygen and nutrients to tissue, clearing waste and pathogens, and recruiting humoral and cellular host defenses in a timely manner.

Hematology deals with the study of blood cells (e.g., erythrocytes, leukocytes, and platelets) and proteins in the circulatory system. Studies have shown that common blood disorders such as anemia, leukocytosis, and bleeding are indirect consequences of infection, inflammation, malnutrition, and malignancy. While hematologic malignancies could produce more severe bleeding, their prevalence is less common than blood disorders.

Advances in recombinant DNA technology have permitted cloning and production of growth factors and blood coagulation factors for the management of hematologic disorders. Several proteins can act on the growth and maintenance of each type of blood precursor cell. Each growth factor may elicit differentiation of a blood cell precursor into many distinct types of

Biotechnology and Biopharmaceuticals, by Rodney J. Y. Ho and Milo Gibaldi
ISBN 0-471-20690-3   Copyright © 2003 by John Wiley & Sons, Inc.

leukocytes. Therefore it is important to understand the fundamental physiological principles of blood cells and growth factors. There follows a brief review of hematology with an emphasis on the therapeutic use of hematopoietic and coagulation factors.

■ 6.1. OVERVIEW

Hematopoietic development of blood cells begins mainly in the spleen and liver of the fetus during early pregnancy. By the seventh month, however, the marrow of a fetus becomes the primary site of blood cell formation [1]. During childhood, the marrow of the central axial skeleton such as the pelvis, spinal cord, and ribs, and of the extremities, such as the wrist and ankle, provides the key site of hematopoiesis. Hematopoiesis at the periphery (also known as extramedullary hematopoiesis) slowly decreases with age. Chronic administration of hematopoietic growth factors can reverse this decline. Severe hemolytic anemia and hematopoietic malignancies can also reverse the process.

Production of blood cells in bone marrow of the central axial skeleton is referred to as medullary hematopoiesis. Hematopoietic tissue in adult bone marrow is well perfused and contains fat cells (adipocytes), and various types of blood and blood precursor cells encased within a protein matrix. Fibroblast, stromal and endothelial cells within bone marrow, serve as sources of matrix proteins as well as a factory for growth factors and chemokines that regulate blood cell production and release matured cells into the circulation [2,3]. Chemokines act as signal lamps for trafficking of lymphocytes in and out of lymphoid tissues. Erythroblasts, neutrophils, lymphoblasts, macrophages, megakaryocytes, and pluripotent stem cells are also found within the calcified lattice crisscrossing the marrow space.

All hematopoietic cells are believed to derive from a common precursor, hematopoietic stem cells [4]. Although stem cells constitute only about 0.05% of bone marrow, this population is maintained through a self-renewal system. Pluripotent stem cells undergo irreversible differentiation into daughter cells that are committed to lineages of unique hematopoietic cell types (Figure 6.1) [5]. While the mechanisms of early stages of lineage commitment by bone marrow to a particular type of blood cell remain elusive, the late stage of this process (differentiation and maturation process) is driven by hematopoietic growth factors. Many of these growth factors are now cloned and recombinant proteins are available for biologic and therapeutic studies. Some have had significant impact in treating hematologic disorders and managing neutropenia-associated infections.

Some hematopoietic growth factors exhibit overlapping specificities for cells of different lineages, particularly in the early stages of differentiation [6]. Studies with recombinant hematopoietic growth factors—including erythropoietin (EPO), thrombopoietin (TPO), granulocyte colony-stimulating factor (CSF), and macrophage colony-stimulating factor (M-CSF)—have shown that many of these proteins exhibit lineage-specific effects during the later stages of cell differentiation. In other words, lineage-specific growth factors tend to act on maturation and deployment of a given type of blood cell.

The cellular source and functional activity of hematopoietic growth factors are listed in Table 6.1. While the names "colony-stimulating factor," "-poietin," and "interleukin" are not obvious, they can be rationalized. The suffix "-poietin" as in erythropoietin and thrombopoietin derives from the Greek term *poieis*, meaning "to make." Colony-stimulating factors are so named because of their ability to stimulate target cells to divide and grow into colonies of cells in culture. The proteins produced by leukocytes that act on neighboring leukocytes are called interleukins. While most

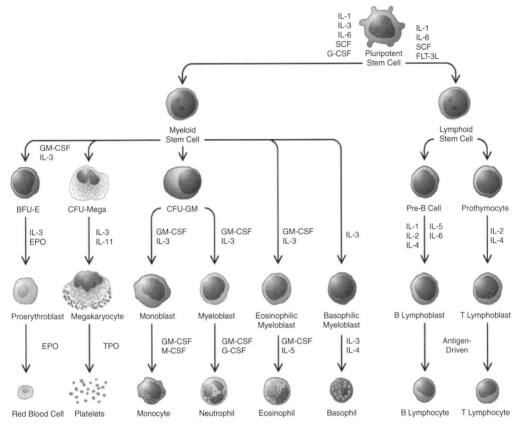

Figure 6.1. The pathways used and hematopoietic growth factors required by pluripotent stem cells in bone marrow to differentiate and develop into distinct type of blood cells. The abbreviations of hematopoietic growth factors are found in Table 6.1.

interleukins support leukocyte growth or lymphocytopoiesis, many also exhibit broad or pleotropic effects on cells of many lineages.

While it is assumed that hematopoietic cell formation is regulated through a concerted and coordinated release of factors in a cascade and series of reactions, the details of how such processes occur remain elusive. It is clear, however, that there are important differences in the growth and turnover (or removal) rates of blood cells. A new population of neutrophils is generated every 24 hours to replace aging cells, which are cleared with a half-life of about 6 to 8 hours [7]. On the other hand, erythrocytes exhibit a well-documented life

span of about 100 days. Despite the differences in capacity, turnover rate, and functions of blood cells, overall, hematopoiesis is tightly regulated to provide sufficient, but not excessive, numbers of different types of blood cells that are needed to sustain life.

The existence of some blood cells, such as erythrocytes and platelets, with long lifespans make cell transfusion therapy practical. Cell transfusion therapy cannot be developed for short-lived cells such as neutrophils with turnover rates of less than 8 hours. Fortunately, for neutrophils, colony stimulating factors can be used to recruit the needed number in blood within 24 hours after administration of these factors.

■ **TABLE 6.1.** Hematopoietic growth factors: Cellular source of synthesis, molecular characteristics, and functions

Hematopoiesis Factor	Cell Source	Functions	Molecular Characteristics
Granulocyte colony-stimulating factor; G-CSF; filgrastim; lenograstim	Monocytes, fibroblasts, endothelial cells	Stimulates formation and function of neutrophils	18.8 kDa protein containing 175 aa
Granulocyte-macrophage colony-stimulating factor; GM-CSF; *Leukine*; sagramostatin	Monocytes, T cells, and fibroblasts	Stimulates formation and function of neutrophils, monocytes, and eosinophils	15.5–19.5 kDa protein containing 127 aa
Macrophage colony-stimulating factor; M-CSF; colony-stimulating factor-1 (CSF-1)	Macrophages, endothelial cells, fibroblasts	Stimulates formation and function of monocytes	18.5 kDa protein containing 159 aa that form a 37 kDa dimmer in its active form
Interleukin-1α and β: IL-1α and IL1-β; endogenous pyrogen hemopoietin-1	Monocytes, endothelial cells, keratinocytes	Support proliferation of T, B, and other cells; at high dose, it induces fever and catabolism	17 kDa proteins containing 153 (IL1-α) and 159 (IL1-β) aa in their mature form
T cell growth factor; IL-2; aldesleukin; *Proleukin*	T cells, large granular lymphocytes or natural killer (NK) cells	T-cell proliferation, antitumor and antimicrobial effects; at high dose, it induces fever and capillary leak syndrome	15.5 kDa protein
Multi-colony-stimulating factor; mast cell growth factor; IL-3	Activated T cells, NK cells	Proliferation of early hematopoietic cells	17.5 kDa protein containing 133–134 aa
B cell growth factor; T cell growth factor II; mast cell growth factor II; IL-4	T lymphocytes	Proliferation of B and T cells; enhances cytotoxic T-cell activities	14 kDa protein containing 130 aa
Eosinophil differentiation factor; eosinophil colony-stimulating factor; IL-5	T lymphocytes	Stimulates eosinophil formation; stimulates T-cell and B-cell functions	32–34 kDa glycoprotein containing 155 aa for each unit of homodimer

B cell stimulatory factor II; hepatocyte stimulatory factor; IL-6	B and T cells, monocytes, fibroblasts, endothelial cells	Promotes B cell differentiation	22–27 kDa protein with 184 aa in its mature form
Lymphopoietin 1; pre-B cell growth factor; IL-7	Lymphoid tissues	Promotes growth of B and T cells	17 kDa protein containing 153 aa
Plasmacytoma-stimulating factor; IL-11; oprelvekin; *Neumega*	Fibroblasts, trophoblastic placental cells	Proliferation of early hematopoietic cells; induces acute-phase protein synthesis	19 kDa protein containing 177 aa
Natural killer cell-stimulating factor; IL-12	B cells, macrophages	Stimulates T cell expansion and interferon-gamma release; the combination of both agents synergistically promotes early hematopoietic cell proliferation	75 kDa heterodimeric protein containing 306 aa p40 and 197 aa p 35 subunits
Erythropoietin; EPO; epoetin	Endothelial cells of kidney glomerular tubules	Stimulates erythrocyte formation and release from marrow	30.4 kDa glycoprotein containing 166 aa; 18.4 kDa unglycosylated
Thrombopoietin; TPO; megakaryocyte growth and development factor (MGDF)	Renal and endothelial cells, hepatocytes, fibroblasts	Stimulates megakaryocyte proliferation and platelet formation	18–70 kDa highly glycosylated protein containing 332 aa
Leukemia inhibitory factor; LIF	Monocytes and lymphocytes, stromal cells	Stimulates hematopoeitic cell differentiation	100–110 kDa glycoprotein containing 789 aa in its mature form
Stem cell factor; kit ligand; steel factor; SCF	Hepatocytes, endothelial cells	Stimulates proliferation of early hematopoietic cells and mast cells	18.5 kDa protein containing 164 aa
fms-like tyrosine kinase 3; FLT-3 ligand; STK-1	T-lymphocytes, stromal cells, fibroblasts	Stimulates early hematopoietic cell differentiation; increases blood dendrite cells	160 kDa protein with 993 aa

6.1.1. Therapeutic Use of Hematopoietic Factors

The finely tuned hematopoietic system, however, sometimes breaks down. Neutropenia—decreased blood neutrophil numbers below 1800 cells/mm^3—is a clinical condition in which neutrophilic leukocytes in the blood fall two standard deviations below normal. Leukopenia and granulocytopenia are terms that are generally used interchangeably with neutropenia. Recurrent infections which typically are inconsequential in healthy subjects, could produce significant consequences in neutropenia, and can be devastating for immunocompromised patients. There are many acquired causes of neutropenia, with infections, drugs, and immune disorders being the most common. The list of drugs that are associated with neutropenia is long and covers many therapeutic areas including antiinfectives, sedatives, diurectics, antiinflammatory agents, and cardiovascular drugs (Table 6.2).

The administration of recombinant granulocyte colony-stimulating factor (G-CSF, filgrastim) can correct neutropenia and reduce infectious morbidity in infected patients with a variety of causes of severe neutropenia. Granulocyte macrophage colony-stimulating factor (GM-CSF, sargramostim) has also been used successfully in neutropenic patients.

Beginning with pioneering studies in the early 1960s, it has been recognized that

■**TABLE 6.2.** Drugs reported to induce neutropenia

Antibiotics		*Anticonvulsants*	*Antihistamines*
Cephalosporins	Lincomycin	Carbamazepine	Brompheniramine
Chloramphenicol	Metronidazole	Ethosuximide	Cimetidine
Clindamycin	Nitrofurantoin	Mephenytoin	Ranitidine
Doxycycline	Penicillins	Phenytoin	Tripelennamine
Flucytosine	Rifampin	Primidone	
Gentamicin	Streptomycin	Trimethadione	
Griseofulvin	Sulfonamides		
Isoniazid	Vancomycin		
Anti-inflammatory agents	*Antimalarials*	*Antithyroid drugs*	*Cardiovascular agents*
Fenoprofen	Amodiaquine	Carbimazole	Captopril
Gold salts	Dapsone	Methimazole	Diazoxide
Ibuprofen	Hydroxychloroquine	Potassium perchlorate	Hydralazine
Indomethacin	Pyrimethamine	Propylthiouracil	Methyldopa
Phenylbutazone	Quinine		Pindolol
			Procainamide
			Propranolol
			Quinidine
Diuretics	*Hypoglycemic agents*	*Phenothiazines*	*Sedatives and neurologic agents*
Acetazolamide	Chlorpropamide	Chlorpromazine	Chlordiazepoxide
Bumetanide	Tolbutamide	Prochlorperazine	Clozapine
Chlorothiazide		Promazine	Desipramine
Chlorthalidone		Thioridazine	Diazepam
Hydrochlorothiazide		Trifluoperazine	Imipramine
Methazolamide		Trimeprazine	Meprobamate
Spironolactone			Metoclopramide

blood progenitor cells can be propagated in culture in the presence of soluble growth factors. These factors were originally termed colony-stimulating factors (CSFs) because of their ability to support the formation of colonies of blood cells by bone marrow cells plated in semisolid medium. During the 1970s and 1980s it became clear that there were multiple types of CSFs based on the different types of colonies that grew in the presence of the different factors. This observation led to the hypothesis that the growth and differentiation of blood cells were controlled, at least in part, by exposure of progenitor cells to CSFs having different lineage specificities.

Following the molecular cloning of the genes for many of these factors and their receptors during the 1980s and 1990s, it became possible to study in detail the structure, function, and biology of the recombinant CSFs as well as the molecular biology of their respective genes. The first indication that GM-CSF can broadly stimulate hematopoiesis, resulted from studies in which GM-CSF, produced in nonhuman primate kidney cells, was administered to monkeys. Recombinant human GM-CSF (rhGM-CSF), when infused intravenously, produced large increments in all classes of leukocytes, including eosinophils, lymphocytes, and reticulocytes [8].

In comparison to GM-CSF, G-CSF is more lineage specific. A primary effect of G-CSF is to promote the conversion of granulocyte colony-forming units into polymorphonuclear leukocytes. Mice that lack G-CSF have chronic neutropenia and reduced bone marrow myeloid precursors and progenitors. They also have a markedly impaired capacity to increase neutrophil and monocyte counts after infection.

The efficacy of human G-CSF was initially evaluated in simian preclinical trials. Monkeys treated with subcutaneous injections of G-CSF showed a dose-related increase in polymorphonuclear neutrophils. These initial studies also evaluated two cyclophosphamide-treated animals. G-CSF was given either before or after cyclophosphamide. The neutrophil count in the treated animal increased dramatically by day six to seven after cyclophosphamide. On the other hand, the control animal remained pancytopenic for three to four weeks after treatment. This type of observation provided the rationale for the administration of G-CSF or GM-CSF to patients with chemotherapy-induced neutropenia [8].

Neutropenia and infection are major dose-limiting side effects of chemotherapy. Febrile neutropenia is especially worrisome. While infectious mortality resulting from febrile neutropenia is low, avoiding its occurrence might be expected to enhance patient quality of life, and allow scheduled chemotherapy. Primary prophylaxis with a colony-stimulating factor can reduce the incidence of febrile neutropenia by as much as 50%. In 2000, the American Society of Clinical Oncology Growth Factors Expert Panel updated *Recommendations for the Use of Hematopoietic Colony-Stimulating Factors* [9]. The panel noted that few common chemotherapy regimens produce significant complications related to neutropenia, but neutropenia is more likely when chemotherapy is given for treatment of relapsed or refractory disease.

After a meticulous review of the biomedical literature, the panel concluded that routine use of colony-stimulating factors for primary prophylaxis of febrile neutropenia in previously untreated patients is not justified by existing data. There is evidence, however, that colony-stimulation factor administration can decrease the probability of febrile neutropenia in subsequent cycles of chemotherapy after a documented occurrence in an earlier cycle. However, dose reduction after a severe episode, rather than administration of colony-stimulating factors, should be considered as a primary therapeutic option. In the absence of clinical evidence or other compelling reasons to maintain chemotherapy dose intensity, physicians should consider chemotherapy dose reduction after febrile neutropenia or severe or prolonged neutropenia after the previous cycle of treatment [9].

Amgen has recently introduced a modified version of filgrastim with a longer duration of action. A single fixed dose of pegfilgrastim (pegylated filgrastim) is as effective as a course of daily filgrastim injections in preventing neutropenia in patients undergoing chemotherapy.

Pegfilgrastim is a covalent conjugate of recombinant methionyl human G-CSF (filgrastim) and monomethoxypolyethylene glycol. Filgrastim is a water-soluble 175 amino-acid protein with a molecular weight of about 19 kDa. Filgrastim is obtained from the bacterial fermentation of a strain of *E. coli* transformed with a genetically engineered plasmid containing the human G-CSF gene. To produce pegfilgrastim, a 20 kDa monomethoxypolyethylene glycol is covalently bound to the N-terminal methionyl residue of filgrastim. The average molecular weight of pegfilgrastim is approximately 39 kDa.

Maintenance of red cell volume is critical to having an adequate oxygen supply to the tissues [10]. Healthy individuals finely balance erythropoiesis and erythrocyte loss and maintain constant hematocrit. The glycoprotein hormone erythropoietin is the principal controller of the homeostatic mechanism that links tissue oxygen delivery to red cell production. While hypothesized as early as 1863, unequivocal evidence of erythropoietin was first published in 1953. A few years later, scientists showed that animals subjected to bilateral nephrectomy were unable to mount an erythropoietin response to hypoxia. Indeed, the kidneys produce about 90% of circulating erythropoietin.

Twenty years passed before researchers purified small amounts of erythropoietin and elucidated the primary amino-acid structure. Erythropoietin has 166 amino-acid residues and a molecular weight of 18.4 kDa. The overall molecular weight is 30.4 kDa because of three N-linked carbohydrate chains. Two internal disulfide linkages between cysteine residues are known to be necessary for biologic activity. In 1985 the cDNA encoding for human erythropoietin was isolated and expressed in mammalian cells [10]. Subsequently recombinant erythropoietin (epoetin alfa) has become an important therapeutic entity and the most successful biotechnology product ever produced.

An inverse log-linear relationship has been found between hematocrit and plasma erythropoietin in anemic patients with normal renal function. Patients with chronic renal failure have inappropriately low erythropoietin levels for their degree of anemia [10]. The severity of anemia correlates with the extent of renal dysfunction. Intravenous or subcutaneous recombinant human erythropoietin given three times a week is the treatment of choice. Some patients seem to do well on only one injection per week. One version of erythropoietin, *Epocrit*, marketed outside the United States, has been associated with pure red cell aplasia.

Anemia of chronic disease is the second most common form of anemia. Recombinant erythropoietin frequently corrects the anemia with few or no transfusions. Anemia may also develop during radiotherapy or chemotherapy. Recombinant erythropoietin has been shown to increase hematocrit and decrease transfusion requirements after the first month of therapy in anemic cancer patients undergoing chemotherapy.

Amgen has recently introduced darbepoetin (*Aranesp*), a long-acting version of human erythropoietin. With two additional N-linked carbohydrate chains, darbepoetin contains more sialic acid than epoetin alfa and has an approximately threefold longer half-life, which leads to less frequent dosing than epoetin alfa. Darbepoetin is approved for both the treatment of anemia due to chronic renal failure and the treatment of chemotherapy-induced anemia.

6.1.2. Therapeutic Use of Coagulation Factors

Hemostasis is the process of blood clot formation at the site of vessel injury. When a blood vessel wall breaks, the hemostatic response must be rapid, localized, and carefully regulated. Abnormal bleeding or a propensity to nonphysiologic thrombosis (i.e., thrombosis not required for hemostatic regulation) may occur when specific elements of these processes are missing or dysfunctional. The elements responsible for normal hemostasis have received extensive review.

Hemostasis begins with the formation of the platelet plug, followed by activation of the clotting cascade, and propagation of the clot. One of the major multicomponent complexes in the coagulation cascade consists of activated factor IX (factor IXa) as the protease, activated factor VIII (factor VIIIa), calcium, and phospholipids as the cofactors, and factor X as the substrate. Factor IXa can be generated by either factor Xa activation of the intrinsic pathway or by the tissue factor/factor VIIa complex.

Factor VIII is synthesized as a single chain polypeptide of 2351 amino acids. A 19 amino-acid signal peptide is cleaved by a protease shortly after synthesis so that circulating plasma factor VIII is a heterodimer. Factor VIII circulates in plasma in a noncovalent complex with von Willebrand factor. Cleavage of plasma factor VIII by thrombin or factor Xa at Arg_{372} and Arg_{1689} is necessary to activate factor VIII. The physiologic role of factor VIII is to accelerate the cleavage of factor X by factor IXa. It has been proposed that factor VIIIa has a stabilizing action that forces the catalytic modules, factor IXa and factor X, together.

The hemophilias are a group of related, usually inherited, bleeding disorders. Inherited bleeding disorders include abnormalities of coagulation factors as well as platelet function. When the term "hemo-philia" is used, it most often refers to the following two disorders:

- Factor VIII deficiency or hemophilia A
- Factor IX deficiency or hemophilia B

The management of patients with hemophilia requires the provision of preventive care, the use of replacement therapy during acute bleeding episodes and as prophylaxis, and the treatment of the complications of the disease and its therapy. Clotting factor concentrates have been given to prevent bleeding and to limit existing hemorrhage. They are still used today. Plasma-derived concentrate therapy in the 1970s and 1980s was associated with serious viral complications. Most patients became infected with hepatitis C virus (HCV), and HIV infection occurred in up to 50% [11]. The current use of donor-screening, virucidal techniques, the insistence on safety testing when the product is first used, and the advent of recombinant products have led to a generation of products with extremely low risk of viral transmission.

Clinical trials have demonstrated excellent efficacy with recombinant human factor VIII concentrates available as *Recombinate* and *Kogenate*. These recombinant factor VIII products are purified from the cell culture of plasmids, not viral DNA-transfected hamster cells and therefore do not express viral sequences. The addition of human serum albumin for stabilization, constitutes the sole possible source for human viral contamination. More recently recombinant factor IX has been genetically engineered by insertion of the human factor IX gene into a Chinese hamster ovary cell line. It has been proved to be safe and effective in the treatment of patients with hemophilia B.

6.1.3. Summary

The most important blood product approved up until now is erythropoietin. It has vastly improved the management of renal failure patients and ameliorated the anemia associated with cancer chemotherapy. Recombinant clotting factors are a more modest advance promising greater purity and lower risk of viral contamination than extracted products. The overall benefits of colony-stimulating factors to reduce the risk of neutropenia in patients with cancer who receive chemotherapy remain controversial.

REFERENCES

1. Tavassoli, M., *Embryonic and fetal hemopoiesis: an overview*. Blood Cells, 1991. **17**(2): 269–81; discussion 282–6.
2. Ogawa, M., and T. Matsungaga, *Humoral regulation of hematopoietic stem cells*. Ann NY Acad Sci, 1999. **827**: 17–23; discussion 23–4.
3. Broxmeyer, H.E., and C.H. Kim, Regulation of hematopoiesis in a sea of chemokine family members with a plethora of redundant activities. Exp Hematol, 1999. **27**(7): 1113–23.
4. Weissman, I.L, *Stem cells: units of development, units of regeneration, and units in evolution*. Cell, 2000. **100**(1): 157–68.
5. Metcalf, D., *Lineage commitment and maturation in hematopoietic cells: the case for extrinsic regulation*. Blood, 1998. **92**(2): 345–7; discussion 352.
6. D'Andrea, A.D., *Hematopoietic growth factors and the regulation of differentiative decisions*. Curr Opin Cell Biol, 1994. **6**(6): 804–8.
7. Lord, B.I., L.B. Woolford, and G. Molineux, *Kinetics of neutrophil production in normal and neutropenic animals during the response to filgrastim (r-metHu G-CSF) or filgrastim SD/01 (PEG-r-metHu G-CSF)*. Clin Cancer Res, 2001. **7**(7): 2085–90.
8. Sieff, C.A., *Introduction to recombinant hematopoietic growth factors*. UpToDate Online, 2002. **10**(2).
9. Ozer, H., J.O. Armitage, C.L. Bennett, J. Crawford, G.D. Demetri, P.A. Pizzo, C.A. Schiffer, T.J. Smith, G. Somlo, J.C. Wade, J.L. Wade, 3rd, R.J. Winn, A.J. Wozniak, and M.R. Somerfield, *2000 update of recommendations for the use of hematopoietic colony-stimulating factors: evidence-based, clinical practice guidelines. American Society of Clinical Oncology Growth Factors Expert Panel*. J Clin Oncol, 2000. **18**(20): 3558–85.
10. Kendall, R.G., *Erythropoietin*. Clin Lab Haematol, 2001. **23**(2): 71–80.
11. Hoots, W.K., and A.D. Shapiro, *Treatment of hemophilia*. UpToDate Online, 2002. **10**(2).

■ SECTION TWO ■

■ 6.2. MONOGRAPHS

EPOETIN ALFA

<u>Trade names</u>: *Epogen, Procrit*

<u>Manufacturer</u>: Amgen Inc, Thousand Oaks, CA

<u>Distributed by</u>: Amgen Inc (*Epogen*), Ortho Biotech Products, Raritan, NJ (*Procrit*)

<u>Indications</u>: Treatment of anemia associated with chronic renal failure. Treatment of anemia related to therapy with zidovudine. Treatment of anemia in cancer patients on chemotherapy. Treatment of anemic patients scheduled to undergo elective surgery to reduce the need for allogeneic blood transfusions.

<u>Approval date</u>: June 1989

<u>Type of submission</u>: Biologics license application

A. General description Erythropoietin is a glycoprotein normally produced in the kidneys and is responsible for the stimulation of red blood cell production. Epoetin alfa is derived via recombinant DNA techniques by mammalian cells after the insertion of the human erythropoietin gene; its amino-acid sequence is identical to that of endogenous erythropoietin. Epoetin alfa contains 165 amino acids. The molecule is heavily glycosylated with a molecular weight of 30.4 kD.

B. Indications and use *Epogen* and *Procrit* are indicated for the treatment of adults and children with anemia associated with chronic renal failure, including patients on dialysis (end-stage renal disease) and patients not on dialysis. They are intended to elevate or maintain the red blood cell level (as manifested by hematocrit or hemoglobin determinations) and to decrease the need for transfusions in these patients. To be considered for therapy, nondialysis patients with symptomatic anemia should have a hematocrit less than 30%. These products are not substitutes for emergency transfusion. *Epogen* and *Procrit* are also indicated for the treatment of anemia related to therapy with zidovudine in HIV-infected patients and for the treatment of anemia in adult patients with nonmyeloid malignancies where anemia is due to the effect of concomitantly administered chemotherapy. They are intended to decrease the need for transfusions in patients who will be receiving concomitant chemotherapy for a minimum of 2 months. *Epogen* and *Procrit* have also been approved for the treatment of anemic patients scheduled to undergo elective, noncardiac, nonvascular surgery to reduce the need for allogeneic blood transfusions.

C. Administration
  Dosage form *Epogen* is formulated as a sterile, buffered solution in single-dose, preservative-free vials. Each ml of solution contains 2000, 3000, 4000, or 10,000 units of epoetin alfa. A single-dose vial containing a more concentrated solution, 40,000 units per ml, is also available, as are multidose, preserved vials.
  Route of administration Epoetin alfa is commonly administered as an intravenous bolus to patients on hemodialysis but is sometimes self-administered subcutaneously.

Recommended dosage and monitoring requirements Intravenous dosing of epoetin alfa is initiated at 50 to 100 units/kg and subcutaneous dosing is generally started at 150 units/kg. Doses are administered three times per week. One recommended dosage schedule for patients undergoing surgery is 300 units/kg per day beginning 10 days before surgery, the day of surgery, and 4 days after surgery.

D. Pharmacology and pharmaceutics
  Clinical pharmacology Erythropoietin is instrumental in the production of red cells from the erythroid tissues in the bone marrow. The majority of this hormone is produced in the kidney in response to hypoxia, with an additional 10% to 15% of synthesis occurring in the liver. Erythropoietin functions as a growth factor, stimulating the mitotic activity of the erythroid progenitor cells and early precursor cells. Chronic renal failure patients often manifest the sequelae of renal dysfunction, including anemia. Anemia in cancer patients may be related to the disease itself or the effect of concomitantly administered chemotherapeutic agents.
  Pharmacokinetics The bioavailability of a subcutaneous dose of epoetin alfa, relative to an intravenous bolus, has been estimated at 22% to 31%. The elimination half-life of epoetin alfa after an intravenous dose is 6 to 13 hours in patients with chronic renal failure. Mean clearance ranges from 0.032 to 0.055 ml/min per kg. The apparent half-life after a subcutaneous dose is 27 hours. Volume of distribution estimates range from 0.021 to 0.063 l/kg.
  Disposition According to Micromedex, some metabolic degradation of epoetin alfa occurs, with small amounts recovered in the urine.
  Drug interactions No evidence of interaction of epoetin alfa with other drugs was observed in the course of clinical trials.

E. Therapeutic response Epoetin alfa has been shown to stimulate erythropoiesis in anemic patients with chronic renal failure, including both patients on dialysis and those who do not require regular dialysis. The first evidence of a response to the administration of epoetin alfa is an increase in the reticulocyte count within 10 days, followed by increases in the red cell count, hemoglobin, and hematocrit, usually within 2 to 6 weeks. Once the hematocrit reaches the suggested target range (30–36%), that level can be sustained by epoetin alfa therapy.

F. Role in therapy Epoetin alfa represents a major advance in the treatment of anemia associated with chronic renal failure. The hormone is an alternative to androgen and red blood cell transfusion therapy, which had been the mainstay of treatment. It may also provide an alternative for those patients who previously could not be treated with blood transfusions because of religious reasons.

G. Other applications Epoetin alfa may find a place in the therapy of anemia of other diseases such as those associated with sickle cell disease, rheumatoid arthritis, and prematurity. In addition the potential for use in autologous blood transfusion programs merits study.

H. Other considerations Epoetin alfa has been designated an orphan product for use in the treatment of anemia of end-stage renal disease, HIV infection, or prematurity in preterm infants, or treatment of myelodysplastic syndrome.

FILGRASTIM

Trade name: *Neupogen*

Manufacturer: Amgen Inc., Thousand Oaks, CA

Indications: Treatment of patients with nonmyeloid malignancies receiving myelo-suppressive anticancer drugs to decrease the incidence of infection. Treatment of adults with acute myeloid leukemia, following induction or consolidation chemotherapy, to reduce the time to neutrophil recovery and the duration of fever. Treatment of patients with nonmyeloid malignancies undergoing myeloablative chemotherapy, followed by marrow transplantation to reduce the duration of neutropenia and neutropenia-related clinical sequelae, such as febrile neutropenia. Treatment of symptomatic patients with congenital neutropenia, cyclic neutropenia, or idiopathic neutropenia to reduce the incidence and duration of neutropenia sequelae. *Neupogen* is also indicated for the mobilization of hematopoietic progenitor cells into the peripheral blood for collection by leukapheresis.

Approval date: 20 February 1991

Type of submission: Product license application

A. General description Filgrastim (G-CSF), also known as recombinant methionyl human granulocyte colony-stimulating factor, is a 175 amino-acid protein manufactured by recombinant DNA technology. Filgrastim is produced by *Escherichia coli* bacteria into which has been inserted the human granulocyte colony-stimulating factor gene. Filgrastim has a molecular weight of 18.8 kD. The protein has an amino-acid sequence that is identical to the natural sequence predicted from human DNA sequence analysis, except for the addition of an N-terminal methionine necessary for expression in *E. coli*. Because Filgrastim is produced in *E. coli*, the product is not glycosylated and thereby differs from G-CSF isolated from a human cell.

B. Indications and use Filgrastim is indicated to lessen neutropenia associated with myelosuppressive chemotherapy, bone marrow transplantation, and severe chronic neutropenia. It is also indicated for the mobilization of peripheral blood progenitor cells (PBPC) prior to PBPC collection. After myeloablative chemotherapy, the transplantation of an increased number of progenitor cells can lead to more rapid engraftment, which may result in a decreased need for supportive care.

C. Administration

Dosage form *Neupogen* is a sterile, clear, preservative-free liquid for parenteral administration. The product is available in single-use vials and prefilled syringes. The single use vials contain either 300 µg or 480 µg filgrastim. The single-use prefilled syringes contain either 300 µg or 480 µg filgrastim.

Route of administration Filgrastim may be given as an intravenous infusion or as a subcutaneous bolus.

Recommended dosage and monitoring requirements A starting dose of *Neupogen* 5 mcg/kg per day is used for cancer patients receiving chemotherapy and for patients with idiopathic or cyclic neutropenia, administered as a single daily injection by subcutaneous bolus injection, by short intravenous infusion (15–30 minutes), or by continuous subcutaneous or continuous intravenous infusion for up to 2 weeks. The dose for congenital neutropenia begins at 6 mcg/kg per day. In cancer patients receiving a bone marrow transplant or requiring peripheral blood stem cell mobilization, the dose is 10 mcg/kg per day. *Neupogen* should be administered no earlier than 24 hours after the administration of cytotoxic chemotherapy; it should not be administered in the period 24 hours before the administration of chemotherapy.

D. Pharmacology and pharmaceutics

Clinical pharmacology Colony-stimulating factors are glycoproteins that act on hematopoietic cells by binding to specific cell surface receptors. G-CSF is one of five hematopoietic growth factors that are involved in the development and activation of hematopoietic elements. These glycoproteins are produced naturally in lymphocytes and monocytes, and have been demonstrated to stimulate progenitor cells of different hematopoietic cell lineages to form colonies of recognizable mature blood cells. Filgrastim specifically promotes proliferation and maturation of neutrophil granulocytes.

Pharmacokinetics The volume of distribution of filgrastim averages 150 ml/kg. Its distribution half-life is 30 minutes, and its elimination half-life is 3 to 4 hours with a clearance of 0.5 to 0.7 ml/min per kilogram.

Disposition No information available.

Drug interactions Drug interactions between *Neupogen* and other drugs have not been fully evaluated. Drugs that may potentiate the release of neutrophils from bone marrow, such as lithium, should be used with caution.

E. Therapeutic response *Neupogen* administration results in a dose-dependent increase in circulating neutrophil counts, following a variety of chemotherapy regimens. In clinical studies, *Neupogen* treatment resulted in a clinically and statistically significant reduction in the incidence of infection, as manifested by febrile neutropenia. The requirements for in-patient hospitalization and antibiotic use were also significantly decreased during the first cycle of chemotherapy.

F. Role in therapy According to Micromedex, *Neupogen* is effective in decreasing the incidence of febrile neutropenia in cancer patients receiving myelosuppressive chemotherapy and in reducing the duration of neutropenia in cancer patients receiving myeloablative chemotherapy with subsequent bone marrow transplant. It is also effective for decreasing the complications of neutropenia in patients with severe chronic neutropenia. *Neupogen* is also used to mobilize peripheral blood stem cells for collection

by apheresis prior to autologous stem cell transplantation. A comparison of sargramostim (*Leukine*) and filgrastim (*Neupogen*) in patients with chemotherapy-induced myelosuppression demonstrated no statistically significant difference in mean number of days to reach an absolute neutrophil count of 500/microliter.

G. Other applications *Neupogen* may also be useful in preventing opportunistic diseases in patients infected with the human immunodeficiency virus and for drug-induced agranulocytosis. The use of topical filgrastim may be effective in reducing oral mucositis due to intensive chemotherapy.

H. Other considerations *Neupogen* has been designated an orphan product for use in the treatment of neutropenia associated with bone marrow transplant, AIDS patients with CMV retinitis being treated with ganciclovir, severe chronic neutropenia, and acute myeloid leukemia.

SARGRAMOSTIM

Trade name: *Leukine*

Manufacturer: Betlex, Seattle, WA

Indications: Treatment of older adult patients with acute myelogenous leukemia (AML) following induction chemotherapy to shorten time to neutrophil recovery and to reduce the incidence of severe and life-threatening infections and treatment of patients who have undergone allogeneic or autologous bone marrow transplantation (BMT) in whom engraftment is delayed or has failed. *Leukine* is also indicated for the mobilization of hematopoietic progenitor cells into peripheral blood for collection by leukapheresis, acceleration of myeloid recovery in patients with non-Hodgkin's lymphoma (NHL), acute lymphoblastic

leukemia (ALL), and Hodgkin's disease undergoing autologous bone marrow transplantation (BMT). It is also indicated for acceleration of myeloid recovery in patients undergoing allogeneic BMT.

Approval date: 5 March 1991

Type of submission: Product license application

A. General description Sargramostim (GM-CSF) is a recombinant human granulocyte-macrophage colony-stimulating factor, a hematopoietic growth factor that preferentially stimulates the granulocyte-macrophage progenitor cells. It is derived from the yeast, *Saccharomyces cerevisiae*. Sargramostim is a glycoprotein of 127 amino acids characterized by three primary molecular species with molecular masses of 19.5, 16.8, and 15.5 kDa. The amino-acid sequence of sargramostim differs from the natural human GM-CSF by a substitution of leucine at position 23, and the carbohydrate moiety may be different from the native protein.

B. Indications and use *Leukine* is indicated: (1) to shorten time to neutrophil recovery and reduce the incidence of severe and life-threatening infections following induction chemotherapy in acute myelogenous leukemia; (2) for mobilization of hematopoietic progenitor cells into peripheral blood to allow for increased collection of cells capable of engraftment and accelerated myeloid reconstitution following peripheral blood progenitor cell transplantation; (3) for accelerated myeloid reconstitution after autologous bone marrow transplantation secondary to non-Hodgkin's lymphoma, acute lymphoblastic leukemia, or Hodgkin's disease; (4) for accelerated myeloid reconstitution following allogeneic bone marrow transplantation, and (5) to prolong survival of patients experiencing graft failure or engraftment delay following autologous or allogeneic bone marrow transplantation.

C. Administration

Dosage form *Leukine* is formulated as a sterile, preserved, injectable solution (500 µg/l) in a 1 ml vial. Lyophilized *Leukine* is a sterile, preservative-free powder (250 µg) that requires reconstitution with sterile water for injection or bacteriostatic water for injection.

Route of administration *Leukine* solution and reconstituted lyophilized *Leukine* are suitable for subcutaneous injection or intravenous infusion.

Recommended dosage and monitoring requirements The recommended dose of *Leukine* is 250 µg/m^2 per day. In acute myelogenous leukemia, the protein is infused over 4 hours starting 4 days after the end of induction chemotherapy. After autologous or allogeneic bone marrow transplant, *Leukine* is infused over 2 hours starting 2 to 4 hours after marrow infusion. In bone marrow transplant failure or engraftment delay, *Leukine* is administered as a 2 hour infusion for 14 days.

D. Pharmacology and pharmaceutics

Clinical pharmacology GM-CSF is one of five hematopoietic growth factors that are involved in the development and functional activation of hematopoietic elements. These glycoproteins are produced naturally in lymphocytes and monocytes and have been demonstrated to stimulate progenitor cells of different hematopoietic cell lineages to form colonies of recognizable mature blood cells. GM-CSF stimulates the proliferation and activation of monocyte-macrophages, and induces these cells to produce cytokines, including tumor necrosis factor and interleukin-1. The biological activity of GM-CSF is species specific. *In vitro* exposure of human bone marrow cells to *Leukine* results in the proliferation of hematopoietic progenitors and in the formation of pure granulocyte, pure macrophage, and mixed granulocyte-macrophage colonies.

Pharmacokinetics The elimination half-life of sargramostim following intravenous administration is about 60 minutes, and its clearance is about 420 to 529 ml/min per m^2 The apparent half-life of sargramostim is 2 to 3 hours after subcutaneous administration.

Disposition No information available.

Drug interactions Because of its ability to potentiate myeloproliferative effects, lithium should be used cautiously in patients also receiving sargramostim. Severe atypical peripheral neuropathy was reported to occur significantly more commonly in patients with lymphomas receiving a colony-stimulating factor (sargramostim or filgrastim) with vincristine than vincristine alone.

E. Therapeutic response

In clinical trials *Leukine* was safe and effective in adult patients with acute myeloid leukemia, reducing the duration of neutropenia as well as chemotherapy-associated mortality and morbidity. *Leukine*, administered alone or after myeloablative chemotherapy, enhances the number of circulating peripheral blood progenitor cells, allowing harvest for autologous transplantation.

F. Role in therapy

Leukine has been effective in mobilizing hematopoietic progenitor cells, in enhancing myeloid engraftment, and in stimulating mature monocytes, macrophages, and neutrophils following high-dose chemotherapy. *Leukine* is also useful in BMT; it reduces the duration to re-engraftment, number of antibiotic treatment days, and the period of hospitalization. Both sargramostim and filgrastim (G-CSF) have been effective in the treatment of chemotherapy-induced neutropenia. A comparison of sargramostim and filgrastim in patients with chemotherapy-induced myelosuppression demonstrated no statistically significant difference in mean number of days to reach an absolute neutrophil count (ANC) of 500/μl. If an increase in neutrophil count is desired, the preferable agent would be filgrastim. However, if the goal of treatment is increased numbers and function of mononuclear cells plus neutrophils, sargramostim would be indicated.

G. Other applications *Leukine* has been effective in producing increases in normally functioning neutrophils, eosinophils, and monocytes in AIDS patients and patients with leukopenia. *Leukine* may prolong survival when used as adjuvant therapy in patients with stage III or IV malignant melanoma. *Leukine* has also been effective in abrogating chemotherapy-related neutropenia in cancer patients, with a reduction in the severity and duration of chemotherapy-induced myelosuppression.

H. Other considerations *Leukine* has been designated an orphan product for use in the treatment of neutropenia associated with bone marrow transplantation, leukemia, graft failure and delay of engraftment, promotion of early engraftment, and to decrease the incidence of death due to infection in patients with acute myelogenous leukemia.

OPRELVEKIN

Trade name: *Neumega*

Manufacturer: Genetics Institute, Inc., Andover, MA

Indications: Prevention of severe thrombocytopenia and the reduction of the need for platelet transfusions following myelosuppressive chemotherapy in patients with nonmyeloid malignancies who are at high risk of severe thrombocytopenia

Approval date: 25 November 1997

Type of submission: Biologics license application

A. General description Interleukin-11 is a thrombopoietic growth factor that stimulates the production of hematopoietic stem cells and megakaryocyte progenitor cells, resulting in increased platelet production. Recombinant human IL-11 (rhIL-11) is synthesized in *Escherichia coli*. The recombinant protein has a molecular weight of 19 kDa and is nonglycosylated. It is 177 amino acids in length and differs from the 178 amino-acid length of native IL-11 only in the absence of the amino-terminal proline residue. This variation has not resulted in measurable differences in biologic activity.

B. Indications and use *Neumega* is indicated for the prevention of severe thrombocytopenia and the reduction of the need for platelet transfusions following myelosuppressive chemotherapy in patients with nonmyeloid malignancies who are at high risk of severe thrombocytopenia. It reduces the need for frequent platelet transfusions following high-dose chemotherapy. While platelet transfusions are safe, they carry small risks of infectious disease transmission as well as a small risk of bacterial contamination during storage. Also, transfusions often need to be repeated frequently in an in-patient setting, and may delay chemotherapy.

C. Administration
   Dosage form *Neumega* is available for injection in single-use vials containing 5 mg of oprelvekin as a sterile lyophilized powder. It is reconstituted by the addition of sterile water for injection.
   Route of administration *Neumega* should be administered subcutaneously as a single injection in the abdomen, thigh, or hip (or upper arm if not self-injecting). In

situations where the physician determines that *Neumega* may be used outside of the hospital or office setting, persons who will be administering *Neumega* should be instructed as to the proper dose, and the method for reconstituting and administering the product.

Recommended dosage and monitoring requirements The recommended dose of *Neumega* in adults is 50 µg/kg given once daily. Dosing should be initiated 6 to 24 hours after the completion of chemotherapy. Platelet counts should be monitored periodically to assess the optimal duration of therapy. Dosing should be continued until the postnadir platelet count is at least 50,000 cells/µl. In controlled clinical studies, doses were administered in courses of 10 to 21 days. Dosing beyond 21 days per treatment course is not recommended. Treatment with *Neumega* should be discontinued at least 2 days before starting the next planned cycle of chemotherapy.

D. Pharmacology and pharmaceutics

Clinical pharmacology Interleukin-11 is a member of a family of human growth factors that includes human growth hormone, granulocyte colony-stimulating factor (G-CSF), and other growth factors. It is produced by fibroblasts, as well as by bone marrow stromal cells, and directly stimulates the proliferation of hematopoietic stem cells and megakaryocyte progenitor cells, as well as lymphoid cells. IL-11 also induces the differentiation of megakaryocytes into platelets. Platelets produced in response to IL-11 are normal morphologically and functionally, and possess normal life spans.

Pharmacokinetics According to product label, analysis of data from a study in healthy men and women who received intravenous and subcutaneous *Neumega* revealed that following subcutaneous administration absorption is the rate-limiting step. Hence the elimination rate constants and associated half-lives of oprelvekin are dependent on the rate of absorption. Following intravenous administration in men, the mean elimination half-life of oprelvekin was 1.5 hours; the apparent mean elimination half-life after subcutaneous administration was 7 hours. No significant gender-based differences in pharmacokinetic endpoints were found. In a study in which a single 50 µg/kg subcutaneous dose was administered to 18 men, the peak serum concentration (C_{max}) of 17.4 ± 5.4 ng/ml (mean ± SD) was reached at 3.2 ± 2.4 hours (T_{max}) following dosing. The mean terminal half-life was 6.9 hours. One study estimated the apparent volume of distribution of oprelvekin to be 112 to 152 ml/kg and its total body clearance to be 2.2 to 2.7 ml/min/kg. In a study in which multiple subcutaneous doses were administered to cancer patients receiving chemotherapy, oprelvekin did not accumulate and clearance was not impaired following multiple doses. Clearance of oprelvekin decreases with increasing patient age, and clearance in infants and children (8 months to 11 years) is between 1.2- and 1.6-fold higher than in adults and adolescents (ages 12 and over). Nevertheless, data from 23 pediatric patients indicated that peak serum values and time to peak serum values were comparable to those of adults. Compared with subjects with normal renal function, peak serum concentrations and area-under-the-plasma-concentration-time-curve values were higher in subjects with severe renal impairment.

Metabolism One report suggests that oprelvekin is extensively metabolized in the kidneys.

Drug interactions No adverse effects of oprelvekin on the activity of filgrastim (G-CSF) was seen during clinical trials in patients who were treated with these two agents concurrently. No information is available on the clinical use of sargramostim (GM-CSF) with *Neumega*. Drug interactions between *Neumega* and other drugs have not been fully evaluated. Based on *in vitro* and nonclinical *in vivo* evaluations of *Neumega*, drug–drug interactions with known substrates of P450 enzymes could not be predicted.

E. Therapeutic response The primary end point of one randomized, placebo-controlled trial was whether the patient required one or more platelet transfusions in the subsequent chemotherapy cycle. More patients avoided platelet transfusion in the *Neumega* arm (28%) than in the placebo arm (7%).

F. Role in therapy According to Micromedex, treatment of severe chemotherapy-related thrombocytopenia is limited to platelet transfusions. There is a need for an alternative, especially due to the frequent use of myeloid colony-stimulating factors (G-CSF, GM-CSF) to reduce febrile neutropenia; although effective, their use increases the risk of acute and prolonged thrombocytopenia, and the need for platelet transfusions. Other cytokines, such as interleukin-1 and interleukin-6, have been investigated as a means of ameliorating chemotherapy-induced thrombocytopenia, but results have been equivocal.

G. Other considerations Recombinant human oprelvekin has been designated an orphan product for use in the prevention of chemotherapy-induced thrombocytopenia.

COAGULATION FACTOR IX

Trade name: *BeneFix*

Manufacturer: Genetics Institute, Inc., Andover, MA

Indications: For use in the control and prevention of hemorrhagic episodes in patients with hemophilia B, including the perioperative management of hemophilia B patients undergoing surgery

Approval date: 11 February 1997

Type of submission: Biologics license application

A. General description Coagulation Factor IX (Recombinant) is a DNA-derived clotting factor. This purified recombinant product is free from the risk of transmitting plasma-derived human viruses because it is produced in Chinese hamster ovary cells that have been modified to express the gene for human factor IX. It is a glycoprotein with an approximate molecular mass of 55 kDa, consisting of 415 amino acids in a single chain. Coagulation Factor IX (Recombinant) has a primary amino-acid sequence that is identical to the Ala 148 allelic form of plasma-derived factor IX, and has structural and functional characteristics similar to those of endogenous factor IX.

B. Indications and use *BeneFix* is intended for the prevention and control of excessive, potentially life-threatening bleeding in patients with hemophilia B, including control and prevention of bleeding in the surgical setting. Patients with hemophilia B, also known as Christmas disease, are unable to form blood clots adequately because of a deficiency or defect in clotting factor IX. Treatment with factor IX products corrects the defect temporarily.

C. Administration
   Dosage form *BeneFix* is formulated as a sterile nonpyrogenic, lyophilized preparation, intended for injection after reconstitution with sterile water for injection. It is available in single-use vials containing the labeled amount of factor IX activity, expressed in international units (IU). Each vial contains nominally 250, 500, or 1000 IU of Coagulation Factor IX (Recombinant).

Route of administration *BeneFix* should be injected intravenously over several minutes, within 3 hours of reconstitution.

Recommended dosage and monitoring requirements Dosage and duration of treatment for all factor IX products depend on the severity of the factor IX deficiency, the location and extent of bleeding, the patient's clinical condition and age, and recovery of factor IX. Dose should be titrated to the patient's clinical response. Patients should be monitored for normalization of coagulation parameters and the development of factor IX inhibitors. Dosing of *BeneFix* may differ from that of plasma-derived factor IX products Frequency of doses in bleeding episodes and surgery is 12 to 24 hours, until bleeding stops and healing begins.

D. Pharmacology and pharmaceutics

Clinical pharmacology Activated factor IX in combination with activated factor VIII activates factor X. This results ultimately in the conversion of prothrombin to thrombin. Thrombin then converts fibrinogen to fibrin, and a clot can be formed. Factor IX is the specific clotting factor deficient in patients with hemophilia B and in patients with acquired factor IX deficiencies. The administration of Coagulation Factor IX (Recombinant) increases plasma levels of factor IX and can temporarily correct the coagulation defect in these patients.

Pharmacokinetics After single intravenous doses of 50 IU/kg of *BeneFix* in 36 patients, each given as a 10-minute infusion, the mean elimination half-life of the protein was 19.4 hours (range: 11–36 hours). There was no significant difference in biological half-life in subsequent evaluations at 6 and 12 months. Another study showed that the elimination half-lives of *BeneFix* and plasma-derived factor IX were not significantly different. Elimination half-lives also did not differ signifi-

cantly among patients treated with four drug product lots manufactured from several batches of drug substance produced from two separate inoculum runs.

Disposition No information available.

Drug interactions No information available.

E. Therapeutic response Clinical studies demonstrated that *BeneFix* increased circulating factor IX activity to a desired level and stopped bleeding episodes.

F. Place in therapy Micromedex notes that the major use of factor IX is in the therapy of hemophilia B, but it may also be useful in patients with clotting disorders secondary to hepatic dysfunction and other conditions. Recombinant factor IX (*BeneFix*), however, is used specifically for the prevention and control of bleeding in patients with hemophilia B.

G. Other consideration *Benefix* was designated an orphan drug by the Office of Orphan Products Development on 3 October 1994. *BeneFix* is the third coagulation factor IX product to receive orphan drug designation for the treatment of hemophilia B. The other two products are plasma-derived Coagulation Factors IX (Human): AlphaNine (Alpha Therapeutic), and Mononine (Armour Pharmaceutical). Both products are manufactured by methods that are effective in reducing the risk of transmitting human viruses; however, these risks have not been totally eliminated. *BeneFix* is the recombinant analogue of the two plasma-derived factor IX products. It has the same principal molecular structural features and is intended for the same use. Hence *BeneFix* should be considered the same drug as the plasma-derived factor IX products unless it can be shown to be clinically superior to the previously approved products. Genetics Institute claims the clinical superiority of *BeneFix* because of its greater safety compared to the plasma-derived products. By virtue of its source and manufacturing methods, *BeneFix* is inherently less likely to transmit human blood-borne viruses and other infectious agents, and is also less likely to transmit animal-derived zoonotic agents than are plasma-derived products.

COAGULATION FACTOR VIIA

Trade name: *NovoSeven*

Manufacturer: Novo Nordisk A/S, Princeton, NJ; Novo Alle, Bagsvaerd, Denmark

Indications: Treatment of bleeding episodes in hemophilia A or B patients with inhibitors to factor VIII or factor IX

Approval date: 25 March 1999

Type of submission: Biologic license application

A. General description Coagulation Factor VIIa (Recombinant) is a vitamin K-dependent glycoprotein consisting of 406 amino-acid residues (molecular weight 50 kD). It is structurally similar to human plasma-derived factor VIIa. The gene for human factor VII is cloned and expressed in baby hamster kidney cells. Recombinant factor VII is secreted into the culture medium in its single-chain form and then proteolytically converted by autocatalysis to the active two-chain form, recombinant coagulation factor VIIa.

B. Indications and use *NovoSeven* is indicated for the treatment of bleeding episodes in hemophilia A or B patients with inhibitors to factor VIII or factor IX. It should be administered to patients only under the direct supervision of a physician experienced in the treatment of hemophilia.

C. Administration
Dosage form *NovoSeven* is supplied as a sterile lyophilized powder of coagulation

factor VIIa (recombinant) in single-use vials. Each vial of the lyophilized drug coagulation factor VIIa contains 1200 or 4800 μg.

Route of administration *NovoSeven* is intended for intravenous bolus administration only, after reconstitution with the appropriate volume of sterile water for injection.

Recommended dosage and monitoring requirements The recommended dose of *NovoSeven* for hemophilia A or B patients with inhibitors is 90 μg/kg given every 2 hours until hemostasis is achieved, or until the treatment has been judged to be inadequate. Doses between 35 and 120 μg/kg have been used successfully in clinical trials, and both the dose and administration interval may be adjusted based on the severity of the bleeding and degree of hemostasis achieved.

D. Pharmacology and pharmaceutics
Clinical pharmacology When coagulation factor VIIa is complexed with tissue factor, it can activate coagulation factor X to factor Xa, as well as coagulation factor IX to factor IXa. Factor Xa, in complex with other factors, then converts prothrombin to thrombin, which leads to the formation of a hemostatic plug by converting fibrinogen to fibrin and thereby inducing local hemostasis.

Pharmacokinetics According to the product label, single-dose pharmacokinetics of recombinant factor VII exhibited dose-proportionality in 15 subjects with hemophilia A or B. The median apparent volume of distribution at steady state was 103 ml/kg (range: 78–139). Median clearance was 33 ml/kg per hour (range: 27–49 ml/kg per hour). The median residence time was 3.0 hours (range: 2.4–3.3), and the half-life was 2.3 hours (range: 1.7–2.7).

Disposition No information available.

Drug interactions The label notes that the risk of potential interactions between *NovoSeven* and coagulation factor concentrates has not been adequately evaluated. Simultaneous use of activated and non-activated prothrombin complex concentrates should be avoided. Although the specific drug interaction was not studied in a clinical trial, there have been more than 50 episodes of concomitant use of anti-fibrinolytic therapies (i.e., tranexamic acid, aminocaproic acid) and *NovoSeven*. *NovoSeven* should not be mixed with infusion solutions until clinical data are available to direct this use.

E. Therapeutic response Evaluation of hemostasis should be used to determine the effectiveness of *NovoSeven*. In a comparison trial of two dose levels of *NovoSeven* in the treatment of joint, muscle, and mucocutaneous hemorrhages in hemophilia A and B patients, the respective efficacy rates for the 35 and 70 µg/kg groups were: excellent 59% and 60%; effective 12% and 11%; and partially effective 17% and 20%. The average number of injections required to achieve hemostasis was 2.8 and 3.2 for the 35 and 70 µg/kg groups, respectively.

F. Role in therapy *NovoSeven* is useful for the treatment and prevention of hemorrhagic episodes in hemophilia A and B patients with factor VII and factor IX inhibitors, respectively. This agent has been shown to control acute bleeding in approximately 80% of patients with or without antibody inhibitors, and its efficacy appears at least comparable to that of other treatement strategies in clinical use. The use of coagulation factor VIIa (recombinant) should be limited mainly to treating patients with moderate- to high-inhibitor titers (most of whom are "high responders," in that challenge with clotting factors significantly increases inhibitor antibody production). Other uses for recombinant factor VIIa include in-home use for early intervention in chronic hemophilic patients and in patients with platelet defects.

G. Other applications *NovoSeven* may be effective for acute bleeding or hemostasis during surgery. In one study *NovoSeven* reversed the effects of the oral anticoagulant acenocoumarol on the prothrombin time and International Normalized Ratio (INR) in healthy volunteers, without evidence of systemic coagulation. It may also transiently correct elevated prothrombin time in patients with cirrhosis-induced coagulopathy.

ANTIHEMOPHILIC FACTOR

Trade name: ***ReFacto, Recombinate***

Manufactured by: Pharmacia AB, Stockholm, Sweden (*ReFacto*); manufacturing of *Recombinate* is shared by Baxter Healthcare Corporation (Hyland Division, Glendale, CA) and Genetics Institute, Inc, Cambridge, MA

Distributed by: Genetics Institute, Inc (*ReFacto*); Baxter Healthcare (*Recombinate*)

Indications: Control and prevention of hemorrhagic episodes and for surgical prophylaxis in patients with hemophilia A. Short-term routine prophylaxis to reduce the frequency of spontaneous bleeding episodes (*ReFacto*). Prevention and control of hemorrhagic episodes in hemophilia A; perioperative management of patients with hemophilia A (*Recombinate*).

Approval date: 6 March 2000 (*ReFacto*); 1 August 1998 (*Recombinate*)

Type of submission: Biologics license application

A. General description The active substance in *ReFacto* is antihemophilic factor (recombinant), a 1438 amino-acid glycoprotein (approximately 170 kDa) that is produced in engineered Chinese hamster ovary (CHO) cells. The CHO cell line secretes B-domain deleted recombinant factor VIII into a defined cell culture medium that contains human serum albumin and recombinant insulin, but does not contain any proteins derived from animal sources. The protein is purified by a chromatography purification process that yields a high-purity, active product. *ReFacto* and the 90 + 80 kD form of plasma factor VIII have many comparable structural characteristics. *Recombinate* is also synthesized by a genetically engineered CHO cell line. In culture the CHO cell line secretes recombinant antihemophilic factor (rAHF) into the cell-culture medium. The rAHF is purified from the culture medium utilizing a series of chromatography columns. A key step in the purification process is an immunoaffinity chromatography methodology in which a purification matrix, prepared by immobilization of a monclonal antibody directed to factor VIII, is utilized to selectively isolate the rAHF in the medium. Von Willebrand factor is coexpressed with the antihemophilic factor (recombinant) and helps to stabilize it. The synthesized glycoprotein produced by the CHO cells has the same biological effects as human antihemophilic factor and structurally has a similar combination of heterogenous heavy and light chains.

B. Indications and use *ReFacto* is indicated for the control and prevention of hemorrhagic episodes and for surgical prophylaxis in patients with hemophilia A (congenital factor VIII deficiency or classic hemophilia). *ReFacto* is also indicated for short-term routine prophylaxis to reduce the frequency of spontaneous bleeding episodes. *Recombinate* is indicated in hemophilia A for the prevention and control of hemorrhagic episodes and is also indicated in the perioperative management of patients with hemophilia A. Both products can be of significant therapeutic value

in patients with acquired antihemophilia factor inhibitors.

C. Administration

Dosage form *ReFacto* is a sterile lyophilized powder for injection available in nominal dosage strengths of 250, 500, and 1000 international units (IU) per vial. *Recombinate* is formulated as a sterile lyophilized powder preparation of concentrated recombinant AHF for intravenous injection and is available in single-dose bottles which contain nominally 250, 500, and 1000 IU per bottle. The final product contains an insufficient quantity of von Willebrand Factor to have any clinically effect in patients with von Willebrand's disease.

Route of administration These products are administered only by intravenous infusion (IV) within 3 hours after reconstitution.

Recommended dosage and monitoring requirements Dosage and duration of treatment depend on the severity of the factor VIII deficiency, the location and extent of bleeding, and the patient's clinical condition. Doses should be titrated to the patient's clinical response. Patients using *ReFacto* should be monitored for the development of factor VIII inhibitors. In the presence of an inhibitor, higher doses may be required. For short-term routine prophylaxis to prevent or reduce the frequency of spontaneous musculoskeletal hemorrhage in patients with hemophilia A, *ReFacto* should be given at least twice a week. In some cases, especially pediatric patients, shorter dosage intervals or higher doses may be necessary. Although dosage of *Recombinate* can be estimated by calculations illustrated in product labeling, the manufacturer strongly recommends that whenever possible, appropriate laboratory tests, including serial AHF assays, be performed on the patient's plasma at suitable intervals to ensure that adequate AHF levels have been reached and are maintained.

D. Pharmacology and pharmaceutics

Clinical pharmacology Factor VIII is the specific clotting factor deficient in patients with hemophilia A. The administration of antihemophilic factor (recombinant) increases plasma levels of factor VIII activity and can temporarily correct the *in vitro* coagulation defect in these patients. Activated factor VIII acts as a cofactor for activated factor IX accelerating the conversion of factor X to activated factor X. Activated factor X converts prothrombin into thrombin. Thrombin then converts fibrinogen into fibrin and a clot is formed. Factor VIII activity is greatly reduced in patients with hemophilia A and therefore replacement therapy is necessary.

Pharmacokinetics In a pharmacokinetic study of previously treated patients, the circulating mean half-life for *ReFacto* was 14.5 ± 5.3 hours (range: 7.6–27.7 hours), which was not significantly different from plasma-derived antihemophilic factor (human), which had a mean half-life of 13.7 ± 3.4 hours (range: 8.8–23.7 hours). Pharmacokinetic studies on 66 patients showed the circulating mean half-life for rAHF (*Recombinate*) to be 14.4 ± 4.9 hours, which was not significantly different from a plasma-derived human antihemophilic factor, which had a mean half-life of 14.0 ± 3.9 hours. One study determined a mean clearance of 127 ml/min per kg.

Disposition No information available.

Drug interactions No information available.

E. Therapeutic response

ReFacto therapy has resulted in significant decreases in the incidence of uncontrolled bleeding into joints and soft tissues and long-term effects of hemorrhage. *ReFacto* was well tolerated and elicited an excellent response when administered exclusively for episodic treatment in a multicenter trial in children previously untreated with any factor VIII products. *Recombinate* has been used with success to treat bleeding episodes in previously treated patients (individuals with hemophilia A who had been treated with plasma-derived AHF) and to prevent bleeds.

F. Role in therapy

Antihemophilic factor is indicated for the treatment of bleeding episodes or perioperative treatment in patients with hemophilia A. Prophylactic use has also been advocated for the prevention and/or reduction of bleeding episodes. The largest issue in treatment with antihemophilic factor is the choice of formulations because of the relative risk of viral transmission. Recombinant factor VIII has the lowest risk of transmission of blood-borne viruses, but its use may be limited due to cost and availability.

G. Other applications

Recombinant-derived and monoclonal antibody-purified formulations are not indicated for treatment of von Willebrand's disease.

H. Other considerations

Antihemophilic factor (recombinant) has been designated an orphan product for use in the treatment or prophylaxis of bleeding in hemophilia A, or prophylaxis before surgery.

ANTIHEMOPHILIC FACTOR

Trade name: ***Kogenate, Kogenate FS***

Manufacturer: Bayer Corporation, Pharmaceutical Division, Elkhart, IN (supply may be limited due to manufacturing problems).

Indications: Treatment of classical hemophilia (hemophilia A) in which there is a demonstrated deficiency of activity of the plasma clotting factor, factor VIII

Approval date: 26 June 2000 (*Kogenate FS*)

<u>Type of submission</u>: Biologics license application

A. General description Antihemophilic Factor (Recombinant) (*Kogenate*), a highly purified glycoprotein consisting of multiple peptides including an 80 kD and various extensions of the 90 kD subunit, is produced by baby hamster kidney cells into which the human factor VIII (*FVIII*) gene has been introduced. The final preparation is stabilized with human albumin and lyophilized. *Kogenate* has the same biological activity as factor VIII derived from human plasma. *Kogenate FS* is a newer formulation of Antihemophilic Factor (Recombinant). The cell-culture medium for *Kogenate FS* contains human plasma protein solution and recombinant insulin, but does not contain any proteins derived from animal sources. Compared to its predecessor product *Kogenate*, *Kogenate FS* incorporates a revised purification and formulation process that eliminates the addition of human albumin.

B. Indications and use *Kogenate* and *Kogenate FS* are indicated for the treatment of classical hemophilia (hemophilia A) in which there is a demonstrated deficiency of activity of the plasma clotting factor VIII. They provide a means of temporarily replacing the missing clotting factor in order to correct or prevent bleeding episodes, or in order to perform emergency or elective surgery in hemophiliacs. *Kogenate* can also be used for treatment of hemophilia A in certain patients with inhibitors to factor VIII. Because *Kogenate FS* has similar biological activity to *Kogenate*, it can be used in the same manner.

C. Administration

Dosage form *Kogenate* and *Kogenate FS* are prepared as sterile, stable, purified, dried concentrates, which require reconsti-

tution for injection. The products are supplied in single-use bottles with the total units of factor VIII activity stated on the label of each bottle. A suitable volume of sterile water for injection is provided.

Route of administration The reconstituted product must be administered intravenously by either direct syringe injection or drip infusion. *Kogenate* and *Kogenate FS* must be administered within 3 hours after reconstitution.

Recommended dosage and monitoring requirements The dosage of *Kogenate* and *Kogenate FS* required for hemostasis must be individualized according to the needs of the patient, the severity of the deficiency, the severity of the hemorrhage, the presence of inhibitors, and the factor VIII level desired. It is often critical to follow the course of therapy with factor VIII level assays. If the calculated dose fails to attain the expected factor VIII levels, or if bleeding is not controlled after administration of the calculated dosage, the presence of a circulating inhibitor in the patient should be suspected. Its presence should be substantiated and the inhibitor level quantified by appropriate laboratory tests. When an inhibitor is present, the dosage requirement for Antihemophilic Factor (Recombinant) is extremely variable, and the dosage can be determined only by the clinical response.

D. Pharmacology and pharmaceutics

Clinical pharmacology According to Micromedex, antihemophilic factor is a high-molecular-weight glycoprotein that functions as a cofactor in the blood coagulation cascade. The molecule consists of the procoagulant protein antihemophilic factor:C (factor VIII:C) and the related protein multimers known as antihemophilic factor:R (factor VIII:R; factor VIII:vWF; von Willebrand factor). Antihemophilic factor:C deficiency results in a bleeding disorder (hemophilia A); in von Willebrand's disease, antihemophilic factor:C and antihemophilic factor:R (von Willebrand protein complex) are reduced. Therapy of antihemophilic factor deficiency is accomplished with the use of lyophilized antihemophilic factor (factor VIII) concentrates. Recombinant-derived formulations reduce the danger of viral transmission.

Pharmacokinetics Pharmacokinetic studies found that the mean biologic half-life of *Kogenate* (rAHF) was about 13 hours. The mean biologic half-life of plasma-derived AHF in the same individuals was 13.9 hours. In a comparative pharmacokinetic study, *Kogenate FS* was shown to be similar to its predecessor product *Kogenate*.

Disposition No information available.

Drug interactions No information available.

E. Therapeutic response The clinical effect of *Kogenate* and *Kogenate FS* is the most important element in evaluating the efficacy of treatment. It may be necessary to administer more *Kogenate* or *Kogenate FS* than would be estimated in order to attain satisfactory clinical results. If the calculated dose fails to attain the expected factor VIII levels, or if bleeding is not controlled after administration of the calculated dosage, the presence of a circulating inhibitor in the patient should be suspected. In a study of previously untreated patients, 3254 infusions were administered to 96 patients over a 48-month enrollment period. Hemostasis was successfully achieved in all cases. Elevating plasma antihemophilic factor activity to 30% is considered adequate for hemostasis; higher percentages are required in special circumstances, such as surgery, trauma, or prolonged external bleeding.

F. Role in therapy Antihemophilic factor concentrates have virtually superceded fresh frozen plasma and cryoprecipitate for the treatment of hemophilia A.

Recombinant-derived and monoclonal antibody-prepared formulations have been developed to reduce the danger of viral transmission. However, recombinant-derived and monoclonal antibody-purified formulations do not contain the large multimers of the von Willebrand factor and are not indicated for treatment of von Willebrand's disease.

G. Other considerations Antihemophilic factor (recombinant) has been designated an orphan product for use in the treatment or prophylaxis of bleeding in hemophilia A, or prophylaxis before surgery. Bayer makes a version of Kogenate FS that is sold by Aventis Behring under the name *Helixate Concentrate* (*Helixate FS*).

LEPIRUDIN

Trade name: ***Refludan***

Manufacturer: Hoechst Marion Roussel, Frankfurt am Main, Germany

Distributed by: Hoechst Marion Roussel, Kansas City, MO

Indications: Anticoagulation in patients with heparin-induced thrombocytopenia and associated thromboembolic disease in order to prevent further thromboembolic complications

Approval date: 11 October 1998

Type of submission: New drug application

A. General description Lepirudin is a recombinant hirudin, a highly specific direct inhibitor of thrombin, derived from yeast cells. The polypeptide is composed of 65 amino acids and has a molecular weight of 7 kD. Natural hirudin is produced in trace amounts as a family of highly homologous isopolypeptides by the medicinal leech. Lepirudin is identical to natural hirudin except for the substitution of leucine for isoleucine at the N-terminal end of the molecule and the absence of a sulfate group on the tyrosine at position 63.

B. Indications and use   *Refludan* is indicated for anticoagulation in patients with heparin-induced thrombocytopenia (HIT) and associated thromboembolic disease in order to prevent further thromboembolic complications.

C. Administration

Dosage form   *Refludan* is supplied as a sterile freeze-dried powder for injection or infusion after reconstitution. Each vial of *Refludan* contains lepirudin 50 mg.

Route of administration   Refludan is administered by intravenous infusion after an initial intravenous bolus.

Recommended dosage and monitoring requirements   The initial dosage of *Refludan* for adult patients with HIT and associated thromboembolic disease is 0.4 mg/kg body weight (up to 110 kg) given intravenously over 15 to 20 seconds as a bolus dose. This is followed by a continuous intravenous infusion of 0.15 mg/kg body weight (up to 110 kg) per hour for 2 to 10 days or longer if clinically needed. In general, therapy with *Refludan* is monitored using the activated partial thromboplastin time (aPTT) ratio. A baseline aPTT should be determined prior to initiation of therapy with *Refludan*, since the drug should not be started in patients presenting with a baseline aPTT ratio of 2.5 or more, in order to avoid initial overdosing.

D. Pharmacology and pharmaceutics

Clinical pharmacology   According to Micromedex, lepirudin directly inhibits all actions of thrombin. It inhibits free and clot-bound thrombin without requiring endogenous cofactors. Lepirudin is not inhibited by platelet factor 4 and acts independently of antithrombin III and heparin cofactor II. It has no direct effect on platelet function, except inhibition of thrombin-induced platelet activation. No physiological inhibitor of lepirudin is known.

Pharmacokinetics   Lepirudin is eliminated primarily by renal excretion (renal clearance 65 to 115 ml/min). Dose adjustment based on creatinine clearance is recommended. The total clearance of lepirudin is 195 ml/min, its elimination half-life is 1.3 hours, and its volume of distribution is 12.2 to 18.0 liters. The systemic clearance of lepirudin in women is about 25% lower than in men. In elderly patients the systemic clearance of lepirudin is 20% lower than in younger patients. Distribution is limited to extracellular space. As the intravenous dose is increased over the range of 0.1 to 0.4 mg/kg, the maximum plasma concentration and the area-under-the-curve increase proportionally.

Dispostion   Lepirudin is eliminated primarily by renal excretion. It may be metabolized by release of amino acids via catabolic hydrolysis of the parent drug.

Drug interactions   Concomitant treatment with thrombolytic agents may increase the risk of bleeding complications and considerably enhance the effect of *Refludan* on aPTT prolongation. Concomitant treatment with coumarin derivatives (vitamin K antagonists) and drugs that affect platelet function may also increase the risk of bleeding.

E. Therapeutic response   Thrombin-dependent tests show dose dependency [aPTT rise proportionally to dose of *Refludan*]. The key criteria of efficacy in two pivotal clinical trials from a laboratory standpoint were platelet recovery and effective anticoagulation. Seven days after the start of treatment with *Refludan* in patients with HIT, the cumulative risk of death, limb amputation, or new thromboembolic complication was substantially lower than in a historical control group.

F. Role in therapy   Lepirudin is a thrombin-inhibitor indicated for use in the treatment of adults with heparin-induced thrombocytopenia (HIT) type II; diagnosis should be substantiated by a heparin-induced platelet aggregation assay. Lepirudin provides a more stable level of anticoagulation than heparin, as indicated by stable prolongation of aPTT and activated clotting time due at least in part to the fact that lepirudin is not inhibited by activated platelets or other proteins known to neutralize heparin. Lepirudin may also offer a better safety-to-efficacy ratio than heparin. Recombinant hirudin (lepirudin), argatroban, and danaparoid have been used successfully in HIT. Only lepirudin is available in the United States.

G. Other applications Lepirudin is also under study for the treatment of myocardial infarction and unstable angina and other thromboembolic disorders requiring parenteral antithrombotic therapy such as disseminated intravascular coagulation, in adults.

BIVALIRUDIN

Trade name: *Angiomax*

Manufacturer: BenVenue Laboratories, Bedford, OH

Marketed by: The Medicines Company, Cambridge, MA

Indications: Prevention of ischemic complications in unstable angina patients during coronary angioplasty

Approval date: 1 December 2000

Type of submission: New drug application

A. General description Bivalirudin, a 20 amino-acid synthetic peptide, is a direct thrombin inhibitor. It is an analogue of recombinant hirudin (*Refludan*), a 65 amino-acid anticoagulant derived from the leech. The molecular weight of bivalirudin is about 2.2 kDa (anhydrous free base peptide).

B. Indications and use *Angiomax* is indicated when given with aspirin for the prevention of ischemic complications in unstable angina patients during coronary angioplasty. It has been used successfully as a substitute for heparin. The efficacy of *Angiomax* appears to be similar to heparin with fewer bleeding complications.

C. Administration
Dosage form Angiomax is available as a sterile lyophilized powder for injection after reconstitution with sterile water for injection. Each single-use vial contains bivalirudin 250 mg.
Route of administration *Angiomax* is administered as an intravenous infusion, sometimes preceded by a bolus.
Recommended dosage and monitoring requirements According to Micromedex, patients with unstable angina undergoing coronary angioplasty should receive an intravenous (IV) bolus of *Angiomax* 1 mg/kg followed by a continuous infusion of 2.5 mg/kg per hour for 4 hours, and if needed, an additional IV infusion of 0.2 mg/kg per hour for up to 20 hours. In patients with unstable angina, a continuous intravenous infusion of 0.2 mg/kg per hour has been administered. *Angiomax* is intended for use with aspirin (300–325 mg daily) and has been studied only in patients receiving concomitant aspirin. Treatment should be initiated just prior to coronary angioplasty. The dose of Angiomax may need to be reduced, and anticoagulation status monitored, in patients with renal impairment.

D. Pharmacology and pharmaceutics
Clinical pharmacology Thrombin is a serine proteinase that plays a central role in the thrombotic process, acting to cleave fibrinogen into fibrin monomers and to activate factor XIII to factor XIIIa. The development of bivalirudin was based on a knowledge of the active residues of hirudin required for binding to thrombin. Studies

have demonstrated that bivalirudin can inactivate both soluble and clot-bound thrombin, and the drug has shown effective antithrombotic activity. When bound to thrombin, bivalirudin inhibits all effects of thrombin, including activation of platelets, cleavage of fibrinogen, and activation of the positive amplification reactions of thrombin.

Pharmacokinetics   The plasma half-life of intravenous bivalirudin is 25 minutes but increases with decreasing glomerular filtration rate. According to the manufacturer, the total body clearance of bivalirudin is 3.4 ml/min per kg but decreases with decreasing renal function.

Disposition   Elimination of bivalirudin is predominantly via metabolic clearance; approximately 20% of a dose is excreted unchanged in the urine. It is speculated that metabolism of bivalirudin may occur in the kidneys and via proteolysis at other sites. Bivalirudin binds only to thrombin; it does not bind to plasma proteins or to red blood cells.

Drug interactions   The safety and effectiveness of *Angiomax* have not been established when used in conjunction with platelet inhibitors other than aspirin.

E. Therapeutic response   Intravenous bivalirudin produces a rapid and dose-dependent prolongation of the activated partial thromboplastin time (aPTT), prothrombin time, activated clotting time (ACT), and thrombin time.

F. Role in therapy   The potential advantages of *Angiomax*—a direct thrombin inhibitor—over heparin include activity against clot-bound thrombin, more predictable anticoagulation, and no inhibition by components of the platelet release reaction. The place in therapy of *Angiomax* will be determined by further comparisons with heparin, low-molecular weight heparins, and recombinant hirudin.

G. Other applications   *Angiomax* has shown efficacy in unstable angina and as an adjunct to streptokinase in acute myocardial infarction; it has been at least as effective as high-dose heparin in coronary angioplasty. One study has suggested the efficacy of subcutaneous *Angiomax* in preventing deep-vein thrombosis in orthopedic surgery patients.

EPTIFIBATIDE

Trade name: *Integrilin*

Manufacturer: COR Therapeutics, S. San Francisco, California

Marketed by: COR Therapeutics, Inc., South San Francisco, CA, and

Key Pharmaceuticals, Inc., Kenilworth, NJ

Indications: Treatment of patients with acute coronary syndrome, including patients who are to be managed medically and those undergoing percutaneous coronary intervention (PCI).

Approval date: 18 May 1998

Type of submission: New drug application

A. General description   Eptifibatide is a cyclic heptapeptide containing six amino acids and one mercaptopropionyl residue. An interchain disulfide bridge is formed between the cysteine amide and the mercaptopropionyl moieties. Eptifibatide binds to the platelet receptor glycoprotein (gp) IIb/IIIa of human platelets and inhibits platelet aggregation. The eptifibatide peptide is produced by solution-phase peptide synthesis, and is purified by preparative reverse-phase liquid chromatography and lyophilized.

B. Indications and use   *Integrilin* is approved for the treatment of patients with acute coronary syndrome (ACS), including patients who are to be managed medically and those undergoing percutaneous coronary intervention (PCI). For ACS patients, *Integrilin* has been shown to decrease the rate of a combined end point of death or new myocardial infarction. The peptide is also indicated for the treatment of patients undergoing PCI. For PCI patients, *Integrilin* has been shown to decrease the rate of a combined end point of death, new myocardial infarction, or need for urgent intervention.

C. Administration

Dosage form *Integrilin* is supplied as a sterile solution in 10 ml vials containing 20 mg of eptifibatide and 100 ml vials containing either 75 mg of eptifibatide or 200 mg of eptifibatide.

Route of administration *Integrilin* is administered intravenously by constant rate infusion, preceded by an intravenous bolus dose.

Recommended dosage and monitoring requirements According to Micromedex, the recommended dose of *Integrilin* in adults presenting with an ACS is 180 µg/kg via intravenous bolus given over 1 to 2 minutes, followed by a continuous intravenous infusion of 2 µg/kg per minute until hospital discharge or initiation of coronary artery bypass grafting (CABG) surgery, but no more than 72 hours. In patients undergoing PCI and not presenting with an ACS, the recommended dose is an intravenous bolus of 135 µg/kg administered over 1 to 2 minutes immediately before the initiation of PCI followed by a continuous intravenous infusion of 0.5 µg /kg per minute for 20 to 24 hours. In the clinical studies of *Integrilin*, most patients received heparin and aspirin.

D. Pharmacology and pharmaceutics

Clinical pharmacology Eptifibatide is a glycoprotein (gp) IIb/IIIa receptor (platelet fibrinogen receptor) antagonist. It selectively and reversibly inhibits the final common pathway of platelet aggregation (binding of fibrinogen, von Willebrand factor, and other ligands to gp IIb/IIIa) and, thereby, may be able to reverse ischemic states produced by platelet thrombosis.

Pharmacokinetics According to the product label, the pharmacokinetics of eptifibatide are linear and dose proportional. Plasma elimination half-life is approximately 2.5 hours. The extent of eptifibatide binding to human plasma protein is about 25%; its mean volume of distribution is 185 ml/kg. Clearance in patients with coronary artery disease is 55–58 ml/kg per hour. Clinical studies have included 2418 patients with serum creatinine between 1.0 and 2.0 mg/dl without dose adjustment. No data are available in patients with more severe degrees of renal impairment, but plasma eptifibatide levels are expected to be higher in such patients. Patients in clinical studies were older than the subjects in clinical pharmacology studies, and they had lower total body eptifibatide clearance and higher eptifibatide plasma levels. Men and women showed no important differences in the pharmacokinetics of eptifibatide.

Disposition In healthy subjects, renal clearance of eptifibatide accounts for approximately 50% of total body clearance, with the majority of the drug excreted in the urine as eptifibatide, deamidated eptifibatide, and other, more polar metabolites. No major metabolites have been detected in human plasma.

Drug interactions While *Integrilin* has been studied and used in combination with heparin and aspirin, there is an increased risk of bleeding when eptifibatide is used with drugs that affect hemostasis. Patients should be closely monitored for signs and symptoms of active bleeding. Concurrent administration of clopidogrel, dipyridamole, nonsteroidal anti-inflammatory agents, oral anticoagulants, or ticlopidine and *Integrilin* may produce enhanced anticoagulation, leading to an increased risk of internal and superficial hemorrhage.

Because the limited data do not allow an estimate of the bleeding risk associated with eptifibatide and a thrombolytic agent, systemic thrombolytic agents should be used with caution in patients who have received *Integrilin*.

E. Therapeutic response In human studies, eptifibatide inhibited *ex vivo* platelet aggregation induced by adenosine diphosphate (ADP) and other agonists in a dose- and concentration-dependent manner. The effect of eptifibatide was observed immediately after administration of a 180μg/kg intravenous bolus. In a placebo-controlled study of patients with acute coronary syndrome, *Integrilin* reduced the occurrence of death from any cause or new myocardial infarction. Similar benefits were observed in patients undergoing coronary angioplasty.

F. Role in therapy *Integrilin* appears to be an effective antiplatelet agent and may be applicable in a wider range of patients than tirofiban (*Aggrastat*). However, comparative data are limited, and further data, especially data comparing other glycoprotein IIb/IIIa receptor antagonists, are needed to determine the place of *Integrilin* in therapy.

G. Other applications In combination with streptokinase, *Integrilin* improved coronary patency in patients experiencing an acute myocardial infarction but had no effect on clinical outcome and increased the rate of bleeding.

DARBEPOETIN ALFA

Trade name: *Aranesp*

Manufacturer: Amgen Inc., Thousand Oaks, CA

Indications: Treatment of anemia associated with chronic renal failure, including patients on dialysis or not on dialysis

Approval date: 17 September 2001

Type of submission: Biologics license application

A. General description Darbepoetin alfa is an erythropoiesis-stimulating protein, closely related to erythropoietin, that is produced in Chinese hamster ovary (CHO) cells by recombinant DNA technology. Darbepoetin is a 165 amino-acid protein that differs from recombinant human erythropoietin in containing five N-linked oligosaccharide chains, whereas recombinant human erythropoietin contains three. The two additional glycosylation sites result from amino-acid substitutions in the erythropoietin peptide backbone. The additional carbohydrate chains increase the molecular weight of the glycoprotein from 30,000 to 37,000 daltons.

B. Indications and use *Aranesp* is indicated for the treatment of anemia associated with chronic renal failure, including patients on dialysis and patients not on dialysis.

C. Administration

   Dosage form *Aranesp* is formulated as a sterile, colorless, preservative-free protein solution. Single-dose vials are available containing 25, 40, 60, 100, or 200μg darbepoetin in a 1 ml solution. Two formulations are available: one contains polysorbate solution and the other contains albumin solution.

   Route of administration *Aranesp* is intended for intravenous or subcutaneous administration as a single weekly injection.

   Recommended dosage and monitoring requirements The dose of *Aranesp* should be started and slowly adjusted based on hemoglobin levels. Hemoglobin should be followed weekly until stabilized and monitored at least monthly thereafter. *Aranesp* dosage should be adjusted to maintain a target hemoglobin not to exceed 12 g/dl.

D. Pharmacology and pharmaceutics

Clinical pharmacology Darbepoetin stimulates erythropoiesis in the same way as endogenous erythropoietin, which is made in the kidneys and released into the bloodstream in response to hypoxia. It is a primary growth factor for erythroid development. Production of endogenous erythropoietin is impaired in patients with chronic renal failure, and erythropoietin deficiency is the primary cause of their anemia.

Pharmacokinetics Darbepoetin has a half-life approximately three times longer than recombinant erythropoietin (*Epoetin alfa*) when given either intravenously or subcutaneously. The mean terminal half-life of darbepoetin following intravenous administration is about 21 hours. Following subcutaneous administration, absorption is slow and rate limiting; the terminal half-life is 49 hours; bioavailability is approximately 37%. There is no evidence of nonlinearity in the therapeutic dosing range. With once-weekly dosing, steady-state serum levels are reached within 4 weeks.

Metabolism No information available.

Drug interactions No formal drug interaction studies of darbepoetin with other medications commonly used in chronic renal failure patients have been performed.

E. Therapeutic response Two studies have evaluated the efficacy of darbepoetin for the correction of anemia in adult patients with chronic renal failure. In one study, the hemoglobin target was reached by 72% of dialysis patients treated with darbepoetin and 84% treated with recombinant erythropoietin. In the other, the primary end point was achieved by 93% of predialysis patients treated with darbepoetin and 92% of patients treated with recombinant erythropoietin.

F. Role in therapy According to Micromedex, the advantage of darbepoetin over recombinant human erythropoietin is its longer half-life and less-frequent dosing requirements. The reduced dosing frequency of darbepoetin should enable cost savings in dialysis settings through a decrease in patient visits to the clinic and in nursing time required. A disadvantage of darbepoetin is a higher frequency of subcutaneous injection-site pain. Darbepoetin should be considered over epoetin alfa if its acquisition cost is not prohibitory.

G. Other applications Darbepoetin is under investigation for the treatment of anemia associated with chemotherapy and advanced HIV infection.

PEGFILGRASTIM

Trade name: *Neulasta*

Manufacturer: Amgen Inc., Thousand Oaks, CA

Indications: Reduction of incidence of infections in patients receiving cancer chemotherapy

Approval date: 31 January 2002

Type of submission: Biologics license application

A. General Description Pegfilgrastim is a covalent conjugate of filgrastim—recombinant methionyl human granulocyte colony-stimulating factor (G-CSF)—and monomethoxypolyethylene glycol. Filgrastim is a water-soluble 175 amino-acid protein with a molecular weight of approximately 19 kDa. It is obtained from the bacterial fermentation of a strain of *E. coli* transformed with a genetically engineered plasmid containing the human G-CSF gene. A 20 kDa monomethoxypolyethylene glycol molecule is covalently bound to the N-terminal methionyl residue of filgrastim to produce pegfilgrastim.

B. Indications and use   *Neulasta* is indicated to decrease the incidence of infection, as manifested by febrile neutropenia, in patients with nonmyeloid malignancies receiving myelosuppressive anticancer drugs associated with a clinically significant incidence of febrile neutropenia.

C. Administration

Dosage form   *Neulasta* is supplied in 0.6 ml prefilled syringes. Each syringe contains pegfilgrastim 6 mg (based on protein weight) and excipients in a sterile solution.

Route of administration   *Neulasta* is administered by subcutaneous injection. A physician may determine that a patient or caregiver can safely administer *Neulasta* at home.

Recommended dosage and monitoring requirements   The recommended dosage of *Neulasta* is a single subcutaneous injection administered once per chemotherapy cycle. *Neulasta* should not be administered in the period between 14 days before and 24 hours after administration of cytotoxic chemotherapy.

D. Pharmacology and pharmaceutics

Clinical pharmacology   Filgrastim and pegfilgrastim are colony-stimulating factors that act on hematopoietic cells by binding to specific cell surface receptors. They stimulate proliferation, differentiation, commitment, and end cell function activation. They both have the same mechanism of action. Pegfilgrastim has reduced renal clearance and prolonged persistence in the body as compared with filgrastim.

Pharmacokinetics   The pharmacokinetics of pegfilgrastim were studied in 379 patients with cancer. The pegylated version of filgrastim demonstrated nonlinear pharmacokinetics, and clearance decreased with increases in dose. Neutrophil receptor binding is an important component of pegfilgrastim clearance, and serum clearance is directly related to the number of neu-

trophils. Consequently pegfilgrastim concentrations declined rapidly at the onset of neutrophil recovery that followed myelosuppressive chemotherapy. The half-life of pegfilgrastim varied considerably in cancer patients, ranging from 15 to 80 hours after subcutaneous injection. No gender- or age-related difference in pegfilgrastim pharmacokinetics was discerned.

Metabolism   No information available.

Drug interactions   No formal drug interaction studies between *Neulasta* and other drugs have been performed.

E. Therapeutic response   *Neulasta* was evaluated in two well-controlled studies employing doxorubicin and docetaxel for the treatment of metastatic breast cancer. The first study investigated the utility of a fixed dose of *Neulasta* and the second study employed a weight-adjusted dose. The investigators anticipated that the chemotherapy regimen would result in a 100% incidence of severe neutropenia with a mean duration of 5 to 7 days, and a 30% to 40% incidence of febrile neutropenia. The duration of severe neutropenia was the primary end point in both studies. The efficacy of *Neulasta* was demonstrated by establishing comparability with filgrastim (*Neupogen*). Both studies determined that the mean days of severe neutropenia of *Neulasta*-treated patients did not exceed that of *Neupogen*-treated patients by more than 1 day in cycle 1 of chemotherapy. The mean days of severe neutropenia was less than two in each study, irrespective of treatment.

F. Role in therapy   According to the manufacturer, until now, filgrastim was the only prescription drug shown to decrease risk of infection and hospitalization as a result of chemotherapy-induced neutropenia. The burden of daily dosing (up to 14 consecutive days), however, led many physicians to delay intervention until a patient developed a neutropenic infection.

The less-frequent dosing of *Neulasta* means that patients will require fewer painful injections and fewer office visits. A clinical trial investigator observed, "This approval means that hundreds of thousands of chemotherapy patients at risk for infection may now receive *Neulasta* as protection at the onset of each treatment cycle.

7

INTERFERONS AND CYTOKINES FOR ANTI-INFECTIVE AND CANCER THERAPY

■ **SECTION ONE** ■

Mamta Chawla-Sarkar, Paul Masci, and Ernest Borden

■ 7.1. INTERFERONS IN CANCER THERAPY

Forty-four years have passed since the initial discovery of interferon (IFN) [1]. Chick chorioallantoic membranes incubated with heat-inactivated influenza virus produced a substance that conferred resistance to infection by live virus [2,3]. Termed "interferon," this substance was later shown to have properties of a protein and to consist of a family of distinct proteins, many of which are structurally related in both amino acid sequence and three-dimensional structure [3,4]. Based on numerous investigations with recombinant IFNs, it is now known that IFNs are pleiotropic cytokines that demonstrate diverse immunomodulatory and antiproliferative properties in addition to their antiviral effects [5,6]. These various biological properties suggested different potential clinical applications. Recombinant biotechnology facilitated full definition of

Biotechnology and Biopharmaceuticals, by Rodney J. Y. Ho and Milo Gibaldi
ISBN 0-471-20690-3 Copyright © 2003 by John Wiley & Sons, Inc.

161

■**TABLE 7.1.** Family of human interferon molecules

Family	Human Chromosome	Number of Genes	AA Number[a]	Homology[b]
Alpha	9	13	165–166	75–85%
Beta	9	1	166	29
Gamma	12	1	173	50
Omega	9	> 6	172	30
Kappa	9	1	207	30

[a]AA = amino acid.

[b]Amino-acid homology compared with αIFN.

the diversity of the IFN family, molecules now differentiated by their primary amino acid sequences [4] (Table 7.1).

By 1981 recombinant DNA technology enabled the introduction and large-scale testing of interferons in various human clinical trials. The availability of gram quantities of recombinant IFNs has permitted their evaluation as therapeutic agents in systematic trials, initially in cancer and then for chronic viral infections and multiple sclerosis. The outcomes of these trials resulted in FDA approval for the use of interferon in more than 14 malignancies including adjuvant therapy in metastatic melanoma, AIDS-related Kaposi's sarcoma, hairy-cell leukemia, and viral diseases such hepatitis B and C. The use of recombinant interferons for viral infection is discussed in Section 7.2. IFN-α was the first of these biologically active proteins to be cloned and expressed in bacteria and became the first biotherapeutic agent produced by recombinant DNA technology to be approved by the US Food and Drug Administration (FDA) for the treatment of cancer. In addition, it has led to introduction of second-generation interferons in clinical trials.

Studies of IFN expression, structure, and function have provided a number of scientific advances. The study of transcription of IFN-stimulated genes (ISGs) and analysis of IFN-induced proteins have provided not only a tool to observe the complexities of inducible gene expression in eukaryotic cells but also led to identification of important regulators of physiological processes.

However, the various mechanisms of action of human IFNs are still not completely understood [6,7].

This section describes the role of interferons as cancer drug therapies, emphasizing key contributions of recombinant biotechnology that made these therapeutic advances a reality. Additionally, some of the mechanisms by which the IFNs elicit antiproliferative or antitumor effects will be explored.

7.1.1. The Interferon Proteins and Genes

Natural IFNs

In 1961 it was reported that human leukocytes were capable of producing IFN in response to viral infections [8,9]. This viral stimulation of white blood cells was initially used to produce "leukocyte IFN" for clinical applications. Identification of a number of varied IFN inducers such as mycoplasma or other microorganisms in cell cultures, lipopolysaccharides (LPS, derived from bacteria membranes), tumor-derived or virus-transformed cells, and synthetic chemical compounds such as polyanions and poly I:C (poly inosine-cytosine) suggested that different IFN mixtures could be derived from interaction of various inducing agents and appropriate target cells [10–16]. Another pH-labile, nonvirus-induced IFN termed "immune-IFN" (induced by immune effector cells) was discovered in 1965. It was produced by human T cells in response to mitogens and other foreign antigens [17]. Similarly another IFN, fibroblast-derived IFN (FIF), was identified and purified from fibroblasts following infection with virus or exposure to poly I:C [10,18]. Since each of the various IFNs that were derived from different cells exhibited antiviral activity, they were initially considered to be the same protein. However, the finding that leukocyte IFN could not be neutralized by antisera against FIF suggested heterogeneity of IFN proteins [19,20].

The IFN used in early clinical trials was a crude protein fraction containing less than 1% IFN by weight [21]. IFNs were purified to homogeneity in solution in amounts sufficient to allow chemical and physical characterization in 1978. The presence of multiple species of IFN became evident during purification of leukocyte IFN using high-pressure liquid chromatography (HPLC). Initially, the observed differences in crude human leukocyte IFN preparations were thought to be due to differing amounts of protein glycosylation, but subsequent amino acid analysis demonstrated 10 separate species of proteins with molecular weights ranging from 15 to 23 kDa. No carbohydrate was detected on any subspecies, hence confirming molecular heterogeneity to be a result of multiple IFN proteins [22–24].

A new nomenclature was therefore designed to describe the subtypes of IFN rather than naming them for the source from which they were originally isolated. The designation of "huIFN-α" was suggested for the major class of IFNs present in leukocyte or lymphoblastoid preparations. "huIFN-β" was used to denote IFN produced by fibroblasts, and "huIFN-γ" was assigned to human immune IFN produced by activated T-lymphocytes [25] (Table 7.2).

Recombinant Human IFNs

Until the development and adaptation of recombinant DNA technology, less than one milligram of pure IFN had been previously isolated. Recombinant DNA technology offered an economic method to produce large amounts of purified huIFNs, once the purification and amino-acid composition of the various species of IFNs was elucidated. The first biologically active interferon-like protein was produced in 1980 from bacteria, after sifting through 20,000 different genetic fragments [26]. The IFNs produced by these clones were designated α1 and α2. The availability of recombinant leukocyte IFN prompted collaboration between biotechnology companies and universities and enabled more complete and precise definition of IFNs as a family of more than a dozen proteins [22–24,27]. Elucidation of the coding regions of IFN-α genes revealed that the heterogeneity in huIFN-α was not due to variations in glycosylation but was due to presence of distinct genes representing each expressed huIFN-α sequence [4]. The complete nucleotide sequences of huIFN-α2 and huIFN-β were determined and cloned in 1981 [27–29].

The antiviral and antiproliferative effects of individual IFN-α subspecies differed from buffy-coat IFN-α [30,31]. Unlike recombinant purified IFN-α2, which is species specific, buffy-coat huIFN-α has broad species effects, producing responses in humans, bovines, and felines. The clinically observed side effects of purified recombinant IFN-α2 and naturally produced buffy-coat IFN, however, were noted to be similar [32,33].

7.1.2. Classification and Biological Properties

Mammalian IFNs are broadly classified into two structurally discrete categories. The type I IFNs include the ubiquitous IFN-α, IFN-β, and IFN-ω subtypes, and the species-restricted IFN-τ. The only type II IFN is IFN-γ. Type I and II IFNs differ in their primary amino-acid sequences, evolu-

■TABLE 7.2. Nomenclature of human interferons

Types	Gene	Protein	AA Residues Distinguishing the Subtype	Primary Cell of Origin
Type I IFN[a]	*IFNA1*	IFN-α1a	Ala-114	Dendritic cells
		IFN-α1b	Val-114	
	IFNA2[b]	IFN-α2a	Lys-23, His-34	
		IFN-α2b	Arg-23, His-34	
		IFN-α2c	Arg-23, Arg-34	
	IFNA4	IFN-α4a	Ala-51, Glu-114	
		IFN-α4b	Thr-51, Val-114	
	IFNA5	IFN-α5		
	IFNA6	IFN-α6		
	IFNA7	IFN-α7a	Met-132, Lys-159, Gly-161	
		IFN-α7b	Met-132, Gln-159, Arg-161	
		IFN-α7c	Thr-132, Lys-159, Gly-161	
	IFNA8	IFN-α8a	Val-98, Leu-99, Cys-100, Asp-101, Arg-161	
		IFN-α8b	Ser-98, Cys-99, Val-100, Met-101, Arg-161	
		IFN-α8c	Ser-98, Cys-99, Val-100, Met-101, Asp-161	
	IFNA10	IFN-α10a	Ser-8, Leu-89	
		IFN-α10b	Thr-8, Ile-89	
	IFNA13	IFN-α13		
	IFNA14	IFN-α14a	Phe-152, Gln-159, Arg-161	
		IFN-α14b	Phe-152, Lys-159, Gly-161	
		IFN-α14c	Leu-152, Gln-159, Arg-161	
	IFNA16	IFN-α16		
	IFNA17	IFN-α17a	Pro-34, Ser-55, Ile-161	
		IFN-α17b	His-34, Ser-55, Ile-161	
		IFN-α17c	His-34, Ser-55, Arg-161	
		IFN-α17d	His-34, Pro-55, Arg-161	
	IFNA21	IFN-α21a	Met-96	
		IFN-α21b	Leu-96	
		IFN-α24		
	IFNB	IFN-β		Fibroblasts
	IFNW1	IFN-ω		
Type II IFN	*IFNG*	IFN-γ		T cells

Adapted from Ref. [25].

[a]IFN = Interferon.

[b]IFN-α2 has 165 amino acids and differs from the other α-IFNs by a deletion of an aspartate residue at position 44. IFN-α2a (Hoffman-LaRoche) and IFN-α2b (Schering) are the two recombinant human IFNs licensed for clinical use in oncology. They differ by a single amino-acid substitution at position 23 (lysine in IFN-α2a and arginine in IFN-α2b).

tionary relationships, receptor cross reactivity, sites of production, and inducibility. Nonetheless, type I and II IFNs induce overlapping sets of responsive genes and activate similar signaling cascades.

Type I Interferons

Type I IFN genes have evolved from a common primordial progenitor and are arrayed on the short arm of chromosome

9 in humans [34–36]. The IFN-β gene has remained as a single copy gene in most species, whereas the IFN-α prototypic gene has subsequently been amplified to a higher number of copies. Comparison of the sequences of IFN-α and IFN-β has demonstrated approximately 45% homology of the encoding nucleotides and 29% homology of subsequently translated amino acids. Type I IFNs are part of the body's first line of defense against viruses, foreign cells (including tumor cells), microbes, and other pathogens [36]. Cells are capable of producing substantial quantities of IFNs in response to virus, bacteria, and other cytokines [37]. The exact IFN subtype produced depends on the inducer and the cell type. Unlike acquired immune responses, the type I IFN response is nonspecific and very rapid, usually peaking within several hours of initial infection [38]. Type I IFNs can also activate components of the immune system, enhance T-cell toxicity, and up-regulate class I major histocompatibility complex (MHC) antigens [39].

IFN-α Family. There are 14 human genes that comprise the IFN-α family. These genes generally encode mature polypeptides with either 165 or 166 amino acids and have a 15% to 25% variation in primary amino-acid sequence. Minor variants consisting of one or two amino-acid differences account for multiple alleles. Thirteen separate proteins are expressed from these 14 genes (Table 7.1), each with a different profile of antiviral and antiproliferative activity [40,41]. Evidence for cell specificity of IFN-α subtypes has been reported. For example, IFN-α1 has greater antiproliferative effects on B-cell neoplasms, whereas IFN-α2 is more specific for other malignancies [42]. When assayed on human cells, IFN-α1 has less antiviral activity than IFN-α2 [43]. Nevertheless, IFN-α1 has potent antiviral activity in bovine-, feline-, and primate-derived cells [44]. Fur-

thermore IFN-α7 can act as an antagonist of natural killer (NK) cell stimulatory activity [3,4].

The recombinant IFN licensed internationally for use in oncology is IFN-α2. The two recombinant types of IFN-α2 used clinically are IFN-α2a (*RoferonA*) and IFN-α2b (*IntronA*). They differ by a single amino-acid substitution at residue 23— lysine in IFN-α2a and arginine in IFN-α2b. A second major interferon-α termed "IFN-α1" is a recombinant subtype used clinically in China. A randomized clinical trial has found IFN-α1 and IFN-α2 to be equipotent in stimulating NK cell cytotoxicity as well as in increasing intracellular 2–5A synthetase activity. Patients receiving IFN-α1, however, experienced less fever, fatigue, and gastrointestinal disturbance when compared to patients receiving IFN-α2 [45].

Hybrid and Consensus IFNs. After the sequences of all IFN-α subtypes and their activities were deduced, recombinant hybrids were designed to improve the activity of natural subtypes. It is possible to radically change the properties of an IFN by making hybrids. For example, a hybrid made from the first 61 amino-acid residues of IFN-α1 and the last 104 amino-acid residues of IFN-α2 exhibited high antiviral activity on mouse cells, while another hybrid produced from the first 91 amino-acid residues of IFN-α1 and the last 72 amino-acid residues of IFN-α4 had no antiviral effect but displayed enhanced antitumor activity in mice, a property neither of the parent molecules exhibit. Structure-function analysis of hybrid molecules has indicated that the NH_2 terminus portion of the IFN-α molecule is important for its biological activity [46].

Consensus IFN (IFN-con1) is a synthetic nonnaturally occurring IFN-α hybrid designed by assigning each of the most frequently observed amino acids in the Hu-IFN-α subtypes. For the most part it is a hybrid between IFN-α2 and IFN-α21 with only five amino acid differences that were introduced for the purposes of cloning. The host range of IFN-con1 was found to be similar to IFN-α2b. The biological activity of consensus IFN (2×10^9 U/mg) is reported to be 5- to 20-fold higher than other IFN-α species ($1-4 \times 10^8$ U/mg). Human clinical trials of IFN-con1 for the treatment of hepatitis C revealed similar efficacy to that of IFN-α2a and IFN-α2b when compared on an antiviral unit basis. However, when compared on a mass basis, IFN-con1 had greater activity in terms of its antiviral, antiproliferative, NK cell activation ability, and interferon-induced gene up-regulation activity when compared to IFN-α2. This more potent activity of IFN-con1 was attributed to its stronger affinity for receptors [47–49].

Pegylated IFN. Polyethylene glycol (PEG) conjugated biomolecules have shown superior clinical properties compared to their unconjugated counterparts [50]. This has been attributed to higher thermal and physical stability, greater protection against proteolytic degradation, higher solubility, longer *in vivo* circulation half-lives, and lower clearance of PEG conjugated molecules. Additional qualities of pegylated proteins are reduced immunogenicity and antigenicity. Since the reported half-life of IFN-α ranges from 4 to 8 hours, and little or no IFN-α is detected in serum 24 hours following administration, daily injections of IFN-α are required to achieve sustained efficacy. Since the early 1990s attempts have been made to develop a pegylated form of IFN-α. Initially 5 kDa PEG was linked to IFN-α2a, but it was abandoned when early phase II trials demonstrated inferior efficacy when compared to IFN-α2a [51]. Recently two more PEG-IFNs have been developed, 40 kDa branched PEG conjugated to IFN-α2a (*Pegasys*: Hoffman-LaRoche) and a 20 kDa PEG conjugated to IFN-α2b (*PEG-Intron*: Schering) [52,53]. Preclinical studies have shown a promising enhanced human pharmacological profile for PEG-IFNα as compared to IFN-α2 [52,53].

IFN-β. Human IFN-β is the predominant subtype expressed by most somatic cells, including fibroblasts and epithelial cells. IFN-β gene is a single copy gene in humans. Natural IFN-β is a 166 amino-acid protein with a single N-linked glycosylation at amino-acid residue 80. This glycosylation is not necessary for most of its biological activity.

A recombinant IFN-β, IFN-β1a (*Rebif*: Serono and *Avonex*: Biogen) is produced for commercial use in Chinese hamster ovary (CHO) cells. A synthetic mutant produced in bacteria has a substitution of serine for cysteine at amino acid 17, yielding IFN-β1b (*Betaseron*: Berlex). Both IFN-β1a and IFN-β1b are approved by the FDA for treatment of multiple sclerosis (MS) and have shown comparable biological activity (see Section 7.3). *In vivo* IFN-α2 and IFN-α1b show comparable biological activity as well as similar side effects [54,55]. However, IFN-β is eliminated faster, resulting in no detectable serum peak levels [56]. The clinical consequence of this is not known. Objective responses, whether partial or complete tumor regression, have been documented in patients with carcinoma of the breast, hairy-cell leukemia, and non-small-cell lung cancer [57,58].

Although IFN-α and IFN-β have approximately 30% amino acid homology, antisera to IFN-α and IFN-β do not cross-react. Interestingly both compete for the same cellular receptor. *In vitro* studies demonstrate greater antiproliferative effects of IFN on different nonhematopoietic cell lines and fibroblasts compared to IFN-α2 [59,60]. However, on lymphoid-derived (Daudi) cell lines or other myeloid progenitor cells, IFN-α displayed greater antiproliferative effects [59]. The antiproliferative effects of IFN-β have been shown to be in part due to induction of apoptosis in melanomas and myelomas [60,61]. The apoptotic potential of IFN-β correlated with preferential induction of the "tumor necrosis factor related apoptosis inducing ligand" molecule (TRAIL/Apo2L) by IFN-β when compared to IFN-α2 in melanoma cell lines [60]. Clinically IFN-α is largely ineffective in treating patients with MS [62]. The mechanisms behind these differential effects of IFN-α2 and IFN-β are still poorly defined.

Other IFNs (IFN-ω, IFN-τ, and IFN-κ). IFN-ω evolved from IFN-α approximately 100 million years ago, probably before amplification of prototypic IFNα species [63]. Like IFN-α, IFN-ω genes have duplicated to high copy numbers in ruminants but not in humans [64]. Unlike IFN-α, IFN-ω protein is 172 amino-acid residues long and is capable of eliciting similar biological responses *in vitro* as compared to other type I IFNs [36].

The IFN-τ genes first identified in ruminants probably arose about 35 million years ago as a result of duplication of IFN-ω. IFN-τ is secreted by early placenta tissue (i.e., trophoblasts) where it is proposed to function in establishing pregnancy. It also possesses biological activities of other type I IFNs, including antiviral and immunomodulation [64]. Like IFN-ω, mature IFN-τ protein is 172 amino acids in length and interacts with the common type I receptor [65]. A recombinant ovine IFN-τ has been reported effective in treatment of multiple melanoma and AIDS virus [66,67]. Clearly, a hu-IFN-τ, lacking potential immunogenicity of ovine protein, would be of therapeutic value; however, no such human gene has yet been identified [68].

A novel IFN (IFN-κ) with 30% homology to other type I proteins has been identified in epidermal keratinocytes. IFN-κ gene is located on chromosome 9, and its protein consists of 207 amino acids. It is induced in keratinocytes upon viral infection, exposure to double stranded RNA, IFN-γ, or IFN-β. It imparts cellular protection against viral infections in a species- and tissue-specific manner, utilizing the same receptors and signaling pathways as other type I IFNs [69].

Type II IFNs: IFN-γ

The type II *IFN* gene evolved from a distinct progenitor that seems to have appeared in more recent times. Unlike type I *IFN* genes, *IFN-γ* is more typical of eukaryotic genes in that it consists of four exons and three 3 introns (*IFN-α* and *IFN-β* have no introns in their respective gene transcripts). It is located on chromosome 12 [70]. The mature protein is 143 amino acids in length, glycosylated, and shares very little homology with the type I IFNs [3,36]. IFN-γ induces macrophages to express cytokines (IL-12, TNF-α), class I and II MHC molecules, and Fc receptors, thereby modulating the host defense system [6,71]. Therefore IFN-γ is often described as an immunomodulatory cytokine and has no direct antiviral-inducing activity [72].

7.1.3. Mechanisms of Action

Receptors

Cellular responses to IFNs require interactions between a small number of stimulatory molecules and high-affinity, multimeric cell surface receptors that exhibit species selectivity. As such, IFN-α and IFN-β share and compete for the same receptor, although IFN-β is capable of binding with greater affinity [73]. One or more species-specific protein components cooperate with IFN receptors to confer species restrictions and mediate biological effects [74,75]. To elicit a cellular response, the IFN-α and IFN-β receptors require two subunits, termed IFNAR-1 and IFNAR-2. IFN-γ, however, binds to a different receptor complex consisting of two subunits, IFNGR-1 and IFNGR-2. Signals are transmitted through the membrane to the cytoplasmic domains of the receptors to

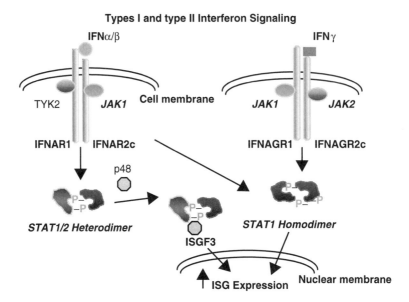

Figure 7.1. Intracellular pathway of signal transduction by interferons (IFN). Type I and II IFNs bind to specific transmembrane receptors. This leads to activation of receptor-associated tyrosine kinases, JAK-1, JAK-2 and Tyk-2 (JAK = Janus kinase; Tyk = tyrosine kinase). Activation of these tyrosine kinases leads to the phosphorylation of inactive transcription factors known as STATs (STAT = signal transducers and activators of transcription). STAT-1 and STAT-2 may form hetero or homodimers. The STAT-1 homodimer translocates to the nucleus and binds to specific gamma activated sequences of the DNA and initiates transcription of IFN-γ stimulated genes (ISGs). In the case of type I IFNs, the STAT-1/2 heterodimer is joined by p48 protein to become ISGF-3 (Interferon stimulated gene factor 3). ISGF-3 binds to a specific DNA sequence or interferon response element (ISRE) to initiate transcription. The IFN receptors are transmembrane paired proteins (IFNAR = IFN alpha receptor; IFNGR = IFN gamma receptor).

mediate differential response (Figure 7.1). IFN-β appears to interact with a receptor heterodimer in a different way than IFN-α [76]. This can result in activation of a selective subset of genes, even though IFN-β binds and activates the same receptor [7,77].

Signal Transduction

Following binding of IFN-α or IFN-β to the receptor, a specific tyrosine kinase, tyk2, as well as tyrosine kinases JAK-1 and JAK-2, are phosphorylated. These activated tyrosine kinases modulate signal-transducing peptides (Figure 7.1) and induce the formation of a complex of protein subunits (ISGF3-α) consisting of STAT (signal transducers and activators of transcription)-1α or STAT-1β, STAT2, and p48

(IRF-9). The phosphorylated ISGF3 complex translocates to the nucleus where it binds to the IFN-stimulated response element (ISRE) of DNA, resulting in transcription of IFN-specific genes. IFN-γ, IFN-α2, or IFN-β form another gene activation complex, STAT1 homodimer, which binds to consensus sequences of DNA termed "gamma-activated sites" (GAS) [6,78]. Other signaling cascades are also activated by IFNs, including phosphatidylinositol-3-kinase and mitogen-activated protein (MAP) kinase cascade [79,80].

7.1.4. Overview of Antitumor Effects in Humans

Two properties of IFNs—potent antiviral and antitumor activities—make them very

■**TABLE 7.3.** Commercially available preparations of human interferon

Interferon Type	Commercial Name	FDA Approved Indications
IFN-α2b	*Intron-A* (Schering)	Melanoma, hepatitis C, CML, hairy-cell leukemia, Kaposi's sarcoma, cutaneous T-cell lymphoma
IFN-α2a	*Roferon-A* (Roche)	Chronic hepatitis C
IFN-β1a	*Avonex* (Biogen Inc.); *Rebif* (Serono)	Multiple sclerosis
IFN-β1b	*Betaseron* (Berlex)	Multiple sclerosis
IFN-γ1b	*Actimmune* (InterMune Pharmaceuticals, Inc.)	Delaying time to disease progression in patients with severe, malignant osteopetrosis.
IFN-αcon-1 (consensus IFN)	*Infergen* (Amgen Inc.)	Chronic hepatitis C

attractive candidates for the treatment of disease. The degree of activity and improvement of quality of life of patients with hairy-cell leukemia led to the first licensed approval of IFN in the United States. Clinically beneficial therapeutic activity of IFN-α2 as a single agent has been demonstrated in hairy-cell leukemia, AIDS-related Kaposi's sarcoma, chronic myelogenous leukemia (CML), B- and T-cell lymphomas, melanoma, multiple myeloma, and renal cell carcinoma. These findings have led to regulatory approvals around the world, with IFN representing the first human protein demonstrated to increase survival of cancer patients (Table 7.3).

Hematologic Malignancies

Hairy-cell leukemia (HCL) is an infrequent B-cell neoplasm characterized by the presence of small lymphocytes with numerous cytoplasmic projections. These cells are found in the peripheral blood and in bone marrow. HCL patients undergoing IFN-α2 therapy may show a decrease in bone marrow infiltration and normalization of peripheral hematologic parameters. More than 85% of patients have objective evidence of partial or complete hematologic response when treated with IFN-α2 [81,82]. Although the mechanism of action of IFN-α against HCL is incompletely understood, it most likely relates to IFN-α-induced B-cell differentiation, inhibition of hairy-cell responsiveness to B-cell growth factors, or activation of antineoplastic immune cell function [83].

Chronic myelogenous leukemia is characterized by the Philadelphia translocation—a chromosomal rearrangement of portions of chromosomes 9 and 22 [t(9:22)], leading to the formation of a chimeric protein product [84]. The disease is clinically characterized by a chronic phase in which patients exhibit thrombocytosis, leukocytosis, and splenomegaly. Invariably patients will progress to an acute or accelerated phase during which a fatal myeloid or lymphoid blast crisis ensues [83]. The median time from diagnosis to acute blastic transformation is 36 to 40 months. While the traditional alkylating agent busulfan can induce complete clinical remission (i.e., normalization of blood count and resolution of palpable splenomegaly) in 60% to 80% of patients, cytogenetic remission is rare. IFN-α2, however, has been shown to result in sustained therapeutic responses in 80% of patients. Moreover 58% of these patients achieved some degree of cytogenetic response [85]. Those with complete response had a five-year survival of 90% [85,86]. Thrombocytosis and leukocytosis can be effectively controlled by IFN-α2 in almost all patients [87,88].

IFN-α2 may have a therapeutic role in both B- and T-cell lymphomas. IFN-α2 (50 × 10^6 units/m^2 administered three times a week) was effective in 45% of patients with advanced cutaneous T-cell lymphomas, with responses lasting from 3 to 25 months [89]. For B-cell lymphomas, the significant single-agent activity of IFN-α2 has been integrated into effective combination therapy for non-Hodgkin's lymphomas [90]. Combinations of IFN-α2 and doxorubicin have been shown to prolong both disease-free and overall survival [91]. A high frequency of IFN-α2 gene deletions has been reported in post-transplant lymphoproliferative disease accompanied by a high response rate to IFN-α2 [92,93].

Solid Tumors

IFNs have clinical activity against a variety of solid tumors such as melanoma, bladder carcinoma, ovarian carcinoma, and renal cell carcinoma [94]. Response rates of different tumor types to IFN-α are variable but often equivalent to the standard-of-care chemotherapeutic agents for these neoplasms. The cumulative response from three series of patients ($n = 124$) treated with recombinant IFN-α2 was 15% (10% partial, 5% complete), a level comparable to results with cytotoxic agents [95]. Significant prolongation of disease-free survival with a trend toward increased overall survival from melanoma has been demonstrated for IFN-α2 when used as an adjuvant therapy in patients at high risk of recurrence following definitive surgery (i.e., stage IIb or III) [96]. Combinations of IFNs with hormones, chemotherapy, and/or IL-2 increase response rates in metastatic melanoma [97–100].

Response rates of 4% to 26% (mean 15%) have been achieved with single-agent

recombinant IFN-α2 in the treatment of metastatic renal cell carcinoma [101,102]. Similarly studies of IFN-γ or combinations of IFN-α2 and IFN-γ have generated response rates of 9% to 30%. A 7.5 month prolongation of overall survival was observed in patients treated with the combination of IFN-α2 and vinblastine compared with those receiving vinblastine alone [103,104]. Other combination therapies with IFNs and chemotherapeutic agents such as 5-fluorouracil (5-FU) and retinoids have shown promise in preclinical antitumor models [105].

Kaposi's sarcoma is a malignant cutaneous neoplasm believed to arise from the lymphatic endothelium [106]. Systemic IFN-α2 therapy can lead to tumor regression in 40% to 50% of patients, with 20% to 30% of patients achieving a complete response [107]. A dose-dependent effect has been observed, and responses have been seen in patients with visceral and nodal disease [108]. Clinical observations on the effectiveness of IFN-α2 in life-threatening hemangiomas of infancy further confirm the potential role of IFNs as antitumor proteins. Systemic IFN-α2 treatment not only resulted in regression of hemangiomas with a response rate of 80%, but life-threatening complications including consumptive coagulopathy and high-output cardiac failure were effectively controlled [109].

7.1.5. Mechanism of Antitumor Action

Gene Modulation

IFNs are pleiotropic cellular modulators. Antitumor effects are postulated to result from either a direct effect on functional capacity or antigenic composition of tumor cells, or from an indirect effect on the modulation of immune effector cell populations that interact with tumor cells. IFNs regulate gene expression, modulate expression of proteins on the cell surface, and activate enzymes that modulate cellular growth. On

■TABLE 7.4. Proteins induced in humans by IFN-α

Cell Surface Proteins	Protein Kinase R
HLA complex	Indoleamine dioxygenase
Class 1 (A,B,C)	Guanylate-binding proteins
Class 2 (DR,DP,DQ)	GTP cyclohydrolase
β_2-Microglobulin	Enzymes and other proteins
Tumor-associated antigens	2–5A synthetase
CEA	Cell attraction/recognition/adhesion
Apoptosis-associated proteins	K12
TRAIL	BST-2
XIAP-associated factor-1	Glectin 9 a
IL1B-converting enzyme	HM145
Plasminogen activator inhibitor	AICL
Signaling/nuclear proteins	Antigen recognition/processing/presentation
Estrogen transcription factor	Toll-like receptor
MyD88	SP-110B
Amphiphysin	HLA-E
TLS/Chop	C1 inhibitor
JunB	B cell activation

a cellular basis, these effects translate into alterations of the state of differentiation, rate of proliferation, apoptosis, and functional activity of many cell types. Induced proteins and their products can be identified on cells and in serum of treated patients (Table 7.4) Biological effects of IFNs peak at 24 to 48 hours after administration. The time to peak effect does not coincide with the time to peak concentration of IFNs in serum [110].

Oligonucleotide gene array studies of melanoma (WM9) and fibrosarcoma (HT1080) cell lines treated with IFN-α2 and IFN-β identified >100 induced genes [7,77]. This impressive impact on transcription makes it difficult to attribute the effects of IFNs to any particular gene product. A number of genes induced by type I IFNs are, however, involved directly or indirectly in apoptosis, including TRAIL/Apo2L, XAF-1, phosphokinase receptor (PKR), PML, 2–5A synthetase, and activated RNAse-L (ribonuclease-L) [77]. Apoptosis is suppressed in RNAse-L-knockout mice treated with different apoptotic agents. Expression of enzymatically inactive RNase-L inhibits antiviral and antiproliferative effects of IFNs. Similarly PKR levels correlate inversely with proliferative activity in different tumors and cell lines [111].

Antiproliferative and Apoptotic Effects

A prerequisite for unregulated growth in cancer cells is an acquired defect in one or more proteins that serve as check points for normal cell cycle progression. IFN-α and IFN-β can affect all phases of the cell cycle: M, G1, and G2 [112]. The cumulative prolongation of the cell cycle can result in cytostasis, an increase in cell size, and apoptosis [113]. IFN-α treatment results in down-regulation of cyclin D3 and cyclin D-cdk4 and cyclin D-cdk6 kinase complexes, and inhibition of hyperphosphorylation of retinoblastoma (Rb) protein, thus suppressing cell cycle progression [114–116]. IFN-α has been shown to induce apoptosis by up-regulating Fas/Fas ligand expression in multiple myeloma cells, chronic myelogenous leukemia, and gliomas [117–119]. IFN-β, although not capable of inducing Fas expression, can induce apoptosis in multiple myeloma and melanoma by induction of TRAIL/Apo2L [60,61]. Both pathways involve recruitment of death domain-containing protein, FADD, activation of the caspase cascade, and DNA fragmentation.

Immune Effector Cells

IFNs augment the effectiveness of all immune effector cell types (cytotoxic T cells, natural killer cells, and monocytes) that have the potential to eliminate tumor target cells. The ability of IFNs to augment NK-cell activity and monocyte function has been demonstrated both *in vitro* and *in vivo* [120]. One of the mechanisms by which IFNs augment T-cell and NK-cell cytotoxicity is by induction and increased expression of TRAIL/Apo2L ligand on cellular surfaces [121]. IFNs enhance antibody-dependent cell-mediated cytotoxicity (ADCC) [122]. They also increase cell surface expression of major histocompatibility (MHC) antigens, tumor-associated antigens (TAA), and Fc receptors [33,123]. An increase in MHC expression may enhance monocyte/macrophage antigen-presenting functions. IFN-activated monocytes also secrete monokines, including colony-stimulating factors (CSFs), TNF and IL-1, as well as plasminogen activator, complement, and other cytotoxic mediators [124].

Angiogenesis Inhibition

Another component of IFN-mediated antitumor effects is inhibition of angiogenesis [5]. Following treatment with IFN, tumor endothelial cells exhibit microvascular injury and a pattern of coagulation necrosis. Systemic administration of IFN-α reduces tumor cell growth in IFN-sensitive cells by directly regulating expression of the angiogenic protein, bFGF [125]. Interleukin-8 (IL-8), a potent mediator of angiogenesis, is inhibited by IFNs [126,127]. IFN-γ up-regulates the expression of an angiostatic protein, IP-10, a CXC

chemokine with the ability to repress the angiogenic activities of bFGF and IL-8 [128]. Clinically IFN-α has been successful in inducing regression of bulky hemangiomas.

7.1.6. Perspective

As the first DNA recombinant technology products effective in cancer treatment, IFNs opened an era of biologic therapies for cancer, multiple sclerosis, and infectious diseases. Although more than 40 years have passed since IFNs were first described, scientific and clinical interest in these multifunctional cytokines has not diminished. The mechanisms of antitumor effects, optimal dose, schedule, and type of IFN for specific clinical indications have yet to be fully described. Progress in cloning IFN receptors and other signaling components has allowed further identification and description of molecular pathways by which IFNs mediate their effects. Further characterization and identification of interferon-stimulated genes or factors that mediate IFN-dependent antiproliferative, apoptotic, and immunomodulatory effects will lead to better utilization of IFNs in the treatment of cancer. Innovative approaches such as the introduction of second-generation IFNs with different biological effects and pharmacokinetic profiles may broaden the therapeutic applications of IFNs over the next decade.

REFERENCES

1. Issacs, A., and J. Lindenmann, *Virus infernce. I. The interferon.* Proc R Soc Lond, Ser B, 1957. **147**: 258–67.
2. Lindenmann, J., *Induction of chick interferon: procedures of the original experiments.* Methods Enzymol, 1981. **78**(Pt): 181–8.
3. Pestka, S., J.A. Langer, K.C. Zoon, and C.E. Samuel, *Interferons and their actions.* Annu Rev Biochem, 1987. **56**: 727–77.

4. Pestka, S., *The human interferon-alpha species and hybrid proteins.* Semin Oncol, 1997. **24**(3 Suppl 9): S9-4–S9-17.

5. Borden, E.C., and B.R.G. Williams, *Interferons.* 5th ed. Cancer Medicine, 5th ed. R.C. Bast, Jr., J.F. Holland, and T.S. Gansler, eds. 2000, Toronto: B.C. Decker.

6. Stark, G.R., I.M. Kerr, B.R. Williams, R.H. Silverman, and R.D. Schreiber, *How cells respond to interferons.* Annu Rev Biochem, 1998. **67**: 227–64.

7. Leaman, D.W., *Mechanisms of interferon action.* Prog Mol Subcell Biol, 1998. **20**: 101–42.

8. Gresser, I., *Production of interferon by suspension of human leukocytes.* Proc Soc Biol Med, 1961. **108**: 799–83.

9. Strander, H., and K. Cantell, *Production of interferon by human leukocytes in vitro.* Ann Med Exp Biol Fenn, 1966. **44**(2): 265–73.

10. Field, A.K., A.A. Tytell, G.P. Lampson, and M.R. Hilleman, *Inducers of interferon and host resistance. II. Multistranded synthetic polynucleotide complexes.* Proc Natl Acad Sci USA, 1967. **58**(3): 1004–10.

11. Maehara, N., and M. Ho, *Cellular origin of interferon induced by bacterial lipopolysaccharide.* Infect Immun, 1977. **15**(1): 78–83.

12. Baron, S., V. Howie, M. Langford, E.M. Macdonald, G.J. Stanton, J. Reitmeyer, and D.A. Weigent, *Induction of interferon by bacteria, protozoa, and viruses: defensive role.* Tex Rep Biol Med, 1981. **41**: 150–7.

13. Lebon, P., M.J. Commoy-Chevalier, B. Robert-Galliot, and C. Chany, *Different mechanisms for alpha and beta interferon induction.* Virology, 1982. **119**(2): 504–7.

14. Taylor, J.L., and S.E. Grossberg, *Chemical induction of interferon: caroxymethylacridanone and other low molecular weight chemicals.* Tex Rep Biol Med, 1982. **41**: 158–62.

15. Tamura, M., and S. Sasakawa, *Induction of human leukocyte interferon by heat-treated poly I: poly C.* Biochem Biophys Res Commun, 1983. **110**(3): 851–8.

16. Naruo, K., Y. Ichimori, M. Koyama, and K. Tsukamoto, *Correlation of interferon gamma inducing ability to sugar binding specificities of various lectins.* J Interferon Res, 1984. **4**(2): 235–41.

17. Wheelock, E.F., *Interferon likevirus inhibitor induced in human leukocytes by phytohemagglutinin.* Science, 1965. **139**: 310–12.

18. Pestka, S., and S. Baron, *Definition and classification of the interferons.* Methods Enzymol, 1981. **78**(PtA): 3–14.

19. Duc-Goiran, P., B. Galliot, and C. Chany, *Studies on virus-induced interferons produced by the human amniotic membrane and white blood cells.* Arch Gesamte Virusforsch, 1971. **34**(3): 232–43.

20. Branca, A.A., and C. Baglioni, *Evidence that types I and II interferons have different receptors.* Nature, 1981. **294**(5843): 768–70.

21. Cantell, K., S. Hirvonen, K.E. Mogensen, and L. Pyhala, *Human leukocyte interferon: production, purification, stability, and animal experiments.* In vitro Monogr, 1974. **3**: 35–8.

22. Rubinstein, M., S. Rubinstein, P.C. Familletti, M.S. Gross, R.S. Miller, A.A. Waldman, and S. Pestka, *Human leukocyte interferon purified to homogeneity.* Science, 1978. **202**(4374): 1289–90.

23. Rubinstein, M., S. Rubinstein, P.C. Familletti, R.S. Miller, A.A. Waldman, and S. Pestka, *Human leukocyte interferon: production, purification to homogeneity, and initial characterization.* Proc Natl Acad Sci USA, 1979. **76**(2): 640–4.

24. Rubinstein, M., W.P. Levy, J.A. Moschera, C.Y. Lai, R.D. Hershberg, R.T. Bartlett, and S. Pestka, *Human leukocyte interferon: isolation and characterization of several molecular forms.* Arch Biochem Biophys, 1981. **210**(1): 307–18.

25. Diaz, M.O., S. Bohlander, and G. Allen, *Nomenclature of the human interferon genes.* J Interferon Cytokine Res, 1996. **16**(2): 179–80.

26. Weissmann, C., S. Nagata, W. Boll, et al., *Structure and expression of human IFN-alpha genes.* Philos Trans R Soc Lond B Biol Sci, 1982. **299**(1094): 7–28.

27. Goeddel, D.V., E. Yelverton, A. Ullrich, et al., *Human leukocyte interferon produced by E. coli is biologically active.* Nature, 1980. **287**(5781): 411–16.

28. Taniguchi, T., L. Guarente, T.M. Roberts, D. Kimelman, J. Douhan, 3rd, and M. Ptashne,

Expression of the human fibroblast interferon gene in Escherichia coli. Proc Natl Acad Sci USA, 1980. **77**(9): 5230–3.

29. Maeda, S., M. Gross, and S. Pestka, *Screening of colonies by RNA-DNA hybridization with mRNA from induced and uninduced cells.* Methods Enzymol, 1981. **79**(Pt): 613–18.

30. Rubinstein, M., *The structure of human interferons.* Biochim Biophys Acta, 1982. **695**(1): 5–16.

31. Rehberg, E., B. Kelder, E.G. Hoal, and S. Pestka, *Specific molecular activities of recombinant and hybrid leukocyte interferons.* J Biol Chem, 1982. **257**(19): 11497–502.

32. Gutterman, J.U., S. Fine, J. Quesada, S.J. Horning, J.F. Levine, R. Alexanian, L. Bernhardt, M. Kramer, H. Spiegel, W. Colburn, P. Trown, T. Merigan, and Z. Dziewanowski, *Recombinant leukocyte A interferon: pharmacokinetics, single-dose tolerance, and biologic effects in cancer patients.* Ann Intern Med, 1982. **96**(5): 549–56.

33. Borden, E.C., *Interferons: in pursuit of the promise.* Prog Clin Biol Res, 1983. **132**(E): 287–96.

34. Wilson, V., A.J. Jeffreys, P.A. Barrie, P.G. Boseley, P.M. Slocombe, A. Easton, and D.C. Burke, *A comparison of vertebrate interferon gene families detected by hybridization with human interferon DNA.* J Mol Biol, 1983. **166**(4): 457–75.

35. Weissmann, C., and H. Weber, *The interferon genes.* Prog Nucleic Acid Res Mol Biol, 1986. **33**: 251–300.

36. De Maeyer, E., and J. De Maeyer-Guignard, *Interferons and other regulatory cytokines.* 1988, New York: John Wiley.

37. Sen, G.C., and P. Lengyel, *The interferon system. A bird's eye view of its biochemistry.* J Biol Chem, 1992. **267**(8): 5017–20.

38. Baron, S., S.K. Tyring, W.R. Fleischmann, Jr., D.H. Coppenhaver, D.W. Niesel, G.R. Klimpel, G.J. Stanton, and T.K. Hughes, *The interferons. Mechanisms of action and clinical applications.* Jama, 1991. **266**(10): 1375–83.

39. Fleischmann, W.R., Jr., and C.M. Fleischmann, *Mechanisms of interferon's antitumor actions.* 1992, Galveston: University of Texas at Galveston, Medical Branch. 299–309.

40. Fish, E.N., K. Banerjee, and N. Stebbing, *Human leukocyte interferon subtypes have different antiproliferative and antiviral activities on human cells.* Biochem Biophys Res Commun, 1983. **112**(2): 537–46.

41. Foster, G.R., O. Rodrigues, F. Ghouze, E. Schulte-Frohlinde, D. Testa, M.J. Liao, G.R. Stark, L. Leadbeater, and H.C. Thomas, *Different relative activities of human cell-derived interferon-alpha subtypes: IFN-alpha 8 has very high antiviral potency.* J Interferon Cytokine Res, 1996. **16**(12): 1027–33.

42. Salmon, S.E., B.G. Durie, L. Young, R.M. Liu, P.W. Trown, and N. Stebbing, *Effects of cloned human leukocyte interferons in the human tumor stem cell assay.* J Clin Oncol, 1983. **1**(3): 217–25.

43. Streuli, M., A. Hall, W. Boll, W.E. Stewart, 2nd, S. Nagata, and C. Weissmann, *Target cell specificity of two species of human interferon-alpha produced in Escherichia coli and of hybrid molecules derived from them.* Proc Natl Acad Sci USA, 1981. **78**(5): 2848–52.

44. Weck, P.K., S. Apperson, L. May, and N. Stebbing, *Comparison of the antiviral activities of various cloned human interferon-alpha subtypes in mammalian cell cultures.* J Gen Virol, 1981. **57**(Pt 1): 233–7.

45. Hawkins, M.J., E.C. Borden, J.A. Merritt, B.S. Edwards, L.A. Ball, E. Grossbard, and K.J. Simon, *Comparison of the biologic effects of two recombinant human interferons alpha (rA and rD) in humans.* J Clin Oncol, 1984. **2**(3): 221–6.

46. Hannigan, G.E., D.R. Gewert, E.N. Fish, S.E. Read, and B.R. Williams, *Differential binding of human interferon-alpha subtypes to receptors on lymphoblastoid cells.* Biochem Biophys Res Commun, 1983. **110**(2): 537–44.

47. Blatt, L.M., J.M. Davis, S.B. Klein, and M.W. Taylor, *The biologic activity and molecular characterization of a novel synthetic interferon-alpha species, consensus interferon.* J Interferon Cytokine Res, 1996. **16**(7): 489–99.

48. Ozes, O.N., Z. Reiter, S. Klein, L.M. Blatt, and M.W. Taylor, *A comparison of inter-*

feron-Con1 with natural recombinant interferons-alpha: antiviral, antiproliferative, and natural killer-inducing activities. J Interferon Res, 1992. **12**(1): 55–9.

49. Geng, Y., D. Yu, L.M. Blatt, and M.W. Taylor, *Tumor suppressor activity of the human consensus type I interferon gene.* Cytokines Mol Ther, 1995. **1**(4): 289–300.

50. Katre, N.V., *The conjunction of proteins with polyethylene glycol and other polymers; altering properties to enhance their therapeutic potential.* Advanced Drug Delivery Systems, 1993. **10**: 91–114.

51. Monkarsh, S.P., Y. Ma, A. Aglione, P. Bailon, D. Ciolek, B. DeBarbieri, M.C. Graves, K. Hollfelder, H. Michel, A. Palleroni, J.E. Porter, E. Russoman, S. Roy, and Y.C. Pan, *Positional isomers of monopegylated interferon alpha-2a: isolation, characterization, and biological activity.* Anal Biochem, 1997. **247**(2): 434–40.

52. Bailon, P., A. Palleroni, C.A. Schaffer, C.L. Spence, W.J. Fung, J.E. Porter, G.K. Ehrlich, W. Pan, Z.X. Xu, M.W. Modi, A. Farid, W. Berthold, and M. Graves, *Rational design of a potent, long-lasting form of interferon: a 40 kDa branched polyethylene glycol-conjugated interferon alpha-2a for the treatment of hepatitis C.* Bioconjug Chem, 2001. **12**(2): 195–202.

53. Grace, M., S. Youngster, G. Gitlin, W. Sydor, L. Xie, L. Westreich, S. Jacobs, D. Brassard, J. Bausch, and R. Bordens, *Structural and biologic characterization of pegylated recombinant IFN-alpha2b.* J Interferon Cytokine Res, 2001. **21**(12): 1103–15.

54. Lucero, M.A., H. Magdelenat, W.H. Fridman, P. Pouillart, C. Billardon, A. Billiau, K. Cantell, and E. Falcoff, *Comparison of effects of leukocyte and fibroblast interferon on immunological parameters in cancer patients.* Eur J Cancer Clin Oncol, 1982. **18**(3): 243–51.

55. Wallach, D., M. Fellous, and M. Revel, *Preferential effect of gamma interferon on the synthesis of HLA antigens and their mRNAs in human cells.* Nature, 1982. **299**(5886): 833–6.

56. Bocci, V., A. Pacini, L. Bandinelli, G.P. Pessina, M. Muscettola, and L. Paulesu, *The role of liver in the catabolism of human*

alpha- and beta-interferon. J Gen Virol, 1982. **60**(Pt 2): 397–400.

57. Wiernik, P.H., B. Schwartz, J.P. Dutcher, N. Turman, and C. Adinolfi, *Successful treatment of hairy cell leukemia with beta-ser interferon.* Am J Hematol, 1990. **33**(4): 244–8.

58. Byhardt, R.W., L. Vaickus, P.L. Witt, A.Y. Chang, T. McAuliffe, J.F. Wilson, C.A. Lawton, J. Breitmeyer, M.E. Alger, and E.C. Borden, *Recombinant human interferon-beta (rHuIFN-beta) and radiation therapy for inoperable non-small cell lung cancer.* J Interferon Cytokine Res, 1996. **16**(11): 891–902.

59. Borden, E.C., T.F. Hogan, and J.G. Voelkel, *Comparative antiproliferative activity in vitro of natural interferons alpha and beta for diploid and transformed human cells.* Cancer Res, 1982. **42**(12): 4948–53.

60. Chawla-Sarkar, M., D.W. Leaman, and E.C. Borden, *Preferential induction of apoptosis by interferon (IFN)-beta compared with IFN-alpha2: correlation with TRAIL/Apo2L induction in melanoma cell lines.* Clin Cancer Res, 2001. **7**(6): 1821–31.

61. Chen, Q., B. Gong, A.S. Mahmoud-Ahmed, A. Zhou, E.D. Hsi, M. Hussein, and A. Almasan, *Apo2L/TRAIL and Bcl-2-related proteins regulate type I interferon-induced apoptosis in multiple myeloma.* Blood, 2001. **98**(7): 2183–92.

62. Knobler, R.L., J.I. Greenstein, K.P. Johnson, F.D. Lublin, H.S. Panitch, K. Conway, S.V. Grant-Gorsen, J. Muldoon, S.G. Marcus, J.C. Wallenberg, et al., *Systemic recombinant human interferon-beta treatment of relapsing- remitting multiple sclerosis: pilot study analysis and six-year follow-up.* J Interferon Res, 1993. **13**(5): 333–40.

63. Henco, K., J. Brosius, A. Fujisawa, J.I. Fujisawa, J.R. Haynes, J. Hochstadt, T. Kovacic, M. Pasek, A. Schambock, J. Schmid, et al., *Structural relationship of human interferon alpha genes and pseudogenes.* J Mol Biol, 1985. **185**(2): 227–60.

64. Roberts, R.M., J.C. Cross, and D.W. Leaman, *Interferons as hormones of pregnancy.* Endocr Rev, 1992. **13**(3): 432–52.

65. Hansen, T.R., M. Kazemi, D.H. Keisler, P.V. Malathy, K. Imakawa, and R.M.

Roberts, *Complex binding of the embryonic interferon, ovine trophoblast protein-1, to endometrial receptors.* J Interferon Res, 1989. **9**(2): 215–25.

66. Johnson, H.M., F.W. Bazer, B.E. Szente, and M.A. Jarpe, *How interferons fight disease.* Sci Am, 1994. **270**(5): 68–75.

67. Dereuddre-Bosquet, N., P. Clayette, M. Martin, A. Mabondzo, P. Fretier, G. Gras, J. Martal, and D. Dormont, *Anti-HIV potential of a new interferon, interferon-tau (trophoblastin).* J Acquir Immune Defic Syndr Hum Retrovirol, 1996. **11**(3): 241–6.

68. Roberts, R.M., L. Liu, Q. Guo, D. Leaman, and J. Bixby, *The evolution of the type I interferons.* J Interferon Cytokine Res, 1998. **18**(10): 805–16.

69. LaFleur, D.W., B. Nardelli, T. Tsareva, et al., *Interferon-kappa, a novel type I interferon expressed in human keratinocytes.* J Biol Chem, 2001. **276**(43): 39765–71.

70. Gray, P.W., and D.V. Goeddel, *Structure of the human immune interferon gene.* Nature, 1982. **298**(5877): 859–63.

71. Talmadge, K.W., H. Gallati, F. Sinigaglia, A. Walz, and G. Garotta, *Identity between human interferon-gamma and "macrophage-activating factor" produced by human T lymphocytes.* Eur J Immunol, 1986. **16**(12): 1471–7.

72. Hardy, K.J., and H.A. Young, *IFN-γ, gene structure and regulation.*

73. Ruzicka, F.J., M.E. Jach, and E.C. Borden, *Binding of recombinant-produced interferon beta ser to human lymphoblastoid cells. Evidence for two binding domains.* J Biol Chem, 1987. **262**(33): 16142–9.

74. Novick, D., P. Orchansky, M. Revel, and M. Rubinstein, *The human interferon-gamma receptor. Purification, characterization, and preparation of antibodies.* J Biol Chem, 1987. **262**(18): 8483–7.

75. Uze, G., G. Lutfalla, and K.E. Mogensen, *Alpha and beta interferons and their receptor and their friends and relations.* J Interferon Cytokine Res, 1995. **15**(1): 3–26.

76. Der, S.D., A. Zhou, B.R. Williams, and R.H. Silverman, *Identification of genes differentially regulated by interferon alpha, beta, or gamma using oligonucleotide arrays.* Proc Natl Acad Sci USA, 1998. **95**(26): 15623–8.

77. Abramovich, C., L.M. Shulman, E. Ratovitski, S. Harroch, M. Tovey, P. Eid, and M. Revel, *Differential tyrosine phosphorylation of the IFNAR chain of the type I interferon receptor and of an associated surface protein in response to IFN-alpha and IFN-beta.* Embo J, 1994. **13**(24): 5871–7.

78. Uddin, S., L. Yenush, X.J. Sun, M.E. Sweet, M.F. White, and L.C. Platanias, *Interferon-alpha engages the insulin receptor substrate-1 to associate with the phosphatidylinositol 3'-kinase.* J Biol Chem, 1995. **270**(27): 15938–41.

79. David, M., E. Petricoin, 3rd, C. Benjamin, R. Pine, M.J. Weber, and A.C. Larner, *Requirement for MAP kinase (ERK2) activity in interferon alpha- and interferon beta-stimulated gene expression through STAT proteins.* Science, 1995. **269**(5231): 1721–3.

80. Darnell, J.E., Jr., I.M. Kerr, and G.R. Stark, *Jak-STAT pathways and transcriptional activation in response to IFNs and other extracellular signaling proteins.* Science, 1994. **264**(5164): 1415–21.

81. Quesada, J.R., J. Reuben, J.T. Manning, E.M. Hersh, and J.U. Gutterman, *Alpha interferon for induction of remission in hairy-cell leukemia.* N Engl J Med, 1984. **310**(1): 15–18.

82. Gutterman, J.U., *Cytokine therapeutics: lessons from interferon alpha.* Proc Natl Acad Sci USA, 1994. **91**(4): 1198–205.

83. Platanias, L.C., and H.M. Golomb, *Clinical use of interferon: Hairy cell, chronic myelegenous and other leukemias,* in *Interferon: principles and medical applications.* 1992, Galveston: University of Texas, Galveston Medical Branch. 427–32.

84. Leibowitz, D., and K.S. Young, *The molecular biology of CML: a review.* Cancer Invest, 1989. **7**(2): 195–203.

85. Kantarjian, H.M., T.L. Smith, S. O'Brien, M. Beran, S. Pierce, and M. Talpaz, *Prolonged survival in chronic myelogenous leukemia after cytogenetic response to interferon-alpha therapy. The Leukemia Service.* Ann Intern Med, 1995. **122**(4): 254–61.

86. Talpaz, M., *Use of interferon in the treatment of chronic myelogenous leukemia.* Semin Oncol, 1994. **21**(6 Suppl 14): 3–7.

87. Ludwig, H., A. Cortelezzi, B.G. Van Camp, E. Polli, W. Scheithauer, R. Kuzmits, W. Linkesch, H. Gisslinger, H. Sinzinger, E. Fritz, et al., *Treatment with recombinant interferon-alpha-2C: multiple myeloma and thrombocythaemia in myeloproliferative diseases.* Oncology, 1985. **42**(Suppl 1): 19–25.

88. Gilbert, H.S., *Long term treatment of myeloproliferative disease with interferon-alpha-2b: feasibility and efficacy.* Cancer, 1998. **83**(6): 1205–13.

89. Bunn, P.A., Jr., K.A. Foon, D.C. Ihde, D.L. Longo, J. Eddy, C.F. Winkler, S.R. Veach, J. Zeffren, S. Sherwin, and R. Oldham, *Recombinant leukocyte A interferon: an active agent in advanced cutaneous T-cell lymphomas.* Ann Intern Med, 1984. **101**(4): 484–7.

90. Borden, E.C., *Innovative treatment strategies for non-Hodgkin's lymphoma and multiple myeloma.* Semin Oncol, 1994. **21**(6 Suppl 14): 14–22.

91. Smalley, R.V., J.W. Andersen, M.J. Hawkins, V. Bhide, M.J. O'Connell, M.M. Oken, and E.C. Borden, *Interferon alfa combined with cytotoxic chemotherapy for patients with non-Hodgkin's lymphoma.* N Engl J Med, 1992. **327**(19): 1336–41.

92. Wood, A., B. Angus, P. Kestevan, J. Dark, G. Notarianni, S. Miller, M. Howard, S. Proctor, and P. Middleton, *Alpha interferon gene deletions in post-transplant lymphoma.* Br J Haematol, 1997. **98**(4): 1002–3.

93. Faro, A., *Interferon-alpha and its effects on post-transplant lymphoproliferative disorders.* Springer Semin Immunopathol, 1998. **20**(3–4): 425–36.

94. Strander, H., and K. Oberg, *Clinical uses of interferons: solid tumors.* 1992, Galveston: University of Texas, Galveston Medical Branch. 533–61.

95. Robinson, W.A., T.I. Mughal, M.R. Thomas, M. Johnson, and R.J. Spiegel, *Treatment of metastatic malignant melanoma with recombinant interferon alpha 2.* Immunobiology, 1986. **172**(3–5): 275–82.

96. Kirkwood, J.M., M.H. Strawderman, M.S. Ernstoff, T.J. Smith, E.C. Borden, and R.H. Blum, *Interferon alfa-2b adjuvant therapy of high-risk resected cutaneous melanoma: the Eastern Cooperative Oncology Group Trial EST 1684.* J Clin Oncol, 1996. **14**(1): 7–17.

97. Thompson, J.A., P.J. Gold, D.R. Markowitz, D.R. Byrd, C.G. Lindgren, and A. Fefer, *Updated analysis of an outpatient chemo-immunotherapy regimen for treating metastatic melanoma.* Cancer J Sci Am, 1997. **3**(Suppl 1): S29–34.

98. Bukowski, R.M., *Natural history and therapy of metastatic renal cell carcinoma: the role of interleukin-2.* Cancer, 1997. **80**(7): 1198–220.

99. Legha, S.S., S. Ring, O. Eton, A. Bedikian, A.C. Buzaid, C. Plager, and N. Papadopoulos, *Development of a biochemotherapy regimen with concurrent administration of cisplatin, vinblastine, dacarbazine, interferon alfa, and interleukin-2 for patients with metastatic melanoma.* J Clin Oncol, 1998. **16**(5): 1752–9.

100. Falkson, C.I., J. Ibrahim, J.M. Kirkwood, A.S. Coates, M.B. Atkins, and R.H. Blum, *Phase III trial of dacarbazine versus dacarbazine with interferon alpha-2b versus dacarbazine with tamoxifen versus dacarbazine with interferon alpha-2b and tamoxifen in patients with metastatic malignant melanoma: an Eastern Cooperative Oncology Group study.* J Clin Oncol, 1998. **16**(5): 1743–51.

101. Kirkwood, J.M., J.E. Harris, R. Vera, S. Sandler, D.S. Fischer, J. Khandekar, M.S. Ernstoff, L. Gordon, R. Lutes, P. Bonomi, et al., *A randomized study of low and high doses of leukocyte alpha-interferon in metastatic renal cell carcinoma: the American Cancer Society collaborative trial.* Cancer Res, 1985. **45**(2): 863–71.

102. Krown, S.E., *Therapeutic options in renal-cell carcinoma.* Semin Oncol, 1985. **12**(4 Suppl 5): 13–17.

103. Quesada, J.R., L. Evans, S.R. Saks, and J.U. Gutterman, *Recombinant interferon alpha and gamma in combination as treatment for metastatic renal cell carcinoma.* J Biol Response Mod, 1988. **7**(3): 234–9.

104. Pyrhonen, S., E. Salminen, M. Ruutu, T. Lehtonen, M. Nurmi, T. Tammela, H. Juusela, E. Rintala, P. Hietanen, and P.L. Kellokumpu-Lehtinen, *Prospective ran-*

domized trial of interferon alfa-2a plus vinblastine versus vinblastine alone in patients with advanced renal cell cancer. J Clin Oncol, 1999. **17**(9): 2859–67.

105. Lindner, D.J., E.C. Borden, and D.V. Kalvakolanu, *Synergistic antitumor effects of a combination of interferons and retinoic acid on human tumor cells in vitro and in vivo.* Clin Cancer Res, 1997. **3**(6): 931–7.

106. Myskowski, P.L., and R. Ahkami, *Advances in Kaposi's sarcoma.* Dermatol Clin, 1997. **15**(1): 177–88.

107. Borden, E.C., and S. Wadler, *Interferons as biochemical modulators.* J Clin Oncol, 1996. **14**(10): 2627–30.

108. de Wit, R., *AIDS-associated Kaposi's sarcoma and the mechanisms of interferon alpha's activity; a riddle within a puzzle.* J Intern Med, 1992. **231**(4): 321–5.

109. Ezekowitz, R.A., J.B. Mulliken, and J. Folkman, *Interferon alfa-2a therapy for life-threatening hemangiomas of infancy.* N Engl J Med, 1992. **326**(22): 1456–63.

110. Merritt, J.A., L.A. Ball, K.M. Sielaff, D.M. Meltzer, and E.C. Borden, *Modulation of 2',5'-oligoadenylate synthetase in patients treated with alpha-interferon: effects of dose, schedule, and route of administration.* J Interferon Res, 1986. **6**(3): 189–98.

111. Savinova, O., B. Joshi, and R. Jagus, *Abnormal levels and minimal activity of the dsRNA-activated protein kinase, PKR, in breast carcinoma cells.* Int J Biochem Cell Biol, 1999. **31**(1): 175–89.

112. Balkwill, F., and J. Taylor-Papadimitriou, *Interferon affects both G1 and S + G2 in cells stimulated from quiescence to growth.* Nature, 1978. **274**(5673): 798–800.

113. Otsuki, T., O. Yamada, H. Sakaguchi, A. Tomokuni, H. Wada, Y. Yawata, and A. Ueki, *Human myeloma cell apoptosis induced by interferon-alpha.* Br J Haematol, 1998. **103**(2): 518–29.

114. Kumar, A., and I. Atlad, *Interferon-a induces the expression of retinoblastoma gene product in human burkitt lymphoma Daudi cells: role in growth regulation.* Proc Natl Acad Sci USA, 1992. **89**: 6659–6603.

115. Tiefenbrun, N., D. Melamed, N. Levy, D. Resnitzky, I. Hoffman, S.I. Reed, and A. Kimchi, *Alpha interferon suppresses the*

cyclin D3 and cdc25A *genes, leading to a reversible G0-like arrest.* Mol Cell Biol, 1996. **16**(7): 3934–44.

116. Subramaniam, P.S., P.E. Cruz, A.C. Hobeika, and H.M. Johnson, *Type I interferon induction of the Cdk-inhibitor p21WAF1 is accompanied by ordered G1 arrest, differentiation and apoptosis of the Daudi B-cell line.* Oncogene, 1998. **16**(14): 1885–90.

117. Weller, M., K. Frei, P. Groscurth, P.H. Krammer, Y. Yonekawa, and A. Fontana, *Anti-Fas/APO-1 antibody-mediated apoptosis of cultured human glioma cells. Induction and modulation of sensitivity by cytokines.* J Clin Invest, 1994. **94**(3): 954–64.

118. Spets, H., P. Georgii-Hemming, J. Siljason, K. Nilsson, and H. Jernberg-Wiklund, *Fas/APO-1 (CD95)-mediated apoptosis is activated by interferon-gamma and interferon- in interleukin-6 (IL-6)-dependent and IL-6-independent multiple myeloma cell lines.* Blood, 1998. **92**(8): 2914–23.

119. Selleri, C., J.P. Maciejewski, F. Pane, L. Luciano, A.M. Raiola, I. Mostarda, F. Salvatore, and B. Rotoli, *Fas-mediated modulation of Bcr/Abl in chronic myelogenous leukemia results in differential effects on apoptosis.* Blood, 1998. **92**(3): 981–9.

120. Edwards, B.S., M.J. Hawkins, and E.C. Borden, *Comparative in vivo and in vitro activation of human natural killer cells by two recombinant alpha-interferons differing in antiviral activity.* Cancer Res, 1984. **44**(7): 3135–9.

121. Sato, K., S. Hida, H. Takayanagi, T. Yokochi, N. Kayagaki, K. Takeda, H. Yagita, K. Okumura, N. Tanaka, T. Taniguchi, and K. Ogasawara, *Antiviral response by natural killer cells through TRAIL gene induction by IFN-alpha/beta.* Eur J Immunol, 2001. **31**(11): 3138–46.

122. Kleinerman, E.S., R. Kurzrock, D. Wyatt, J.R. Quesada, J.U. Gutterman, and I.J. Fidler, *Activation or suppression of the tumoricidal properties of monocytes from cancer patients following treatment with human recombinant gamma- interferon.* Cancer Res, 1986. **46**(10): 5401–5.

123. Basham, T.Y., M.F. Bourgeade, A.A. Creasey, and T.C. Merigan, *Interferon*

increases *HLA synthesis in melanoma cells: interferon-resistant and -sensitive cell lines.* Proc Natl Acad Sci USA, 1982. **79**(10): 3265–9.

124. Nathan, C.F., T.J. Prendergast, M.E. Wiebe, E.R. Stanley, E. Platzer, H.G. Remold, K. Welte, B.Y. Rubin, and H.W. Murray, *Activation of human macrophages. Comparison of other cytokines with interferon-gamma.* J Exp Med, 1984. **160**(2): 600–5.

125. Dinney, C.P., D.R. Bielenberg, P. Perrotte, R. Reich, B.Y. Eve, C.D. Bucana, and I.J. Fidler, *Inhibition of basic fibroblast growth factor expression, angiogenesis, and growth of human bladder carcinoma in mice by systemic interferon-alpha administration.* Cancer Res, 1998. **58**(4): 808–14.

126. Koch, A.E., P.J. Polverini, S.L. Kunkel, L.A. Harlow, L.A. DiPietro, V.M. Elner, S.G. Elner, and R.M. Strieter, *Interleukin-8 as a macrophage-derived mediator of angiogenesis.* Science, 1992. **258**(5089): 1798–801.

127. Reznikov, L.L., A.J. Puren, G. Fantuzzi, H. Muhl, L. Shapiro, D.Y. Yoon, D.L. Cutler, and C.A. Dinarello, *Spontaneous and inducible cytokine responses in healthy humans receiving a single dose of IFN-alpha2b: increased production of interleukin-1 receptor antagonist and suppression of IL-1-induced IL-8.* J Interferon Cytokine Res, 1998. **18**(10): 897–903.

128. Strieter, R.M., S.L. Kunkel, D.A. Arenberg, M.D. Burdick, and P.J. Polverini, *Interferon gamma-inducible protein 10 (IP-10), a member of the C-X-C chemokine family, is an inhibitor of angiogenesis.* Biochem Biophys Res Commun, 1995. **210**(1): 51–7.

■ SECTION TWO ■

GARY L. DAVIS

■ 7.2. INTERFERONS IN VIRAL HEPATITIS

In 1957 Isaacs and Lindemann noted that virally infected cells produced a protein that conferred viral resistance to naïve cells [1]. They called the substance "interferon." The properties of interferons immediately captured the imagination of virologists. However, clinical research was limited by the inability to produce interferon in sufficient quantity and to adequately purify the protein. The first major step in overcoming these problems was the discovery by Gresser that human leukocytes could produce interferons when stimulated in culture by Sendai virus [2].

Subsequently Cantell at the Finnish Blood Bank used this strategy to develop a large-scale process of preparing leukocyte interferon from virus-stimulated buffy-coat lymphocytes pooled from blood donors [3]. This allowed production of comparatively large amounts of crude interferon, but efficiency was low. Despite access to more than 90,000 units of blood, only 6×10^{11} units of crude interferon were produced in 1980 [4,5], allowing only limited experimentation in humans.

Greenberg in 1976 [6] was the first to report that interferon has activity against a hepatitis virus. Four patients with chronic hepatitis B received short courses of low-dose interferon (6×10^3 to 17×10^4 units per kilogram per day). Three of the four patients demonstrated a reduction in viral replication, and two appeared to have permanently lost hepatitis B *e* antigen (HBeAg). Unfortunately, the limited production and purity of the Cantell interferon did not allow for continuation of these experiments. Based on current treatment recommendations, the entire production of Cantell interferon in 1980 would have allowed treatment of only about 200 to 250 patients with hepatitis B or C.

The problems of quantity and purity were resolved by the application of recombinant technology to interferon production. The gene encoding interferon-α was cloned and inserted into *Escherichia coli* by Nagata in 1980 [7]. Later studies have shown that interferon-α is a multigene family with 10% to 30% sequence diversity [8]. These studies confirmed the existence of multiple subtypes of α-interferon [5].

Genentech employed mRNA isolated from Cantell interferon-producing buffy-coat lymphocytes and cell lines to develop clones that were used for large-scale production of α-interferon by recombinant technology [5]. The recombinant interferons that were subsequently approved for use in viral hepatitis were versions of interferon-α-2, but they differed by a single amino acid (α-2a has a lysine at position 23, while α-2b has an arginine at that position) [9].

Expression of both of these proteins in *E. coli* was facilitated by prokaryotic promoters (a tryptophan promoter for α-2a and a β-lactamase promoter for α-2b), improved vectors, and optimized fermentation conditions, such that production levels as high as 10^{10} units per liter could be reached [9]. Elaborate procedures for purifying interferon from the cultivated biomass were developed wherein multiple steps of centrifugation, filtration, chromatography, and precipitation were included [9]. The resulting products were highly reproducible and provided a high degree of purity. Despite the manipulations involved in recombinant cultivation and purification, recombinant α-interferons retained their antiviral activity and appeared to be as well tolerated in humans as natural interferons [10–12]. The ability to produce large quantities of purified interferon resulted in the initiation in the mid-1980s of large-scale trials of recombinant interferons in patients with viral hepatitis.

7.2.1. Interferon Treatment of Chronic Hepatitis B

Hepatitis B virus (HBV) is a DNA virus that causes one of the most common infectious diseases in the world, affecting about two billion people at some time during their lives [13]. About 15% of infected individuals develop chronic infection, characterized by persistence of hepatitis B surface antigen (HBsAg) in serum and some degree of HBV replication. Infected patients with active liver disease usually have detectable hepatitis B e antigen (HBeAg) and high levels of serum HBV DNA, while those without active liver disease typically lack HBeAg but have antibody to HBeAg (anti-HBe), low alanine aminotransferase (ALT) levels, and undetectable or low-titer HBV DNA replication [14].

Approximately 350 million people worldwide are chronic carriers of HBV, with the majority living in Asia and Africa. In the United States approximately one million people have chronic HBV infection. Although chronically infected, individuals may remain asymptomatic for long periods. Spontaneous loss of HBeAg occurs in 7% to 20% of patients each year, but spontaneous loss of HBsAg occurs in only 1% to 2% per year [14]. Health experts estimate that 2% of patients with chronic HBV infection develop cirrhosis each year, and that 15% to 25% of patients with chronic HBV infection will die prematurely from cirrhosis or hepatocellular carcinoma (HCC).

The primary goal of treatment is to convert patients with chronic HBV infection from a high to a low level of HBV replication as evidenced by HBeAg seroconversion to anti-HBe with loss of detectable serum HBV DNA as measured by relatively insensitive tests such as solution DNA hybridization [15]. This serologic change parallels biochemical and histologic improvement that are usually long lasting and probably associated with increased survival and a lower risk of developing hepatocellular carcinoma [16,17].

As noted, Greenberg was the first to demonstrate the antiviral activity of a leukocyte-derived interferon against HBV infection [6]. Subsequent studies, using recombinant interferon-α, demonstrated durable HbeAg to anti-HBe seroconversion in 25% to 40% of patients receiving a 4 to 6 month course of therapy at a dose of 30 to 35 million units (MU) per week given as either 5 MU daily or 10 MU three times per week. The three-times-per-week dosing schedule may be better tolerated [18–22]. A meta-analysis of 15 clinical trials showed that HBeAg loss occurred in 33% of patients treated with interferon compared with 12% of untreated or placebo-treated controls [23]. Longer treatment duration may result in an even higher response rate [24]. Response to treatment correlates with low pretreatment HBV DNA levels, high pretreatment serum ALT levels, and active inflammation on liver biopsy [25]. Side effects of interferon are common but usually manageable and include flu-like symptoms, cytopenia, fatigue, and neuropsychiatric problems. Side effects disappear on dose reduction or discontinuation of therapy.

The US Food and Drug Administration (FDA) approved recombinant α-interferons for the treatment of chronic hepatitis B in 1992. These agents are currently recommended for patients with compensated chronic hepatitis B and detectable HBsAg, HBeAg, and HBV DNA in serum. Other subsets of patients with chronic hepatitis B who may benefit from treatment include those with HBeAg-negative infection (the so-called precore mutation of the virus) [26] or hepatitis D superinfection [27,28], though prolonged therapy is required and durable responses are less likely. Interferon treatment may also benefit patients with extrahepatic features of HBV infection such as glomerulonephritis or vasculitis, though this experience is largely anecdotal [29].

Interferon therapy should be used with great caution in patients with decompensated cirrhosis since treatment may flare their disease, resulting in hepatic failure, and is often associated with significant cytopenia or infection [30,31]. Other absolute or relative contraindications to interferon treatment include a history of hypersensitivity to interferon, immune suppression associated with organ transplantation, active autoimmune disease, and significant psychiatric disease, especially depression.

Despite the success of recombinant interferon therapy in patients with chronic hepatitis B, it has been largely replaced by treatment with nucleoside analogues that appear to be at least as effective but can be given orally with few side effects. Lamivudine is an oral nucleoside analogue that is highly active against the hepatitis B virus. One year of treatment with lamivudine results in loss of HBeAg in more than 30% of cases, HbeAg-to-anti-HBe seroconversion in almost 20%, and histologic improvement in more than half of cases [32]. Longer courses of lamivudine further increase the chance of seroconversion (28% at 2 years, 40% at 3 years, 48% at 4 years) but also result in a high likelihood of developing a YMDD mutation in the polymerase gene that confers lamivudine resistance [33]. Other potent antiviral agents for hepatitis B are in clinical trials and will continue to make interferon a less attractive therapeutic option.

7.2.2. Interferon Treatment of Chronic Hepatitis C

Hepatitis C virus (HCV) is an RNA virus that is a common cause of parenterally acquired viral hepatitis; chronic infection follows acute infection in 80% to 85% of cases. Although liver disease resulting from chronic HCV infection is only slowly progressive, HCV is the most common cause of chronic liver disease in the United States, the most common etiology for hepatocellular carcinoma, and the leading indication for liver transplantation [34–36].

Because of the risk of progressive liver disease, efforts toward controlling HCV infection have been vigorous and were underway even before the virus was identified in 1989 [37]. Indeed, the beneficial effects of interferon in HCV infection were shown in the very first study of recombinant α-interferon for chronic non-A, non-B hepatitis (subsequently renamed chronic hepatitis C after the discovery of the virus). That study showed improvement with normalization of serum ALT levels in 8 of 10 interferon-treated subjects as well as histological improvement on liver biopsy in patients who received a longer duration of therapy [37].

Subsequent analysis of stored serum samples showed reduction of HCV RNA levels during treatment and durable eradication of virus in some cases [38]. These early results were confirmed by randomized controlled trials in patients with chronic hepatitis C [39–41]. Durable viral eradication (termed "sustained virologic response," or SVR) was achieved in 6% to 15% of patients after six months of treatment with recombinant interferon at doses of 3 to 6 MU administered subcutaneously three times per week. SVR increased to 13% to 25% if treatment was extended to 12 months [41]. The combination of the oral nucleoside analogue ribavirin with recombinant interferon increased SVR to 41% [42–44]. Ribavirin, however, is potentially embryotoxic and induces a dose-dependent hemolytic anemia, a situation that calls for close monitoring during therapy.

Long-acting pegylated interferons are likely to replace standard recombinant interferons because they allow once-weekly dosing and provide more predictable viral suppression [45,46]. Pegylation involves the covalent attachment of polyethylene glycol to the recombinant interferon molecule without otherwise altering its structure [47]. Pegylated interferons have reduced renal clearance that results in drug half-lives of 50 to 77 hours compared with 4 to 5 hours for standard interferons [47]. The combination of pegylated interferons with ribavirin has resulted in SVR rates of 54% to 56% [48,49]. Pegylated interferon-α-2b was approved by the FDA in 2001, and pegylated-α-2a has now also been approved.

Response of HCV infection to interferon-based therapies is highly dependent on viral factors. The most important factor in determining response to treatment is the viral genotype. HCV has six major genotypes, representing nucleotide sequence diversity of 30% to 45% [50]. The major genotypes in the United States and Europe are 1, 2, and 3, representing approximately 75%, 13%, and 6% of isolates [51]. Trials with standard recombinant interferon and ribavirin showed that patients with genotype 1 were less likely to respond to therapy than those with genotype 2 or 3 (SVR: 29% vs. 65%) and required 12 months of treatment, as compared with only 6 months for genotype 2 or 3 [42–43]. Response to the combination of pegylated interferon and ribavirin is seen in only 42% to 46% of patients with genotype 1, but in 76% to 82% of those with genotypes 2 and 3 [48,49]. The level of HCV RNA and the presence of fibrosis on liver biopsy also influence treatment response but to a much lower degree than genotype [49].

The contraindications to and tolerance of interferon-based therapies in the treatment of HCV infection are similar to those described for patients with HBV infection. As for HBV infection, considerable efforts are being made to develop new agents to improve response rates in patients with HCV infection. Direct antiviral strategies with antisense oligonucleotides, ribozymes, and inhibitors of the viral enzymes—polymerase, helicase, and protease—are under investigation. However, it is likely that interferons will continue to serve as the foundation of therapy for HCV infections, with new agents serving as adjuncts.

7.2.3. Summary

Although early studies suggested that α-interferon was active against hepatitis viruses, human trials were prohibited by a meager supply of drug and its lack of purity. Recombinant technology allowed the production of the large amounts of high-quality interferon that were needed to study interferon's effects in patients with viral hepatitis. Subsequent studies demonstrated efficacy in both chronic hepatitis B and C. Although the use of α-interferons in hepatitis B infections has largely been replaced, these agents remain the standard of care for treatment of patients with chronic hepatitis C infections. HCV infection is eradicated in more than half of patients treated with a long-acting form of recombinant α-interferon in combination with the oral nucleoside analogue ribavirin. The eradication of HCV infection is associated with a reduction in hepatic inflammation and fibrosis on liver biopsy [52,53]. These data coupled with an understanding of the natural course of the infection predict that viral eradication translates into reductions in morbidity and mortality in future years [54,55]. The section that follows concerns the use of interferons in the treatment of multiple sclerosis, an immunologic rather than infectious disorder, illustrating the remarkable breadth of biologic activity offered by interferons.

REFERENCES

1. Isaacs, A., and J. Lindenmann, *Virus interference: I. The interferon*. Proc R Soc Lond, 1957. B**147**: 258–67.
2. Gresser, I., *Production of interferon by suspensions of human leukocytes*. Proc Soc Exp Biol Med, 1961. **108**: 799–803.
3. Cantell, K., H. Strander, G.Y. Hadhazy, and H.R. Nevanlinna, *How much interferon can be prepared in human leukocyte suspensions?* The Interferons. A. Rita, ed. New York: Academic Press, 223–32.
4. Cantell, K., *Why is interferon not in clinical use today?* Interferon, 1979. **1**: 1–28.
5. Dunnick, J.K., R.C. Merigan, and G.J. Galasso, *From the National Institute of Allergy and Infectious Diseases. Report on a workshop of DNA recombinant technology in interferon cloning*. J Infect Dis, 1981. **143**: 297–300.
6. Greenberg, H.B., R.B. Pollard, L.I. Lutwick, P.B. Gregory, W.S. Robinson, and T.C. Merigan, *Effect of human leukocyte interferon on hepatitis B virus infection in patients with chronic active hepatitis*. New Engl J Med, 1976. **295**: 517–22.
7. Nagata, S., H. Taira, A. Hall, et al. *Synthesis in e. coli of a polypeptide with human leukocyte interferon activity*. Nature, 1980. **284**: 316–19.
8. Streuli, M., S. Nagata, and C. Weissman, *At least three human type α interferons: structure of α2*. Science, 1980. **209**: 1343–49.
9. Baron, E., and S. Narula, *From cloning to a commercial realization: human alpha interferon*. Crit Reviews in Biotech, 1990. **10**: 179–90.
10. Goeddel, D.V., E. Yelverton, A. Ullrich, et al., *Human leukocyte interferon produced by E. coli is biologically active*. Nature, 1980. **287**: 411–67.
11. Trown, P.W., R.J. Wills, and J.J. Kamm, *The preclinical development of Roferon-A*. Cancer, 1986. **57**: 1648–56.
12. Gutterman, J.U., S. Fein, J. Quesada, et al., *Recombinant leukocyte A interferon: pharmacokinetics, single-dose tolerance, and biological effects in cancer patients*. Ann Intern Med, 1982. **96**: 549–56.
13. Lee, W.M., *Hepatitis B virus infection*. N Engl J Med, 1997. **337**: 1733–45.
14. Hoofnagle, J.H., and A.M. Di Bisceglie, *Serologic diagnosis of acute and chronic viral hepatitis*. Semin Liver Dis, 1991. **11**: 73–83.
15. Hoofnagle, J.H., and A.M. Di Bisceglie, *The treatment of chronic viral hepatitis*. N Engl J Med, 1997. **336**: 347–56.
16. Niederau, C., T. Heintges, S. Lange, et al., *Long-term follow-up of HBeAg-positive patients treated with interferon alfa for chronic hepatitis B*. N Engl J Med, 1996. **334**: 1422–27.
17. Lau, D.T., J. Everhart, D.E. Kleiner, et al., *Long-term follow-up of patients with chronic hepatitis B treated with interferon alpha*. Gastroenterology, 1997. **113**: 1660–67.
18. Alexander, G.J.M., J. Brahm, E.A. Fagan, et al., *Loss of HBsAg with interferon therapy*

in chronic hepatitis B virus infection. Lancet, 1987. **2**: 66–69.

19. Hoofnagle, J.H., M.G. Peters, K.D. Mullen, et al., *Randomized, controlled trial of recombinant human a-interferon in patients with chronic hepatitis B.* Gastroenterology, 1988. **95**: 1318–25.

20. Perrillo, R.P., E.R. Schiff, G.L. Davis, et al., *A randomized, controlled trial of interferon alfa-2b alone and after prednisone withdrawal for the treatment of chronic hepatitis B.* N Engl J Med, 1990. **323**: 295–301.

21. Lok, A.S.F., P.C. Wu, C.L. Lai, et al., *A controlled trial of interferon with or without prednisone priming for chronic hepatitis B.* Gastroenterology, 1992. **102**: 2091–97.

22. Di Bisceglie, A.M., T.L. Fong, M.W. Fried, et al., *A randomized controlled trial of recombinant alpha-interferon therapy for chronic hepatitis B.* Am J Gastroenterol, 1993. **88**: 1887–92.

23. Wong, D.K.H., A.M. Cheung, K. O 'Rourke, C.D. Naylor, A.S. Detsky, and J. Heathcote, *Effect of alpha-interferon treatment in patients with hepatitis B e antigen-positive chronic hepatitis B: a meta-analysis.* Ann Intern Med, 1993. **119**: 312–23.

24. Janssen, H.L., G. Gerken, V. Carreno, et al., *Interferon alfa for chronic hepatitis B infection: increased efficacy of prolonged treatment. The European Concerted Action on Viral Hepatitis (EUROHEP).* Hepatology, 1999. **30**: 238–43.

25. Brook, M.G., P. Karayiannis, and H.C. Thomas, *Which patients with chronic hepatitis B virus infection will respond to alpha-interferon therapy? A statistical analysis of predictive factors.* Hepatology, 1989. **10**: 761–63.

26. Brunetto, M.R., F. Oliveri, G. Rocca, et al., *Natural course and response to interferon of chronic hepatitis B accompanied by antibody to hepatitis B e antigen.* Hepatology, 1989. **10**: 198–202.

27. Rosina, F., C. Pintus, C. Meschievitz, and M. Rizzetto, *A randomized controlled trial of a 12-month course of recombinant human interferon-α in chronic delta (type D) hepatitis: a multicenter Italian study.* Hepatology, 1991. **13**: 1052–56.

28. Farci, P., A. Mandas, A. Coiana, et al., *Treatment of chronic hepatitis D with interferon alfa-2a.* N Engl J Med, 1994. **330**: 88–94.

29. Conjeevaram, H.S., J.H. Hoofnagle, H.A. Austin, Y. Park, M.W. Fried, and A.M. Di Bisceglie, *Long-term outcome of hepatitis B virus-related glomerulonephritis after therapy with interferon alfa.* Gastroenterology, 1995. **109**: 540–46.

30. Hoofnagle, J.H., A.M. Di Bisceglie, J.G. Waggoner, and Y. Park, *Interferon alfa for patients with clinically apparent cirrhosis due to chronic hepatitis B.* Gastroenterology, 1993. **104**: 1116–21.

31. Perrillo, R., C. Tamburro, F. Regenstein, et al., *Low-dose titratable interferon alfa in decompensated liver disease caused by chronic infection with hepatitis B virus.* Gastroenterology, 1995. **109**: 908–16.

32. Dienstag, J.L., E.R. Schiff, T.L. Wright, et al., *Lamivudine as initial treatment for chronic hepatitis B in the United States.* New Engl J Med, 1999. **341**: 1256–63.

33. Ling, R., D. Mutimer, M. Ahmed, et al., *Selection of mutations in the hepatitis B virus polymerase during therapy of transplant recipients with lamivudine.* Hepatology, 1996. **24**: 711–13.

34. Meyer, R.A., and S.C. Gordon, *Epidemiology of hepatitis C virus infection in a suburban Detroit community.* Am J Gastroenterol, 1991. **86**: 1224–6.

35. El-Serag, H.B., and A.C. Mason, *Rising incidence of hepatocellular carcinoma in the United States.* New Engl J Med, 1999. **340**: 745–50.

36. *Primary Liver Disease of Liver Transplant Recipients 1991 and 1992* (From the UNOS Scientific Registry). UNOS Update, 1993. **9**(8): 27.

37. Hoofnagle, J.H., K.D. Mullen, D.B. Jones, et al., *Treatment of chronic non-A, non-B hepatitis with recombinant human alpha interferon: a preliminary report.* New Engl J Med, 1986. **315**: 1575–8.

38. Shindo, M., A.M. Di Bisceglie, L. Cheung, et al., *Decrease in serum hepatitis C viral RNA during alpha-interferon therapy for chronic hepatitis C.* Ann Intern Med, 1991. **115**: 700–4.

39. Davis, G.L., L.A. Balart, E.R. Schiff, et al., *Treatment of chronic hepatitis C with recombinant interferon alfa: a multicenter randomized controlled study.* New Engl J Med, 1989. **321**: 1501–6.

40. Di Bisceglie, A.M., P. Martin, C. Kassianides, et al., *Recombinant interferon alfa therapy*

for chronic hepatitis C: a randomized, double-bind, placebo-controlled trial. New Engl J Med, 1989. **321**: 1506–10.

41. Poynard, T., V. Leroy, M. Cohard, et al., *Meta-analysis of interferon randomized trials in the treatment of viral hepatitis C: effects of dose and duration.* Hepatology, 1996. **24**: 778.

42. McHutchison, J.G., S. Gordon, E.R. Schiff, et al., *Interferon alfa-2b montherapy versus interferon alfa-2b plus ribavirin as initial treatment for chronic hepatitis C: results of a U.S. multicenter randomized controlled study.* New Engl J Med, 1998. **339**: 1485–92.

43. Poynard, T., P. Marcellin, S. Lee, et al., *Randomised trial of interferon alpha2b plus ribavirin for 48 weeks or for 24 weeks versus interferon alpha-2b plus placebo for 48 weeks for treatment of chronic infection with hepatitis C virus.* Lancet, 1998. **352**: 1426–32.

44. Schalm, S.W., O. Weiland, B.E. Hansen, et al., *Interferon-ribavirin for chronic hepatitis C with and without cirrhosis: analysis of individual patient data of six controlled studies.* Gastroenterology, 1999. **117**: 408–13.

45. Zeuzem, S., S.V. Feinman, J. Rasenack, et al., *Peginterferon alfa-2a in patients with chronic hepatitis C.* N Engl J Med, 2000. **343**: 1666–72.

46. Lindsay, K.L., C. Trepo, T. Heintges, et al., *A randomized, double-blind trial comparing pegylated interferon alfa-2b to interferon alfa-2b as initial treatment for chronic hepatitis C.* Hepatology, 2001. **34**: 395–403.

47. Kozlowski, A., and J.M. Harris, *Improvements in protein pegyylation: pegylated interferons for treatment of hepatitis C.* J Controlled Release, 2001. **72**: 217–224.

48. Fried, M.W., M.L. Shiffman, R.K. Reddy, et al., *Pegylated interferon alfa-2a in combination with ribavirin: efficacy and safety results*

from a phase III, randomized, actively controlled, multicenter study (abstract). Gastroenterology, 2001. **120**: A55.

49. Manns, M.P., J.G. McHutchison, S. Gordon, et al., *Peginterferon alfa-2b plus ribavirin compared to interferon alfa-2b plus ribavirin for the treatment of chronic hepatitis C: a randomized trial.* Lancet, 2001. **358**: 958–65.

50. Simmonds, P., *Variability of hepatitis C virus.* Hepatology, 1995. **21**: 570–83.

51. Lau, J.Y.N., M. Mizokami, J.A. Kolberg, et al., *Application of six hepatitis C virus subtyping systems to sera of patients with chronic hepatitis C in the United States.* J Infect Dis, 1994. **171**: 281–89.

52. Marcellin, P., N. Boyer, C. Degott, M. Martinot-Peignoux, V. Duchatelle, E. Giostra, et al., *Long-term histologic and viral changes in patients with chronic hepatitis C who responded to alpha interferon.* Liver, 1994. **14**: 302–7.

53. Poynard, T., J. McHutchison, G.L. Davis, Esteban, R. Mur, Z. Goodman, P. Bedossa, and J. Albrecht, *Impact of interferon alfa-2b and ribavirin on progression of liver fibrosis in patients with chronic hepatitis C.* Hepatology, 2000. **32**: 1131–7.

54. Davis, G.L., J.E. Albright, S.F. Cook, and D.M. Rosenberg, *Projecting the future healthcare burden from hepatitis C in the United States* (abstract). Hepatology, 1998. **28**: 390A.

55. Wong, JB., G.M. McQuillan, J.G. McHutchison, and T. Poynard, *Estimating future hepatitis C morbidity, mortality, and costs in the United States.* Am J Public Health, 2000. **90**: 1562–69.

56. Davis, G.L., J.E. Albright, S.F. Cook, and D.M. Rosenberg, *Projecting the future healthcare burden from hepatitis C in the United States.* Hepatology, 1998. **28**: 390A.

■ SECTION THREE ■

■ 7.3. INTERFERON-β IN THE TREATMENT OF MULTIPLE SCLEROSIS

Multiple sclerosis (MS) is a chronic, inflammatory, demyelinating disease of the central nervous system (CNS), affecting 250,000 to 350,000 people in the United States. MS derives its name from the multiple scarred areas or lesions in brain tissues visible on macroscopic examination [1,2].

These demyelinated lesions or plaques vary in size from 1 or 2 mm to several cen-

■TABLE 7.5. Multiple sclerosis disease presentation and the location of CNS lesions

Lesion Location	Symptoms
Cerebrum and cerebellum	Balance problems, speech problems, coordination, tremors
Motor nerve tracts	Muscle weakness, spasticity paralysis, vision problems, bladder, bowel problems
Sensory nerve tract	Altered sensation, numbness, prickling, burning sensation

timeters and are characterized by the presence of both activated $CD4^+$ T-lymphocytes and myelin-laden macrophages, both of which may orchestrate demyelination. At sites of inflammation, the blood-brain barrier is disrupted, but the vessel wall itself is often preserved, distinguishing the MS lesion from other vasculitis. Poor correlation exists between the plaque burden and the severity of disease.

Viral and autoimmune etiologies, as well as genetic and environmental contributions, have been proposed; the specific cause of this disease, however, is yet to be identified. Depending on the location of the lesion within the CNS, MS patients may experience varying signs and symptoms, as outlined in Table 7.5. The disorder has a typical onset in early adulthood and largely affects women. Fifteen years after onset, about 50% of patients require help in walking. Advanced magnetic resonance imaging (MRI) may allow clinicians to follow the progression of the disease and monitor response to treatment.

About 80% of MS patients have an episodic form of the disease called relapsing-remitting multiple sclerosis. An episode typically starts with sensory disturbances, limb weakness, and clumsiness; symptoms evolve over several days, stabilize, and then often improve spontaneously or in response to corticosteroids. In the initial phase of the disease, relapses are generally followed by complete or nearly complete clinical recovery. However, persistent signs of CNS dysfunction may develop after a relapse, and the disease may progress between relapses (secondary progressive MS).

The recent addition of interferon-β-1a and interferon-β1b to the therapeutic arsenal for the treatment of MS aims to shut down inflammation at the blood-brain barrier and thereby reduce the rate of relapse and decrease frequency and severity of MS disease symptoms. Both β-interferons have demonstrated benefits in the treatment of patients with established MS, including slowing the progression of physical disability, reducing the rate of clinical relapses, and reducing the development of brain lesions, as assessed by MRI. Several trials have found that interferon-β1b (*Betaseron*) reduced the frequency of relapse by approximately 30% [3–5]. These studies also suggested a trend toward a delay in the progression of disability. Interferon-β1a (*Avonex*) was subsequently found to reduce the frequency of relapse [6–8].

While the specific mechanisms of action of interferon-β1a and interferon-β1b in MS are not fully understood, each interferon has a number of immune-mediating activities (see Section 7.1). A recent review article on multiple sclerosis observed: "The interferons reduce the proliferation of T cells and the production of tumor necrosis factor α, decrease antigen presentation, alter cytokine production to favor ones governed by type 2 helper T (Th₂) cells, increase the secretion of interleukin-10, and reduce the passage of immune cells across the blood-brain barrier by means of their effects on adhesion molecules, chemokines, and proteases" [2].

Opinions vary on when to initiate treatment with interferon-β. According to the National Multiple Sclerosis Society, these agents should be considered in patients with relapsing-remitting MS who have had recent relapses. Some neurologists delay treatment until there is a more prolonged history of recurrent relapses, because, in their experience, patients may have a benign early course.

Two recent trials report that treatment of patients earlier in the course of MS—after a first acute clinical demyelinating event—is of value [9,10]. The results of these studies suggest that interferon-β-1a may delay the development of a second bout of demyelination. These reports may influence decisions regarding the timing of interferon therapy, but the inconvenience of medication requiring injection, treatment-related side effects, cost, and the lack of evidence of an important long-term benefit of interferon-β will deter many specialists from starting treatment earlier in the course of the disease.

Other drug therapies for the treatment of MS are corticosteroids, glatiramer acetate (*Copaxone*), high-dose interferon-β1a (*Rebif*), and mitoxantrone (*Novantrone*). Corticosteroids (e.g., intravenous methylprednisolone) are often used to treat clinically significant relapses in an effort to hasten recovery. Glatiramer, also called copolymer-1, is a mixture of synthetic polypeptides containing the L-amino acids, such as glutamic acid, alanine, lysine, and tyrosine. It may promote the proliferation of immune cells that produce Th_2 cytokines; compete with myelin basic protein for presentation of MHC class II molecules and thereby inhibit antigen-specific T-cell activation; alter the function of macrophages; and induce antigen-specific suppressor T cells [2].

Like the β-interferons, glatiramer reduces the frequency of relapse. A high-dose formulation of interferon-β-1a, *Rebif*, has recently been approved in the United States, Canada, and Europe. The Food and Drug Administration has also approved *Novantrone*, a cytotoxic agent with associated anti-inflammatory activities, for the treatment of patients with secondary progressive multiple sclerosis based on a phase III trial that provided clinical and MRI evidence of reduced disease activity.

REFERENCES

1. Harrison, T.R., and A.S. Fauci, *Harrison's principles of internal medicine*. 14th ed. 1998, New York: McGraw-Hill Health Professions Division. 2 v. (xxvi, 2569, 11, 170).

2. Noseworthy, J.H., C. Lucchinetti, M. Rodriguez, and B.G. Weinshenker, *Multiple Sclerosis*. N Engl J Med, 2000. **343**(13): 938–52.

3. Paty, D.W., and D.K. Li, *Interferon beta-1b is effective in relapsing-remitting multiple sclerosis. II. MRI analysis results of a multicenter, randomized, double-blind, placebo-controlled trial. UBC MS/MRI Study Group and the IFNB Multiple Sclerosis Study Group*. Neurology, 1993. **43**(4): 662–7.

4. *Interferon beta-1b in the treatment of multiple sclerosis: final outcome of the randomized controlled trial. The IFNB Multiple Sclerosis Study Group and The University of British Columbia MS/MRI Analysis Group*. Neurology, 1995. **45**(7): 1277–85.

5. *Interferon beta-1b is effective in relapsing-remitting multiple sclerosis. I. Clinical results of a multicenter, randomized, double-blind, placebo-controlled trial. The IFNB Multiple Sclerosis Study Group*. Neurology, 1993. **43**(4): 655–61.

6. Simon, J.H., L.D. Jacobs, M. Campion, et al., *Magnetic resonance studies of intramuscular interferon beta-1a for relapsing multiple sclerosis. The Multiple Sclerosis Collaborative Research Group*. Ann Neurol, 1998. **43**(1): 79–87.

7. Jacobs, L.D., D.L. Cookfair, R.A. Rudick, et al., *Intramuscular interferon beta-1a for disease progression in relapsing multiple sclerosis. The Multiple Sclerosis Collaborative Research Group (MSCRG)*. Ann Neurol, 1996. **39**(3): 285–94.

8. Rudick, R.A., J.A. Cohen, B. Weinstock-Guttman, R.P. Kinkel, and R.M. Ransohoff, *Management of multiple sclerosis.* N Engl J Med, 1997. **337**(22): 1604–11.

9. Jacobs, L.D., R.W. Beck, J.H. Simon, R.P. Kinkel, C.M. Brownscheidle, T.J. Murray, N.A. Simonian, P.J. Slasor, A.W. Sandrock, and The CHAMPS Study Group, *Intramuscular Interferon Beta-1A Therapy Initiated during a First Demyelinating Event in Multiple Sclerosis.* N Engl J Med, 2000. **343**(13): 898–904.

10. Comi, G., M. Filippi, F. Barkhof, L. Durelli, G. Edan, O. Fernandez, H. Hartung, P. Seeldrayers, P.S. Sorensen, M. Rovaris, V. Martinelli, and O.R. Hommes, *Effect of early interferon treatment on conversion to definite multiple sclerosis: a randomised study.* Lancet, 2001. **357**(9268): 1576–82.

■ SECTION FOUR ■

■ 7.4. MONOGRAPHS

INTERFERON ALFACON-1

Trade name: *Infergen*

Manufacturer: Amgen, Inc., Thousand Oaks, CA

Indications: Treatment of chronic hepatitis C virus (HCV) infection in patients 18 years of age or older with compensated liver disease who have anti-HCV serum antibodies and/or the presence of HCV RNA

Approval date: 6 October 1997

Type of submission: Biologics license application

A. General description Interferon alfacon-1 is a recombinant nonnaturally occurring type-I interferon derived from a synthetically constructed gene composed primarily of the consensus amino acids at each position in the human alpha interferon family. The 166 amino-acid (19.4 kDa) sequence of interferon alfacon-1 was derived by scanning the sequences of several natural interferon-alpha subtypes and assigning the most frequently observed amino acid in each corresponding position. Four additional amino-acid changes were made to facilitate the molecular construction, and a corresponding synthetic DNA sequence was constructed using chemical synthesis methodology. Interferon alfacon-1 is produced in *E. coli* cells that have been genetically altered by insertion of the synthetically constructed sequence that codes for interferon alfacon-1. Interferon alfacon-1 differs from interferon alfa-2 at 20/166 amino acids (88% homology), and has approximately 34% homology with human interferon-beta.

B. Indications and use *Infergen* is indicated for treating chronic hepatitis C virus (HCV) infection in adults with compensated liver disease who have anti-HCV serum antibodies and/or the presence of HCV RNA. It is also effective in the subsequent treatment of patients who did not respond or relapsed after initial interferon therapy. In some patients with chronic HCV infection, *Infergen* normalizes serum alanine aminotransferase (ALT) concentrations, reduces serum HCV RNA concentrations to undetectable quantities (<100 copies/ml), and improves liver histology.

C. Administration
  Dosage form *Infergen* is a sterile, clear, preservative-free liquid. The product is available in single-use vials and prefilled syringes. *Infergen* vials and prefilled syringes contain 0.03 mg/ml of interferon alfacon-1.
  Route of administration *Infergen* is to be administered undiluted by subcutaneous injection.
  Recommended dosage and monitoring requirements The recommended dose of

Infergen for treatment of chronic HCV infection is 9 μg three times weekly, administered subcutaneously for 24 weeks. At least 48 hours should elapse between doses of *Infergen*. Patients who tolerated previous interferon therapy and did not respond or relapsed following its discontinuation may be subsequently treated with 15 μg of Infergen thrice weekly for 6 months. If home use is determined to be desirable by the physician, instruction on appropriate use should be given by a health care professional. There are significant differences in specific activities among interferons. Health care providers should be aware that changes in interferon brand may require adjustments of dosage and/or change in route of administration. Patients should be warned not to change brands of interferon without medical consultation.

D. Pharmacology and pharmaceutics

Clinical pharmacology According to the product label, interferons are a family of naturally occurring proteins that are produced and secreted by cells in response to viral infections or to various synthetic and biological inducers. Type-I interferons include a family of more than 25 interferon alphas as well as interferon beta and interferon omega. All type-I interferons share common biological activities generated by binding of interferon to the cell-surface receptor, leading to the production of several interferon-stimulated gene products. Type-I interferons induce biologic responses that include antiviral, antiproliferative and immunomodulatory effects. While all alpha interferons have similar biological effects, not all the activities are shared by each alpha interferon, and in many cases the extent of activity varies substantially for each interferon subtype. The antiviral, antiproliferative NK-cell activation and gene-induction activities of *Infergen* have been compared with other recombinant alfa interferons in *in vitro* assays and have demonstrated similar ranges of activity. *Infergen* exhibited at least five times higher specific activity *in vitro* than interferon alfa-2a (*Roferon*) and interferon alfa-2b (*Intron*), both of which are used in the treatment of HCV infection.

Pharmacokinetics The pharmacokinetic properties of *Infergen* have not been evaluated in patients with chronic hepatitis C. Pharmacokinetic profiles were evaluated in normal, healthy volunteer subjects after subcutaneous injection of interferon alfacon-1 at doses up to 9 μg. Plasma levels of interferon alfacon-1 at any dose were too low to be detected. However, analysis of *Infergen*-induced cellular products—induction of 2′5′ oligoadenylate synthetase and (beta)-2 microglobulin—after treatment in these subjects revealed a statistically significant, dose-related increase in the area under the curve (AUC).

Disposition No information available.

Drug interactions No formal drug interaction studies have been conducted with *Infergen*. However, *Infergen* should be used cautiously in patients who are receiving agents that are metabolized via the cytochrome P-450 pathway. Patients taking drugs that are metabolized by this pathway should be monitored closely for changes in the therapeutic and/or toxic levels of concomitant drugs. Interferons have been reported to inhibit cytochrome P-450-mediated drug metabolism.

E. Therapeutic response Efficacy of *Infergen* therapy was determined by measurement of serum alanine aminotransferase (ALT) concentrations at the end of therapy (24 weeks) and following 24 weeks of observation after the end of treatment of adults with chronic HCV infection. Serum HCV RNA was also assessed using a quantitative reverse transcriptase polymerase chain reaction (RT-PCR). At the end of 24 weeks of treatment, ALT normalization was observed in 39% of patients on *Infergen* and in 35% of patients on interferon alfa-2b (*Intron* A). Only 17% of patients in each group had normal ALT levels after 24 weeks of follow-up. After treatment, 33% of patients on *Infergen* had HCV RNA below detectable limits, compared with 25% of patients on interferon alfa-2b. After another 24 weeks, however, less than 10% of patients in either group had undetectable HCV RNA.

F. Role in therapy Interferon alfacon-1 is an option for the initial treatment of patients with HCV infection and for subsequent treatment in patients who did not respond or relapsed after initial interferon therapy. For initial treatment, however, other therapies are usually more effective.

G. Other applications Limited data show possible effectiveness of *Infergen* in the treatment of persistent genital papillomavirus disease and hematologic improvement in patients with hairy-cell leukemia.

INTERFERON ALFA-2A

Trade name: ***Roferon-A***

Manufacturer: Roche Laboratories, Nutley, NJ

Indications: Treatment of chronic hepatitis C, hairy-cell leukemia and AIDS-related Kaposi's sarcoma, as well as for the treatment of chronic phase, Philadelphia chromosome-positive chronic myelogenous leukemia (CML)

Approval date: 21 November 1998

Type of submission: Biologic license application

A. General description *Roferon-A* (interferon alfa-2a recombinant) is manufactured by recombinant DNA technology that employs a genetically engineered *Escherichia coli* bacterium containing DNA that codes for the naturally occurring human protein. Interferon alfa-2a recombinant is a highly purified protein containing

165 amino acids, and it has an approximate molecular weight of 19 kDa.

B. Indications and use *Roferon-A* is indicated for the treatment of chronic hepatitis C, hairy-cell leukemia, and AIDS-related Kaposi's sarcoma in patients 18 years of age or older. In addition it is indicated for chronic phase, Philadelphia chromosome (Ph) positive chronic myelogenous leukemia (CML) patients who are minimally pretreated (within 1 year of diagnosis).

C. Administration

Dosage form *Roferon-A* is available as single-use injectable solutions, single-use prefilled syringes, and multidose injectable solutions.

Route of administration *Roferon-A* can be administered intramuscularly or subcutaneously. The route of administration for the prefilled syringe is subcutaneous only.

Recommended dosage and monitoring requirements The recommended dose of *Roferon-A* for chronic hepatitis C is 3 million international units (MIU) three times a week for 12 months. The recommended dose for CML is 9 MIU daily; optimal duration of therapy has not been determined. The recommended dose for hairy cell leukemia is 3 MIU daily for 16 to 24 weeks. The recommended dose for AIDS-related Kaposi's sarcoma is 36 MIU daily for 10 to 12 weeks. There are five types of interferon alfa currently available in the United States: interferon alfa-2a, peginterferon alfa-2a, interferon alfa-2b, interferon alfacon-1, and interferon alfa n3. Doses are not consistent for these preparations.

D. Pharmacology and pharmaceutics

Clinical pharmacology The mechanism by which interferon alfa-2a recombinant, or any other interferon, exerts antitumor or antiviral activity is not clearly understood. However, it is believed that direct antiproliferative action against tumor cells, inhibition of virus replication, and modulation of the host immune response play important roles in antitumor and antiviral activity. Interferon alfa-2a recombinant is similar to interferon alfa-2b recombinant in that both are pure clones of single interferon subspecies; they differ by the sequence of two amino acids.

Pharmacokinetics A mean elimination half-life of approximately 5 hours has been reported after intravenous doses of *Roferon-A*. Pharmacokinetic parameters are similar in healthy subjects and cancer patients after intramuscular doses. Dose-proportionate increases in serum levels occur with doses up to 198 MIU. The bioavailability of interferon alfa-2a after intramuscular administration is 80% to 83%, and its volume of distribution is 0.223 to 0.748 liter/kg. The total body clearance of interferon alfa-2a has been reported to range from 2.14 to 3.62 ml/min per kg.

Disposition Like other alpha interferons, interferon alfa-2a undergoes extensive proteolytic degradation in the kidney (during tubular reabsorption); hepatic metabolism is minimal.

Drug interactions Alfa interferons may affect the oxidative metabolic process by reducing the activity of hepatic microsomal cytochrome enzymes in the P450 group. Although the clinical relevance is still unclear, this suppression should be taken into account when prescribing concomitant therapy with drugs metabolized in this manner. *Roferon-A* has been reported to reduce the clearance of theophylline. Caution should also be exercised when administering *Roferon-A* in combination with other potentially myelosuppressive agents. The neurotoxic, hematotoxic, or cardiotoxic effects of previously or concurrently administered drugs may be increased by interferons. Interactions could occur following concurrent administration of centrally acting drugs. Use of *Roferon-A* in conjunction with interleukin-2 may potentiate risks of renal failure.

E. Therapeutic response According to the product label, studies have shown that *Roferon-A* can normalize serum transaminases, improve liver histology, and reduce viral load in patients with chronic hepatitis C virus (HCV) infection. Other studies have shown that *Roferon-A* can produce clinically meaningful tumor regression or disease stabilization in patients with hairy-cell leukemia or in patients with AIDS-related Kaposi's sarcoma. In Ph-positive CML, *Roferon-A* supplemented with intermittent chemotherapy has been shown to prolong overall survival and to delay disease progression compared to patients treated with chemotherapy alone. In addition *Roferon-A* has been shown to produce sustained complete cytogenetic responses in a small subset of patients with CML in chronic phase.

F. Role in therapy According to Micromedex, *Roferon-A* is recommended as the drug of choice in renal carcinoma and the chronic phase of chronic myelogenous leukemia. The role of *Gleevec* (Imatinib) in CML is yet to be determined, but it may replace the use of interferon alfa-2a. *Roferon-A* is an alternative (for unresponsive/intolerant patients) to current regimens of choice in hairy-cell leukemia, multiple myeloma, metastatic melanoma, and AIDS-related Kaposi's sarcoma. Other alpha interferons appear to have similar efficacy and can be used in lieu of *Roferon-A* in some instances. In particular, interferon alfa-2b (*Intron*) can be considered to have the same role as interferon alfa-2a in chronic hepatitis C, Kaposi's sarcoma, and hairy-cell leukemia.

G. Other applications *Roferon-A* has also been used in renal cell carcinoma, multiple myeloma, metastatic melanoma, and primary resected cutaneous melanoma.

H. Other considerations   Interferon alfa-2a has been designated an orphan drug product for the treatment of chronic myelogenous leukemia, AIDS-related Kaposi's sarcoma, renal cell carcinoma, metastatic malignant melanoma, and esophageal and colorectal cancer.

INTERFERON ALFA-2B

Trade name: *Intron A*

Manufacturer: Schering Division of Schering-Plough, Kenilworth, NJ

Indications: Treatment of patients with hairy-cell leukemia, malignant melanoma, follicular non-Hodgkin's lymphoma (NHL), AIDS-related Kaposi's sarcoma, chronic hepatitis C, and the skin disorder, condylomata acuminata

Approval date: 16 November 1997

Type of submission: Biologic license application

A. General description   *Intron A* (interferon alfa-2b recombinant) is a recombinantly derived water-soluble protein with a molecular weight of 19.3 kDa. It is obtained from the bacterial fermentation of a strain of *Escherichia coli* bearing a genetically engineered plasmid containing an interferon alfa-2b gene from human leukocytes.

B. Indications and use   *Intron A* is indicated for the treatment of adult patients with hairy-cell leukemia, as an adjuvant to surgical treatment for adult patients with malignant melanoma who are free of disease but at high risk for recurrence, for initial treatment of clinically aggressive follicular NHL in conjunction with anthracycline-containing combination chemotherapy in adult patients, for intralesional treatment of selected adult patients with condylomata acuminata (venereal or genital warts) involving external surfaces of the genital and perianal areas, and for

the treatment of selected adult patients with AIDS-related Kaposi's sarcoma. *Intron A* is also indicated for the treatment of chronic hepatitis C in adult patients with compensated liver disease who have a history of blood or blood-product exposure and/or are HCV antibody positive and for the treatment of chronic hepatitis B in children and adults with compensated liver disease.

C. Administration
    Dosage form   *Intron A* is supplied as a sterile lyophilized powder for injection and a solution for injection. *Intron A* solution for injection in multidose pens contains a prefilled, multidose cartridge for subcutaneous administration.
    Route of administration   *Intron A* is intended for intramuscular, subcutaneous, intralesional, or intravenous injection.
    Recommended dosage and monitoring requirements   The recommended dose of *Intron A* for hairy-cell leukemia is 2 million international units (MIU)/m^2 intramuscularly or subcutaneously three times a week. The recommended dose for malignant melanoma includes a 4-week induction phase of 20 MIU intravenously 5 days a week, followed by maintenance therapy of 10 MIU subcutaneously three times a week for a total of 48 weeks. The recommended dose for follicular lymphoma is 5 MIU subcutaneously three times a week for up to 18 months (with anthracycline-containing chemotherapy). The recommended dose for condylomata acuminata is 1 MIU intralesionally three times a week for 3 weeks. The recommended dose for AIDS-related Kaposi's sarcoma is 30 MIU/m^2 intramuscularly or subcutaneously three times a week. The recommended dose for chronic hepatitis C is 3 MIU intramuscularly or subcutaneously three times a week. The recommended dose for chronic hepatitis B in adults is 5 MIU intramuscularly or subcutaneously daily or 10 MIU three times a week for 16 weeks. The recommended dose for chronic hepatitis B in children is

$3\,\text{MIU/m}^2$ subcutaneously three times a week for the first week, then $6\,\text{MIU/m}^2$ for a total of 16 to 24 weeks.

D. Pharmacology and pharmaceutics

Clinical pharmacology The interferons are a family of naturally occurring small proteins and glycoproteins produced and secreted by cells in response to viral infections and to synthetic or biological inducers. Interferons as a class possess anti-tumor, antiviral, and immunomodulating activity. Interferon alfa-2b recombinant is similar to interferon alfa-2a recombinant in that both are pure clones of single interferon subspecies; they differ by the sequence of two amino acids.

Pharmacokinetics The elimination half-life of interferon alfa-2b after subcutaneous or intramuscular administration is 2 to 3 hours.

Dispostion Interferon alfa-2b undergoes extensive proteolytic degradation in the kidney (during tubular reabsorption); hepatic metabolism is minimal.

Drug Interactions According to the product label, interactions between *Intron A* and other drugs have not been fully evaluated. Caution should be exercised when administering *Intron A* therapy in combination with other potentially myelo-suppressive agents such as zidovudine. Concomitant use of alfa interferon and theophylline decreases theophylline clearance, resulting in a 100% increase in serum theophylline levels.

E. Therapeutic response The percentage of patients with hairy-cell leukemia who required red blood cell or platelet transfusions decreased significantly during treatment with *Intron A*, and the percentage of patients with confirmed and serious infections declined as granulocyte counts improved. *Intron A* therapy produced a significant increase in relapse-free and overall survival in patients with malignant melanoma. Patients receiving the combination of *Intron A* therapy plus che-

motherapy had a significantly longer progression-free survival than patients who received chemotherapy alone. *Intron A* treatment of condylomata was significantly more effective than placebo, as measured by disappearance of lesions, decreases in lesion size, and by an overall change in disease status. Studies in patients with chronic hepatitis C demonstrated that *Intron A* therapy can produce meaningful effects on this disease, manifested by normalization of serum alanine aminotransferase and reduction in liver necrosis and degeneration.

F. Role in therapy According to Micromedex, *Intron A* is the agent of choice for the treatment of malignant melanoma (surgical adjuvant) and chronic hepatitis B. It is also the drug of choice for chronic hepatitis C in combination with ribavirin. The combination is available under the name *Rebetron*. *Intron A* is an alternative (unresponsive/intolerant patients) to pentostatin in hairy-cell leukemia, topical podophyllin regimens in condyloma acuminata, and standard regimens in multiple myeloma.

G. Other applications Intron A has been used in the treatment of many viral infections and malignancies, but documentation of safety and efficacy is poor.

H. Other considerations Interferon alfa-2b has been designated an orphan product for use in the treatment of Kaposi's sarcoma.

INTERFERON BETA-1A

Trade name: *Avonex*

Manufacturer: Biogen, Inc., Cambridge, MA

Indications: Treatment of relapsing-remitting forms of multiple sclerosis (MS) to slow the accumulation of physical disability and decrease the frequency of clinical exacerbations

<u>Approval date</u>: 17 May 1996

<u>Type of submission</u>: Product license application

A. General description   Interferon beta-1a is produced by Chinese hamster ovary (CHO) cells into which the human interferon beta gene has been introduced. The amino-acid sequence of the recombinant protein produced by these cells is identical to naturally occurring interferon beta. Interferon beta-la is a single-chain glycosylated polypeptide, 166 amino-acid residues in length, with an approximate molecular weight of 22.5 kDa.

B. Indications and use   *Avonex* is indicated for the treatment of relapsing forms of multiple sclerosis to slow the accumulation of physical disability and decrease the frequency of clinical exacerbations. Safety and efficacy in patients with chronic progressive multiple sclerosis have not been evaluated.

C. Administration

   Dosage form   *Avonex* is formulated as a sterile, lyophilized powder for injection after reconstitution with supplied diluent or sterile water for injection. Each 1.0 ml of reconstituted *Avonex* contains 30 μg of interferon beta-1a.

   Route of administration   *Avonex* is administered by intramuscular injection. Subcutaneous administration should not be substituted for intramuscular administration. Subcutaneous and intramuscular administration have been observed to have nonequivalent pharmacokinetic and pharmacodynamic parameters following administration to healthy volunteers.

   Recommended dosage and monitoring requirements   The recommended dosage of *Avonex* for the treatment of relapsing forms of multiple sclerosis is 30 μg injected intramuscularly once a week. *Avonex* is intended for use under the supervision of a physician. Patients may self-inject if their physician determines that it is appropriate, after proper training in intramuscular injection technique. In addition to those laboratory tests normally required for monitoring patients with multiple sclerosis, complete blood and differential white blood cell counts, platelet counts, and blood chemistries, including liver function tests, are recommended during *Avonex* therapy. During clinical trials, these tests were performed at least every 6 months.

D. Pharmacology and pharmaceutics

   Clinical pharmacology   *Avonex* is indicated for the treatment of relapsing-remitting forms of MS. MS is a chronic inflammatory disorder of the central nervous system (CNS) that results in injury to the myelin and to myelin-producing cells. Two-thirds to three-quarters of affected individuals are women. Clinical onset is usually between the ages of 20 and 40 years. Four different clinical forms of MS have been recognized. The first, relapsing-remitting type of MS (RRMS), accounts for approximately 85% to 90% of MS cases at onset. It is characterized by discrete attacks of neurologic dysfunction. These attacks generally evolve acutely over days to weeks. Over the ensuing several weeks to months, the large majority of patients experience a substantial recovery of function (but not necessarily complete). The specific interferon-induced proteins and mechanisms by which *Avonex* exerts its effects in multiple sclerosis have not been fully defined. Clinical studies conducted in MS patients showed that interleukin 10 (IL-10) levels in cerebrospinal fluid (CSF) were increased in patients treated with *Avonex* compared to placebo. However, no relationship has been established between absolute levels or changes in levels of IL-10 and clinical outcome in multiple sclerosis.

Pharmacokinetics After completion of phase III clinical trials that supported FDA approval, Biogen developed a new CHO cell line that carried the interferon beta gene and began production of a product referred to as BG9216. These CHO cells were adapted for suspension culture. Data supporting the use of this cell line were submitted to the FDA and showed that the specific activity of BG9216 was greater than the material used in the clinical trials, BG9015. In addition pharmacokinetic bioequivalence studies in humans showed that BG9216 was not equivalent to BG9015. Based on the biochemical and pharmacokinetic differences, Biogen was informed that BG9216 was not comparable to BG90 15. Biogen then developed another interferon beta-1a cell line, and the product produced by this cell line was designated BG9418. BG9418 has been extensively characterized and compared in side-by-side analyses with BG9015. Biological, biochemical, and biophysical analyses have shown that the two molecules are comparable. Biological activities of each molecule are similar using several different assays. Finally, pharmacokinetic studies in humans using the two molecules revealed a pattern of clearance from the blood that was determined to be equivalent by rigorous statistical analyses. For these reasons the agency determined that BG9015 and BG9418 are comparable and that clinical data derived from the use of BG9015 can support the approval of the BG9418 molecule.

After an intramuscular dose of *Avonex*, serum levels of interferon beta-1a peak between 3 and 15 hours and then decline at a rate consistent with a 10-hour elimination half-life. The terminal half-life of interferon beta-1a after intravenous administration has been estimated at between 3 and 4 hours. Serum levels of interferon beta-1a may be sustained after intramuscular administration due to prolonged absorption from the injection site. Systemic exposure, as determined by area under the curve and peak plasma concentration, is greater following intramuscular than subcutaneous administration. According to one report the volume of distribution of interferon beta-1a is about 62 liters, and its total body clearance is 334 ml/h per kg.

Disposition No information available

Drug interactions No drug interaction studies have been conducted with *Avonex*. Other interferons have been found to reduce cytochrome P-450-mediated drug metabolism. Hepatic microsomes isolated from *Avonex*-treated rhesus monkeys showed no influence on hepatic P-450 enzyme metabolism activity.

E. Therapeutic response The primary outcome assessment for the pivotal clinical trial supporting the approval of *Avonex* was time to progression in disability, measured as an increase in the Expanded Disability Status Scale (EDSS) of at least 1.0 point that was sustained for at least 6 months. An increase in EDSS score reflects accumulation of disability. This end point was used to ensure that progression reflected a permanent increase in disability rather than a transient effect due to an exacerbation. Time to onset of sustained progression in disability was significantly longer in patients treated with *Avonex* than in patients receiving placebo. The Kaplan-Meier estimate of the percentage of MS patients progressing by the end of 2 years was 34.9% for placebo-treated patients and 21.9% for *Avonex*-treated patients, indicating a slowing of the disease process.

F. Place in therapy Avonex has shown significant advantages over interferon beta-1b (*Betaseron*) in the treatment of multiple sclerosis. It is administered once a week rather than every other day, and it is not associated with the high incidence of injection site skin necrosis reported with interferon beta-1b in some studies. It appears (based on indirect comparison) at least as effective as interferon beta-1b in reducing the signs of physical disability, and in decreasing the frequency of exacerbations and increasing the periods of stability between exacerbations of the disease.

G. Other considerations In 1993 the FDA approved an interferon beta (interferon beta-1b, *Betaseron*) and granted the product orphan drug status in the treatment of MS. A clinical trial with *Betaseron* showed efficacy in reducing the rate of exacerbations by approximately one third. The side effects, however, were considerable. *Betaseron* differs structurally from natural human interferon-beta; it lacks the amino terminus methionine, the cysteine at position 17 is replaced by a serine, and it is not glycosylated. The Biogen product, *Avonex* (interferon beta-1a), retains the terminal methionine, has no substitutions, and is glycosylated. Under the regulations of the Orphan Drug Act, *Avonex* was determined to be a different product from *Betaseron*, because of a difference in safety profile involving the occurrence of injection site skin necrosis. Analyses of safety data submitted in the *Avonex* Product License Application (PLA) showed that no injection site necrosis was reported in 158 patients treated with interferon beta-1a. In contrast, the incidence of injection site necrosis reported in the *Betaseron* PLA was 5% in 124 patients treated with interferon beta-1b. Further supportive evidence for a difference in skin necrosis incidence is suggested by the 85% incidence of injection site reactions in the *Betaseron* phase III study versus only 4% in the *Avonex* phase III study.

INTERFERON BETA-1B

Trade name: ***Betaseron***

Manufacturer: Chiron Corporation, Emeryville, CA

Distributed by: Berlex Laboratories, Richmond, CA

Indications: Treatment of ambulatory patients with relapsing-remitting multiple sclerosis to reduce the frequency of clinical exacerbations

Approval date: 23 July 1993

Type of submission: Product license application

A. General description Interferon beta-1b is a protein product produced by recombinant DNA techniques. It is manufactured by bacterial fermentation of a strain of *Escherichia coli* that bears a genetically engineered plasmid containing the gene for human interferon beta$_{ser17}$. The native gene was obtained from human fibroblasts and altered in a way that substitutes serine for the cysteine residue found at position 17. Interferon beta-1b has 165 amino acids and an approximate molecular weight of 18.5 kDa. It does not include the carbohydrate side chains found in the natural material. Natural interferon beta is a glycoprotein produced by animal or human cells primarily in response to viruses. It contains 166 amino-acid residues and has a molecular weight of approximately 20 kDa.

B. Indications and use *Betaseron* is indicated for use in ambulatory patients with relapsing-remitting multiple sclerosis to reduce the frequency of clinical exacerbations. The safety and efficacy of *Betaseron* in chronic-progressive MS has not been fully evaluated.

C. Administration

    Dosage form *Betaseron* is available as a sterile, lyophilized powder for injection. After reconstitution with accompanying diluent, *Betaseron* vials contain 0.25 mg interferon beta-1b per ml of solution.

Route of administration *Betaseron* is intended for subcutaneous injection. Patients should be instructed in injection techniques to assure the safe self-administration of *Betaseron*. Sites for self-injection include arms, abdomen, hips, and thighs.

Recommended dosage and monitoring requirements For prevention of exacerbations in patients with relapsing-remitting multiple sclerosis, the recommended dose of *Betaseron* is 0.25 mg injected subcutaneously every other day.

D. Pharmacology and pharmaceutics

Clinical pharmacology Relapsing-remitting MS is characterized by recurrent attacks of neurologic dysfunction followed by complete or incomplete recovery. The mechanisms by which *Betaseron* exerts its actions in multiple sclerosis (MS) are not clearly understood. However, it is known that the biologic response-modifying properties of interferon beta-1b are mediated through its interactions with specific cell receptors found on the surface of human cells. The binding of interferon beta-1b to these receptors induces the expression of a number of interferon-induced gene products that are believed to be the mediators of the biological actions of interferon beta-1b.

Pharmacokinetics The bioavailability of subcutaneous interferon beta-1b is 50%. In patients receiving single intravenous *Betaseron* doses up to 2.0 mg, increases in serum concentrations of interferon beta-1b were dose proportional. Mean serum clearance values ranged from 9.4 ml/min per kg to 28.9 ml/min per kg and were independent of dose. Mean terminal elimination half-life values ranged from 8.0 minutes to 4.3 hours and mean steady-state volume of distribution values ranged from 0.25 liter/kg to 2.88 liters/kg. Three-times-a-week intravenous dosing for 2 weeks resulted in no accumulation of interferon beta-1b in the serum of patients.

Disposition According to Micromedex, data on the metabolism of interferon beta are lacking. However, this compound is similar to interferon alfa, which is rapidly inactivated in body fluids and tissue.

Drug interactions Interactions between *Betaseron* and other drugs have not been fully evaluated.

E. Therapeutic response One study in MS patients demonstrated a 31% reduction in annual exacerbation rate, from 1.31 in the placebo group to 0.9 in the *Betaseron* 0.25 mg group. The proportion of patients free of exacerbations was 16% in the placebo group, compared with 25% in the *Betaseron* group. Over the 2-year period, there were 25 MS-related hospitalizations in the *Betaseron*-treated group compared to 48 hospitalizations in the placebo group.

F. Role in therapy *Betaseron* is useful for reducing symptomatic exacerbation in multiple sclerosis (MS) patients with relapsing-remitting disease. The drug should be considered in patients with clinically definite or laboratory-supported definite disease. It is not indicated in those patients with primary progressive MS. Interferon beta-1a (*Avonex*) has also demonstrated activity in MS patients.

G. Other applications Treatment with interferon beta-1b has demonstrated therapeutic effect in patients with secondary progressive multiple sclerosis by delaying sustained neurological deterioration. It may also be effective in the treatment of Kaposi's sarcoma and malignant glioma.

H. Other considerations Interferon beta-1b was designated an orphan product for use in the treatment of multiple sclerosis. Its marketing exclusivity, however, was cut short by the approval of *Avonex*.

INTERFERON GAMMA-1B

Trade name: *Actimmune*

Manufacturer: InterMune Pharmaceuticals, Inc., Burlingame, CA

Indications: Reducing the frequency and severity of serious infections associated with chronic granulomatous disease and delaying time to disease progression in patients with severe, malignant osteopetrosis

Approval date: 10 February 2000

Type of submission: Biologics license application

A. General description Interferon gamma-1b recombinant, a biologic response modifier, is a single-chain polypeptide containing 140 amino acids. It is produced by fermentation of a genetically engineered *Escherichia coli* bacterium containing the DNA that encodes for the human protein. *Actimmune* is a highly purified sterile solution consisting of noncovalent dimers of two identical 16.5 kDa monomers.

B. Indications and use *Actimmune* is indicated for reducing the frequency and severity of serious infections associated with chronic granulomatous disease. It is also indicated for delaying time to disease progression in patients with severe, malignant osteopetrosis, a rare genetic disease characterized by abnormally dense bone, due to defective resorption of immature bone.

C. Administration

Dosage form *Actimmune* is a sterile solution filled in single-dose vials. Each 0.5 ml of *Actimmune* contains interferon gamma-1b 100 μg.

Route of administration *Actimmune* is given by subcutaneous injection. Optimal injection sites are the right and left deltoid and anterior thigh.

Recommended dosage and monitoring requirements In patients with chronic granulomatous disease or malignant osteopetrosis (body surface area greater than $0.5\,m^2$), the recommended dose of *Actimmune* is $50\,\mu g/m^2$ subcutaneously three times weekly. Subcutaneous administration of 1.5 μg/kg three times weekly is recommended for smaller patients. Acetaminophen may be administered with *Actimmune* to reduce headache, myalgia, and fever. It should be started 4 hours before injecting *Actimmune* and continued for 24 hours after the injections. Furthermore bedtime administration of *Actimmune* may help to minimize the flu-like symptoms of fever, headache, chills, myalgia, and fatigue.

D. Pharmacology and pharmaceutics

Clinical pharmacology According to the product label, the most striking differences between interferon-gamma and other classes of interferon concern the immunomodulatory properties of this molecule. While gamma, alpha, and beta interferons share certain properties, interferon-gamma has potent phagocyte-activating effects not seen with other interferon preparations. These effects include the generation of toxic oxygen metabolites within phagocytes *in vitro*, which are capable of mediating the intracellular killing of selected microorganisms. To the extent that interferon-gamma is produced by antigen-stimulated T-lymphocytes and regulates the activity of immune cells, it is appropriate to characterize interferon-gamma as a lymphokine of the interleukin type. With respect to chronic granulomatous disease (an inherited disorder characterized by deficient phagocyte oxidative metabolism), clinical trials of the systemic administration of *Actimmune* provided evidence for a treatment-related enhancement of phagocyte function including elevation of superoxide levels and improved killing of *Staphylococcus aureus*. In severe, malignant osteopetrosis (another inherited disorder characterized by an osteoclast defect leading to bone overgrowth and deficient phagocyte oxidative metabolism), a treatment-related enhancement of superoxide production by phagocytes was observed *in situ*. Interferon gamma-1b was also found to enhance osteoclast function *in vitro*.

Pharmacokinetics According to product label, interferon gamma-1b is rapidly cleared after intravenous administration (1.4 liters/min) and slowly absorbed after intramuscular or subcutaneous injection. After intramuscular or subcutaneous injection, the apparent fraction of dose absorbed was greater than 90%. The mean elimination half-life after intravenous administration in healthy male subjects was 38 minutes. The mean elimination half-lives after intramuscular or subcutaneous dosing were 2.9 and 5.9 hours, respectively. Pharmacokinetic studies in patients with chronic granulomatous disease have not been performed.

Disposition No information available.

Drug interactions Interactions between *Actimmune* and other drugs have not been fully evaluated. Caution should be exercised when administering *Actimmune* in combination with other potentially myelosuppressive agents.

E. Therapeutic response A randomized, double-blind, placebo-controlled study of *Actimmune* in patients with chronic granulomatous disease was terminated early following demonstration of a highly statistically significant benefit of *Actimmune* therapy compared to placebo with respect to time to serious infection, the primary end point of the investigation. A controlled, randomized study in patients with severe, malignant osteopetrosis compared *Actimmune* plus calcitriol with calcitriol alone. The median time to disease progression was significantly delayed in the *Actimmune* plus calcitriol arm versus calcitriol alone.

F. Role in therapy According to Micromedex, the primary place in therapy of *Actimmune* is as prophylaxis against infection in patients with chronic granulomatous disease, and as an adjunct to conventional therapies (i.e., antimicrobial agents, bone marrow transplantation, leukocyte infusions) in these patients.

G. Other applications *Actimmune* may have application in the treatment of a variety of cancers (e.g., malignant melanoma, ovarian cancer), AIDS, rheumatoid arthritis, hepatitis B, and cutaneous leishmaniasis, atopic dermatitis, and keloidal scarring. There is preliminary evidence that *Actimmune* may benefit patients with pulmonary fibrosis.

H. Other considerations Interferon gamma-1b has been designated an orphan product for use in the treatment of chronic granulomatous disease, renal cell carcinoma, and severe congenital osteoporosis.

ALDESLEUKIN

Trade name: *Proleukin*

Manufacturer: Chiron Corporation, Emeryville, CA

Indications: Treatment of adults with metastatic renal cell carcinoma and adults with metastatic melanoma

Approval date: 5 May 1992

Type of submission: Product license application

A. General description Aldesleukin, a human recombinant interleukin-2 product, is a highly purified protein with a molecular weight of approximately 15.3 kDa. It is produced by recombinant DNA technology using a genetically engineered *Escherichia coli* strain containing an analogue of the human interleukin-2 gene. Genetic engineering techniques are used to modify the human IL-2 gene, and the resulting expression clone encodes a modified human interleukin-2. This recombinant form differs from native interleukin-2 in the following ways: (a) it is not glycosylated, because it is derived from *E. coli*; (b) the molecule has no N-terminal alanine; the codon for this amino acid was deleted during the genetic engineering procedure; and (c) the molecule has serine substituted for cysteine at amino acid position 125; this was accomplished by site-specific manipulation during the genetic engineering procedure.

B. Indications and use *Proleukin* is indicated for the treatment of adults with metastatic renal cell carcinoma. *Proleukin* is also indicated for the treatment of adults with metastatic melanoma. Careful patient selection is mandatory prior to the administration of the cytokine.

C. Administration
  Dosage form *Proleukin* is supplied as a sterile, lyophilized cake in single-use vials intended for injection on reconstitution with sterile water for injection. When reconstituted, each ml contains aldesleukin 18 million IU (1.1 mg).
  Route of administration *Proleukin* is given as an intravenous infusion over 15 minutes.
  Recommended dosage and monitoring requirements The dose of *Proleukin* in metastatic renal cell carcinoma is 600,000 units/kg every 8 hours for a maximum of 14 doses. Following 9 days of rest, the schedule is repeated for another 14

doses, for a maximum of 28 doses per course, as tolerated. During clinical trials, doses were frequently withheld for toxicity. The dose is the same for metastatic melanoma patients. Several medications have been shown to be useful in the management of patients receiving *Proleukin*, including nonsteroidal antiinflammatory agents, meperidine, antihistamines, and antiemetics and antidiarrheals.

D. Pharmacology and pharmaceutics
  Clinical pharmacology Aldesleukin is an antineoplastic agent and biological response modifier, a highly purified human recombinant interleukin-2 (IL-2) lymphokine. This modified IL-2 mimics the biological activities of native nonrecombinant IL-2. The exact mechanism by which aldesleukin mediates antitumor activity is unknown. The *in vivo* administration of aldesleukin in animals and humans produces multiple immunological effects in a dose-dependent manner. These effects include activation of cellular immunity with profound lymphocytosis, eosinophilia, and thrombocytopenia, and the production of cytokines, including tumor necrosis factor, IL-1 and gamma interferon. Interleukin-2 is one of several lymphocyte-produced messenger regulatory molecules that mediate immunocyte interactions and functions. The interaction of interleukin-2 with target cells occurs via a specific cell surface receptor present on T-lymphocytes and certain malignant lymphocytes.
  Pharmacokinetics Aldesleukin displays biphasic pharmacokinetics, with an alpha (distribution) half-life of 13 minutes and a beta (terminal) half-life of 85 minutes. In cancer patients, the mean clearance rate of aldesleukin is 268 ml/min.
  Disposition Aldesleukin is primarily eliminated by metabolism in the kidney. It is cleared from the circulation by both glomerular filtration and peritubular extraction with little or no bioactive protein excreted in the urine.

Drug interactions *Proleukin* may affect central nervous system function. Therefore interactions could occur following concomitant administration of psychotropic drugs. Concurrent administration of drugs possessing nephrotoxic, myelotoxic, cardiotoxic, or hepatotoxic effects with *Proleukin* may increase toxicity in these organ systems. Reduced kidney and liver function secondary to *Proleukin* treatment may delay elimination of concomitant medications and increase the risk of adverse events from those drugs. Beta-blockers and other antihypertensives may potentiate the hypotension seen with *Proleukin*.

E. Therapeutic response In renal cell carcinoma studies, an objective response was seen in 15% of patients, with 7% complete and 8% partial responders. In metastatic melanoma studies, an objective response was seen in 16% of patients, with 6% complete and 10% partial responders. In patients with HIV infection, *Proleukin* increases the CD4 count, with no effect on viral load.

F. Role in therapy The role of *Proleukin* in therapeutics has yet to be defined. Response rates of less than 20% have been reported in patients with metastatic renal cell carcinoma. Despite low response rates with *Proleukin* in the treatment of renal cell carcinoma and melanoma, the cytokine is nonetheless the drug of choice for both of these difficult cancers, because afflicted patients are without good treatment options. Therefore the risk–benefit of using this highly toxic agent should be weighed. According to Micromedex, most investigators agree that patients who are asymptomatic or who are symptomatic, but who are still ambulatory, are candidates for therapy. *Proleukin* may play a role in the treatment of HIV infection. In several trials aldesleukin increased the number of CD4 cells without an increase in plasma HIV load.

G. Other applications *Proleukin* may also be useful in the management of HIV infection.

H. Other considerations *Proleukin* has been designated an orphan drug for use in the treatment of metastatic renal cell carcinoma, metastatic melanoma, non-Hodgkin's lymphoma, and primary immunodeficiency disease associated with T-cell defects.

DENILEUKIN DIFTITOX

Trade name: *Ontak*

Manufacturer: Seragen, Inc., Hopkinton, MA

Marketed by: Ligand Pharmaceuticals, San Diego, CA

Indications: Treatment of patients with persistent or recurrent cutaneous T-cell lymphoma (CTCL) whose malignant cells express the CD25 component of the IL-2 receptor

Approval date: 5 February 1999

Type of submission: Biologics license application

A. General description Denileukin diftitox is a recombinant, DNA-derived, interleukin-2 receptor specific ligand, cytotoxic fusion protein consisting of diphtheria toxin fragments A and B fused to interleukin-2. It is produced by expression of a recombinant fusion protein in *Escherichia coli* that contains nucleotide sequences for human interleukin-2, and sequences for the enzymatically active fragment A of diphtheria toxin and the membrane-translocating portion of diphtheria toxin fragment B. Denileukin diftitox has a molecular weight of 58 kDa.

B. Indications and use *Ontak* is indicated in recurrent or refractory CTCL in patients whose malignant cells express the CD25 component of the interleukin-2 receptor. The safety and efficacy of denileukin diftitox in patients with CTCL whose malignant cells do not express the CD25 component of the IL-2 receptor have not been examined.

C. Administration

Dosage form *Ontak* is supplied in single-use vials as a frozen solution for injection on thawing at room temperature or in a refrigerator. Each 2 ml vial of *Ontak* contains recombinant denileukin diftitox 300 µg. The thawed solution must not be refrozen.

Route of administration *Ontak* is intended for intravenous use only and should be infused over at least 15 minutes. It is not intended for bolus administration.

Recommended dosage and monitoring requirements For the treatment of refractory cutaneous T-cell lymphoma, the recommended intravenous dose of *Ontak* is 9 or 18 µg/kg daily for 5 days every 3 weeks has been administered for up to 6 months. Tumor expression of the interleukin-2 receptor by immunohistological assay (for p55 or p75 subunit) is a requirement for the treatment of refractory CTCL with *Ontak*. Because *Ontak* also affects normal lymphocytes, and because patients with CTCL are prone to infections, patients should be monitored carefully during treatment.

D. Pharmacology and pharmaceutics

Clinical pharmacology Denileukin diftitox is a fusion protein designed to direct the cytocidal action of diphtheria toxin to cells that express the IL-2 receptor. It binds avidly to the high-affinity interleukin-2 receptor on target-cell surfaces, where it is rapidly internalized via receptor-mediated endocytosis; the enzymatically active fragment A portion of diphtheria toxin is subsequently released into the cytosol and inhibits protein synthesis, resulting in cell death. The potential for denileukin diftitox to exhibit selectivity for interleukin-2 receptor-expressing malignancies and other cells bearing high-affinity interleukin-2 receptors led to its clinical development

Pharmacokinetics Denileukin diftitox has a terminal half-life of 70 to 80 minutes, a volume of distribution ranging from 0.06 to 0.08 liter/kg, and a total body clearance of 1 to 2 ml/min per kg. Clearance rates have been significantly affected by the development of antibodies to denileukin diftitox, reducing mean systemic exposure by approximately 75%.

Disposition No information available.

Drug interactions No clinical drug interaction studies have been conducted.

E. Therapeutic response *Ontak* was tested in a phase 3 clinical trial of 71 patients with advanced CTCL who had failed at least one other treatment, such as interferon, chemotherapy, or radiation. In this study 30% of the patients' tumors were reduced 50% or more, with the response lasting an average of 4 months. Ten percent of the patients achieved a complete clinical remission that lasted an average of 9 months. Such complete remissions would not be expected in an untreated group.

F. Role in therapy *Ontak* is a biologic treatment for cutaneous T-cell lymphoma, (CTCL), a rare slow-growing form of non-Hodgkin's lymphoma. *Ontak* is approved to treat certain patients with advanced or recurrent CTCL, when other treatments have failed. *Ontak* targets cells with receptors for IL-2 on their surfaces, including malignant cells and some normal lymphocytes, resulting in cell death. Approximately 60% of patients with CTCL were found to have tumors with these IL-2 receptors and only these patients have been shown to benefit from *Ontak*. Approximately 1000 people in the United States are diagnosed with CTCL each year. *Ontak* is recommended for use in CTCL as an alternative to other modalities (e.g., topical carmustine, electron beam radiotherapy, interferon alfa, systemic chemotherapy) in refractory or relapsed patients.

G. Other applications *Ontak* is being evaluated in other non-Hodgkin's lymphomas, rheumatoid arthritis, psoriasis, and HIV infection.

H. Other considerations *Ontak* was granted orphan drug status in 1996 for the treatment of patients with persistent or recurrent cutaneous T-cell lymphoma whose malignant cells express the CD25 component of the IL-2 receptor.

ANAKINRA

Trade name: *Kineret*

Manufacturer: Amgen Inc., Thousand Oaks, CA

Indications: Reduction of signs and symptoms of rheumatoid arthritis

Approval date: 14 November 2001

Type of submission: Biologics license application

A. General description Anakinra is a recombinant nonglycosylated form of the human interleukin-1 receptor antagonist (IL-1Ra). It differs from native human IL-1Ra by the addition of a methionine residue at its amino terminus. Anakinra consists of 153 amino acids and has a molecular weight of 17.3 kDa. It is produced

by recombinant DNA technology using an *Escherichia coli* bacterial expression system.

B. Indications and use *Kineret* is indicated for the reduction in signs and symptoms of moderately to severely active rheumatoid arthritis in patients 18 years of age or older who have failed one or more disease-modifying antirheumatic drugs (DMARDs). *Kineret* can be used alone or in combination with DMARDs other than tumor necrosis factor (TNF) blocking agents.

C. Administration

Dosage form *Kineret* is supplied in single-use, 1 ml prefilled glass syringes with needles. Each prefilled syringe contains anakinra 100 mg.

Route of administration *Kineret* is intended for administration by means of subcutaneous injection.

Recommended dosage and monitoring requirements The recommended dose of *Kineret* for the treatment of patients with rheumatoid arthritis is 100 mg administered subcutaneously once daily. Higher doses did not result in a greater response. Patients or care providers should not be allowed to administer *Kineret* until they have demonstrated a thorough understanding of procedures and an ability to inject the product. Treatment with *Kineret* should not be initiated in patients with active infections and should be discontinued if a patient develops a serious infection. Neutrophil counts should be assessed prior to initiating *Kineret*, and while receiving the drug, monthly for 3 months, and thereafter quarterly for up to 1 year, because some patients may experience a decreased count.

D. Pharmacology and pharmaceutics

Clinical pharmacology Anakinra blocks the biologic activity of IL-1 by competitively inhibiting IL-1 binding to the interleukine-1 type I receptor (IL-1RI), which is expressed in a variety of tissues and organs. IL-1 production is induced in response to inflammatory stimuli and mediates various physiologic and immunological responses. IL-1 has a broad range of activity, including cartilage degradation, through rapid loss of proteoglycans, as well as stimulation of bone resorption. The levels of naturally occurring IL-1Ra in synovium and synovial fluid from patients with rheumatoid arthritis are not sufficient to compete with the elevated amount of locally produced IL-1.

Pharmacokinetics The absolute bioavailability after a 70 mg subcutaneous dose of *Kineret* in healthy subjects is 95%. In subjects with rheumatoid arthritis, maximum plasma concentrations of anakinra occurred 3 to 7 hours after subcutaneous administration. The terminal half-life ranged from 4 to 6 hours. No unusual pharmacokinetics was observed on repeated daily dosing.

Disposition No information available.

Drug interactions No formal drug interaction studies in human subjects have been conducted. Studies in rats did not demonstrate changes in the clearance of methotrexate or anakinra when the two drugs were administered together.

E. Therapeutic response *Kineret*'s Package Insert presents data from two well-controlled trials in which the improvement in signs and symptoms of active rheumatoid arthritis was assessed using the American College of Rheumatology (ACR) response criteria (ACR$_{20}$, ACR$_{50}$, ACR$_{70}$). In one study, *Kineret* or placebo was given to patients who were also receiving methotrexate. In the other, *Kineret* was compared with placebo in patients not receiving a DMARD. In both studies, patients treated with *Kineret* were likely to achieve an ACR$_{20}$ or higher magnitude

of response than patients treated with placebo. Significant improvement occurred more quickly in patients treated with both *Kineret* and methotrexate than in patients who received *Kineret* monotherapy.

F. Role in therapy According to Micromedex, *Kineret* has shown moderate efficacy in severe rheumatoid arthritis when given alone. When combined with methotrexate, *Kineret* improved responses compared with previous methotrexate alone. However, the ability of this agent to significantly modify disease progression has not been established. An additional study is needed to confirm slowing of joint erosions. Direct comparisons with other cytokine modulators (e.g., etanercept, infliximab) with disease-modifying activity will be essential to establish its role in therapy. A recent review of autoimmune diseases [N Engl J Med, 2001; **345**: 340–50] suggests that in general, antagonism of interleukin-1 receptors appears less effective than blockade of tumor necrosis factor in rheumatoid arthritis. Safety permitting, combination regimens of *Kineret* with tumor necrosis factor-alpha blockers should be investigated. Preliminary data suggest a higher rate of serious infections when *Kineret* and etanercept are used in combination compared with *Kineret* used alone, and their combined use is strongly discouraged except in special circumstances. The requirement of daily subcutaneous doses of anakinra is a disadvantage (inconvenience and local side effects). Micromedex concludes, "At present, clinical data for anakinra are insufficient to recommend it over other agents for symptomatic benefit or as disease-modifying therapy in patients with severe disease, or as early therapy to prevent bone erosions. Methotrexate remains the disease-modifying agent of choice."

PEGINTERFERON ALFA-2B

Trade name: *PEG-Intron*

Manufacturer: Schering Corp., Kenilworth, NJ

Indications: Treatment of chronic hepatitis C

Approval date: 19 January 2001

Type of submission: Biologics license application

A. General Description Peginterferon alfa-2b is a covalent conjugate of recombinant alfa interferon with monomethoxy polyethylene glycol (PEG). The molecular weight of the PEG portion of the molecule is 12 kDa. The average molecular weight of the peginterferon alfa-2b molecule is approximately 31 kDa. Interferon alfa-2b, the starting material used to manufacture peginterferon alfa-2b, is a water-soluble protein with a molecular weight of 19.3 kDa produced by recombinant DNA techniques. It is obtained from the bacterial fermentation of a strain of *Escherichia coli* bearing a genetically engineered plasmid containing an interferon gene from human leukocytes.

B. Indications and use *PEG-Intron* monotherapy is indicated for the treatment of chronic hepatitis C in patients not previously treated with interferon alfa who have compensated liver disease and are at least 18 years of age.

C. Administration

Dosage form *PEG-Intron* (powder for injection) is a lyophilized powder supplied in 2-ml vials. Each vial contains a lyophilized mixture of peginterferon alfa-2b and excipients. Following reconstitution with 0.7 ml of supplied diluent, each vial contains *PEG-Intron* at strengths of 100 μg/ml, 160 μg/ml, 240 μg/ml, or 300 μg/ml.

Route of administration *PEG-Intron* is intended for subcutaneous administration. A patient can self-inject *PEG-Intron* if the physician determines that it is appropriate, if the patient agrees, and if training in proper injection technique has been given to the patient.

Recommended dosage and monitoring requirements *PEG-Intron* is administered once weekly for 1 year. The dose should be administered on the same day of each week. Initial dosing should be based on body weight. Serum hepatitis C virus (HCV) should be assessed after 24 weeks of treatment. Discontinuation of treatment should be considered in patients who have not achieved HCV RNA below the limit of detection after 24 weeks of therapy. Patients should be monitored for neuropsychiatric events, bone marrow toxicity, endocrine disorders, cardiovascular events, colitis, pancreatitis, autoimmune disorders, pulmonary disorders, and hypersensitivity.

D. Pharmacology and pharmaceutics

Clinical pharmacology The biological activity of peginterferon alfa-2b is derived from its interferon alfa-2b moiety. Interferons exert their cellular activities by binding to specific membrane receptors on the cell surface and initiate a complex sequence of intracellular events. These include the induction of certain enzymes, suppression of cell proliferation, immunomodulating activities such as enhancement of the phagocytic activity of macrophages and augmentation of the cytotoxicity of lymphocytes for target cells, and inhibition of virus replication in virus-infected cells.

Pharmacokinetics Pegylation of interferon alfa-2b produces a product whose clearance is substantially lower that than of nonpegylated interferon alfa-2b. The mean absorption half-life of peginterferon alfa-2b following a single subcutaneous injection of *PEG-Intron* is 4.6 hours. Time to peak concentration is variable and may range from 15 to 44 hours after injection. Serum levels are sustained for up to 2 to 3 days. Serum levels following single doses of peginterferon alfa-2b increase in a dose-related manner. Some unexpected accumulation occurs on repeated dosing. The mean half-life of peginterferon alfa-2b is approximately 40 hours; apparent clearance is about 22 ml/h per kg; renal elimination accounts for about 30% of the clearance. The clearance of peginterferon alfa-2b is reduced by about half in patients with impaired renal function.

Disposition No information available.

Drug interactions It is not known if *PEG-Intron* causes clinically significant drug–drug interactions with drugs metabolized by the liver in patients with chronic hepatitis C infection.

E. Therapeutic response A randomized trial compared treatment with *PEG-Intron*, once weekly, to treatment with *Intron A* (interferon alfa-2b), three times weekly, in 1219 previously untreated adults with chronic hepatitis from HCV infection. Patients were treated for 48 weeks. Response to treatment was defined as undetectable HCV RNA (less than 100 copies per ml) and normalization of the liver enzyme alanine aminotransferase (ALT) at 24 weeks post-treatment. Response rates to the 1.0 µg/kg *PEG-Intron* doses and to *Intron A* were 24% and 12%. Patients receiving *PEG-Intron* with viral genotype 1 had a response rate of 14%, while patients with other viral genotypes had a 45% response rate.

F. Role in therapy According to Micromedex, current agents of choice in the treatment of chronic hepatitis C are interferon alfa-2a, alfa-2b, alfacon-1, peginterferon alfa-2b, or alfa-2b plus ribavirin. Efficacy of these interferons is similar, although response rates have been higher with the addition of oral ribavirin. In a recent randomized trial of patients with chronic hepatitis C, combination therapy with peginterferon alfa-2b plus ribavirin was shown to be more effective than interferon alfa-2b plus ribavirin. The benefit of peginterferon alfa-2b plus ribavirin was most apparent in patients with hepatitis C virus genotype 1 infection (42%), a substantial improvement over the current response rate (approximately 29–33%) that can be achieved with standard interferon alfa-2b/ribavirin therapy.

INTERFERON BETA-1A

Trade name: *Rebif*

Manufacturer: Serono, Inc., Randolph, MA

Indications: Treatment of patients with multiple sclerosis

Approval date: 7 March 2002

Type of submission: Biologics license application

A. General Description *Rebif* (interferon beta-1a) is a purified 166 amino-acid glycoprotein with a molecular weight of approximately 22.5 kDa. It is produced by recombinant DNA technology using genetically engineered Chinese hamster ovary cells into which the human interferon beta gene has been introduced. The amino-acid sequence of *Rebif* is identical to that of natural fibroblast-derived human interferon beta. Natural interferon beta and interferon beta-1a (*Rebif*) are glycosylated, with each having a single N-linked complex carbohydrate moiety.

B. Indications and use *Rebif* is indicated for the treatment of patients with relapsing forms of multiple sclerosis to decrease the frequency of clinical exacerbations and delay the accumulation of physical disability.

C. Administration

Dosage form *Rebif* is formulated as a sterile solution in a graduated prefilled syringe. Each 0.5 ml of *Rebif* contains interferon beta-1a, either 44 µg or 22 µg, and excipients.

Route of administration *Rebif* is intended for subcutaneous injection. Appropriate instruction for self-injection or injection by a caregiver should be provided. If a patient is to self-administer *Rebif*, the ability of that patient to self-administer and properly dispose of syringes should be assessed. The initial injection should be performed under supervision. Patients should be advised of the importance of rotating sites of injection with each dose, to minimize the likelihood of severe injection reactions or necrosis.

Recommended dosage and monitoring requirements The recommended dosage of *Rebif* is 44 µg injected three times per week. It should be administered at the same time—preferably in the late afternoon or evening—on the same three days at least 48 hours apart each week. Patients are generally started at 8.8 µg three times per week, and the dose is increased over a 4-week period to 44 µg three times per week. Following the administration of each dose, any residual product remaining in the syringe should be discarded. Patients should be advised to report immediately any symptoms of depression and/or suicidal ideation. Dose reduction should be considered if hepatic transaminases are raised above five times the upper limit of normal.

D. Pharmacology and pharmaceutics

Clinical pharmacology Interferons are a family of naturally occurring proteins that are produced by eukaryotic cells in response to viral infections and other biological inducers. They possess immunomodulatory, antiviral, and antiproliferative biological activity and exert their effects by binding to specific receptors on the surface of cells. Interferon beta is produced naturally by various cell types including fibroblasts and macrophages. Binding of interferon beta to its receptors initiates a complex cascade of intracellular events that lead to the expression of interferon-induced gene products and markers, which may mediate some of the biological activities.

Pharmacokinetics The pharmacokinetics of interferon beta-1a (*Rebif*) in patients with multiple sclerosis have not been evaluated. In healthy subjects, a single injection resulted in a peak serum concentration at about 16 hours after administration. The mean serum elimination half-life was 69 hours but varied widely. Following every-other-day subcutaneous injections, an increase in the area under the serum concentration versus time curve (AUC) of approximately 240% was observed. Clearance has been estimated at 33 to 55 liters/h.

Disposition No information available.

Drug interactions No formal drug interaction studies have been conducted with *Rebif*. Monitoring patients for neutropenia and lymphopenia is required if *Rebif* is given in combination with myelosuppressive agents.

E. Therapeutic response Two studies evaluated the safety and efficacy of *Rebif* in patients with relapsing-remitting multiple sclerosis. Study 1 was a placebo-controlled, 2-year trial in 560 patients with multiple sclerosis for at least 1 year. The primary efficacy end point was the number of clinical exacerbations. The mean numbers of exacerbations per patient over 2 years was 1.82 and 1.73 in those who received *Rebif* either 22 µg or 44 µg three times per week compared with a mean number of 2.56 in those who received placebo. The decrease in the number of exacerbations in the *Rebif* groups compared with the placebo group was statistical significant, but the difference between the two doses of *Rebif* was not. Study 2 was a randomized open-label trial comparing *Rebif* with *Avonex*, which also contains interferon beta-1a, in 677 multiple sclerosis patients. The primary efficacy end point was the proportion of patients who remained free of exacerbations at 24 weeks. Patients treated with subcutaneous *Rebif* 44 µg three times per week were more likely to remain relapse-free during the 24-week treatment period than were patients treated with intramuscular *Avonex* 30 µg once weekly (75% vs. 63%). The risk of relapse on *Rebif* relative to the risk on *Avonex* was 0.68.

F. Role in therapy   After reviewing the study comparing *Rebif* and *Avonex*, the FDA's Division of Clinical Trial Design and Analysis determined that the efficacy difference is meaningful because it signifies that a patient on *Rebif* is 32% less likely to experience an exacerbation, which can substantially lower his or her quality life for weeks or months, and that *Rebif* is clinically superior to *Avonex*. Adverse reactions were generally similar between the two treatment groups, but those receiving *Rebif* had a higher incidence of injection site reactions (80% vs. 24%), hepatic function

disorder (14% vs. 7%), and leukopenia (3% vs. <1%) than did those receiving *Avonex*.

G. Other considerations   Serono conducted the comparative study to demonstrate that *Rebif* is clinically superior to *Avonex*. No other path was available for Serono to have the product licensed by the FDA before mid-2003 because *Avonex* carried an orphan drug designation based on its superior safety profile over *Betaseron*, the first interferon beta to be licensed for multiple sclerosis. Based on the results of the *Rebif* versus *Avonex* study, the FDA determined that *Rebif* is clinically superior to *Avonex*. While recognizing that the safety profile associated with *Rebif* is not as favorable as the safety profile of *Avonex*, the FDA has determined that the severity and frequency of such adverse events do not render *Rebif* unlicensable under Section 35 of the Public Health Service Act. Further, under the orphan drug regulations, if Serono demonstrates that *Rebif* is superior to *Avonex* based on efficacy, Serono does not have to show that *Rebif* is safer than, or as safe as, *Avonex*.

8

HORMONES

■ **SECTION ONE** ■

SANDRA BLETHEN

■ 8.1. PROTEIN HORMONES AS THERAPEUTICS: YESTERDAY, TODAY, AND TOMORROW

8.1.1. Peptide Hormones in the Prerecombinant DNA Era (Yesterday)

Twenty-five years ago, the peptide and protein hormones used clinically were purified from animal or human sources. In a few cases, where a smaller fragment had

biological activity, synthetic forms were also available. If the amino-acid sequence of the animal protein was such that the protein had biological activity in humans, its purification from animals slaughtered for food was possible. Common examples were insulin (where only one amino acid is different between porcine and human insulin and two amino acids are different between bovine and human insulin), glucagon, and parathyroid hormone (PTH). Although effective, these prepara-

Biotechnology and Biopharmaceuticals, by Rodney J. Y. Ho and Milo Gibaldi
ISBN 0-471-20690-3   Copyright © 2003 by John Wiley & Sons, Inc.

■TABLE 8.1. Clinically useful peptide and protein hormones in the pre-recombinant DNA era

Compound	Source(s)	Clinical Use
Insulin	Bovine and porcine pancreata	Treatment of insulin-dependent diabetes mellitus
Glucagon	Bovine and porcine pancreata	Treatment of acute hypoglycemia
Parathyroid hormone (PTH)	Bovine parathyroid glands	Differential diagnosis of hypocalcemia
Thyrotropin stimulating hormone (TSH)	Bovine pituitary	Identification of thyroid cancer metastases in patients previously treated for thyroid cancer
Adrenocortitropic hormone (ACTH)	Bovine pituitary	Diagnosis of adrenal disorders Treatment of gelastic seizures
Cosyntropin	Chemically synthesized active peptide fragment	Diagnosis of adrenal disorders
Antidiuretic hormone (ADH)	Bovine posterior pituitaries	Treatment of diabetes insipidus
Desmopressin Acetate (DDAVP)	Chemically synthesized analogue of arginine vasopressin	Treatment of diabetes insipidus
Somatotropin or growth hormone (GH)	Human cadaver pituitaries	Treatment of severe GH deficiency in children
Human menopausal gonadotropin (hMG)	Urine of postmenopausal women	Induction of ovulation
Human chorionic somatrotropin (hCG)	Human placenta	Analogue of pituitary luteinizing hormone (LH) used for evaluation of testicular function and treatment of cryptorchidism

tions were often contaminated with other animal proteins, and their use could lead to the development of antibodies. In some cases, where high antibody titers were accompanied by high binding capacities, the drug could become ineffective.

For protein hormones where there was less sequence homology with the equivalent bovine or porcine protein, supplies for clinical use were derived directly from human tissues. Somatotropin or growth hormone (GH) was purified from cadaver pituitaries. The limitations on GH use were due to its scarcity, which is well described in a review by Frasier [1]. Other peptide hormones purified from human sources

include human chorionic gonadotropin—the luteinizing (LH) analogue isolated from human placenta—and human menopausal gonadotropins—a mixture of LH and follicle stimulating hormone (FSH) isolated from urine. The supply of these hormones was sufficient for clinical application.

While small peptides, such as somatostatin, a tetradecapeptide, could be synthesized chemically, synthesis of larger molecules resulted in low yields of desired peptide, usually contaminated with peptides of similar structure. Examples of clinically useful peptides and protein hormones, their sources, and uses are listed in Table 8.1 and shown in Figure 8.1.

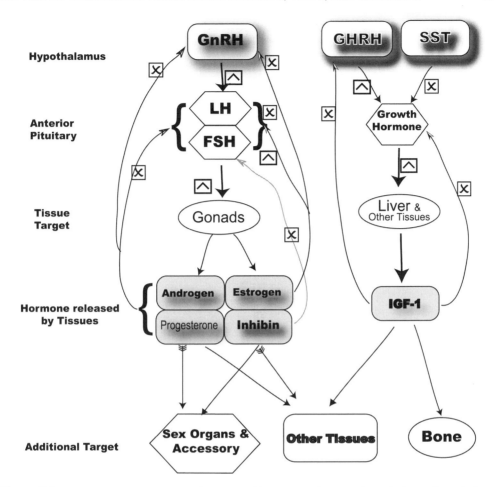

Figure 8.1. Schematic representation of integrated hormone release and regulation at hypothalamus, anterior pituitary, and at the primary and secondary target tissues. Hypothalmic releasing factors gonadotropin-releasing hormone (GnRH), growth hormone release hormone (GHRH), and somatostatin (SST) stimulate or inhibit release of luteinizing hormone (LH; lutropin), follicle-stimulating hormone (FSH: follitropin), or growth hormone (H) at anterior pituitary. Release of these hormones by anterial pituitary influences the primary target tissues, gonads, liver, and other tissues to release effector hormones such as androgen, estrogen, progesterone inhibin, or hepatic hormones such as insulin-like growth factor-1 (IGF1). These hormones, which regulate sex organs and accessory tissues, bone, and other tissues, may also directly or indirectly down regulate hormones of hypothalamus and anterior pituitary. ⌂ Indicates induction; ⌧ indicates inhibition.

8.1.2. The Recombinant DNA Era (Today)

The situation changed with development of methods for synthesizing proteins by recombinant DNA techniques. In 1979 Goeddel et al. [2,3] published two papers describing the introduction of the genes for human insulin and human GH into *Escherichia coli*. This was followed by ex-pression of the respective proteins, which were then taken into clinical trials. In 1982 insulin became the first recombinant protein hormone marketed for clinical use in the United States.

Originally, recombinant human GH (hGH) was expressed with an additional methionine at the N-terminal because all of the amino-acid sequences of *E. coli* pro-teins begin with an S-adenosyl methionine.

■TABLE 8.2. Properties of protein and peptide hormones derived from hypothalamus, pituitary, and other tissues

Hormone	Molecular Weight (Da)	Number of Peptide Chains	Amino-Acid Length	Carbohydrate (%)
Somatotropic hormones				
Growth hormone (GH)	22,000	1	191	0
Prolactin (Prl)	22,500	1	198	0
Placental lactogen (PL)	22,300	1	191	0
Glycoprotein hormone[a]				
Luteinizing hormone (LH)	29,400	2	α-92 β-115	23
Follicle-stimulating hormone (FSH)	32,600	2	α-92 β-115	28
Chorionic gonadotropin (CG)	38,600	2	α-92 β-145	33
Thyroid-stimulating hormone (TSH)	30,500	2	α-92 β-112	22
Pro-opiomelanocortin (POMC)-derived hormones				
Corticotropin (ACTH)	4,500	1	39	0
α-Melanocyte-stimulating hormone	1,650	1	13	0
β-Melanocyte-stimulating hormone (β-MSH)	2,100	1	18	0
β-Lipotropin (β-LPH)	9,500	1	91	0
γ-Lipotropin (γ-LPH)	5,800	1	58	0

Source: Adapted from Ref. [21].
[a]The amino-acid sequences of the α subunits of LH, FSH, TSH, and CG are identical. Specificity resides in β subunits.

The additional methionine did not affect the biological activity of the product [4], and clinical trials were soon underway. Recombinant GH increased levels of insulinlike-growth factor-I (IGF-I) in adults [5] and increased growth rates in GH-deficient (GHD) children [6]. In 1985, concerns about contamination of human-derived pituitary hGH with the infectious agent causing Creutzfeldt-Jakob disease resulted in the discontinuation of its use in children with growth failure due to lack of endogenous GH secretion [7]. It was replaced by recombinant methionyl GH (Table 8.2). As we look back at the era of using proteins derived from human tissues as therapeutic modalities, it is amazing that many more problems related to contamination with infectious agents did not occur.

The next step in the saga of recombinant hGH was production of a molecule that did not have the extra methionine. In 1987 Chang et al. prepared a hybrid gene that included the secretion signal-coding region of *E. coli* heat-stable enterotoxin II. When cells containing this hybrid were grown in culture, 90% of the hGH produced was secreted into the periplasm. This also simplified the purification process [8].

With increased availability of GH, other clinical uses could be explored, such as growth failure due to Turner syndrome [9], chronic renal insufficiency [10], Prader-

must precede a meal. Hence, attempts achieve euglycemia result in wide excu sions of insulin levels and sometimes pi cipitate clinical hypoglycemia.

The introduction of new versions rapid-acting recombinant insulin prepar by introducing modifications in the insu sequence has decreased the time fro injection to insulin action, allowing for insulin injection at the same time as a me [17]. Insulin lispro and insulin aspart a rapidly acting forms of the hormone th can be given at mealtime or even after meal. The latter approach is especia important in managing an infant or you child with diabetes as the caregiver c adjust the insulin dose to the child's actu food intake [18]. A new analogue, insul glargine, appears to solve the problem peaks and troughs of insulin action follo ing NPH or UltraLente injections [19]; provides nearly constant levels of insul action, similar to physiologic insulin secr tion [20].

8.1.4. Summary

The ability to synthesize recombina protein hormones has allowed great access to treatment and the exploration new uses for familiar hormones. The deve opment of designer hormones represen the next step with alterations in amino-ac sequence resulting in changes in pharm cokinetics and differences in biologic activity.

REFERENCES

1. Frasier, S.D., *The not-so-good old day working with pituitary growth hormone North America, 1956 to 1985.* J Pedia 1997. **131**(1 Pt 2): S1–4.
2. Goeddel, D.V., H.L. Heyneker, T. Hozun R. Arentzen, K. Itakura, D.G. Yansun M.J. Ross, G. Miozzari, R. Crea, and P.l Seeburg, *Direct expression in Escherich coli of a DNA sequence coding for hum*

logues and their potential in the management of diabetes mellitus. Diabetologia, 1999. **42**(10): 1151–67.
3. Rutledge, K.S., H.P. Chase, G.J. Klingensmith, P.A. Walravens, R.H. Slover, and S.K. Garg, *Effectiveness of postprandial Humalog in toddlers with diabetes.* Pediatrics, 1997. **100**(6): 968–72.
4. Bolli, G.B., *The pharmacokinetic basis of insulin therapy in diabetes mellitus.* Diabetes Res Clin Pract, 1989. **6**(4): S3–15; discussion S15–6.
5. Lepore, M., S. Pampanelli, C. Fanelli, F. Porcellati, L. Bartocci, A. Di Vincenzo, C. Cordoni, E. Costa, P. Brunetti, and G.B. Bolli, *Pharmacokinetics and pharmacodynamics of subcutaneous injection of long-acting human insulin analog glargine, NPH insulin, and ultralente human insulin and continuous subcutaneous infusion of insulin lispro.* Diabetes, 2000. **49**(12): 2142–8.
6. Ascoli, M., and D.L. Segaloff, *Adenohypophyseal hormones and their hypothalamic releasing factors.* In Goodman & Gilman's 9th edition. J.E. Hardman and L.E. Limbird, eds. McGraw-Hill, NY. NY, 1996. pp. 1363–82.

TWO ■

A. General description Insulin is se reted by the beta cells of the pancreas. It s a protein with a molecular weight of .6kDa, composed of two amino-acid hains connected by disulfide linkages. Recombinant human insulins are struc urally identical to the insulin produced by he pancreas in the human body. Insulin preparations differ in concentration, onset nd duration, purity, and species of origin. Insulins are divided into rapid-acting preparations, intermediate-acting prepara ions, and long-acting preparations. Rapid onset preparations include regular, crys alline zinc, and prompt insulin zinc (semi ente), and have an average onset of 0.5 o 1 hour and a duration of effect of 5 to 8 hours (regular, crystalline zinc) or 12 to 16 hours (semilente). Intermediate-acting preparations [isophane insulin (NPH) and insulin zinc suspension (lente)] have an approximate onset of 1.5 to 3 hours and a duration of 22 to 48 hours. Long-acting preparations [protamine zinc suspension and extended insulin zinc suspension (ultralente)] have a general onset of 4 to 6 hours and a duration of more than 36 hours.

Humulin is synthesized in a laboratory strain of *Escherichia coli* bacteria that has been genetically altered by the addition of the gene for human insulin. *Novolin* is produced by recombinant DNA technology utilizing *Saccharomyces cerevisiae* (bakers' yeast) as the production organism. *Velosulin BR* is also obtained by recombinant-DNA technology utilizing *S. cerevisiae* as the production organism.

B. Indications and use Insulin is indicated for all patients with type 1 diabetes mellitus and for patients with type 2 diabetes mellitus who are inadequately controlled by oral medication.

C. Administration

Dosage forms All human insulin products contain 100 units/ml. *Humulin* is available in six formulations — Regular (R), NPH (N), Lente (L), Ultralente (U), *Humulin 50/50*, and *Humulin 70/30*. *Humulin 50/50* is a sterile suspension containing a mixture of 50% Human Insulin Isophane Suspension and 50% Human Insulin (rDNA origin). It is an intermediate-acting insulin combined with the more rapid onset of action of regular insulin. The duration of activity may last up to 24 hours following injection. *Humulin 70/30* is a mixture of 70% Human Insulin Isophane Suspension and 30% Human Insulin (rDNA origin). It is also an intermediate-acting insulin combined with the more rapid onset of action of regular insulin. The duration of activity may last up to 24 hours following injection. *Humulin 70/30* is also available in the form of a prefilled cartridge and pen. *Humulin R* and *N*

are also available in cartridges and *Humalin N* is also available in a pen. *Humulin L* is a sterile, amorphous, and crystalline suspension of recombinant human insulin with zinc providing an intermediate-acting insulin with a slower onset and a longer duration of activity (up to 24 hours) than regular insulin. *Humulin N* is a sterile, crystalline suspension of human insulin with protamine and zinc providing an intermediate-acting insulin with a slower onset of action and a longer duration of activity (up to 24 hours) than that of regular insulin. *Humulin R* consists of zinc-insulin crystals dissolved in a sterile solution. *Humulin R* has had nothing added to change the speed or length of its action. It takes effect rapidly and has a relatively short duration of activity (4–12 hours) as compared with other insulins. *Humulin R (U-500)*, like *Humulin R*, consists of zinc-insulin crystals dissolved in a sterile solution but contains 500 units/ml rather than 100 units/ml. *Humulin R (U-500)* takes effect rapidly but has a relatively long duration of activity following a single dose (up to 24 hours) as compared with other Regular insulins. It is available by prescription only. *Humulin U* is a sterile crystalline suspension of human insulin with zinc providing a slower onset and a longer and less intense duration of activity (up to 28 hours) than regular insulin or the intermediate-acting insulins (NPH and Lente).

Novolin is available in four formulations: Regular (R), NPH (N), Lente (L), and *Novolin 70/30*. They correspond to the several formulations of *Humulin*. *Novolin 70/30*, *Novolin N*, and *Novolin R* are supplied in prefilled cartridges and in prefilled disposable syringes. *Velosulin BR* is a sterile solution of human insulin (rDNA origin) in a phosphate buffer.

Route of administration *Humulin 50/50*, *Humulin 70/30*, *Humulin L*, *Humulin N*, *Humulin R*, *Humulin R (U-500)*, and *Humulin U* are for subcutaneous injection only. The same applies to formulations containing *Novolin*. *Velosulin BR* is indicated for use with external insulin infusion pumps (subcutaneous).

Recommended dosage and monitoring requirements Insulin dosage should be titrated to achieve near-normal levels of glucose, while avoiding hypoglycemia. For newly diagnosed type 1 diabetes mellitus, the initial daily dosage of insulin is 0.6 to 0.75 units/kg. For patients with type 2 diabetes mellitus, the starting dosage of insulin is usually 10 units and may be increased weekly in increments of 5 to 10 units. The label for all products that contain human insulin (recombinant DNA origin) warns: HUMAN INSULIN PRODUCTS DIFFER FROM ANIMAL-SOURCE INSULINS BECAUSE THEY ARE STRUCTURALLY IDENTICAL TO THE INSULIN PRODUCED BY YOUR BODY'S PANCREAS AND BECAUSE OF THEIR UNIQUE MANUFACTURING PROCESS. ANY CHANGE OF INSULIN SHOULD BE MADE CAUTIOUSLY AND ONLY UNDER MEDICAL SUPERVISION. CHANGES IN STRENGTH, MANUFACTURER, TYPE (E.G., REGULAR, NPH, LENTE), SPECIES (BEEF, PORK, BEEF-PORK, HUMAN), OR METHOD OF MANUFACTURE (RDNA VERSUS ANIMAL-SOURCE INSULIN) MAY RESULT IN THE NEED FOR A CHANGE IN DOSAGE.

D. Pharmacology and pharmaceutics

Clinical pharmacology Insulin is secreted by the beta cells of the pancreas. Insulin deficiency results in marked reduction in the rate of transport of glucose across cell membranes, resulting in hyperglycemia; there is also a reduction in the enzyme system that catalyzes the conversion of glucose to glycogen, and an abnormally high rate of conversion of protein to glucose. Insulin stimulated endogenously or administered exogenously stimulates carbohydrate metabolism and facilitates transfer of glucose into cardiac muscle, skeletal muscle, and adipose tissue and converts glucose to glycogen.

Pharmacokinetics The onset of action, peak effect, and duration of action depend primarily on the insulin product and site of administration. For patients with a glomerular filtration rate (GFR) between 10 and 50 ml/min, a dosage reduction of approximately 25% may be indicated; a dosage reduction of approximately 50% may be indicated in patients with a GFR less than 10 ml/min. Insulin is distributed primarily in extracellular fluids and has an apparent volume of distribution of 0.15 liter/kg.

Disposition Insulin is metabolized in the liver and kidney.

Drug interactions Acarbose co-administered with insulin is likely to cause a lowered blood glucose concentration, thereby increasing the hypoglycemic potential of insulin. At adult doses of greater than 650 mg daily, aspirin may enhance the hypoglycemic effect of insulin through an intrinsic hypoglycemic effect. Corticosteroid therapy tends to increase blood glucose in diabetic patients, and higher doses of insulin may be required. Ethanol use in patients with diabetes or cirrhosis impairs the recovery from insulin-induced hypoglycemia.

E. Therapeutic response Near-normalization of blood glucose and glycosylated hemoglobin levels.

F. Role in therapy All patients with type 1 diabetes mellitus require insulin therapy to survive. For optimal control, the patient should perform self-monitoring of blood glucose. The majority of patients diagnosed with type 2 diabetes mellitus do not require immediate treatment with insulin. Intravenous regular insulin is recommended for treating patients with diabetic ketoacidosis. Patients with diabetes mellitus are at risk for microvascular (i.e., retinopathy, nephropathy, and neuropathy) and macrovascular (i.e., myocardial infarction) complications that lead to morbidity and mortality. The Diabetes Control and Complications Trial demonstrated that intensive insulin therapy, designed to achieve a near-normal glycosylated hemoglobin, reduced microvascular complications in patients with type 1 diabetes. The increased use of human recombinant insulin and highly purified insulins has led to a decline in the incidence of local reactions, insulin allergy, immune resistance, and lipoatrophy in patients with type 1 diabetes. *Humulin R (U-500)* is especially useful for the treatment of diabetic patients with marked insulin resistance (daily requirements more than 200 units), since a large dose may be administered subcutaneously in a reasonable volume.

G. Other applications Delivery of insulin via an implantable insulin pump has resulted in good glycemic control with fewer episodes of severe hypoglycemia.

INSULIN LISPRO

Trade name: *Humalog*

Manufacturer: Eli Lilly, Indianapolis, IN

Indications: Treatment of diabetes mellitus

Approval date: 4 April 2000

Type of submission: New drug application

A. General description *Humalog* (insulin lispro, rDNA origin) is a rapid-acting human insulin analog. Chemically it is Lys(B28), Pro(B29) human insulin analogue (molecular weight 5.8 kDa), created when the amino acids at positions 28 and 29 on the insulin B-chain are reversed. *Humalog* is synthesized in a laboratory strain of *Escherichia coli* bacteria that has been genetically altered by the addition of the gene for insulin lispro.

B. Indications and use *Humalog* is an insulin analogue that is indicated in the treatment of patients with diabetes mellitus for the control of hyperglycemia. It has a more rapid onset and shorter duration of action than human regular insulin.

C. Administration

Dosage forms *Humalog* injection consists of zinc-insulin lispro crystals dissolved in a sterile solution. Each 10 ml vial of *Humalog* injection contains insulin lispro 100 units. *Humalog Mix 75/25* [75% insulin lispro protamine suspension and 25% insulin lispro injection (rDNA origin)] is a mixture of insulin lispro solution, a rapid-acting blood glucose-lowering agent and insulin lispro protamine suspension, an intermediate-acting blood glucose-lowering agent.

Route of administration *Humalog* and *Humalog Mix 75/25* are intended for subcutaneous injection only. *Humalog* was absorbed at a consistently faster rate than human regular insulin in healthy male volunteers given 0.2 U/kg human regular insulin or *Humalog* at abdominal, deltoid, or thigh sites, the three sites most often used by patients with diabetes.

Recommended dosage and monitoring requirements The dosage of insulin lispro alone or mixed with insulin lispro protamine suspension is individualized for each patient and depends on the patient's metabolic needs, eating habits, and other lifestyle variables. When used as a mealtime insulin, *Humalog* should be given within 15 minutes before or immediately after a meal. Human regular insulin is best given 30 to 60 minutes before a meal. To achieve optimal glucose control, the amount of longer-acting insulin being given may need to be adjusted when using *Humalog*.

D. Pharmacology and pharmaceutics

Clinical pharmacology According to Micromedex, insulin lispro differs from human insulin by inversion of the beta-chain amino-acids proline and lysine at positions 28 and 29, respectively; these changes result in a reduced capacity for self-association compared with insulin. Self-association of insulin to higher molecular forms has been a rate-limiting step in achieving a biologic response after subcutaneous injection. The modifications inherent in insulin lispro confer monomeric behavior, resulting in more rapid absorption after subcutaneous injection, an earlier peak effect, and a shorter and more consistent duration of action relative to human regular insulin; these characteristics mimic plasma insulin dynamics in nondiabetic subjects in response to meals.

Pharmacokinetics Insulin lispro is rapidly absorbed after subcutaneous administration; its bioavailability (55–77%) is similar to that of regular human insulin. Peak serum insulin levels after subcutaneous insulin lispro occur earlier (30–90 minutes) than after subcutaneous human regular insulin (50–120 minutes), are higher, and are of shorter duration. The pharmacokinetic profiles of *Humalog* and human regular insulin are comparable to one another when administered to normal volunteers by the intravenous route. *Humalog* was absorbed at a consistently faster rate than human regular insulin in healthy male volunteers given human regular insulin or *Humalog* at abdominal, deltoid, or thigh subcutaneous sites. After abdominal administration of *Humalog*, serum drug levels are higher and the duration of action is slightly shorter than after deltoid or thigh administration. *Humalog* has less intra- and interpatient variability compared to human regular insulin. Insulin lispro has a volume of distribution of 0.26 to 0.36 liter/kg. Its elimination half-life is dose dependent (26–52 minutes) after subcutaneous doses. *Humalog Mix 75/25* has an early peak corresponding to the rapid onset of action of insulin lispro and a late peak corresponding to insulin lispro protamine suspension.

Disposition Human metabolism studies of *Humalog* have not been conducted. Studies in animals indicate that the metabolism of insulin lispro is identical to that of regular human insulin.

Drug interactions Insulin requirements may be increased by medications with hyperglycemic activity such as corticosteroids, isoniazid, certain lipid-lowering drugs (e.g., niacin), estrogens, oral contraceptives, phenothiazines, and thyroid replacement therapy. Insulin requirements may be decreased in the presence of drugs with hypoglycemic activity, such as oral hy-poglycemic agents, salicylates, sulfa antibiotics, and certain antidepressants (monoamine oxidase inhibitors), certain angiotensin-converting-enzyme inhibitors, beta-adrenergic blockers, inhibitors of pancreatic function (e.g., octreotide), and alcohol. Beta-adrenergic blockers may mask the symptoms of hypoglycemia in some patients.

E. Therapeutic response A decrease in postprandial glucose levels and near-normalization of blood glucose and glycosylated hemoglobin levels when combined with a long-acting insulin or a long-acting oral hypoglycemia agent.

F. Role in therapy *Humalog* is an insulin analogue with improved pharmacokinetic characteristics compared with regular human insulin. *Humalog* more effectively reduces peak concentrations of glucose after a meal, but is short acting and poses a low risk of hypoglycemia. It is indicated for patients with type 1 or type 2 diabetes. When *Humalog* is used for patients with type 1 diabetes, a longer-acting insulin should be added to the regimen; however, *Humalog* may be used with a sulfonylurea rather than a longer-acting insulin for patients with type 2 diabetes.

G. Other considerations *Humalog* is available by prescription only.

INSULIN GLARGINE

Trade name: *Lantus*

Manufacturer: Hoechst Marion Roussel, Frankfurt am Main, Germany

Distributed by: Aventis Pharmaceuticals Inc., Kansas City, MO

Indications: Treatment of adult and pediatric patients with type 1 diabetes mellitus or adult patients with type 2 diabetes mellitus who require basal (long-acting) insulin for the control of hyperglycemia

Approval date: 1 April 2000

Type of submission: New drug application

A. General description Insulin glargine is a recombinant human insulin analogue with a long duration of action (up to 24 hours). It is produced by recombinant DNA technology utilizing a laboratory strain of *Escherichia coli* as the production organism. Insulin glargine differs from human insulin in that the amino acid asparagine at position A21 is replaced by glycine and two arginines are added to the C-terminus of the B-chain. It has a molecular weight of about 6.1 kDa.

B. Indications and use *Lantus* is indicated for once-daily subcutaneous administration at bedtime in the treatment of adult and pediatric patients with type 1 diabetes mellitus who are at least 6 years old or adult patients with type 2 diabetes mellitus who require basal (long-acting) insulin for the control of hyperglycemia.

C. Administration
  Dosage form *Lantus* is a sterile solution of insulin glargine for use as an injection. Each milliliter of *Lantus* contains 100 IU insulin glargine. *Lantus* is available in 5 ml and 10 ml vials and cartridges for use only in a proprietary insulin delivery device (pen).
  Route of administration *Lantus* is administered by subcutaneous injection once daily at bedtime. Intravenous administration of the usual subcutaneous dose could result in severe hypoglycemia. As with all insulins, injection sites within an injection area (abdomen, thigh, or deltoid) must be rotated from one injection to the next.
  Recommended dosage and monitoring requirements *Lantus*'s potency is approximately the same as human insulin. Individualized *Lantus* regimens should be developed for each patient based on self-

monitoring of blood glucose. In a clinical study with insulin-naive patients with type 2 diabetes already treated with oral antidiabetic drugs, *Lantus* was started at an average dose of 10 IU once daily, and subsequently adjusted according to the patient's need to a total daily dose ranging from 2 to 100 IU. In clinical studies, when patients were changed over from once-daily NPH human insulin or ultralente human insulin to once-daily *Lantus*, the initial dose was usually not changed. However, when patients were switched from twice-daily NPH human insulin to *Lantus* once daily at bedtime, to reduce the risk of hypoglycemia, the initial dose was usually reduced by approximately 20% (compared to total daily dose of NPH human insulin) within the first week of treatment and then adjusted based on patient response.

D. Pharmacology and pharmaceutics
  Clinical pharmacology The primary activity of insulin, including insulin glargine, is regulation of glucose metabolism. Insulin glargine is less soluble than native human insulin at physiological pH, and precipitates in skin following subcutaneous injection, resulting in delayed absorption. At pH 4, as in the *Lantus* injection solution, it is completely soluble. After injection into the subcutaneous tissue, the acidic solution is neutralized, leading to formation of microprecipitates from which small amounts of insulin glargine are slowly released, resulting in a relatively constant concentration/time profile over 24 hours with no pronounced peak. This profile allows once-daily dosing as a patient's basal insulin.
  Pharmacokinetics Insulin glargine is characterized by slow absorption from the subcutaneous injection site and a flat plasma insulin profile. Absorption patterns are similar after subcutaneous injection into the arm, abdomen, or thigh. A study in patients with type 1 diabetes found that the median time between injection and the end

of pharmacological effect was 14.5 hours (range: 9.5–19.3 hours) for NPH human insulin, and 24 hours (range: 10.8 to >24.0 hours) (24 hours was the end of the observation period) for insulin glargine. The effect of renal impairment on the pharmacokinetics of insulin lispro has not been studied. However, some studies with human insulin have shown increased circulating levels of insulin in patients with renal failure. Careful glucose monitoring and dose adjustments of insulin, including insulin lispro, may be necessary in patients with renal dysfunction.

Disposition Insulin glargine is partially metabolized at the carboxyl terminus of the B chain to two active metabolites with activity similar to insulin.

Drug interactions According to the product label, a number of substances affect glucose metabolism and may require insulin dose adjustment and particularly close monitoring. Examples of drugs that may increase the blood-glucose-lowering effect and susceptibility to hypoglycemia are oral antidiabetic products, ACE inhibitors, fibrates, fluoxetine, MAO inhibitors, propoxyphene, salicylates, somatostatin analogues (e.g., octreotide), and sulfonamides. Examples of substances that may reduce the blood-glucose-lowering effect of insulin are corticosteroids, diuretics, sympathomimetic agents (e.g., epinephrine, albuterol, terbutaline), isoniazid, and phenothiazine derivatives, somatropin, thyroid hormones, and estrogens, and progestins (e.g., in oral contraceptives). Beta-blockers, clonidine, lithium salts, and alcohol may either potentiate or weaken the blood-glucose-lowering effect of insulin. In addition, under the influence of sympatholytic medicinal products such as beta-blockers, clonidine, guanethidine, and reserpine, the signs of hypoglycemia may be reduced or absent.

E. Therapeutic response Near normalization of blood glucose and glycosylated hemoglobin is the goal of insulin therapy. In clinical studies *Lantus* achieved a level of glycemic control similar to NPH human insulin as measured by glycosylated hemoglobin.

F. Role in therapy Insulin glargine is as effective as human isophane (NPH) insulin suspension in patients with type 1 or 2 diabetes and is associated with a lower incidence of nocturnal hypoglycemia in some patients.

G. Other considerations *Lantus* is available by prescription only.

INSULIN ASPART

Trade name: ***NovoLog***

Manufacturer: Novo Nordisk, Princeton, NJ

Indications: Treatment of adult patients with diabetes mellitus, for the control of hyperglycemia

Approval date: 1 July 2000

Type of submission: New drug application

A. General description Insulin aspart (rDNA origin) is a rapid-acting human insulin analogue. It is homologous with regular human insulin with the exception of a single substitution of the amino acid proline by aspartic acid in position B28, and is produced by recombinant DNA technology utilizing *Saccharomyces cerevisiae* (baker yeast) as the production organism. It has a molecular weight of 5.8 kDa.

B. Indication *NovoLog* is indicated for the treatment of adult patients with diabetes mellitus, for the control of hyperglycemia. Because *NovoLog* has a more rapid onset and a shorter duration of action than human regular insulin, it should normally be used in regimens together with intermediate or long-acting insulin.

C. Administration

Dosage form *NovoLog* is a sterile, aqueous, clear, and colorless solution that contains insulin aspart 100 units/ml. *NovoLog* is available in 10 ml vials and 3 ml *PenFill* cartridges, each containing 100 units of insulin aspart per ml. *NovoLog PenFill* cartridges are for use with proprietary delivery devices and disposable needles.

Route of administration *NovoLog* should be administered by subcutaneous injection in the abdominal wall, the thigh, or the upper arm. Injection sites should be rotated within the same region. As with all insulins, the duration of action will vary according to the dose, injection site, blood flow, temperature, and level of physical activity.

Recommended dosage and monitoring requirements The dosage of *NovoLog* should be individualized and determined in accordance with the needs of the patient. The total daily individual insulin requirement is usually between 0.5 and 1.0 units/kg per day. In a meal-related treatment regimen, *NovoLog* may provide 50% to 70% of this requirement and the remainder by an intermediate-acting or long-acting insulin. Patients may require more basal insulin in relation to bolus insulin and more total insulin when using *NovoLog* compared to regular human insulin to prevent premeal hyperglycemia. Additional basal insulin injections may be necessary. Because of the fast onset of action, the injection of *NovoLog* should immediately be followed by a meal. Because of the short duration of action of *NovoLog*, patients with type 1 diabetes also require a longer-acting insulin to maintain adequate glucose control.

D. Pharmacology and pharmaceutics

Clinical pharmacology The primary activity of insulin aspart is the regulation of glucose metabolism. In humans, the effect of the insulin analogue is more rapid in onset and of shorter duration, compared to regular human insulin after subcuta-neous injection. The single substitution of the amino acid proline with aspartic acid at position B28 in insulin aspart reduces the molecule's tendency to form hexamers as observed with regular human insulin. Insulin aspart is therefore more rapidly absorbed after subcutaneous injection compared to regular human insulin.

Pharmacokinetics The relative bio-availability of *NovoLog* compared to regular human insulin indicates that the two insulins are absorbed to a similar extent. In studies in healthy human subjects and patients with type 1 diabetes, *NovoLog* consistently reached peak serum concentrations approximately twice as fast as regular human insulin. The median time to maximum concentration in these trials was 40 to 50 minutes for *NovoLog* versus 80 to 120 minutes for regular human insulin. In a clinical trial in patients with type 1 diabetes, *NovoLog* and regular human insulin, both administered subcutaneously at a dose of 0.15 U/kg body weight, reached mean maximum concentrations of 82.1 and 35.9 mU/liter, respectively. Pharmacokinetic/pharmacodynamic characteristics of insulin aspart have not been established in patients with type 2 diabetes. The intraindividual variability in time to maximum serum insulin concentration for healthy male subjects was significantly less for *NovoLog* than for regular human insulin. In a clinical study in healthy nonobese subjects, the pharmacokinetic differences between *NovoLog* and regular human insulin were observed independent of the injection site. *NovoLog* has a low binding to plasma proteins, less then 10%, similar to regular human insulin. After subcutaneous administration in healthy male subjects, *NovoLog* seemed to be more rapidly eliminated than regular human insulin with an average apparent half-life of 81 minutes compared to 141 minutes for regular human insulin.

Disposition No information available.

Drug interactions See **Insulin Glargine**.

E. Therapeutic response Near normalization of hyperglycemia as measured by glycosylated hemoglobin (HbA1c)

F. Role in therapy Limited evidence suggests that better control of postprandial glucose may lead to long-term benefits. Both insulin lispro and insulin aspart, because of their rapid absorption after subcutaneous administration, are effective in this regard, but they have not been compared.

G. Other considerations *NovoLog* is available by prescription only.

GLUCAGON

Trade names: *GlucaGen*, *Glucagon*

Manufacturer: Novo Nordisk A/S, Bagsvaerd, Denmark (*GlucaGen*); Eli Lilly, Indianapolis, IN (*Glucagon*)

Distributed by: Bedford Laboratories, Division of Ben Venue Laboratories, Bedford, OH (*GlucaGen*)

Indications: Treatment for severe hypoglycemia

Approval date: 22 June 1998

Type of submission: New drug application

A. General description Glucagon (rDNA origin) is a polypeptide hormone identical to human glucagon. Glucagon is produced in the pancreas and consists of 29 amino acids (MW: 3.5 kDa). It increases blood glucose and relaxes smooth muscle of the gastrointestinal tract. *Glucagon* (Eli Lilly) is synthesized in a special nonpathogenic laboratory strain of *Escherichia coli* bacteria that has been genetically altered by the addition of the gene for glucagon. *GlucaGen* is produced by expression of recombinant DNA in a *Saccharomyces cerevisiae* vector with subsequent purification.

B. Indications and use Glucagon (rDNA origin) is indicated as a treatment for severe hypoglycemic reactions that may occur in patients with diabetes treated with insulin. Because Glucagon (rDNA origin) depletes glycogen stores, the patient should be given supplemental carbohydrates as soon as he/she awakens and is able to swallow, especially children or adolescents with type 1 diabetes. Glucagon (rDNA origin) is also indicated for use during radiologic examinations to temporarily inhibit movement of the gastrointestinal tract. It is as effective for this examination as are anticholinergic drugs.

C. Administration

Dosage form *Glucagon* (Eli Lilly) is available for use intravenously, intramuscularly, or subcutaneously in a kit that contains a vial of sterile glucagon and a syringe of sterile diluent. The vial contains glucagon (rDNA origin) 1 mg. *GlucaGen* is supplied as a sterile, lyophilized white powder in a 2 ml vial that contains glucagon (rDNA origin) 1 mg, accompanied by sterile water for reconstitution.

Route of administration Glucagon for injection (rDNA origin) may be administered intravenously, intramuscularly, or subcutaneously.

Recommended dosage and monitoring requirements For the treatment of hypoglycemia, the dose of *GlucaGen* for adult and pediatric patients weighing more than 25 kg is 1 mg by injection. Half the adult dose (0.5 mg) is recommended for pediatric patients weighing less than 25 kg or younger than 6 to 8 years old. *Glucagon* (Eli Lilly) 1 mg is recommend for all patients weighing more than 20 kg. Emergency assistance should be sought if the patient fails to respond within 15 minutes after subcutaneous or intramuscular injection of glucagon. The glucagon injection may be repeated while waiting for emergency assistance. Intravenous glucose must be administered if the patient fails to respond to glucagon. When the patient has responded to the treatment, oral carbohydrate should be given to restore the liver glycogen and prevent recurrence of hypoglycemia.

D. Pharmacology and pharmaceutics

Clinical pharmacology Glucagon is produced by alpha cells in the islets of Langerhans in the pancreas. In the liver, both gluconeogenesis and glycogenolysis are stimulated, resulting in an increase in blood sugar. Glucagon acts only on liver glycogen, converting it to glucose.

Pharmacokinetics According to the product label, *Glucagon* (Eli Lilly) has been studied following intramuscular, subcutaneous, and intravenous administration in adult volunteers. Administration of intravenous *Glucagon* (Eli Lilly) showed dose proportionality between 0.25 and 2.0 mg. Calculations from a 1 mg dose indicated a mean volume of distribution of 0.25 liter/kg and a mean clearance of 13.5 ml/min per kg); half-life ranged from 8 to 18 minutes. Maximum plasma concentrations of 7.9 ng/ml were achieved approximately 20 minutes after subcutaneous administration. With intramuscular dosing, maximum plasma concentrations of 6.9 ng/ml were attained approximately 13 minutes after dosing. Intramuscular injection of *GlucaGen* demonstrated a mean apparent half-life of 45 minutes, probably reflecting prolonged absorption from the injection site.

Disposition Glucagon (rDNA origin) is inactivated in the liver, kidney, and plasma.

Drug interactions No information available

E. Therapeutic response

Glucagon (rDNA origin) is usually administered intramuscularly or subcutaneously. After injection, patients in hypoglycemia usually recover within 15 minutes. In a study of fasting healthy human subjects, a subcutaneous dose of *Glucagon* (Eli Lilly) 1 mg resulted in a mean peak glucose concentration of 136 mg/dl 30 minutes after injection. Similarly, following intramuscular injection, the mean peak glucose level was 138 mg/dl, which occurred at 26 minutes after injection.

F. Role in therapy

According to Micromedex, glucagon (rDNA origin) is effective in the treatment of hypoglycemia due to severe insulin reactions and idiopathic hypoglycemia of neonates. In the treatment of insulin reactions it may be used alone or in combination with intravenous glucose. Glucagon has the advantage that it may be given intramuscularly or subcutaneously and may be administered in cases where it is difficult or impossible to establish venous access. Hypoglycemia from excessive doses of a sulfonylurea may not respond adequately to glucagon because the oral agent may cause the secretion of insulin. Although glucose is the preferred agent for treating hypoglycemia when an intravenous line is available, glucagon is an effective alternative.

G. Other applications

Micromedex notes that glucagon (rDNA origin) may also be used to reduce smooth muscle tone in the gastrointestinal tract. It has also shown to be effective in the treatment of anaphylaxis, biliary tract pain, beta-adrenergic blocker overdose, esophageal obstruction, and as premedication in endoscopic procedures.

SOMATREM

Trade name: ***Protropin***

Manufacturer: Genentech Inc., South San Francisco, CA

Indications: Treatment of growth hormone deficiency in children

Approval date: 17 October 1985

Type of submission: Product license application

A. General description Somatrem is a polypeptide hormone produced by DNA technology. It has 192 amino-acid residues and a molecular weight of 22 kDa. Somatrem contains the sequence of 191 amino acids constituting pituitary-derived human growth hormone plus an additional amino acid, methionine, on the N-terminus of the molecule. Somatrem is synthesized in a strain of *Escherichia coli* bacteria modified by the addition of the gene for human growth hormone production.

B. Indications and use *Protropin* is administered to children with documented lack of human growth hormone.

C. Administration
Dosage form *Protropin* is supplied as a sterile, lyophilized powder for reconstitution with bacteriostatic water for injection. Each vial contains somatrem 5 mg or 10 mg.
Route of administration *Protropin* is intended for subcutaneous or intramuscular administration.
Recommended dosage and monitoring requirements *Protropin* dosage must be individualized for each patient. The recommended dose is up to 0.1 mg/kg administered three times weekly by intramuscular or subcutaneous injection. Therapy should be discontinued when epiphyses have fused, when final height is achieved, or when the patient ceases to respond to *Protropin*.

D. Pharmacology and pharmaceutics
Clinical pharmacology Deficient secretion of growth hormone during years of active body growth results in pituitary dwarfism. The nomenclature for the various biosynthetic growth hormone preparations reflects the source and the chemical composition of the product. Somatropin refers to GH of the same amino-acid sequence as that in naturally occurring growth hormone. Somatropin from human pituitary glands is abbreviated hGH or pit-hGH; recombinant origin somatropin is termed recombinant GH or

rGH. Somatrem refers to the methionine derivative of recombinant GH, and is abbreviated met-rGH. Somatrem is a more antigenic preparation, but despite the presence of anti-GH antibodies, growth responses to met-rGH are similar to those seen in patients treated with rGH.
Pharmacokinetics The same area under the curve (AUC) and disappearance rates are observed following injection of pituitary-derived hGH or Somatrem.
Disposition No information available.
Drug interactions Excessive glucocorticoid therapy may inhibit the growth-promoting effects of human growth hormone.

E. Therapeutic response *Protropin* has been found to be effective in achieving growth rate increases in children with growth hormone deficiency.

F. Role in therapy Micromedex notes that like somatropin (rDNA origin), somatrem is effective in the treatment of children with documented deficiency of growth hormone; its effectiveness in constitutional short stature remains to be determined. Somatrem can be considered to be equal to somatropin in efficacy, but because of the increased incidence of immunogenic effects resulting in neutralizing antigrowth hormone antibodies seen with somatrem, it will likely take a secondary role to recombinant somatropin products in therapy.

G. Other applications According to Micromedex, adolescents (as well as children) with Turner's syndrome have benefited from *Protropin*. It may be useful as well for children with growth retardation secondary to renal failure. The effectiveness of growth hormone for children with constitutional short stature has not been adequately determined.

H. Other considerations *Protropin* has been designated an orphan product for use in the treatment of growth failure due to lack of endogenous growth hormone secretion or short stature secondary to Turner's syndrome.

SOMATROPIN (RECOMBINANT)

Trade names: ***Humatrope, Genotropin, Norditropin, Nutropin, Saizen, Serostim***

Manufacturer: Eli Lilly, Indianapolis, IN (*Humatrope*); Pharmacia, Stockholm, Sweden (*Genotropin*); Novo Nordisk A/S, Denmark (*Norditropin*); Genentech, Inc., South San Francisco, CA (*Nutropin*); Serono Laboratories, Inc., Randolph, MA (*Saizen, Serostim*)

Indications: Long-term treatment of pediatric patients who have growth failure due to an inadequate secretion of endogenous growth hormone (*Genotropin, Humatrope, Norditropin, Nutropin, Saizen*) or to Prader-Willi syndrome (*Genotropin*). Treatment of short stature associated with Turner's syndrome (*Humatrope, Nutropin*). Treatment of growth failure in children associated with chronic renal insufficiency up to the time of renal transplantation (*Nutropin*). Long-term replacement therapy in adults with growth hormone deficiency (*Genotropin, Humatrope, Nutropin*). Treatment of AIDS wasting or cachexia (*Serostim*).

Approval dates: 1 March 1997 (Humatrope); 1 August 2000 (Serostim); 1 September 2000 (Saizen); 1 June 2000 (Genotropin)

Type of submission:

A. General description Somatropin (rDNA origin) is a polypeptide hormone of recombinant DNA origin. It has 191 amino-acid residues and a molecular weight of about 22 kDa. The amino-acid sequence of the product is identical to that of human growth hormone of pituitary origin. *Humatrope* and *Genotropin* are synthesized in a strain of *Escherichia coli* that has been mod-ified by the addition of the gene for human growth hormone. *Norditropin* is synthesized by a special strain of *E.coli* bacteria that has been modified by the addition of a plasmid that carries the gene for human growth hormone. *Nutropin* is synthesized by a specific laboratory strain of *E. coli* as a precursor consisting of the rhGH molecule preceded by the secretion signal from an *E. coli* protein. This precursor is directed to the plasma membrane of the cell. The signal sequence is removed and the native protein is secreted into the periplasm so that the protein is folded appropriately as it is synthesized. *Saizen* is produced by a mammalian cell line (mouse C127) that has been modified by the addition of the human growth hormone gene. *Saizen*, with the correct three-dimensional configuration, is secreted directly through the cell membrane into the cell-culture medium for collection and purification. *Serostim* is produced by a mammalian cell line (mouse C127) that has been modified by the addition of the human growth hormone gene. *Serostim* is secreted directly through the cell membrane into the cell-culture medium for collection and purification.

B. Indications and use *Humatrope* and *Nutropin* are indicated for the long-term treatment of pediatric patients who have growth failure due to an inadequate secretion of normal endogenous growth hormone and for the treatment of short stature associated with Turner syndrome in patients whose epiphyses are not closed. They are also indicated for replacement of endogenous growth hormone in adults with growth hormone deficiency. *Genotropin* is indicated for the long-term treatment of pediatric patients who have growth failure due to an inadequate secretion of endogenous growth hormone and the long-term treatment of pediatric patients who have growth failure due to Prader-Willi syndrome (PWS). The diagnosis of PWS should be confirmed by appropriate genetic testing. Other causes of short stature in pediatric patients should be excluded. *Genotropin* is also indicated for the long-term replacement therapy in adults with growth hormone deficiency of either childhood- or adult-onset etiology. *Norditropin* and *Saizen* are indicated for the long-term treatment of children who have growth failure due to inadequate secretion of endogenous growth hormone. *Serostim* is indicated for the treatment of AIDS wasting or cachexia.

C. Administration

Dosage form Somatropin (rDNA origin) is a sterile, lyophilized powder intended for injection after reconstitution. Most products are available in vials. *Humatrope* pen cartridges are also available. *Genotropin* is dispensed in a two-chamber cartridge. The front chamber contains recombinant somatropin and excipients; the rear chamber contains a preservative, mannitol, and water for injection. *Genotropin* and *Norditropin* are also dispensed as a single-use syringe device containing a two-chamber cartridge. *Nutropin* is also available as a sterile solution (*Nutropin AQ*) and as a long-acting injectable suspension (*Nutropin Depot*). The *Nutropin Depot* formulation consists of micronized particles of rhGH embedded in biocompatible, biodegradable polylactide-coglycolide microspheres. Before administration, the powder is suspended in a sterile aqueous solution.

Route of administration *Humatrope* and *Saizen* are intended for subcutaneous or intramuscular administration. *Genotropin*, *Nutropin*, and *Serostim* are intended for subcutaneous use. *Norditropin* is intended for subcutaneous injection in the thighs after reconstitution. *Nutropin Depot* in administered by subcutaneous injection once or twice a month.

Recommended dosage and monitoring requirements The recommended dosage of somatropin for growth-hormone deficiency varies with the product and should be individualized. Therapy for children

should not be continued if epiphyseal fusion has occurred. During therapy, dosage should be titrated if required by the occurrence of side effects or to maintain the Insulin-like Growth Factor (IGF-I) response below the upper limit of normal IGF-I levels, matched for age and sex.

D. Pharmacology and pharmaceutics

Clinical pharmacology Somatropin (rDNA origin) stimulates linear growth in pediatric patients who lack adequate normal endogenous growth hormone. Preclinical and clinical testing have demonstrated that products containing somatropin (rDNA origin) are therapeutically equivalent to human growth hormone of pituitary origin in normal adults. Treatment of growth hormone-deficient pediatric patients and patients with Turner syndrome produces increased growth rate and IGF-I (Insulin-like Growth Factor-I/Somatomedin-C) concentrations similar to those seen after therapy with human growth hormone of pituitary origin. In pediatric patients who have growth hormone deficiency (GHD) or Prader-Willi syndrome, treatment stimulates linear growth and normalizes concentrations of IGF-I. In adults with GHD, treatment results in reduced fat mass, increased lean body mass, metabolic alterations that include beneficial changes in lipid metabolism, and normalization of IGF-I concentrations. Treatment of patients with chronic renal insufficiency results in an increase in growth rate and an increase in IGF-I levels similar to that seen with pituitary-derived hGH. Treatment of patients with AIDS-associated wasting results in an increase in lean body mass (LBM) and a decrease in body fat with a significant increase in body weight due to the dominant effect of LBM gain.

Pharmacokinetics The absolute bioavailability of *Humatrope* after subcutaneous injection is about 75%; absolute bioavailability is somewhat lower after intramuscular administration. According to label, the volume of distribution of somatropin (rDNA origin) after intravenous injection is about 0.07 liter/kg, and the mean clearance is 0.14 liter/h/kg. The mean half-life of intravenous somatropin (rDNA origin) is 0.36 hours, whereas subcutaneously and intramuscularly administered somatropin have mean half-lives of 3.8 and 4.9 hours, respectively. The longer half-life observed after subcutaneous or intramuscular administration is due to slow absorption from the injection site. *Humatrope* achieves pharmacokinetic profiles equivalent to human growth hormone in normal adults. Following a subcutaneous (thigh) injection of *Genotropin* to adult growth hormone-deficient patients, approximately 80% of the dose was systemically available as compared with that available following intravenous dosing. Results were comparable in both male and female patients. Following an intravenous infusion of *Norditropin*, investigators determined a mean clearance of approximately 139 (± 105) ml/min for hGH and a terminal elimination half-life of approximately 21.1 (± 5.1) minutes. After a subcutaneous dose, the mean apparent terminal half-life values were estimated to be approximately 7 to 10 hours. The mean terminal $t_{1/2}$ of hGH after subcutaneous administration of *Nutropin* is significantly longer than that seen after intravenous administration (2.1 ± 0.43 hours vs. 19.5 ± 3.1 minutes), indicating that the subcutaneous absorption of the compound is slow and rate-limiting. Clearance of hGH after intravenous administration of *Nutropin* in healthy adults and children is reported to be in the range of 116 to 174 ml/hr per kg. Estimates of relative bioavailability in GHD children for a single subcutaneous dose of *Nutropin Depot* ranged from 33% to 38% when compared to a single subcutaneous dose of *Nutropin AQ*.

Disposition Somatropin is metabolized by the liver and kidney.

Drug interactions Excessive glucocorticoid therapy may prevent optimal response to somatropin. If glucocorticoid replacement therapy is required, the glucocorticoid dosage and compliance should be monitored carefully to avoid either adrenal insufficiency or inhibition of growth promoting effects. Limited published data indicate that growth hormone treatment increases cytochrome P450 (CYP450) mediated antipyrine clearance in humans. These data suggest that GH administration may alter the clearance of compounds known to be metabolized by CYP450 liver enzymes. Careful monitoring is advisable when GH is administered in combination with other drugs known to be metabolized by CYP450 liver enzymes.

E. Therapeutic response In clinical trials in adults, the primary efficacy measure was body composition (lean body mass and fat mass). *Humatrope*-treated adult onset patients, as compared with those treated with placebo, experienced an increase in lean body mass and a decrease in body fat. Similar changes were seen in childhood onset growth hormone deficient patients. In some studies, Turner's syndrome patients treated to final adult height achieved statistically significant average height gains. Beneficial changes in body composition were observed at the end of the 6-month treatment period for adult patients receiving *Genotropin* as compared with the placebo patients. Patients with Prader-Willi syndrome who received *Genotropin* showed significant increases in linear growth during the first year of study. After 12 weeks of treatment, 74% of the AIDS patients treated with *Serostim* gained weight while only 48% of the placebo-treated patients gained weight.

F. Role in therapy Somatropin (rDNA origin) is a safe alternative to growth hormone derived from human pituitary glands, which carries the risk of viral contamination. Pituitary-derived growth hormone has been withdrawn from the market because of the associated risk of Creutzfeldt-Jacob disease. Recombinant growth hormone products are now the mainstay of treatment in growth hormone deficiency in children and other appropriate patients.

H. Other considerations   *Humatrope* has been designated an orphan product for use in the long-term treatment of children who have growth failure due to inadequate secretion of normal endogenous growth hormone and in the treatment of Turner's syndrome. *Genotropin* has been designated an orphan product for use in the treatment of adults with growth hormone deficiency, in the treatment of growth failure in children who were born small for gestational age, and in the treatment of short stature in patients with Prader-Willi syndrome. *Norditropin* has been designated an orphan product for the treatment of short stature associated with Turner's syndrome. *Nutropin* has been designated an orphan product for use in the long-term treatment of children who have growth failure due to a lack of adequate endogenous growth hormone secretion. *Saizen* holds an orphan drug designation for enhancement of nitrogen retention in hospitalized patients suffering from severe burns and treatment of idiopathic or organic growth hormone deficiency in children with growth failure. *Serostim* is designated an orphan drug for treatment of children with AIDS-associated failure-to-thrive including AIDS-associated wasting.

FOLLITROPIN ALFA

Trade name: ***Gonal-F***

Manufacturer: Serono Labs, Randolph, MA

Indications: Induction of ovulation and pregnancy; development of multiple follicles in the anovulatory patient; induction of spermatogenesis in men

Approval date: 1 June 2000

Type of submission: New drug application

A. General description   Follitropin alfa for injection is a human follicle-stimulating hormone (FSH) preparation of recombinant DNA origin. It consists of two noncovalently linked, nonidentical glycoproteins designated as the (alpha)- and (beta)-subunits. The (alpha)- and (beta)-subunits have 92 and 111 amino acids, respectively, and their primary and tertiary structures are indistinguishable from those of human follicle-stimulating hormone. Recombinant FSH is produced in genetically modified Chinese hamster ovary (CHO) cells cultured in bioreactors. Based on available data derived from physicochemical tests and bioassays, follitropin alfa and follitropin beta are indistinguishable.

B. Indications and use   *Gonal-F* is indicated for the induction of ovulation and pregnancy in anovulatory infertile patients in whom the cause of infertility is functional and not due to primary ovarian failure. It is also indicated for the development of multiple follicles in an anovulatory patient participating in an assisted reproductive technology program. In men with primary and secondary hypogonadotropic hypogonadism in whom the cause of infertility is not due to primary testicular failure, *Gonal-F* is indicated for the induction of spermatogenesis.

C. Administration
   Dosage form   *Gonal-F* is a sterile, lyophilized powder intended for subcutaneous injection after reconstitution with sterile water for injection. Each ampule contains recombinant FSH 37.5 IU, 75 IU, or 150 IU.
   Route of administration   *Gonal-F* is to be administered as a subcutaneous injection.

Recommended dosage and monitoring requirements For ovulation induction, the usual starting dose of *Gonal-F* is 75 international units (IU) subcutaneously per day. Over the course of treatment, doses of *Gonal-F* may range up to 300 IU per day depending on the individual patient response. *Gonal-F* should be administered until adequate follicular development is indicated by serum estradiol and vaginal ultrasonography. A response is generally evident after 5 to 7 days. For women participating in assisted reproductive technologies, the usual dose is 150 to 225 IU/d, initiated in the early follicular phase (cycle day 2 or 3). For inducing spermatogenesis in men, the usual dose of *Gonal-F* is 150 IU subcutaneously three times per week. *Gonal-F* must be given in conjunction with human chorionic gonadotropin (hCG).

D. Pharmacology and pharmaceutics

Clinical pharmacology According to Micromedex, FSH is a complex glycoprotein secreted by the anterior pituitary that influences the processes involved with gonadal germ cell development in both genders. Preparations of human FSH are used in the treatment of human infertility. In women, these preparations stimulate follicular development, and in males, they are used in combination with human chorionic gonadotropin to initiate and maintain spermatogenesis in hypogonadotropic hypogonadism patients.

Pharmacokinetics The absorption rate of *Gonal-F* following subcutaneous or intramuscular administration was found to be slower than the elimination rate. Hence the pharmacokinetics of *Gonal-F* are absorption rate-limited. After intravenous administration to pituitary down-regulated, healthy female volunteers, the serum profile of FSH appears to be described by a two-compartment open model with a distribution half-life of about 2 to 2.5 hours. Steady-state serum levels were reached after 4 to 5 days of daily administration.

Total clearance after intravenous administration in healthy females was 0.6 liter/h; mean residence time was 17 to 20 hours; volume of distribution was about 10 liters.

Disposition FSH metabolism following administration of *Gonal-F* has not been studied in humans. FSH renal clearance was 0.07 liter/h after intravenous administration representing approximately 1/8 of total clearance.

Drug interactions The label states that no drug–drug interaction studies have been performed.

E. Therapeutic response

In anovulatory women, the goal of therapy is adequate follicular development as determined by ultrasound in combination with measurement of serum estradiol levels. In men, the aim of treatment is maintenance of serum testosterone levels within the normal range.

F. Role in therapy

According to Micromedex, follitropin alfa is derived from recombinant technology and is devoid of contaminants. This agent can therefore be administered by subcutaneous injection. Recombinant technology may also improve batch-to-batch consistency and unlike previous exogenous preparations, follitropin alfa does not contain any luteinizing hormone activity. Results from initial studies have failed to detect the formation of antibodies. No significant differences in safety or efficacy were observed between follitropin alfa (*Gonal-F*) and follitropin beta (*Follistim*) in a prospective, randomized trial. Ovarian and endometrial response rates, pregnancy rates, and ongoing pregnancy rates were all similar, indicating that the two recombinant products are comparable in clinical efficacy. In studies comparing the efficacy of follitropin alfa to that of highly purified urinary follicle-stimulating hormone (uFSH) in women undergoing ovulation induction for *in vitro* fertilization or intracytoplasmic sperm injection and subsequent embryo transfer, follitropin alfa was as good as or superior to uFSH for inducing multiple follicle development.

G. Other considerations Recombinant follitropin alfa has been designated an orphan product for the induction of spermatogenesis in men with primary and secondary hypogonadotropic hypogonadism in whom the cause of infertility is not due to primary testicular failure.

FOLLITROPIN BETA

Trade name: *Follistim*

Manufacturer: Organon Inc., West Orange, NJ

Indications: Development of multiple follicles in ovulatory patients and the induction of ovulation and pregnancy in anovulatory infertile patients

Approval date: 1 December 1998

Type of submission: New drug application

A. General description Follitropin beta contains human follicle-stimulating hormone (hFSH), a glycoprotein hormone that is manufactured by recombinant DNA technology. Follitropin beta has a dimeric structure containing two glycoprotein subunits (alpha and beta). Both the 92 amino-acid alpha-chain and the 111 amino-acid beta-chain have complex heterogeneous structures arising from two N-linked oligosaccharide chains. Follitropin beta is synthesized in a Chinese hamster ovary cell line that has been transfected with a plasmid containing the two subunit DNA sequences encoding for hFSH. The compound is considered to contain no LH activity. The amino-acid sequence and tertiary structure of the product are indistinguishable from that of human follicle-stimulating hormone (hFSH) of urinary source. Also, based on available data derived from physicochemical tests

and bioassay, follitropin beta and follitropin alfa are indistinguishable.

B. Indications and use *Follistim* is indicated for the development of multiple follicles in ovulatory patients participating in an assisted reproductive technology program. It is also indicated for the induction of ovulation and pregnancy in anovulatory infertile patients in whom the cause of infertility is functional and not due to primary ovarian failure.

C. Administration

Dosage form *Follistim* consists of a sterile, freeze-dried cake intended for injection after reconstitution with sterile water for injection. Each vial of *Follistim* contains follitropin beta 75 international units (IU).

Route of administration *Follistim* is intended for subcutaneous or intramuscular administration.

Recommended dosage and monitoring requirements According to Micromedex, the usual starting dose of *Follistim* is 150 to 225 IU per day injected intramuscularly or subcutaneously, adjusted accordingly on an individual basis. The initial dose of 75 IU per day for 7 to 14 days is employed in anovulatory women who have not responded to clomiphene citrate treatment. The initial dose of 150 IU per day, followed by a maintenance dose of 75 to 375 IU per day for 6 to 12 days, is employed for controlled ovarian hyperstimulation followed by assisted reproduction.

D. Pharmacology and pharmaceutics

Clinical pharmacology FSH, the active component of *Follistim*, is required for normal follicular growth, maturation, and gonadal steroid production. In the female the level of FSH is critical for the onset and duration of follicular development, and consequently for the timing and number of follicles reaching maturity. In order to effect the final phase of follicle maturation, resumption of meiosis and rupture of the follicle in the absence of an endogenous LH surge, human chorionic gonadotropin (hCG) must be given following the administration of *Follistim* when patient monitoring indicates that appropriate follicular development parameters have been reached.

Pharmacokinetics The bioavailablity of follitropin beta following subcutaneous and intramuscular administration of *Follistim* was 77.8% and 76.4%, respectively, but absorption was faster after intramuscular injection than after subcutaneous injection. There is dose proportionality in serum levels over a range of doses following subcutaneous or intramuscular injection. The volume of distribution of follitropin beta in healthy, pituitary-suppressed, female subjects following intravenous administration of a 300 IU dose was approximately 8 liters. The elimination half-life following a single intramuscular dose (300 IU) of *Follistim* in female subjects was 43.9 ± 14.1 hours (mean ± SD). The elimination half-life following a 7-day intramuscular treatment with 75 IU, 150 IU, or 225 IU was 26.9 ± 7.8, 30.1 ± 6.2, and 28.9 ± 6.5 hours (mean ± SD), respectively.

Disposition Recombinant FSH is biochemically very similar to urinary FSH, and it is therefore anticipated that it is metabolized in the same manner.

Drug interactions The label states that no drug/drug interaction studies have been performed.

E. Therapeutic response Follicle maturation during follitropin beta therapy is best monitored by daily ultrasonography and daily measurement of serum and/or urine estradiol levels. Ovulation may be clinically confirmed by an increase in serum progesterone as well as an elevation in basal body temperature.

F. Role in therapy According to Micromedex, recombinant follitropin beta is preferred over urofollitropin (uFSH). It lacks contaminants, provides for greater

batch-to-batch consistency, and can be injected either subcutaneously or intramuscularly. Preliminary studies have not detected any antifollitropin beta antibodies following repeated injections. Two large *in vitro* fertilization trials have shown that follitropin beta is more efficacious than urofollitropin as assessed by the number of oocytes retrieved or by a higher embryo development rate. No significant differences in safety or efficacy were observed between follitropin alfa and follitropin beta in a prospective randomized trial.

G. Other applications Micromedex notes that follitropin beta, in combination with human chorionic gonadotropin treatment results in near normal sperm concentration and motility, but sperm morphology remained subnormal in a hypogonadotropic male.

NAFARELIN

Trade name: *Synarel*

Manufacturer: G.D. Searle & Co., Skokie, IL

Indications: Management of endometriosis and treatment of central precocious puberty

Approval date: 1 October 1999

Type of submission: New drug application

A. General description Nafarelin acetate, the active component of *Synarel Nasal Solution*, is a synthetic decapeptide, an analogue of the naturally occurring gonadotropin-releasing hormone (GnRH) in which the sixth amino acid of the native molecule (glycine) has been replaced by D-napthylalanine.

B. Indications and use *Synarel* is indicated for management of endometriosis, including pain relief and reduction of endometriotic lesions. Experience with the drug for the management of endometriosis has been limited to women 18 years of age and older treated for 6 months. *Synarel* is also indicated for treatment of central precocious puberty (CPP) (gonadotropin-dependent precocious puberty) in children of both sexes.

C. Administration

Dosage form Two versions of *Synarel* are available: *Synarel Nasal Solution for Central Precocious Puberty* and *Synarel Nasal Solution for Endometriosis*. They appear to be the same except for package and labeling. Each 0.5 ounce bottle contains 8 ml *Synarel Nasal Solution* 2 mg/ml (as nafarelin base). Each bottle is supplied with a metered spray pump that delivers 200 µg of nafarelin per spray.

Route of administration The appropriate dose of *Synarel Nasal Solution* is sprayed into one or both nostrils.

Recommended dosage and monitoring requirements For the management of endometriosis, the recommended daily dose of *Synarel* is 400 µg. This is achieved by one spray (200 µg) into one nostril in the morning and one spray into the other nostril in the evening. Treatment should be started between days 2 and 4 of the menstrual cycle. In an occasional patient, the 400 µg daily dose may not be sufficient to produce amenorrhea. The recommended duration of administration is 6 months. Sneezing during or immediately after dosing may impair drug absorption. For the treatment of central precocious puberty (CPP), the recommended daily dose of *Synarel* is 1600 µg. The dose can be increased to 1800 µg daily if adequate suppression cannot be achieved. The 1600 µg dose is achieved by two sprays (400 µg) into each nostril in the morning and two sprays into each nostril in the evening. The patients head should be tilted back slightly, and 30 seconds should elapse between sprays. If the prescribed therapy has been well tolerated by the patient, treatment of

CPP with *Synarel* should continue until resumption of puberty is desired.

D. Pharmacology and pharmaceutics

Clinical pharmacology According to Micromedex, nafarelin is a potent synthetic decapeptide analogue of GnRH in which the sixth amino acid of the native molecule (glycine) has been replaced by D-napthyl-alanine. The resulting product is approximately 200 times more potent than GnRH. Nafarelin also appears to have a quicker onset and longer duration of action than GnRH. The effects of nafarelin on the pituitary gland and sex hormones are dependent on its length of administration. Continued administration suppresses gonadotrope responsiveness to endogenous GnRH, resulting in reduced secretion of LH and FSH and, secondarily, decreased ovarian and testicular steroid production. In women, nafarelin inhibits ovulation and reduces serum estrogen concentrations in a dose-dependent fashion, to approximately postmenopausal levels. Ovulatory menses return about 50 days following discontinuation of therapy.

Pharmacokinetics Nafarelin acetate is rapidly absorbed into the systemic circulation after intranasal administration. Maximum serum concentrations (measured by RIA) were achieved between 10 and 40 minutes. Bioavailability from a 400 µg dose averaged 2.8% (range: 1.2–5.6%). The average serum half-life of nafarelin following intranasal administration is approximately 3 hours. Twice daily intranasal administration of 200 or 400 µg of *Synarel* in healthy women for 22 days did not lead to significant accumulation of the drug. In children, nafarelin acetate was also rapidly absorbed into the systemic circulation after intranasal administration. The average serum half-life of nafarelin following intranasal administration of a 400 µg dose was approximately 2.5 hours.

Dispostion Nafarelin is primarily degraded by peptidase.

Drug interactions The label states that no pharmacokinetic-based drug-drug interaction studies have been conducted with *Synarel*. However, because nafarelin acetate is a peptide that is primarily degraded by peptidase and not by cytochrome P-450 enzymes, drug interactions would not be expected to occur.

E. Therapeutic response *Synarel* administration leads to decreased secretion of gonadal steroids; consequently tissues and functions that depend on gonadal steroids for their maintenance become quiescent.

F. Role in therapy Nafarelin administered by nasal spray is as effective as oral danazol in the treatment of endometriosis and is not associated with the adverse effects on blood lipids that are seen with danazol, another antigonadal tropic agent. The therapeutic benefit from *Synarel*, however, tends to provide only temporary relief. Currently, the only treatment for endometriosis that is not associated with a high recurrence rate is bilateral oophorectomy. In patients concerned about future fertility, *Synarel* is an excellent alternative to oophorectomy. It produces a reversible chemical castration that is as effective as oophorectomy or orchiectomy in reducing sex steroid production while maintaining the possibility of future fertility. When used regularly in girls and boys with central precocious puberty (CPP) at the recommended dose, *Synarel* suppresses luteinizing hormone and sex steroid hormone levels to prepubertal levels, affects a corresponding arrest of secondary sexual development, and slows linear growth and skeletal maturation.

G. Other applications Nafarelin may be an effective contraceptive in women because it suppresses ovulation. Subcutaneous administration of nafarelin is partially effective in suppressing spermato-genesis. Nafarelin may also be effective in hirsutism due to ovarian androgen hypersecretion, and for *in vitro* fertilization. It may provide symptomatic relief for benign prostatic hyperplasia.

LEUPROLIDE

Trade name: *Lupron*

Manufacturer: Takeda Chemical Industries, Ltd., Osaka, Japan

Distributed by: TAP Pharmaceuticals Inc., Lake Forest, IL

Indications: Palliative treatment of advanced prostate cancer, management of endometriosis, and treatment of central precocious puberty

Approval date: 16 April 1993

Type of submission: New drug application

A. General description Leuprolide is a synthetic nonapeptide analogue of naturally occurring gonadotropin releasing hormone (GnRH). The analogue possesses greater potency than the natural hormone.

B. Indications and use *Lupron Injection*, *Lupron Depot*, *Lupron Depot-3 Month*, and *Lupron Depot-4 Month* are indicated in the palliative treatment of advanced prostate cancer. These products offer an alternative treatment when orchiectomy or estrogen administration is either not indicated or unacceptable to the patient. *Lupron Injection Pediatric* and *Lupron Depot-PED* are indicated in the treatment of children with central precocious puberty. *Lupron Depot* and *Lupron Depot-3 Month* are indicated for the management of endometriosis, including pain relief and reduction of endometriotic lesions.

C. Administration
Dosage forms *Lupron Injection* and *Lupron Injection Pediatric* are sterile,

aqueous solutions intended for subcutaneous injection. They are available in a 2.8 ml multiple-dose vial containing 5 mg/ml of leuprolide acetate. *Lupron Depot 3.75 mg* and *7.5 mg* are available in prefilled dual-chamber syringes containing sterile lyophilized microspheres that, when mixed with diluent, become a suspension intended as a monthly intramuscular injection. *Lupron Depot-3 Month 11.25 mg* and *22.5 mg* are available in prefilled dual-chamber syringes containing sterile lyophilized microspheres that, when mixed with diluent, become a suspension intended as an intramuscular injection to be given once every 3 months. *Lupron Depot-4 Month 30 mg* is available in a prefilled dual-chamber syringe containing sterile lyophilized microspheres that, when mixed with diluent, become a suspension intended as an intramuscular injection to be given once every 4 months. *Lupron Depot-PED*, intended for children, is available in a prefilled dual-chamber syringe containing sterile lyophilized microspheres that, when mixed with diluent, become a suspension intended as a single intramuscular injection. The front chamber of *Lupron Depot-PED* prefilled dual-chamber syringe contains leuprolide acetate 7.5 mg, 11.25 mg, or 15 mg.

Route of administration *Lupron Injection* is intended for subcutaneous injection. *Lupron Depot* is intended for intramuscular injection.

Recommended dosage and monitoring requirements According to Micromedex, the effective doses of *Lupron* in prostate cancer are 1 mg subcutaneously daily, or in the depot formulation 7.5 mg intramuscularly (IM) monthly, 11.25 mg every 3 months, or 30 mg every 4 months. *Lupron* Depot-3 month 22.5 mg is used in the treatment of advanced prostate cancer. In endometriosis, the effective dose is 3.75 mg depot IM monthly or 11.25 mg every 3 months for 6 months. In central precocious puberty, the recommended starting dose of *Lupron Injection Pediatric* is 50 µg/kg

administered as a single subcutaneous injection daily. The recommended starting dose of the pediatric depot formulations is 0.3 mg/kg every 4 weeks (minimum 7.5 mg) administered as a single intramuscular injection. The starting dose will be dictated by the child's weight.

D. Pharmacology and pharmaceutics

Clinical pharmacology According to the product label, leuprolide, a GnRH agonist, acts as a potent inhibitor of gonadotropin secretion when given continuously in therapeutic doses. Studies indicate that following an initial stimulation, chronic administration of leuprolide results in suppression of ovarian and testicular steroidogenesis. This effect is reversible upon discontinuation of drug therapy. Administration of leuprolide has resulted in inhibition of the growth of certain hormone-dependent tumors as well as atrophy of the reproductive organs. In humans, subcutaneous administration of single daily doses of leuprolide results in an initial increase in circulating levels of luteinizing hormone (LH) and follicle-stimulating hormone (FSH), leading to a transient increase in levels of the gonadal steroids (testosterone and dihydrotestosterone in males, and estrone and estradiol in premenopausal females). However, continuous daily administration of leuprolide results in decreased levels of LH and FSH in all patients. In males, testosterone is reduced to castrate levels. In premenopausal females, estrogens are reduced to postmenopausal levels. These decreases occur within 2 to 4 weeks after initiation of treatment, and castrate levels of testosterone in prostate cancer patients have been demonstrated for periods of up to 5 years. In children with central precocious puberty, stimulated and basal gonadotropins are reduced to prepubertal levels. Testosterone in males and estradiol in females are reduced to prepubertal levels. Reduction of gonadotropins will allow for normal physical and psychological growth and development. Natural maturation occurs when gonadotropins return to pubertal levels following discontinuation of leuprolide acetate.

Pharmacokinetics Leuprolide is not active when given orally. Bioavailability by subcutaneous administration is comparable to that by intravenous administration. The mean steady-state volume of distribution of leuprolide following intravenous bolus administration to healthy male subjects was 27 liters, mean systemic clearance was 7.6 liters/h, with a terminal elimination half-life of approximately 3 hours based on a two-compartment model, and *in vitro* binding to human plasma proteins ranged from 43% to 49%. Following a single injection of *Lupron Depot-3 Month 11.25 mg* in female subjects, leuprolide appeared to be released at a constant rate following the onset of steady-state levels during the third week after dosing and mean levels then declined gradually to near the lower limit of detection by 12 weeks. Following a single injection of *Lupron Depot-4 Month 30 mg* in orchiectomized prostate cancer patients, leuprolide appeared to be released at a constant rate following the onset of steady-state levels during the fourth week after dosing, providing steady plasma concentrations throughout the 16-week dosing interval. A single dose of *Lupron Depot 3.75 mg* was administered by intramuscular injection to healthy female volunteers. Following the initial rise, leuprolide concentrations in plasma started to plateau within 2 days after dosing and remained relatively stable for about 4 to 5 weeks.

Disposition No information available.

Drug interactions None have been reported.

E. Therapeutic response In prostate cancer, a complete response is considered to be disappearance of tumor masses, normalization of elevated acid phosphatase, disappearance of osteoblastic lesions, recalcification of osteolytic lesions, and normalization of hepatomegaly and abnormal liver function tests. Disease progression is suggested by an increase in size of tumor mass of greater than 25% and new lesions. Testosterone levels of less than 25 to 50 ng/dl have been associated with objective responses in prostate cancer.

F. Role in therapy According to Micromedex, leuprolide appears to be as effective as diethylstilbestrol, but with a lower order of toxicity. In combination with flutamide, leuprolide is often considered first-line therapy for prostate cancer. The effectiveness of leuprolide in the treatment of endometriosis is similar to other GnRH analogues, and is equivalent to danazol, thus making it an effective alternative for treating endometriosis. Comparative evaluation with other GnRH analogues is lacking, so leuprolide's ultimate place in the treatment of central precocious puberty has not been determined.

G. Other applications Limited data show some beneficial effects of leuprolide in the treatment of breast cancer. According to Micromedex, there is good documentation that leuprolide is effective for bowel pain and nausea associated with irritable bowel syndrome. Leuprolide has been used for controlled ovarian hyperstimulation to enhance the *in vitro* fertilization-embryo transfer procedure. In endometriosis, the goal of treatment is pain relief and reduction of endometriotic lesions. In children with central precocious puberty, stimulated and basal gonadotropins are reduced to prepubertal levels. Testosterone and estradiol are reduced to prepubertal levels in males and females, respectively.

H. Other considerations Leuprolide has been designated an orphan product for use in the treatment of central precocious puberty.

GOSERELIN

Trade name: ***Zoladex***

Manufacturer: Zeneca Limited, Macclesfield, England

Distributed by: Zeneca Pharmaceuticals, Wilmington, DE

Indications: Palliative treatment of advanced carcinoma of the prostate and in combination with flutamide for the management of locally confined Stage T2b-T4 (Stage B2-C) carcinoma of the prostate, and the management of endometriosis. *Zoladex* is also used as an endometrial-thinning agent

Approval date: 9 April 1998

Type of submission: New drug application

A. General description Goserelin acetate is a potent synthetic decapeptide analogue of gonadotropin-releasing hormone (GnRH), with a molecular weight of 1.3 kDa (free base).

B. Indications and use *Zoladex* is indicated in the palliative treatment of advanced carcinoma of the prostate and in combination with flutamide for the management of locally confined Stage T2b-T4 (Stage B2-C) carcinoma of the prostate. *Zoladex* is also indicated for the management of endometriosis, including pain relief and reduction of endometriotic lesions, and for use in the palliative treatment of advanced breast cancer in pre- and peri-menopausal women. *Zoladex* is also used as an endometrial-thinning agent prior to endometrial ablation for dysfunctional uterine bleeding. *Zoladex 3-Month* is indicated in the palliative treatment of advanced carcinoma of the prostate, and in combination with flutamide, for the management of locally confined Stage T2b-T4 (Stage B2-C) carcinoma of the prostate.

C. Administration

Dosage forms *Zoladex* is supplied as a sterile, biodegradable implant containing goserelin acetate equivalent to 3.6 mg of goserelin. The product is designed for subcutaneous injection with continuous release over a 28-day period. Goserelin acetate is dispersed in a matrix of D,L-lactic and glycolic acids copolymer and presented as a sterile, 1 mm diameter cylinder, pre-loaded in a special single use syringe with a 16-gauge needle. *Zoladex 3-Month* is supplied as a sterile, biodegradable product containing goserelin acetate equivalent to 10.8 mg of goserelin. It is designed for subcutaneous implantation with continuous release over a 12-week period.

Route of administration *Zoladex* is injected as a subcutaneous implant. The injection site of choice is the upper abdominal wall, but the midline of the lower abdomen has also been used.

Recommended dosage and monitoring requirements The 3.6 mg implant administered every 28 days is indicated for prostate cancer, endometriosis, breast cancer, and endometrial thinning. The 10.8 mg implant administered every 12 weeks is only indicated in prostate cancer. For the management of endometriosis, the recommended duration of administration is 6 months.

D. Pharmacology and pharmaceutics

Clinical pharmacology Goserelin acts as a potent inhibitor of pituitary gonadotropin secretion. Following initial administration in males, the implant causes an initial increase in serum luteinizing hormone (LH) and follicle-stimulating hormone (FSH) levels with subsequent increases in serum levels of testosterone. With time, however, goserelin leads to sustained suppression of pituitary gonadotropins, and serum levels of testosterone consequently fall into the range normally seen in surgically castrated men. This leads to accessory sex organ regression. In clinical trials with follow-up of more than 2 years, suppression of serum testosterone to castrate levels has been maintained for the duration of therapy. In females, a similar down-regulation of the pituitary gland by chronic exposure to goserelin leads to suppression of gonadotropin secretion, a decrease in serum estradiol to levels consistent with the postmenopausal state, and would be expected to lead to a reduction of ovarian size and function, reduction in the size of the uterus and mammary gland, as well as a regression of sex hormone-responsive tumors, if present. Serum estradiol is suppressed to levels similar to those observed in postmenopausal women. Estradiol, LH, and FSH levels return to pretreatment values within 12 weeks following the last implant administration in all but rare cases.

Pharmacokinetics Following administration of *Zoladex*, goserelin is released from the depot at a much slower rate initially for the first 8 days, and then there is more rapid and continuous release for the remainder of the 28-day dosing period. Despite the change in the releasing rate of goserelin, administration of *Zoladex* every 28 days resulted in testosterone levels that were suppressed and maintained in the range normally seen in surgically castrated men.

Disposition Metabolism of goserelin, by hydrolysis of the C-terminal amino acids, is the major clearance mechanism. Clearance of goserelin following subcutaneous administration of a solution formulation is very rapid and occurs via a combination of hepatic metabolism and urinary excretion. More than 90% of a subcutaneous radiolabeled solution formulation dose of goserelin is excreted in urine. Approximately 20% of the dose in urine is accounted for by unchanged goserelin.

Drug interactions No formal drug–drug interaction studies have been performed.

E. Therapeutic response According to the product label, in studies of patients with advanced prostate cancer that compared *Zoladex* with orchiectomy, the long-term endocrine responses and objective responses were similar between the two treatment arms. Additionally, duration of survival was similar between the two treatment arms in a comparative trial. In Stage B2-C prostatic carcinoma, *Zoladex* (3.6 mg depot) and flutamide prior to and during radiation was associated with a significantly lower rate of local failure compared with radiation alone. In endometriosis, controlled clinical studies using the 3.6 mg formulation every 28 days for 6 months, showed that *Zoladex* was as effective as danazol therapy in relieving clinical symptoms (dysmenorrhea, painful coitus, and pelvic pain) and signs (pelvic tenderness, pelvic induration) of endometriosis and in decreasing the size of endometrial lesions as determined by laparoscopy.

F. Role in therapy Goserelin is useful as an alternative to estrogen therapy or orchiectomy for the palliation of advanced carcinoma of the prostate. It produces chemical castration, which is as effective as orchiectomy or oophorectomy in terms of reducing sex steroid production. Preliminary data indicate that goserelin is as effective and better tolerated than diethylstilbestrol therapy. In patients who refuse orchiectomy, especially young men, goserelin may be a useful alternative. In endometriosis, goserelin has been shown to be at least as effective as danazol, a standard pharmacologic treatment, and was significantly better tolerated.

G. Other applications Goserelin is useful for *in vitro* fertilization and is possibly effective in controlling precocious puberty and early puberty. Goserelin may promote fibroid regression, but rapid regrowth results upon discontinuation.

GANIRELIX

Trade name: *Antagon*

Manufacturer: Vetter Pharma-Fertigung GmbH & Co. KG, Ravensburg, Germany

<u>Distributed by</u>: Organon Inc, West Orange, NJ

<u>Indications</u>: Inhibition of premature luteinizing hormone surges in women undergoing controlled ovarian hyperstimulation

<u>Approval date</u>: 1 July 1999

<u>Type of submission</u>: New drug application

A. General description Ganirelix is a synthetic decapeptide with high antagonistic activity against naturally occurring gonadotropin-releasing hormone (GnRH). It is derived from native GnRH with substitutions of amino acids at positions 1, 2, 3, 6, 8, and 10. The molecular weight for ganirelix acetate is 1.6 kDa as an anhydrous free base.

B. Indications and use *Antagon* is indicated for the inhibition of premature luteinizing hormone (LH) surges in women undergoing controlled ovarian hyperstimulation.

C. Administration
Dosage form *Antagon* is supplied as a sterile, aqueous solution intended for subcutaneous injection only. Each sterile, prefilled syringe contains ganirelix acetate 250 μg per 0.5 ml.
Route of administration *Antagon* is intended for subcutaneous administration.
Recommended dosage and monitoring requirements Once-daily subcutaneous doses of *Antagon* 250 μg have been effective for prevention of premature LH surges during ovarian stimulation in *in vitro* fertilization procedures.

D. Pharmacology and pharmaceutics
Clinical pharmacology According to the product label, the pulsatile release of GnRH stimulates the synthesis and secretion of luteinizing hormone (LH) and follicle-stimulating hormone (FSH). The frequency of LH pulses in the mid- and late-follicular phase is approximately 1 pulse per hour. These pulses can be detected as transient rises in serum LH. At midcycle, a large increase in GnRH release results in an LH surge. The midcycle LH surge initiates several physiologic actions, including ovulation, resumption of meiosis in the oocyte, and luteinization. Luteinization results in a rise in serum progesterone with an accompanying decrease in estradiol levels. Ganirelix acetate acts by competitively blocking the GnRH receptors on the pituitary gonadotroph and subsequent transduction pathway. It induces a rapid, reversible suppression of gonadotropin secretion. The suppression of pituitary LH secretion by ganirelix is more pronounced than that of FSH. An initial release of endogenous gonadotropins has not been detected with ganirelix, which is consistent with an antagonist effect. Upon discontinuation of ganirelix, pituitary LH and FSH levels are fully recovered within 48 hours.

Pharmacokinetics The pharmacokinetics of ganirelix are dose proportional in the dose range of 125 to 500 μg. Ganirelix is rapidly absorbed following subcutaneous injection; the mean absolute bioavailability following a single 250 μg subcutaneous injection to healthy female volunteers was 90%. The mean was volume of distribution of ganirelix in healthy females following intravenous administration of a single 250 μg dose is 44 liters. *In vitro* protein binding to human plasma is about 80%. Mean half-life after subcutaneous administration of a single dose is 12.8 hours, and 16.2 hours after multiple dosing.

Disposition No information available.
Drug interactions No formal drug–drug interaction studies have been performed.

E. Therapeutic response *Antagon* induces a rapid, reversible suppression of gonadotropin secretion.

F. Role in therapy Subcutaneous ganirelix has been effective as an adjunct to recombinant follicle-stimulating hormone (rFSH) for preventing premature LH surges and facilitating a good clinical outcome in women undergoing ovarian stimulation for *in vitro* fertilization. According to Micromedex, the principal advantage of ganirelix over gonadotropin-releasing hormone (GnRH) agonists (e.g., leuprolide, goserelin) is the avoidance of a stimulatory effect on pituitary cells. At present, ganirelix should be considered an alternative to GnRH agonists for preventing premature LH surges during ovarian hyperstimulation.

OXYTOCIN

Trade name: *Pitocin*

Manufacturers: Monarch Pharmaceuticals, Bristol, TN (*Pitocin*); Wyeth-Ayerst, Philadelphia, PA (*Oxytocin for Injection*); Eon Labs, Laurelton, NY (*Oxytocin Solution for Injection*); Fujisawa, Deerfield, IL (*Oxytocin Solution for Injection*)

Indications: Induction and augmentation of labor; management of postpartum bleeding; and adjunctive therapy in the management of incomplete or inevitable abortion

Approval date: 27 July 1997 (*Pitocin*)

Type of submission: New drug application

A. General description Oxytocin is a nonapeptide (molecular weight 1.0 kDa) that differs by two amino acids from vasopressin; it is stored and released from the posterior pituitary gland. Commercially available oxytocin is prepared synthetically.

B. Indications and use Oxytocin is widely used for induction and augmentation of labor. It is also indicated for the control of postpartum uterine bleeding and as adjunctive therapy in the management

of incomplete or inevitable abortion in the second trimester.

C. Administration
   Dosage form *Pitocin* is available in a 1 ml ampule, a 1 ml prefilled syringe, a 1 ml vial, and a 10 ml multiple-dose vial. Each preparation contains oxytocin 10 units/ml. Oxytocin Solution for Injection (Eon, Fujisawa) is supplied as a 1 ml ampule or vial and a 10 ml vial; each contains oxytocin 10 units/ml. Oxytocin for injection also contains oxytocin 10 units/ml.
   Route of administration The use of oxytocin for the induction of labor requires intravenous administration. According to Micromedex, pulsatile administration achieves the same results as continuous administration with a significantly lower dose and infusion volume. Intravenous infusion of oxytocin is routinely used postpartum to control uterine bleeding.
   Recommended dosage and monitoring requirements According to Micromedex, oxytocin dosage and rate of administration are determined by uterine response. The usual dosage for induction and augmentation of labor is 0.5 to 2 milliunits (mU)/min. To reduce postpartum bleeding, 10 U of oxytocin may be infused at a rate of 20 to 40 mU/min after delivery of the placenta. Fetal heart rate, resting uterine tone, and the frequency, duration, and force of contractions should be monitored during oxytocin therapy. Oxytocin should be discontinued if uterine hyperactivity or fetal distress develop.

D. Pharmacology and pharmaceutics
   Clinical pharmacology The uterine myometrium contains receptors specific to oxytocin. Oxytocin promotes uterine contractions by increasing intracellular concentrations of calcium. It stimulates both electrical and contractile activity. Plasma prostaglandin levels rise significantly during oxytocin infusion in women with

successful induction of labor when compared to women in whom induction failed.

Pharmacokinetics Oxytocin is distributed throughout the extracellular fluid and is eliminated with a half-life of 3 to 5 minutes.

Dispostion Oxytocin is subject to enzymatic hydrolysis, primarily by tissue oxytocinase.

Drug interactions Co-administration of dinoprostone in patients receiving intravenous oxytocic agents such as oxytocin may result in uterine hyperstimulation and is therefore contraindicated. Severe hypertension has been reported when oxytocin was given 3 to 4 hours following prophylactic administration of a vasoconstrictor in conjunction with spinal anesthesia. Oxytocin may enhance the neuromuscular blockade of succinylcholine.

E. Therapeutic response Oxytocin promotes uterine contractions.

F. Role in therapy According to Micromedex, oxytocin is routinely used for the induction of labor at term and postpartum for the control of uterine bleeding. Oxytocin is not the drug of choice for induction of labor for abortion. Oxytocin infusion has been used following prostaglandin or hypertonic abortifacients to shorten the induction to abortion time when inducing second-trimester abortion, inducing abortion when a patient has failed to respond to the abortifacient, or to induce abortion after membranes have ruptured.

OCTREOTIDE

Trade name: *Sandostatin*

Manufacturer: Novartis Pharmaceuticals, East Hanover, NJ

Indications: Reduction of blood levels of growth hormone and IGF-I in acromegaly patients; symptomatic treatment of me-tastatic carcinoid tumors; and treatment of profuse watery diarrhea associated with intestinal tumors

Approval date: 25 November 1998

Type of submission: New drug application

A. General description Octreotide is a long-acting cyclic octapeptide (molecular weight 1.0 kDa) with pharmacologic actions mimicking those of the natural hormone somatostatin. The usefulness of natural somatostatin is limited by its extremely short duration of action, which requires continuous infusion. Octreotide is a molecular form (an octapeptide with two D-amino-acid substitutions) that overcomes the problem of short duration of action.

B. Indications and use *Sandostatin* is indicated to reduce blood levels of growth hormone and IGF-I (somatomedin-C) in acromegaly patients who have had inadequate response to or cannot be treated with surgical resection, pituitary irradiation, or bromocriptine mesylate at maximally tolerated doses. It is also indicated for the symptomatic treatment of patients with metastatic carcinoid tumors, where it suppresses or inhibits the severe diarrhea and flushing episodes associated with the disease, and for the treatment of the profuse watery diarrhea associated with VIP (vasoactive intestinal peptide)-secreting tumors (vipomas).

C. Administration

Dosage forms *Sandostatin* sterile solution is available as 1 ml ampules containing octreotide (as acetate) 50, 100, or 500 µg, and 5 ml multidose vials containing octreotide (as acetate) 200 and 1000 µg/ml. *Sandostatin LAR Depot* is available in 5 ml vials in three strengths: 10 mg, 20 mg, or 30 mg. The contents of the vial, when mixed with diluent, becomes a suspension that is given as a monthly intragluteal injection. The octreotide is uniformly distributed within the microspheres, which are made of biodegradeable polymers.

Route of administration *Sandostatin* may be administered subcutaneously or intravenously. *Sandostatin LAR Depot* is intended for intragluteal administration only.

Recommended dosage and monitoring requirements *Sandostatin* injectable solution is usually administered subcutaneously in initial doses of 50 to 100 μg two or three times daily. Upward dose titration is frequently required. *Sandostatin LAR Depot* injectable suspension should never be administered intravenously or subcutaneously. Patients not currently receiving octreotide acetate should begin therapy with subcutaneous injections and then can be switched to the depot injection of 20 mg intragluteally at 4-week intervals for 3 months.

D. Pharmacology and pharmaceutics

Clinical pharmacology Octreotide exerts pharmacologic actions similar to the natural hormone, somatostatin. It is an even more potent inhibitor of growth hormone, glucagon, and insulin than somatostatin. Like somatostatin, octretide also suppresses luteinizing hormone response to gonadotropin-releasing hormone, decreases splanchnic blood flow, and inhibits release of serotonin, gastrin, vasoactive intestinal peptide, secretin, motilin, and pancreatic polypeptide. By virtue of these pharmacological actions, *Sandostatin* has been used to treat the symptoms associated with metastatic carcinoid tumors (flushing and diarrhea), and VIP-secreting adenomas (watery diarrhea). Octreotide also substantially reduces growth hormone and/or IGF-I levels in patients with acromegaly.

Pharmacokinetics After subcutaneous injection, octreotide is absorbed rapidly and completely from the injection site; intravenous and subcutaneous doses were found to be bioequivalent. Peak concentrations and area-under-the-curve values were dose proportional both after subcutaneous or intravenous single doses up to 400 μg and with multiple doses of 200 μg three times a day. At daily doses of 600 μg/d as compared to 150 μg/d, clearance was reduced by about 66%, suggesting nonlinear kinetics of the drug. In healthy volunteers the distribution of octreotide from plasma was rapid, the volume of distribution was estimated to be 13.6 liters, and the total body clearance was 10 liters/h. The elimination of octreotide from plasma had an apparent half-life of 1.7 hours compared with a half-life of only 1 to 3 minutes with the natural hormone, somatostatin. In an elderly population, dose adjustments may be necessary due to a significant increase in the half-life and a significant decrease in the clearance of the drug. In patients with acromegaly, the pharmacokinetics differ somewhat from those in healthy volunteers. The volume of distribution was estimated to be 21.6 ± 8.5 liters, and the total body clearance was increased to 18 liters/h. In patients with severe renal failure requiring dialysis, clearance was reduced to about half that found in normal subjects (from approximately 10 liters/h to 4.5 liters/h). The effect of hepatic diseases on the disposition of octreotide is unknown. Drug release for octreotide injectable suspension depends on the biodegradation of the microspheres in the muscle.

Dispostion About 32% of the dose of octreotide is excreted unchanged into the urine. Octreotide is also metabolized by the liver.

Drug interactions *Sandostatin* has been associated with changes in nutrient absorption, so it may effect the absorption of orally administered drugs. Concomitant administration of *Sandostatin* with cyclosporine may decrease blood levels of cyclosporine and result in transplant rejection. Patients receiving insulin, oral hypoglycemic agents, beta blockers, calcium channel blockers, or agents to control fluid and electrolyte balance may require dose adjustments of these agents.

E. Therapeutic response Laboratory tests that may be helpful as biochemical markers in determining and following patient response depend on the specific tumor. Based on diagnosis, measurement of the following substances may be useful in monitoring the progress of therapy: growth hormone and IGF-I (somatomedin-C) in acromegaly; 5-HIAA (urinary 5-hydroxyindole acetic acid), plasma serotonin, and plasma substance P in patients with carcinoid tumors; and plasma vasoactive intestinal peptide in patients with vipomas.

F. Role in therapy Octreotide appears to have an established role in both the long-and short-term treatment of symptoms caused by metastatic carcinoid tumors and vipomas. Bromocriptine has been demonstrated to reduce plasma growth hormone levels in patients with acromegaly. In comparison with bromocriptine, octreotide results in a greater reduction of plasma growth hormone levels, and the reduction is more rapid and more prolonged.

G. Other applications Octreotide has been reported to be effective in treating noninfectious diarrhea and diarrhea in patients with AIDS, and as an adjunct to endoscopic variceal ligation, to reduce recurrent bleeding from esophageal varices. Micromedex lists many other potential uses.

9

ENZYMES

■ SECTION ONE ■

■ 9.1. OVERVIEW OF ENZYME THERAPIES

Enzymes are biologic catalysts exhibiting remarkable substrate specificity and high efficiency. They are found throughout biological systems and are located both inside and outside of cells. With few exceptions (e.g., RNA molecules), enzymes are proteins that catalyze unique biochemical reactions with exquisite substrate selectivity. Through lowering the activation energy essential for moving the process forward, a typical enzyme is capable of accelerating a chemical reaction rate by a millionfold or more [1]. Enzyme catalytic activities are essential for viability of cells and tissues because these activities are an integral part of biochemical reactions that carry out DNA, RNA, and protein synthesis, repair, and metabolism. More than a thousand enzyme-catalyzed reactions have been reported in the literature, and a majority of the more than 10,000 proteins synthesized in mammalian cells exhibit enzyme activity [2].

Highly coordinated chemical reactions, carried out simultaneously in response to the continuously changing cellular environment, are mediated by several enzyme-regulating mechanisms. Enzymes are regulated by covalent (e.g., phosphorylation) and proteolytic modifications, binding to stimulatory and inhibitory proteins, and

Biotechnology and Biopharmaceuticals, by Rodney J. Y. Ho and Milo Gibaldi
ISBN 0-471-20690-3   Copyright © 2003 by John Wiley & Sons, Inc.

allosteric interactions. Because not all cellular reactions occur in the same intracellular compartment, spatial segregation within intracellular (e.g., mitochondrion, lysosome, and nucleus) organelles permits enzymes to carry out their specialized tasks in the most efficient manner. A change or perturbation of the dynamic, tightly linked, and highly orchestrated enzyme equilibrium may lead to significant clinical consequences. On the other hand, understanding of mechanisms and the role enzymes play in biologic systems provides opportunity to use enzymes as therapeutic tools for treating diseases, especially those for which traditional small-molecule, chemical drugs are unlikely to be effective.

9.1.1. Enzyme Replacement Therapy

Deficient or aberrant enzyme activity in metabolic pathways may produce clinical symptoms due to shortage or accumulation of key products or substrates. Some of the enzymes displaying aberrant activity and their related clinical consequences are listed in Table 9.1. Several enzymes have been tested as replacement therapy to ameliorate disease symptoms. The effectiveness of chronic enzyme replacement therapy may depend on the tissue source (i.e., human vs. animal tissues) used for enzyme isolation. The protein orthologue purified from an animal source may elicit antibody response to the enzyme, thereby reducing its potency on repeated use. A number of strategies, including production of human recombinant enzyme instead of an animal orthologue, as well as surface modification of proteins with polyethylene glycol as a means to reduce protein immunogenicity, have improved the effectiveness of enzyme replacement therapy.

Adenosine Deaminase for Severe Combined Immunodeficiency Syndrome

Severe combined immunodeficiency (SCID) is an autosomal recessive syn-

drome characterized by the absence of T- and B-cell function from birth. In 1972 Giblett and co-workers serendipitously discovered that in some patients with the syndrome, the disease is due to an inherited deficiency of adenosine deaminase (ADA) [3]. ADA can form multimers with an apparent molecular weight of 298 kDa, composed of 36 to 42 kDa subunits [4]. ADA is ubiquitous, and diagnosis of deficiency can be made with many cell types including erythrocytes, lymphocytes, and fibroblasts; prenatal diagnosis can be performed with chorionic villous, amniotic cells, and fetal blood samples [3].

Infants with SCID have profound immunodeficiency and present with frequent episodes of diarrhea, pneumonia, otitis, sepsis, and cutaneous infections. Persistent infections with opportunistic organisms such as *Pneumocystis carinii*, Epstein-Barr virus, *Candida albicans*, cytomegalovirus, parainfluenzae 3 virus, respiratory syncitial virus, adenovirus, varicella, and bacille Calmette-Guérin (BCG) lead to death within the first or second year of life. ADA deficiency also occurs in adults, but with a much later onset and milder, but clinically discernible, immunodeficiency [3,5].

The abnormal T- and B-cell functions in patients with SCID are the result of ADA deficiency. The *ADA* gene has been mapped to chromosome 20q.13, and a number of point and deletion mutations have been identified in SCID patients [5–7]. ADA catalyses the irreversible deamination of adenosine and 2'-deoxyadenosine to inosine and 2'-deoxyinosine as a part of purine nucleoside metabolism. Adenosine and deoxyadenosine are suicide inactivators of S-adenosylhomocysteine (SAH) hydrolase, and lead indirectly to intracellular accumulation of SAH, which is a potent inhibitor of methylation reactions. Cellular methylation function is essential for detoxification of adenosine and deoxyadenosine. As a result ADA deficiency leads to accumulation to

■TABLE 9.1. Some reported enzyme deficiencies and their clinical presentation

Enzyme Defect	Clinical Features and Symptoms
Glucose-6-phosphatase	Common (type-Ia/von Gierke), severe hypoglycemia, growth retardation, enlarged liver and kidney, hypoglycemia, elevated blood lactate, cholesterol, triglycerides, and uric acid
Glucose-6-phosphate translocase	Type Ib; equivalent to 10% of total type I; similar to type Ia, with additional findings of neutropenia and neutrophil dysfunction
Liver and muscle debranching enzyme	Type IIIa/Cori or Forbes: intermediate severity of hypoglycemia; liver cirrhosis can occur in adulthood
debranching enzyme in liver (normal muscle debrancher activity)	Type IIIb: Liver symptoms same as in type IIIa; no muscle symptoms
Liver phosphorylase	Type VI/Hers: hepatomegaly, mild hypoglycemia, hyperlipidemia, and ketosis; symptoms improve with age
Liver phosphorylase kinase α subunit	Type IX/phosphorylase kinase deficiency: similar to type VI presentation (X-linked)
Glycogen synthase	Fasting hypoglycemia and ketosis, elevated lactic acid and hyperglycemia after glucose load (decreased glycogen storage capacity)
Branching enzyme	Type IV/Andersen: failure to thrive, hypotonia, hepatomegaly, splenomegaly, progressive liver cirrhosis and failure (death usually before fifth year); some without progression
Muscle phosphorylase	Type V/McArdle: exercise intolerance, muscle cramps, myoglobinuria on strenuous exercise, increased creatine kinase (predominantly in male)
Phosphofructokinase-M subunit	Type VII/Tarui: similar to type V, with additional findings of a compensated hemolysis
Phosphoglycerate kinase	Similar to type V, with additional findings of a hemolytic anemia and CNS dysfunction (X-linked, rare)
Phosphoglycerate mutase deficiency (M subunit)	Similar to type V (rare; mostly in African-American population)
Lactate dehydrogenase deficiency (M subunit)	Similar to type V, with additional findings of erythematous skin eruption and uterine stiffness resulting in childbirth difficulty in female (rare)
Fructose 1,6-bisphosphate aldolase A	Similar to type V, plus hemolytic anemia
Pyruvate kinase-muscle isozyme	Muscle cramps and/or fixed muscle weakness
Muscle-specific phosphorylase kinase	Similar to type V with muscle weakness and atrophy (autosomal recessive)
Lysosomal acid α-glucosidase	Type II/Pompe: usually undetectable, or very low level of enzyme activity in infantile form; residual enzyme activity in late-onset. Infant: hypotonia, muscle weakness, cardiac enlargement and failure, fatal early
	Juvenile and adult: progressive skeletal muscle weakness and atrophy, proximal muscle and respiratory muscle are seriously affected

■TABLE 9.1. *Continued*

Enzyme Defect	Clinical Features and Symptoms
Cardiac-specific phosphorylase kinase	Severe cardiomyopathy and early heart failure (very rare)
Galactose 1-phosphate uridyl transferase	Galactosemia, vomiting, hepatomegaly, jaundice, cataracts, amino aciduria, failure to thrive; long-term complications exist even with early diagnosis and treatment
Galactokinase	Cataracts (benign)
Uridine diphosphate galactose 4-epimerase	Galactosemia with additional findings of hypotonia and nerve deafness
Fructose 1-phosphate aldolase B	Hereditary fructose intolerance, vomiting, lethargy, failure to thrive, hepatic failure; good prognosis with early diagnosis and fructose restriction
Fructose 1,6-diphosphatase	Episodic hypoglycemia and lactic acidosis; good prognosis when fasting is avoided
Alkaline phosphatase	Hypophosphatasia due to accumulation of inhibitors of mineralization such as pyrophosphate, and inability to raise phosphorus levels at the site of calcium-phosphorus deposition into hydroxyapatite
Adenosine deaminase	Severe combined immune deficiency
β-glucuronidase	Gaucher's diseases
Superoxide dismutase	Amyotrophic lateral sclerosis
α-glucosidase	Pompe syndrome
α-galactosidase	Farby syndrome

Source: Data adapted, in part from Chen [33].

Note: CK, creatine kinase; M, muscle; CNS, central nervous system.

toxic levels of intracellular purine metabolites and impairment of T- and B-cell functions. The loss of T- and B-cell functions presents clinically as an inability to recover from even a mild course of infection.

The first evidence that supplementation of exogenous ADA may be helpful in SCID patients came from the 1975 report of Polmar et al. [8], demonstrating that addition of bovine-intestinal ADA or human-erythrocyte ADA to cultures of lymphocytes of a SCID patient restored their ability to proliferate when stimulated with mitogens. The ability to respond to mitogens is an indicator of immune function restoration. Therapeutic use of enzyme extracted from calf tissue revealed that this form of ADA has a short half-life, and elicits antibodies, which may further increase clearance of the enzyme on repeated administration. Isolation of sufficient quantities of ADA from human tissue was impractical. However, surface modification of the bovine enzyme with polyethylene glycol (PEG) polymers increases hydration and reduces immunogenicity, thereby making the PEG-modified or pegylated ADA a therapeutic modality [9].

Currently pegylated ADA (*Adagen*) is the only ADA pharmaceutical preparation available for enzyme replacement therapy in patients with SCID [10]. Although enzyme replacement therapy does not fully restore ADA levels, the levels achieved are sufficient to protect patients against opportunistic and life-threatening infections. PEG-ADA therapy is currently recommended as an alternative treatment for patients for whom a MHC-matched bone marrow is not available, or are considered high-risk candidates for bone marrow transplantation [11]. According to one report, mortality and morbidity with PEG-ADA have been less than that for bone marrow transplantation procedures [11] .

Now that the ADA gene has been cloned and expressed, it is essential to determine whether recombinant human ADA can be used to increase safety and efficacy of replacement therapy through significant reduction of the immunogenicity of bovine enzyme without the need of pegylation. *ADA* gene therapeutic strategies have also been proposed and are being tested. As a true "orphan drug" developed to treat a very small patient population, the high cost of PEG-ADA therapy will continue to be a barrier.

Glucocerebrosidase for Gaucher's Disease

Gaucher's disease, a genetic defect in the glucocerebrosidase enzymes, specifically β-glucuronidase, leads to accumulation of glucocerebroside (a glycolipid) in lysosomes and is potentially fatal [12]. Gaucher's disease was first described by P. C. E. Gaucher in 1882. About 50 years later investigator reported that patients with this condition accumulated a sphingoglycolipid called glucocerebroside. Brady and co-workers in 1964 demonstrated that Gaucher's disease was due to reduced activity of a β-glucosidase called glucocerebrosidase that is now known as β-glucuronidase [13].

Patients with Gaucher's disease have been classified into three major categories based on clinical signs and symptoms: type 1, nonneuronopathic (adult); type 2, acute neuronopathic (infantile); and type 3, subacute neuronopathic (juvenile). All types of Gaucher's disease can be demonstrated

to be associated with deficient glucocerebrosidase activity. Patients may experience anemia, bone damage, enlarged liver and spleens, and pulmonary insufficiency; a few develop severe central nervous system damage. With the highest prevalence observed in the Ashkenazi Jewish population, an estimated 10,000 to 20,000 individuals in the United States are affected by this disorder.

Because the β-glucuronidase is localized in lysosomes and the exogenously administered enzyme is capable of accumulating in the lysosomes of most blood cells, including leukocytes, enzyme replacement therapy is a logical strategy [14]. β-glucuronidase extracted from human placental tissues (*Ceredase*) was used initially, and later substituted with a human recombinant form (*Cerezyme*) for treatment of Gaucher's disease [15,16]. Both placental and human recombinant β-glucuronidase were modified to expose terminal mannose residues on the glycosylated enzyme to enhance localization of the enzyme to lysosomes in leukocytes that express a high density of mannose receptors [17]. In symptomatic type-I patients, enzyme replacement therapy significantly ameliorates hematologic and hepatic abnormalities, but it is less effective in reversing pulmonary and skeletal manifestation [17].

According to one estimate, the total dose of enzyme needed for each patient requires extraction of material from approximately 50,000 placentae or 100 millions placentae for the 2000 patients currently on enzyme replacement therapy. With the gene cloned and expressed in large quantities in a fully glycosylated human recombinant form in Chinese hamster ovary (CHO) cells, the availability of the enzyme is less restricted. However, the cost of manufacturing remains high, partly due to the relatively high dose of enzyme needed to produce therapeutic effects. Annual cost of enzyme replacement therapy may be upward of $80,000 for each patient.

Enzyme replacement is a chronic therapy that typically requires administration every week or two. Antibody response is a concern because immune reactions could produce significant morbidity, and mortality in rare cases. According to the manufacturer, about 15% of patients treated to date have developed IgG antibodies to *Cerezyme* within the first year, and most often within six months of initiating therapy. After 12 months of therapy, patients who do not seroconvert are less likely to develop antibodies. Nearly half of patients with detectable IgG antibodies are reported to experience some symptoms of hypersensitivity. Whether exposure of mannose on the enzyme (for targeting the enzyme to leukocytes) plays a role in eliciting or enhancing antibody response to the enzyme is not known and probably worth investigating. About 13% of patients on *Ceredase* also produce antibody response to the enzyme extracted from placenta.

Recombinant β-glucuronidase has been engineered in vectors suitable for somatic gene therapy. A gene therapy human study has been initiated to determine the feasibility of this strategy for reversing Gaucher's disease, and initial data show adequate safety but limited expression of the enzyme [13]. We expect that improvements in gene delivery systems will provide a higher degree and more prolonged expression of the enzyme.

Pancreatic Enzymes for Steatorrhea

Steatorrhea occurs in patients whose lipase output is at 10% or less of normal. Lipase and other pancreatic enzyme insufficiencies are observed in cystic fibrosis and chronic alcoholic pancreatitis. Patients with various liver diseases may also present with steatorrhea [18]. For these patients, pancreatic enzymes—mainly lipase, protease, and amylase—extracted with alcohol from porcine pancreases have been shown to provide amelioration of diarrhea. These enzymes are enriched and formulated in enteric-coated capsules to prevent protein degradation in the acid environment of the stomach. Enteric coating prevents exposure of the lyophilized pancreatic enzymes to tryptic and peptic enzymes in the stomach.

The dose of enzyme required to treat steatorrhea may vary among individuals, and the dose should be individualized to achieve optimal therapeutic effects. In addition less-than-precise enzyme extraction and inconsistent enteric-coating procedures demand careful consideration in selecting from the array of pancreatic enzyme products now available. As much as a 30-fold difference may be seen in pancreatic enzyme activity among products after being exposed to simulated gastric fluid [19].

9.1.2. Enzymes as Therapeutic Agents

Detailed understanding of enzymatic reactions, substrate specificity, and their respective role in integrated physiologic processes has allowed development of therapeutic applications that draw on the unique, highly effective, and specific catalytic functions of enzymes. Traditional drug molecules that are designed as receptor agonists or antagonists can provide some of the effects of enzyme therapy, but lacking catalytic function, they are less efficient in mediating cascade events. Enzyme applications include dissolving blood clots to promote reperfusion and enhancement of cytotoxicity in cancer cells. The first therapeutic enzymes (i.e., streptokinase, asparaginase) were extracted from tissue or bacterial sources.

The advent of recombinant DNA technology has allowed the isolation of genes and expression of proteins that are found in biologic tissues in exceedingly small quantities. This has permitted large-scale production of enzymes such as tissue plasminogen activator (i.e., tPA, alteplase) that cannot be extracted from tissues in quantities required for therapeutic use. Table 9.2

■TABLE 9.2. Other enzymes with therapeutic potential

Enzymes	Potential Therapeutic Use
α-L-iduronidase	Hurler disorder
Superoxide dismutase (SOD)	Amyotrophic lateral sclerosis (ALS)
Phenylalanine ammonia-lyase (PAL)	Phenylketonuria (PKU)
α-glucosidase	Pompe syndrome
α-galactosidase	Fabry syndrome
N-acetylgalactosamine-4-sulfatase	Mucopolysaccharidosis VI
Aspartylglucosaminidase	Aspartylglucosaminuria ??
Fatty aldehyde dehydrogenase (FALDH) and Fatty alcohol:NAD$^+$ oxidoreductase (FAO)	Sjögren Larsson syndrome
Alkyl-dihydroxyacetone phosphate (DHAP) synthase	Isolated alkyl-DHAP synthase deficiency

Sources: Adapted from Rizzo [30] and Russell and Clarke [31].

lists some additional enzymes with potential for therapeutic applications.

Thrombolytic Enzymes

To maintain hemostasis, blood must be retained in the vasculature as fluid. At the same time, blood components must be able to respond rapidly with a clot when a vascular injury occurs. To repair a vascular injury, platelets in blood first adhere as aggregates to the endothelial cells at the affected site and form an initial blood clot. Platelets then stimulate and activate coagulation factors found in plasma to form a more stable fibrin clot. As the injury is resolved and healed, the clot is degraded. Thrombosis is a pathological event wherein a blood clot occludes a blood vessel, resulting in ischemic necrosis of the tissue fed by the blood vessel. Ischemic necrosis involves local anemia and oxygen deprivation. Thrombosis of a coronary artery may lead to myocardial infarction or unstable angina [20].

A fibrinolytic system reverses thrombosis and thereby maintains hemostasis. Fibrinolytic enzymes and other factors work in concert to dissolve the clot using a series of rather nonspecific proteases that act locally by digesting the fibrin deposited at the site of injury. The fibrinolytic system is tightly regulated so that the proteolytic activities of the enzymes do not cause unintended vascular and tissue damage, yet permit the localization of sufficient fibrinolytic activity to prevent extensive blood clotting [21].

Enzymes capable of catalyzing fibrinolytic and thrombolytic activities are listed in Table 9.3. These enzymes are directly or indirectly involved in activation of plasminogen to plasmin to degrade fibrin polymers in the clot (Figure 9.1). Streptokinase does not directly activate plasminogen, but binding of streptokinase to plasminogen (even in the absence of fibrin) exposes the proteolytic cleavage site (arginine 560) on plasminogen, leading to nonfibrin-specific activation of plasmin activity. A similar mechanism is used by urokinase in converting plasminogen to plasmin. Therefore, both of these enzymes mediate non-fibrin-specific thrombolytic activity that may lead to systemic lytic states (bleeding).

On the other hand, tPA and the variant TNKase (tenecteplase) catalyze the plasminogen to plasmin conversion only in a fibrin-dependent manner, thereby, in theory, reducing nonfibrin-dependent systemic lytic states. Clinical studies, however, comparing streptokinase with alteplase

■TABLE 9.3. Enzymes that catalyze fibrinolytic activity leading to thrombolysis

Enzyme	Characteristics	Pharmaceutical Source	Mechanisms of Action
Streptokinase Anistreplase (streptokinase-plasminogen complex)	47 kDa streptococci protein	β-hemolytic streptococci	Binding of streptokinase exposes plaminogen, which is then cleaved to form plasmin. Plasmin mediate fibrinolysis. Systemic lytic state and immunogenicity may limit its use. Because antistreplase is already lined to plaminogen, the onset of fibrinolytic is faster. These enzymes are not fibrin-specific.
Tissue plasminogen activator (tPA: alteplace)	527 aa serine protease	Human recombinant produced in Chinese hamster ovary cells	tPA is a serine protease that catalyzes plasminogen to plasmin. The conversion is enhanced by the presence of fibrin. It is less immunogenic.
TNKase (tenecteplase)	A variant of human tPA–527 aa	Human recombinant produced in Chinese hamster ovary cells.	Similar to that of tPA. The sequence modification may provide additional reduction in nonintracranial major bleeding complications.
Urokinase Prourokinase (single-chain urokinase)	411 aa two-chain serine protease	Cultured human kidney cells	Similar action as streptokinase and also lacks fibrin specificity; prourokinase exhibits fibrin specificity.

Note: aa, amino acid.

have failed to detect a significant difference in bleeding. This is partly due to the low incidence of major bleeding (i.e., 2–6%) even when heparin is included as part of therapeutic management. It took multiple clinical studies enrolling a large number (several thousand) of patients by the GUSTO (*G*lobal *U*tilization of *S*treptokinase and *T*issue plasminogen activator for *O*ccluded coronary arteries) investigators to eventually demonstrate a 1% mortality advantage of alteplase over streptokinase [22,23]. TNKase, a sequence-modified version of tPA, has been shown to produce even fewer nonintracranial major bleeding complications than tPA. The new enzyme derivative reduces major bleeding from 5.9% to 4.7% and decreases the need for blood transfusions. As tPA and TNKase are human recombinant proteins, as opposed to the streptococci protein, streptokinase, immunogenicity is less of an concern.

DNAse Enzyme Therapy

Cystic fibrosis is a life-threatening genetic disease caused by a dysfunctional cystic fibrosis transmembrane regulator, CFTR protein, which modulates salt and water transport into and out of cells. This ion-channel defect leads to poorly hydrated, thick, mucous secretions in the airways and severely impaired mucociliary func-

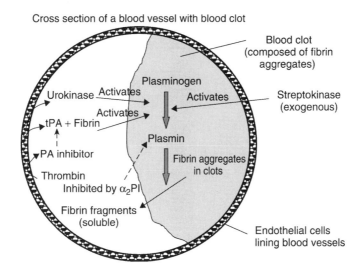

Cross section of a blood vessel with blood clot

Figure 9.1. Schematic representation of a cross-sectional view of a blood vessel with a clot and thrombolytic enzymes that convert fibrin in the blood clot into fibrin fragments. The endogenous enzymes, urokinase and tissue plasminogen activator (tPA), as well as exogenous enzymes such as streptokinase, activate plasminogen into plasmin, which catalyzes fibrin degradation reactions. The degraded fibrin fragments are then removed from the clot. α_2 plasmin inhibitor (α_2PI) can inhibit plasminogen to plasmin conversion.

tion. Impairment of these vital lung defense mechanism leads to progressive pulmonary dysfunction and respiratory failure, which accounts for approximately 90% of the premature deaths in patients with cystic fibrosis.

Retention of viscous purulent secretions, which contain high concentrations of extracellular DNA—released by degenerating leukocytes that accumulate in response to infection [24]—in the airways contributes both to reduced pulmonary function and to exacerbations of infection [24,25]. Digestion of DNA polymers in purulent secretion with DNAse (dornase-α or *Pulmozyme*) has been shown to reduce sputum viscosity in cystic fibrosis patients. The availability of recombinant DNAse has allowed its use in an aerosol formulation to deliver the enzyme into the deep lung alveoli of CF patients. The purified glycoprotein contains 260 amino acids with an approximate molecular weight of 37,000 daltons [26]. The primary amino-acid sequence is identical to that of the native human enzyme. Administration of aerosolized DNAse to CF patients significantly reduces the risk of respiratory infection [27].

L-asparaginase an Oncolytic Agent

L-asparagine aminodohydrolase (type EC-2, EC 3.5.1.1) or L-asparaginase catalyzes hydrolysis of L-asparagine (an essential amino acid) to L-aspartic acid and ammonia. Leukemic cells lacking asparagine synthase are unable to synthesize asparagine and are dependent on an exogenous source of asparagine for their survival. Treatment with L-asparaginase causes a rapid depletion of extracellular asparagines, leading to selective killing of these leukemic cells. Normal cells, with ability to synthesize asparagine intracellularly, are less sensitive to L-asparaginase oncolytic activity. This enzyme therapy takes advantage of a specific metabolic defect in some leukemic cells to provide a therapeutic effect, especially when used in combina-

■TABLE 9.4. Comparisons of plasma half-life of L-asparaginase preparations available for clinical use

	Plasma Half-life	
Enzyme	Patient Number	Mean ± SD (Days)
Pegylated-L-asparaginase[a] (*E. coli*)	10	5.73 ± 3.24
E. coli L-asparaginase	17	1.24 ± 0.17
Erwinia L-asparaginase	10	0.65 ± 0.13

Data source: Asselin [32].

[a]Pegylated-L-asparaginase, PEG-conjugated L-asparaginase (of *Escherichia coli*); Erwinia L-asparginase is extracted from *Erwinia chrysanthemi.*

tion with other chemotherapeutic agents such as vincristine, methotrexate, cytarabine, daunorubicin, and doxorubicin [28].

There are three different L-asparaginase preparations available for clinical use (see Table 9.4). Two of them are unmodified enzymes purified from *Erwinia chrysanthemi* (formerly known as *Erwinia caratovora*) and *Escherichia coli.* Repeated dosing of bacterial enzymes, however, elicits immunogenicity (hypersensitivity). This problem can be minimized by attachment of polyethylenglycol (as discussed for ADA enzyme replacement therapy) to L-asparaginase. A preparation of L-asparginase with polyethylene glycol is commercially available and has been shown to significantly reduce the degree and incidence of hypersensitivity and increase the half-life of the enzyme (Table 9.4) [29].

9.1.3. Summary and Future Prospects

The availability of endogenous enzymes and their variants has allowed the use of enzymes as replacement therapy and as therapeutic agents. With continued refinement in recombinant protein technology, the cost of human recombinant proteins may become more affordable. As more novel human recombinant enzymes are developed, a wide variety of medical disorders will be amenable to enzyme therapies. Additional molecular engineering such as pegylation or creation and modification of

glycosylation sites on the peptide backbone of enzymes may enhance the pharmacokinetic properties of "native" enzymes for pharmaceutical applications.

Development of enzyme variants using strategies such as exposing terminal mannose on β-glucuronidase to target cells and tissues expressing mannose receptors has improved delivery of enzymes. While gene therapy holds promise for long-term treatment of abnormal enzymatic functions due to genetic disorders, the therapeutic use of enzymes will continue to depend on recombinant enzymes that can be produced in a cost-effective manner. The progress made in therapies for in-born enzyme-deficiency errors such as Gaucher's disease and SCID, as well as for treatment of diseases such as myocardial infarction and cystic fibrosis is likely to be applied to other diseases and genetically influenced disorders.

REFERENCES

1. Stryer, L., *Biochemistry.* 4th ed. 1995, New York: W.H. Freeman. xxxiv, 1064.
2. Alberts, B., *Molecular biology of the cell.* 3rd ed. 1994, New York: Garland Pub. xliii, 1294.
3. Giblett, E.R., J.E. Anderson, F. Cohen, B. Pollara, and H.J. Meuwissen, *Adenosine-deaminase deficiency in two patients with severely impaired cellular immunity.* Lancet, 1972. **2**(7786): 1067–9.
4. Kelley, W.N., P.E. Daddona, and M.B. van der Weyden, *Characterization of human*

adenosine deaminase. Ciba Found Symp, **1977**(48): 277–93.

5. Hirschhorn, R., *Adenosine deaminase deficiency.* Immunodefic Rev, 1990. **2**(3): 175–98.

6. Hirschhorn, R., A. Ellenbogen, and S. Tzall, *Five missense mutations at the adenosine deaminase locus (ADA) detected by altered restriction fragments and their frequency in ADA—patients with severe combined immunodeficiency (ADA-SCID).* Am J Med Genet, 1992. **42**(2): 201–7.

7. Hirschhorn, R., *Overview of biochemical abnormalities and molecular genetics of adenosine deaminase deficiency.* Pediatr Res, 1993. **33**(1 Suppl): S35–41.

8. Polmar, S.H., E.M. Wetzler, R.C. Stern, and R. Hirschhorn, *Restoration of in-vitro lymphocyte responses with exogenous adenosine deaminase in a patient with severe combined immunodeficiency.* Lancet, 1975. **2**(7938): 743–6.

9. Beauchamp, C., P.E. Daddona, and D.P. Menapace, *Properties of a novel PEG derivative of calf adenosine deaminase.* Adv Exp Med Biol, 1984. **165**(Pt A): 47–52.

10. Hershfield, M.S., S. Chaffee, and R.U. Sorensen, *Enzyme replacement therapy with polyethylene glycol-adenosine deaminase in adenosine deaminase deficiency: overview and case reports of three patients, including two now receiving gene therapy.* Pediatr Res, 1993. **33**(1 Suppl): S42–7; discussion S47–8.

11. Hershfield, M.S., *PEG-ADA: an alternative to haploidentical bone marrow transplantation and an adjunct to gene therapy for adenosine deaminase deficiency.* Hum Mutat, 1995. **5**(2): 107–12.

12. Harrison, T.R., and K.J. Isselbacher, *Harrison's principles of internal medicine.* 13th / ed. 1994, New York: McGraw-Hill. 2 v. xxxii, 2496.

13. Brady, R.O., *Gaucher's disease: past, present and future.* Baillieres Clin Haematol, 1997. **10**(4): 621–34.

14. Pfeffer, S.R., *Targeting of proteins to the lysosome.* Curr Top Microbiol Immunol, 1991. **170**: 43–65.

15. Grabowski, G.A., N.W. Barton, G. Pastores, J.M. Dambrosia, T.K. Banerjee, M.A. McKee, C. Parker, R. Schiffmann, S.C. Hill, and R.O. Brady, *Enzyme therapy in type 1 Gaucher disease: comparative efficacy of mannose-terminated glucocerebrosidase from natural and recombinant sources.* Ann Intern Med, 1995. **122**(1): 33–9.

16. Niederau, C., *First long-term results of imiglucerase therapy of type 1 Gaucher disease.* Eur J Med Res, 1998. **3**(1/2): 25–30.

17. Brady, R.O., and N.W. Barton, *Enzyme replacement therapy for Gaucher disease: critical investigations beyond demonstration of clinical efficacy.* Biochem Med Metab Biol, 1994. **52**(1): 1–9.

18. Williams, C.N., and J.J. Sidorov, *Steatorrhea in patients with liver disease.* Can Med Assoc J, 1971. **105**(11): 1143–6 passim.

19. Hendeles, L., A. Dorf, A. Stecenko, and M. Weinberger, *Treatment failure after substitution of generic pancrelipase capsules: correlation with in vitro lipase activity.* Jama, 1990. **263**(18): 2459–61.

20. Majerus, P.W., G.J. Broze, Jr, J.P. Miletich, and D.M. Tollefsen, *Anticoagulant, Thrombolytic, and antiplatelet drugs,* in *Goodman & Gilman's the pharmacological basis of therapeutics.* L.S. Goodman et al., eds. 2001, McGraw-Hill: New York. xxvii, 2148.

21. Lijnen, H.R., and D. Collen, *Molecular basis of thrombolytic therapy.* J Nucl Cardiol, 2000. **7**(4): 373–81.

22. Metz, B.K., H.D. White, C.B. Granger, R.J. Simes, P.W. Armstrong, J. Hirsh, V. Fuster, C.M. MacAulay, R.M. Califf, and E.J. Topol, *Randomized comparison of direct thrombin inhibition versus heparin in conjunction with fibrinolytic therapy for acute myocardial infarction: results from the GUSTO-IIb Trial. Global Use of Strategies to Open Occluded Coronary Arteries in Acute Coronary Syndromes (GUSTO-IIb) Investigators.* J Am Coll Cardiol, 1998. **31**(7): 1493–8.

23. Hudson, M.P., C.B. Granger, E.J. Topol, K.S. Pieper, P.W. Armstrong, G.I. Barbash, A.D. Guerci, A. Vahanian, R.M. Califf, and E.M. Ohman, *Early reinfarction after fibrinolysis: experience from the global utilization of streptokinase and tissue plasminogen activator (alteplase) for occluded coronary arteries (GUSTO I) and global use of strategies to open occluded coronary arteries (GUSTO III) trials.* Circulation, 2001. **104**(11): 1229–35.

24. Collins, F.S., *Cystic fibrosis: molecular biology and therapeutic implications.* Science, 1992. **256**(5058): 774–9.

25. Boat, T., *Cystic Fibrosis*, in *Textbook of respiratory medicine*. N.J. Murray JF, ed. 1988, Saunders WB: Philadelphia. p. 1126–52.

26. Shak, S., D.J. Capon, R. Hellmiss, S.A. Marsters, and C.L. Baker, *Recombinant human DNase I reduces the viscosity of cystic fibrosis sputum*. Proc Natl Acad Sci USA, 1990. **87**(23): 9188–92.

27. Fuchs, H.J., D.S. Borowitz, D.H. Christiansen, E.M. Morris, M.L. Nash, B.W. Ramsey, B.J. Rosenstein, A.L. Smith, and M.E. Wohl, *Effect of aerosolized recombinant human DNase on exacerbations of respiratory symptoms and on pulmonary function in patients with cystic fibrosis. The Pulmozyme Study Group*. N Engl J Med, 1994. **331**(10): 637–42.

28. Capizzi, R., and J. Holcenberg, *Asparaginase*, in *Cancer Med*. H.A. Frei, ed. 1993, Philadelphia: Lea and Febiger.

29. Clavell, L.A., R.D. Gelber, H.J. Cohen, S. Hitchcock-Bryan, J.R. Cassady, N.J. Tarbell, S.R. Blattner, R. Tantravahi, P. Leavitt, and S.E. Sallan, *Four-agent induction and intensive asparaginase therapy for treatment of childhood acute lymphoblastic leukemia*. N Engl J Med, 1986. **315**(11): 657–63.

30. Rizzo, W.B., *Inherited disorders of fatty alcohol metabolism*. Mol Genet Metab, 1998. **65**(2): 63–73.

31. Russell, C.S., and L.A. Clarke, *Recombinant proteins for genetic disease*. Clin Genet, 1999. **55**(6): 389–94.

32. Asselin, B.L., J.C. Whitin, D.J. Coppola, I.P. Rupp, S.E. Sallan, and H.J. Cohen, *Comparative pharmacokinetic studies of three asparaginase preparations*. J Clin Oncol, 1993. **11**(9): 1780–6.

33. Chen, Y.-T., *Chapter 350: Glycogen storage diseases and other inherited disorders of carbohydrate metabolism*. Harrison's Text online. 2002, accessed 4/2/02. E. Braunwald, A.S. Fauci, K.J. Isselbacher, D.L. Kasper, S.L. Hauser, D.L. Longo, and J.L. Jameson, eds.

■ SECTION TWO ■

■ 9.2. MONOGRAPHS

ALGLUCERASE
IMIGLUCERASE

Trade name: ***Ceredase*, *Cerezyme***

Manufacturer: Genzyme Corporation, Cambridge, MA

Indications: Replacement therapy for patients with a confirmed diagnosis of type 1 Gaucher's disease

Approval date: 4 May 1991 (*Ceredase*), November 1999 (*Cerezyme*)

Type of submission: Product license application (*Ceredase*), Biologic license application (*Cerezyme*)

A. General description Alglucerase is a monomeric glycoprotein, with a molecular weight of 59.3 kDa, consisting of 497 amino-acid residues. It is a modified form of β-glucocerebrosidase from pooled human placental tissue. Alglucerase differs from the native molecule in that it is deglycosylated so that its terminal carbohydrate residues consist primarily of mannose, which are specifically recognized by carbohydrate receptors on macrophages. Imiglucerase is an analogue of the human enzyme β-glucocerebrosidase produced by recombinant DNA technology using mammalian cell culture (Chinese hamster ovary). (β)-Glucocerebrosidase is a lysosomal glycoprotein enzyme that catalyzes the hydrolysis of the glycolipid glucocerebroside to glucose and ceramide. Purified imiglucerase is a monomeric glycoprotein of 497 amino acids, containing 4 N-linked glycosylation sites. Imiglucerase differs from placental glucocerebrosidase by one amino acid at position 495 where histidine is substituted for arginine. The oligosaccharide chains at the glycosylation sites have been modified to terminate in mannose sugars. The modified carbohydrate structures on imiglucerase are somewhat different from those on placental glucocerebrosidase. These mannose-terminated oligosaccharide chains of imiglucerase are specifically recognized by endocytic carbohydrate receptors on macrophages, the cells that accumulate lipid in Gaucher's disease.

B. Indications and use *Ceredase* and *Cerezyme* are indicated for long-term replacement therapy in type 1 Gaucher's disease, which results in one or more of the following conditions: anemia, thrombocytopenia, bone disease, and hepatomegaly or splenomegaly.

C. Administration

Dosage form *Ceredase* is a sterile solution for injection, supplied in 5 ml bottles containing alglucerase 80 units/ml. *Cerezyme* is supplied as a sterile lyophilized product for injection. Each vial contains imiglucerase 212 or 424 units. Amost all patients with Gaucher's disease use *Cerezyme* for enzyme replacement therapy. *Ceredase* is only available in limited supplies for the treatment of patients who do not tolerate *Cerezyme*.

Route of administration *Ceredase*, on dilution, is administered as an intravenous infusion. *Cerezyme*, after reconstitution with sterile water for injection, is also administered as an intravenous infusion.

Recommended dosage and monitoring requirements The initial dose of alglucerase is 60 units/kg, infused intravenously over 1 to 2 hours. This dose is usually repeated every 2 weeks, but may be given as often as every other day or as infrequently as every 4 weeks depending on response. The initial dose of *Cerezyme* is 2.5 to 60 units/kg infused intravenously over 1 to 2 hours. This dose is usually repeated every 2 weeks, but may be given as often as three times a week or as infrequently as every 4 weeks depending on clinical response.

D. Pharmacology and pharmaceutics

Clinical pharmacology (β)-Glucocerebrosidase catalyzes the hydrolytic cleavage of the glycolipid glucocerebroside to glucose and ceramide within the lysosomes of phagocytic cells in the reticuloendothelial system. This process normally occurs as a catabolic pathway of membrane lipids derived from hematologic cell turnover. A deficiency of β-glucocerebrosidase results in accumulation of glucocerebroside within tissue macrophages, which become engorged with the glycolipid. These cells are termed Gaucher cells and are responsible for the clinical manifestations of Gaucher's disease. Within macrophages, alglucerase and imiglucerase are endocytosed into lysosomes where they cleave glucocerebroside and restore normal metabolic clearance of glucose and ceramide.

Pharmacokinetics After a single intravenous dose of *Ceredase*, the apparent volume of distribution of alglucerase ranged from 49 to 282 ml/kg, the plasma clearance ranged from 6 to 25 ml/min per kg, and the elimination half-life from 4 to 20 minutes. During single 1-hour intravenous infusions of *Cerezyme*, the apparent volume of distribution of imiglucerase was 90 to 150 ml/kg; its plasma clearance ranged from 9.8 to 20.3 ml/min per kg, and its elimination half-life from 3.6 to 10.4 minutes.

Disposition No information available.

Drug interactions Caution may be advisable in administration of *Cerezyme* to patients previously treated with *Ceredase* and who have developed antibodies to *Ceredase* or who have exhibited symptoms of hypersensitivity to *Ceredase*.

E. Therapeutic response Significant reductions in hepatosplenomegaly, improvement in hematologic deficiencies, and decreased cachexia have been reported with long-term *Cerezyme* therapy in patients with type 1 Gaucher's disease. Symptomatic improvement occurs within 6 months of induction of therapy. Serum hemoglobin and platelet values decrease to near-baseline values within 6 months of discontinuation of therapy.

F. Role in therapy *Ceredase* and *Cerezyme* are important advances in the treatment of type 1 Gaucher's disease.

Prior to the development of *Ceredase* and *Cerezyme*, traditional therapy was palliative, consisting primarily of splenectomy, which is also thought to accelerate disease progression. Although the patient population likely to benefit from enzyme replacement is small (approximately 5000 to 10,000 patients), it may prevent the devastating sequelae that occur in patients with Gaucher's disease. *Cerezyme* was developed to overcome supply constraints of placenta-derived *Ceredase*, which was previously the only effective treatment for type 1 Gaucher's disease. The availability of *Cerezyme* may resolve problems with supply completely; the manufacturer may phase out production of *Ceredase*, replacing it with *Cerezyme*. Unlike *Ceredase*, which is a modified form of β-glucocerebrosidase extracted from pooled human placental tissue, *Cerezyme* is produced by recombinant DNA technology using mammalian cell culture; therefore the risk of viral contamination and subsequent infection is reduced. Imiglucerase is at least as effective as alglucerase in the treatment of type 1 Gaucher's disease.

G. Other considerations *Ceredase* and *Cerezyme* have been designated as orphan products for use in the treatment of types 1, 2, and 3 Gaucher's disease.

PEGADEMASE BOVINE

Trade name: *Adagen*

Manufacturer: Enzon, Inc, Piscataway, NJ

Indications: Enzyme replacement therapy for adenosine deaminase (ADA) deficiency in patients with severe combined immunodeficiency disease (SCID) who are not suitable candidates for—or who have failed—bone marrow transplantation

Approval date: 21 March 1990

Type of submission: Product license application

A. General description Pegademase bovine is a modified enzyme. Its chemical name is (monomethoxypolyethylene glycol succinimidyl)$_{n=11-17}$-adenosine deaminase. It is a conjugate of numerous strands of monomethoxypolyethylene glycol (PEG), molecular weight 5 kDa, covalently attached to the enzyme adenosine deaminase (ADA). ADA used in the manufacture of *Adagen* is derived from bovine intestine.

B. Indications and use According to the label, *Adagen* is indicated for enzyme replacement therapy for ADA deficiency in patients with SCID who are not suitable candidates for—or who have failed—bone marrow transplantation. *Adagen* is recommended for use in infants from birth or in children of any age at the time of diagnosis. *Adagen* is not intended as a replacement for bone marrow transplant therapy. *Adagen* is also not intended to replace continued close medical supervision and the initiation of appropriate diagnostic tests and therapy (e.g., antibiotics, nutrition, oxygen, and gammaglobulin) as indicated for intercurrent illnesses.

C. Administration

Dosage form *Adagen* is a sterile solution, for injection. It is supplied in 1.5 ml single-dose vials. Each ml contains pegademase bovine 250 units.

Route of administration *Adagen* should only be given by intramuscular injection and should not be diluted or mixed with other drugs prior to use.

Recommended dosage and monitoring requirements Once-weekly intramuscular injections of *Adagen* are recommended with the following schedule: 10 units/kg, 15 units/kg, and 20 units/kg for the first, second, and third dose, respectively. Maintenance doses are 20 units/kg per week, with incremental increases by 5 units/kg per week to a maximum single dose of 30 units/kg.

D. Pharmacology and pharmaceutics

Clinical pharmacology Adenosine deaminase is an enzyme that catalyzes the deamination of both adenosine and deoxyadenosine. In the absence of the ADA enzyme, the purine substrates adenosine and 2'-deoxyadenosine accumulate, causing metabolic abnormalities that are directly toxic to lymphocytes. Hereditary lack of adenosine deaminase activity results in severe combined immunodeficiency disease, a fatal disorder of infancy characterized by profound defects of both cellular and humoral immunity. Unless children are kept in protective isolation or receive bone marrow transplantation to reconstitute the immune system, death usually intervenes prior to 2 years of age. Specifically, *Adagen* therapy is aimed at reversing the accumulation of total red cell adenine deoxyribonucleotides, depleting total adenine ribonucleotides, and restoring S-adenosylhomocysteine hydrolase activity, which is inactivated by deoxyadenosine.

Pharmacokinetics Pegademase bovine is rapidly absorbed following intramuscular administration of *Adagen*; plasma adenosine deaminase activity generally normalizes after 2 to 3 weeks of weekly intramuscular injections. The half-life of pegademase is 48 to 72 hours.

Disposition No information available.

Drug interactions There are no known drug interactions with *Adagen*. However, the antiviral agent vidarabine (*Vira-A*) is a substrate for ADA and 2'-deoxycoformycin (*Pentostatin*) is a potent inhibitor of ADA. Thus the activities of these drugs and *Adagen* could be substantially altered if they are used in combination with one another.

E. Therapeutic response

Adagen has been effective in reversing biochemical abnormalities in children with adenosine deaminase deficiency and severe combined immunodeficiency disease (SCID). It is imperative that treatment with *Adagen* be carefully monitored by measurement of the level of ADA activity in plasma. Monitoring of the level of deoxyadenosine triphosphate (dATP) in erythrocytes is also helpful in determining that the dose of *Adagen* is adequate.

F. Role in therapy

According to Micromedex, the goal of therapy with *Adagen* is to correct immune function by reversal of the biochemical abnormalities caused by adenosine deaminase deficiency. *Adagen*'s role in therapy at this time would appear to be as an alternative when bone marrow transplantation is not feasible or has been unsuccessful. It may also be considered in lieu of transplantation in milder cases of adenosine deaminase deficiency. *Adagen* is preferable to red cell transfusions in these patients. While regular administration of *Adagen* can improve immune function and reduce the incidence of opportunistic infections in patients with ADA-deficient SCID, it is of no value in patients with immunodeficiency due to other causes.

H. Other considerations

Adagen has been designated an orphan product for use in replacement therapy of ADA deficiency in patients with severe combined immunodeficiency.

DORNASE ALFA

Trade name: ***Pulmozyme***

Manufacturer: Genentech, Inc., South San Francisco, CA

Indications: Management of cystic fibrosis patients to improve pulmonary function and protection against respiratory tract infections

Approval date: 30 December 1993

Type of submission: Product license application

A. General description Recombinant human deoxyribonuclease I (rhDNase), an enzyme that selectively cleaves DNA, is produced by genetically engineered Chinese hamster ovary (CHO) cells containing DNA encoding for the native human protein, deoxyribonuclease I (DNase). The purified glycoprotein contains 260 amino acids with an approximate molecular weight of 37 kDa. The primary amino-acid sequence is identical to that of the native human enzyme.

B. Indications and use Daily administration of *Pulmozyme* in conjunction with standard therapies is indicated in the management of cystic fibrosis patients to improve pulmonary function. In patients with a forced vital capacity (FVC) ≥40% of predicted. Daily administration of *Pulmozyme* has also been shown to reduce the risk of respiratory tract infections requiring parenteral antibiotics.

C. Administration

Dosage form *Pulmozyme Inhalation Solution* is a sterile, preservative-free, highly purified solution of recombinant human deoxyribonuclease I (dornase alfa). Each *Pulmozyme* single-use ampule will deliver 2.5 ml of the solution to the nebulizer bowl. The aqueous solution contains dornase alfa 1.0 mg/ml.

Route of administration *Pulmozyme* is administered by inhalation of an aerosol mist produced by a compressed-air-driven nebulizer system.

Recommended dosage and monitoring requirements The recommended dose of *Pulmozyme* in most cystic fibrosis patients is 2.5 mg, inhaled once daily using a recommended nebulizer. Some patients, however, especially those over 21 years old or with an FVC greater than 85%, may benefit from twice daily administration. *Pulmozyme* should not be diluted or mixed with other agents in the nebulizer.

D. Pharmacology and pharmaceutics

Clinical pharmacology According to Micromedex, normal bronchial mucus is viscid, primarily due to the presence of mucoproteins and mucopolysaccharides. In patients with cystic fibrosis, abnormally thick mucus is secreted, resulting in chronic bronchiolar obstruction and infection. The inflammatory response to infection leads to disintegration of bacteria and polymorphonuclear leukocytes and other tissues, releasing DNA into the environment; this greatly increases the viscosity of the mucus and further impairs its clearance from the lungs. Like DNase, the endogenous enzyme, dornase alfa, cleaves and depolymerizes extracellular deoxyribonucleic acid (DNA) and separates DNA from proteins; this allows endogenous proteolytic enzymes to break down the proteins, and substantially decreases the viscoelasticity and surface tension of purulent sputum.

Pharmacokinetics The pharmacokinetics of dornase alfa have not been characterized. Minimal systemic absorption occurs via inhalation.

Disposition No information available.

Drug interactions No formal drug interaction studies have been carried out. Clinical trials have indicated that *Pulmozyme* can be effectively and safely used in conjunction with standard cystic fibrosis therapies including oral, inhaled, and/or parenteral antibiotics, bronchodilators, enzyme supplements, vitamins, oral or inhaled corticosteroids, and analgesics.

E. Therapeutic response Improvement in labored breathing and sputum clearance are indicative of a therapeutic response to *Pulmozyme*. In a well-controlled trial, *Pulmozyme*, compared with placebo, resulted in significant reductions in the number of patients experiencing respiratory tract infections requiring use of parenteral antibiotics. Within 8 days of the start of treatment with *Pulmozyme*, mean forced expiratory volume in 1 second (FEV1) increased 7.9% in those treated once a day compared to baseline values. Placebo recipients did not show significant mean changes in pulmonary function testing.

F. Role in therapy *Pulmozyme* is a mucolytic enzyme used in the treatment of cystic fibrosis. Although it is not a cure, dornase alfa is an effective mucolytic for adjunctive treatment. All compliant patients with cystic fibrosis, irrespective of disease severity, who produce purulent sputum are potential candidates for *Pulmozyme* therapy. It is useful for liquefying the thick mucus secreted by cystic fibrosis patients, and causes both objective improvement (as measured by pulmonary function testing) and subjective symptomatic improvement. *Pulmozyme* reduces the frequency of respiratory infections requiring parenteral antibiotics and improves pulmonary function.

G. Other applications *Pulmozyme* may also be useful in chronic bronchitis.

H. Other considerations *Pulmozyme* has been designated an orphan product for use in the treatment of excessive mucous viscosity in patients with cystic fibrosis.

PEGASPARGASE

Trade name: *Oncaspar*

Manufacturer: Enzon, Inc., Piscataway, NJ

Distributed by: Aventis Pharmaceuticals Products Inc., Parsippany, NJ

Indications: Treatment of patients with acute lymphoblastic leukemia (ALL) who are hypersensitive to native forms of L-asparaginase

Approval date: 1 February 1994

Type of submission: Product license application

A. General description Pegaspargase is a modified version of the antineoplastic enzyme L-asparaginase. L-asparaginase is derived from cultures of *Escherichia coli*. The covalent attachment of polyethylene glycol (PEG) to L-asparaginase, producing pegaspargase, is a means of overcoming many of the problems associated with L-asparaginase. The chemical name of pegaspargase is monomethoxypolyethylene glycol succinimidyl L-asparaginase. L-asparaginase is modified by covalently conjugating units of monomethoxypolyethylene glycol (PEG), molecular weight of 5 kDa, to the enzyme, forming the active ingredient PEG-L-asparaginase.

B. Indications and use *Oncaspar* is indicated for patients with ALL who require L-asparaginase in their treatment regimen but have developed hypersensitivity to the native forms of the enzyme. *Oncaspar*, like native L-asparaginase, is generally used in combination with other chemotherapeutic agents, such as vincristine, methotrexate, cytarabine, daunorubicin, and doxorubicin.

C. Administration
Dosage form *Oncaspar* is a preservative-free, sterile solution in buffered saline in 5 ml single-dose vials. Each ml contains pegaspargase 750 IU.
Route of administration *Oncaspar* is intended for intramuscular or intravenous administration only. The preferred route of administration, however, is the intramuscular route because of the lower incidence of toxicity compared to the intravenous route of administration.
Recommended dosage and monitoring requirements The recommended dose of *Oncaspar* for older children is 2500 IU/m^2 every 14 days intramuscularly or intravenously (intramuscular is the preferred route), in combination with other cytotoxic agents. When administering *Oncaspar* intramuscularly, the volume at a single injection site should be limited to 2 ml. If

the volume to be administered is greater than 2 ml, multiple injection sites should be used. Only in unusual situations, should *Oncaspar* be administered as the sole induction agent in the treatment of acute leukemia. Patients should be observed for 1 hour after administration for signs of hypersensitivity reactions.

D. Pharmacology and pharmaceutics

Clinical pharmacolog According to Micromedex, leukemic cells are unable to synthesize asparagine due to a lack of asparagine synthetase and are dependent on an exogenous source of asparagine for survival. Rapid depletion of asparagine resulting from treatment with the enzyme L-asparaginase, which catalyzes hydrolysis of L-asparagine to aspartic acid and ammonia, kills the leukemic cells. The absence of L-asparagine in tumor cells results in interference of protein synthesis and subsequent DNA and RNA synthesis. Normal cells, however, are less affected by the rapid depletion because of their ability to synthesize asparagine. This is an approach to therapy based on a specific metabolic defect in some leukemic cells that do not produce asparagine synthetase. L-asparaginase is effective in the treatment of ALL, but severe adverse effects and the immunogenicity of the enzyme have limited its clinical utility.

Pharmacokinetics The elimination half-life of pegaspargase has been estimated to be 5.7 days, compared with a half-life of a few minutes up to 5 hours for L-asparaginase, and its volume of distribution has been estimated to be 2.1 liters/m^2.

Disposition Metabolism of pegaspargase appears to be similar to that of L-asparaginase, which is inactivated by serum proteases, as well as immune and reticulendothelial systems.

Drug interactions Vaccination with a live vaccine in a patient immunocompromised by a chemotherapeutic agent has resulted in severe and fatal infections.

Live virus and bacterial vaccines should not be administered to a patient receiving an immunosuppressive chemotherapeutic agent. At least 3 months should elapse between the discontinuation of chemotherapy and vaccination with a live vaccine. Unfavorable interactions of *Oncaspar* with some antitumor agents have been demonstrated, and *Oncaspar* may interfere with the enzymatic detoxification of other drugs, particularly in the liver. Physicians using a given treatment regimen should be thoroughly familiar with its benefits and risks.

E. Therapeutic response
Oncaspar was evaluated as part of combination therapy in open-label studies of relapsed, previously L-asparaginase hypersensitive acute leukemia patients. The reinduction response rate was 50% (36% complete remissions and 14% partial remissions).

F. Role in therapy
Oncaspar has been effective in relapsed ALL and refractory non-Hodgkin's lymphoma. Polyethylene glycol (PEG) conjugation of L-asparaginase bestows *Oncaspar* with reduced immunogenicity and a prolonged plasma half-life compared to native L-asparaginase. Potential advantages of *Oncaspar* over the native enzyme include a more convenient dosing schedule (every 2 weeks), a lower incidence of toxicity (including anaphylaxis), decreased tendency for resistance development, and efficacy/safety in patients refractory to, or intolerant of, native L-asparaginase.

G. Other applications
Oncaspar has been given to patients with non-Hodgkin's lymphoma intramuscularly every 2 weeks for one to five courses, in a phase II study and has shown some activity.

H. Other considerations
Oncaspar has been designated an orphan product for use in the treatment of ALL.

ALTEPLASE

Trade name: *Activase*

Manufacturer: Genentech, Inc., South San Francisco, CA

Indications: Management of acute myocardial infarction (AMI), acute ischemic stroke, and acute massive pulmonary embolism

Approval date: 18 June 1996

Type of submission: Product license application

A. General description Alteplase, a purified glycoprotein of 527 amino acids, is a tissue plasminogen activator produced by recombinant DNA technology. It is synthesized using the complementary DNA (cDNA) for natural human tissue-type plasminogen activator obtained from a human melanoma cell line. The manufacturing process involves the secretion of the enzyme alteplase into the culture medium by an established mammalian cell line (Chinese hamster ovary cells) into which the cDNA for human tissue-type plasminogen activator has been genetically inserted.

B. Indications and use *Activase* is indicated for use in the management of acute myocardial infarction in adults for the improvement of ventricular function following AMI, the reduction of the incidence of congestive heart failure, and the reduction of mortality associated with AMI. *Activase* is also indicated for the management of acute ischemic stroke in adults for improving neurological recovery and reducing the incidence of disability and for the management of acute massive pulmonary embolism in adults. For stroke management, *Activase* should only be initiated within 3 hours after the onset of stroke symptoms, and after exclusion of intracranial hemorrhage by a cranial computerized tomography (CT) scan or other diagnostic imaging method sensitive for the presence of hemorrhage.

C. Administration

Dosage form *Activase* is a sterile, lyophilized powder for intravenous administration after reconstitution with sterile water for injection. Each vial contains alteplase 50 or 100 mg.

Route of administration *Activase* should be administered only by intravenous infusion or by means of a small bolus followed by infusion.

Recommended dosage and monitoring requirements The recommended dose of *Activase* to achieve patency following myocardial infarction is 100 mg given as a "front-loaded" intravenous infusion; patients weighing greater than 67 kg should receive a 15 mg bolus, followed by a 30-minute infusion of 50 mg, followed by a 60-minute infusion of 35 mg; if the patient weighs less than 67 kg, the 15 mg bolus should be followed by 0.75 mg/kg infused over 30 minutes, followed by 0.5 mg/kg over 60 minutes. The total alteplase dose should not exceed 100 mg. Alternatively, but less convenient, *Activase* may be given as 60 mg, 20 mg, and 20 mg by infusion over the first, second, and third hours, respectively. The dose for stroke is 0.9 mg/kg (max: 90 mg): 10% given as a bolus, followed by the remainder over 60 minutes; the dose for pulmonary embolism is 100 mg infused over 2 hours. Therapy with alteplase should begin as soon as possible after the onset of AMI symptoms.

D. Pharmacology and pharmaceutics

Clinical pharmacology Alteplase is a fibrin-enhanced plasminogen activator (serine protease) that specifically cleaves the Arg-Val bond of plasminogen, resulting in the formation of plasmin. Plasmin is the enzyme responsible for clot dissolution. *Activase* binds to fibrin in a thrombus and converts the entrapped plasminogen to plasmin. This initiates local fibrinolysis with, in theory, limited systemic proteolysis.

Pharmacokinetics Alteplase in AMI patients is rapidly cleared from the plasma with an initial half-life of less than 5 minutes. There is no difference in the dominant initial plasma half-life between the two different dosage regimens. The plasma clearance of alteplase is 380 to 570 ml/min. The clearance of alteplase is mediated primarily by the liver.

Disposition Alteplase is broken down in the liver to its constituent amino acids.

Drug interactions The interaction of *Activase* with other cardioactive or cerebroactive drugs has not been studied. In addition to bleeding associated with heparin and vitamin K antagonists, drugs that alter platelet function (e.g., aspirin, dipyridamole, and abciximab) may increase the risk of bleeding if administered prior to, during, or after *Activase* therapy. Aspirin and heparin have been administered concomitantly with and following infusions of *Activase* in the management of acute myocardial infarction or pulmonary embolism. Because heparin, aspirin, or *Activase* may cause bleeding complications, careful monitoring for bleeding is advised, especially at arterial puncture sites. The concomitant use of heparin or aspirin during the first 24 hours following symptom onset was prohibited in the pivotal stroke trial. The safety of such concomitant use with *Activase* for the management of acute ischemic stroke is unknown.

E. Therapeutic response *Activase*, and other thrombolytic agents, used in a timely manner during an evolving myocardial infarction, decrease mortality and improve left ventricular function. Resolution of chest pain, resolution of baseline EKG changes, reduced total creatine phosphokinase (CPK) release, and preserved left ventricular function are evidence of cardiac reperfusion. *Activase*, administered within the first 3 hours of ischemic stroke onset, has been shown to improve recovery.

F. Role in therapy *Activase* is effective in producing recanalization of occluded coronary arteries following AMI. Alteplase, anistreplase, reteplase, tenecteplase, and streptokinase have similar efficacy in acute myocardial infarction in terms of decreased mortality and improved left ventricular function. However, accelerated infusion with *Activase* has resulted in higher rates of early reperfusion than has streptokinase. It has been as effective as streptokinase, possibly superior in some cases, but an advantage is arguable. *Activase* is also effective in the treatment of acute massive pulmonary embolism and stroke. Thrombolytic agents have been shown to be more effective than heparin in accelerating the resolution of pulmonary embolism.

G. Other applications *Activase* may have utility in other vascular disorders such as deep-vein thrombosis.

RETEPLASE

Trade name: ***Retavase***

Manufacturer: Centocor, Inc. Malvern, PA

Indications: The management of acute myocardial infarction (AMI) in adults for the improvement of ventricular function following a heart attack, the reduction of the incidence of congestive heart failure, and the reduction of mortality associated with AMI

Approval date: 30 October 1996

Type of submission: Biologics license application

A. General description Reteplase (recombinant plasminogen activator) is a nonglycosylated deletion variant of tissue plasminogen activator (t-PA), produced by recombinant DNA technology in *Eschenchia coli*. Recombinant t-PA (alteplase) is widely used as a thrombolytic drug for the treatment of thromboembolic disease, including AMI. Reteplase differs from alteplase in that it lacks three domains but retains the protease domain for enzymatic activity and the kringle 2 domain for fibrin selectivity. Reteplase contains 355 of the 527 amino acids of native t-PA. The protein is isolated from *E. coli*, converted into its active form by an *in vitro* folding process, and purified by chromatographic separation. The molecular weight of reteplase is 39.6 kDa.

B. Indications and use *Retavase* is indicated for use in the management of acute myocardial infarction (AMI) in adults. Benefits include improvement of ventricular function following infarction, reduction of the incidence of congestive heart failure, and reduction of mortality. Treatment should be initiated as soon as possible after the onset of symptoms.

C. Administration

Dosage form *Retavase* is a sterile, lyophilized powder for injection after reconstitution with sterile water for injection.

Route of administration *Retavase* is for intravenous administration only.

Recommended dosage and monitoring requirements *Retavase* is administered as a double bolus injection. Each intravenous bolus delivers reteplase 10 U and is administered over 2 minutes. The second bolus is given 30 minutes after initiation of the first. Each bolus injection should be given via an intravenous line in which no other medication is being simultaneously injected or infused.

D. Pharmacology and pharmaceutics

Clinical pharmacology Plasminogen activators catalyze the cleavage of endogenous plasminogen to generate plasmin. Plasmin, in turn, degrades the fibrin matrix of a thrombus, the proximate cause of an AMI, lyses the thrombus, and thereby restores patency to the affected coronary vessel and salvages myocardial tissue.

Pharmacokinetics The deletions made in tissue plasminogen activator to produce reteplase as well as its lack of glycosylation result in a longer half-life. While the half-life for alteplase is only 4 minutes, that for reteplase is approximately 13 to 16 minutes (based on the measurement of thrombolytic activity). Consequently, the development of *Retavase* has been designed to allow for dosing as a bolus method rather than an infusion. At the time *Retavase* received FDA approval, alteplase (*Activase*) was given as an intravenous infusion over 3 hours. Thus *Retavase* simplified thrombolytic drug administration during an emergent myocardial infarction. Today, alteplase may also be given by means of a less complicated 90-minute accelerated intravenous infusion. Based on measurement of thrombolytic activity, reteplase is cleared from plasma at a rate of 250 to 450 ml/min. The pharmacokinetics of reteplase are governed by multiple parallel processes. Different processes may be characterized as saturable or nonsaturable. Saturable processes include binding to endogenous inhibitors, whereas nonsaturable processes are associated with organ uptake.

Disposition Reteplase is primarily cleared by the liver and kidney. Animal studies suggest that reteplase is inactivated by blood components.

Drug interactions The interaction of *Retavase* with other cardiovascular drugs has not been studied. Heparin and vitamin K antagonists, as well as drugs that alter platelet function (e.g., aspirin, dipyridamole, and abciximab) may increase the risk of bleeding if administered prior to or after *Retavase* therapy. Administration of *Retavase* may cause decreases in plasminogen and fibrinogen. During therapy, if coagulation tests and/or measurements of fibrinolytic activity are performed, the results may be unreliable.

E. Therapeutic response In a comparative study with streptokinase in acute MI patients, the primary end point was mortality at 35 days after initiation of treatment. Mortality was slightly lower in the *Retavase* group, but the difference was not statistically significant. Two studies evaluated coronary artery perfusion through the infarct-related artery 90 minutes after the initiation of therapy as the primary end point. In both trials the percentage of patients with complete flow at 60 minutes was significantly higher with *Retavase* than with alteplase.

F. Role in therapy Reteplase is a novel thrombolytic agent. It has a longer half-life than alteplase, which allows bolus administration. Its administration technique is much simpler than that of alteplase. In addition reteplase has achieved more rapid, complete, and sustained thrombolysis of the infarct-related artery compared to standard doses of alteplase with comparable safety. Reteplase is at least as effective as streptokinase and alteplase in AMI.

G. Other applications *Retavase* has demonstrated efficacy in the management of pulmonary embolism, and performed comparably to alteplase. It has been used successfully as a replacement for urokinase for the treatment of acute limb ischemia.

TENECTEPLASE

Trade name: *TNKase*

Manufacturer: Genentech, Inc., South San Francisco, CA

Indications: Reduction of mortality associated with acute myocardial infarction (AMI)

Approval date: 2 June 2000

Type of submission: Biologics license application

A. General description Tenecteplase, a 527 amino-acid glycoprotein, is a genetically engineered variant of alteplase. Production of tenecteplase involves multiple-point mutations of the alteplase molecule: replacement of threonine 103 by asparagine (addition of glycosylation site), replacement of asparagine 117 by glutamine (removal of glycosylation site), and replacement of four amino acids (lysine 296, histidine 297, arginine 298, arginine 299) by four alanines. These changes reportedly result in a longer plasma half-life, enhanced fibrin specificity, and increased resistance to inactivation by plasminogen activator inhibitor-1 compared with alteplase. Tenecteplase is produced in Chinese hamster ovary cells, which conserve unmodified carbohydrate side chains.

B. Indications and use *TNKase* is indicated for use in the reduction of mortality associated with acute myocardial infarction. Treatment should be initiated as soon as possible after the onset of symptoms.

C. Administration

Dosage form *TNKase* is a sterile, lyophilized powder for injection after reconstitution with sterile water for injection. Each vial of *TNKase* will deliver 50 mg of tenecteplase.

Route of administration *TNKase* is administered by means of a single intravenous bolus over 5 seconds.

Recommended dosage and monitoring requirements In AMI, reperfusion has been achieved with single intravenous bolus doses of *TNKase* 30 to 50 mg, based on body weight. The dose of *TNKase* should not exceed 50 mg. Concomitant aspirin and heparin have usually been given.

D. Pharmacology and pharmaceutics

Clinical pharmacology The most important therapeutic goal currently in the management of AMI is the prompt, complete, and sustained restoration of antegrade perfusion following coronary occlusion. Clinical studies have demonstrated a relation between 90-minute patency of the infarct-related artery (IRA) and mortality. Regardless of the thrombolytic agent used, an occluded IRA at 90 minutes has been associated with a 30-day mortality rate of about 9%, whereas the mortality rate with restoration of "normal" perfusion is about 4%. Tenecteplase, a modified form of human tissue plasminogen activator, binds to fibrin and converts plasminogen to plasmin. Tenecteplase conversion of plasminogen to plasmin is increased in the presence of fibrin. This fibrin specificity decreases systemic activation of plasminogen and the resulting degradation of circulating fibrinogen as compared to a molecule lacking this property. Following administration of *TNKase*, there are decreases in circulating fibrinogen and plasminogen. The clinical significance of fibrin-specificity on safety (e.g., bleeding) or efficacy has not been established.

Pharmacokinetics Compared to alteplase, tenecteplase has a lower plasma clearance (175 ml/min vs. 570 ml/min), longer initial half-life (20–24 minutes vs. 4 minutes), and may be more resistant to inactivation by plasminogen activator inhibitor-1. Its initial volume of distribution is weight related and approximates plasma volume.

Disposition Tenecteplase is metabolized in the liver, but the extent of metabolism is unknown. It is also inactivated in plasma by circulating plasminogen activator inhibitor-1. The degree of inactivation may be less than that of alteplase.

Drug interactions Caution is called for when administering tenecteplase with drugs that affect platelet function such as aspirin, dipyridamole, clopidogrel, and glycoprotein IIb/IIIa inhibitors. Administration of tenecteplase before, during, or after any of these drugs may increase the risk of bleeding. Anticoagulants (e.g., heparin and vitamin K antagonists) may also increase the risk of bleeding if administered prior to, during, or after *TNKase* therapy.

E. Therapeutic response In a clinical trial with nearly 17,000 participants, single-bolus tenecteplase and front-loaded (accelerated infusion) alteplase (*Activase*) exhibited equivalent rates of 30-day mortality (6.18% and 6.15%, respectively) and intracranial hemorrhage in patients within 6 hours of AMI [Lancet, 1999; **354**: 716–22].

F. Role in therapy Thrombolytic agents currently licensed for the treatment of AMI in the United States include streptokinase, tissue plasminogen activator, anistreplase, reteplase, and tenecteplase. *TNKase* and alteplase have similar clinical efficacy for thrombolysis after myocardial infarction (i.e., similar mortality and intracranial hemorrhage rates). However, advantages of *TNKase* include ease and rapidity of administration, longer half-life, greater fibrin specificity, and lower noncerebral bleeding rates. Reteplase shares some characteristics of tenecteplase (e.g., similar half-life, rapid onset, and ease of administration).

DROTRECOGIN ALFA (ACTIVATED)

Trade name: *Xigris*

Manufacturer: Eli Lilly, Indianapolis, IN

Indications: Reduction of mortality in adult patients with severe sepsis

Approval date: 21 November 2001

Type of submission: Biologics license application

A. General description Drotrecogin alfa (activated) is a recombinant form of human Activated Protein C. An established human cell line with the complementary DNA for the inactive human Protein C zymogen secretes the protein into the fermentation medium. Human Protein C is enzymatically activated by cleavage with thrombin and subsequently purified. Drotrecogin alfa (activated) is a serine protease with the same amino-acid sequences as human plasma-derived Activated Protein C. The enzyme is a glycoprotein of approximately 55 kDa molecular weight, consisting of a heavy and light chains linked by a disulfide bond. Drotrecogin alfa (activated) and human plasma-derived Activated Protein C have the same sites of glycosylation, although there are some differences in glycosylation structure.

B. Indications and use   *Xigris* is indicated for the reduction of mortality in adult patients with severe sepsis (sepsis associated with acute organ dysfunction) who have a high risk of death [as determined by the acute physiology and chronic health evaluation (APACHE) score, the most widely used method of assessing the severity of illness in acutely ill patients in intensive care units].

C. Administration
   Dosage form   *Xigris* is supplied as a sterile, lyophilized powder in vials containing 5 and 20 mg drotrecogin alfa (activated) and excipients.
   Route of administration   *Xigris* is administered by means of intravenous infusion.
   Recommended dosage and monitoring requirements   *Xigris* is administered at an infusion rate of 24 µg/kg per hour for a total duration of infusion of 96 hours. If the infusion is interrupted, *Xigris* should be restarted at the 24 µg/kg per hour infusion rate. In the event of clinically important bleeding, the infusion should be stopped immediately. Each patient being considered for therapy with *Xigris* should be carefully evaluated and anticipated benefits weighed against potential risks associated with therapy.

D. Pharmacology and pharmaceutics
   Clinical pharmacology   Activated Protein C exerts an antithrombotic effect by inhibiting Factors Va and VIIIa. *In vitro* data indicate that it has indirect profibrinolytic activity through its ability to inhibit plasminogen activator inhibitor-1 (PAI-1) and to limit production of activated thrombin-activatable-fibrinolysis inhibitor. *In vitro* data also indicate that Activated Protein C may exert an anti-inflammatory effect by inhibiting human tumor necrosis factor production by monocytes, by blocking leukocyte adhesion to selectins, and by limiting thrombin-induced inflammatory responses within the microvascular endothelium.
   Pharmacokinetics   Endogenous plasma protease inhibitors inactivate drotrecogin alfa (activated) and endogenous Activated Protein C. Plasma levels of endogenous Activated Protein C are usually below detection limits in both healthy subjects and patients with severe sepsis. In patients with severe sepsis, *Xigris* infusions produce steady-state concentrations that are proportional to infusion rate within 2 hours after initiation. In a phase III trial, the median clearance of drotrecogin alfa (activated) was 40 liters/h. Plasma clearance of drotrecogin alfa (activated) in patients with severe sepsis is approximately 50% higher than that in healthy subjects.
   Disposition   No information available.
   Drug interactions   Drug interactions with *Xigris* have not been studied in patients with severe sepsis.

E. Therapeutic response   The efficacy of *Xigris* was studied in a well-controlled trial of 1690 patients with severe sepsis. Patients received a 96-hour infusion of *Xigris* or placebo starting within 48 hours after the onset of the first sepsis-induced organ dysfunction. The primary efficacy end point was all-cause mortality at 28 days after the start of treatment. The study was stopped after a planned interim analysis because mortality was significantly lower in patients on *Xigris* than in patients on placebo (25% vs. 31%, $p = 0.005$). Baseline APACHE score was correlated with risk of death. Among patients receiving placebo, those in the lowest quartile of APACHE scores had 12% mortality, whereas those in the fourth quartile had 49% mortality. The observed mortality difference between *Xigris* and placebo was limited to the half of patients with higher risk of death (i.e., APACHE score >25).

F. Role in therapy According to Micromedex, *Xigris* should be considered an adjunct in sepsis patients with evidence of organ dysfunction; it is not recommended in patients with mild sepsis and no organ dysfunction. *Xigris* has been shown to reduce mortality, albeit modestly, in patients with severe sepsis. No other adjunctive treatment directed at underlying pathology has been shown to significantly alter clinical outcome in severe sepsis. Patient selection is an important consideration in the use of *Xigris*.

10

ANTIBODIES AND DERIVATIVES

■ SECTION ONE ■

■ 10.1. OVERVIEW

In 1798 Edward Jenner published the classc memoir, "An Inquiry into the Causes and Effect of the Variolae Vacciniae," documenting how inoculation with cowpox protected humans against smallpox infection [1]. Louis Pasteur's formulation of the germ theory extended the understanding of this kind of protection against infection [1,2]. About 100 years later, isolation of the diphtheria bacillus and description of a protective substance (antitoxin) by Roux and Yersin demonstrated that the protective substance found in the serum of immunized animals can be transferred to susceptible animals and thereby confer passive immunity [4]. The antitoxin or protective substance found in serum, now known as antibodies, is synthesized on stimulation by a pathogen (i.e., an antigen). The process is tightly regulated.

The immune system—the host's defense system against infection by pathogenic microorganisms—consists of distinct cell types with specialized roles at different stages of the immune response [5]. Immune system cells are found in the spleen, lymph nodes, Peyer's patches in the intestine, tonsils, bone marrow, blood, and thymus. White blood cells, including monocytes (macrophages), dendritic and Langerhans cells, natural killer cells, mast cells, and basophils, may interact with lymphocytes in the presentation of an antigen and thereby mediate the immune response. The differ-

Biotechnology and Biopharmaceuticals, by Rodney J. Y. Ho and Milo Gibaldi
ISBN 0-471-20690-3 Copyright © 2003 by John Wiley & Sons, Inc.

■TABLE 10.1. Biological properties and localization of the five classes of antibodies

Class	Heavy Chain	Light Chain	MW (kDa)	Properties	Localization
IgG	γ	κ or λ	150	Toxin neutralizing, agglutinating, opsonizing, bacteriolytic (with aid of complement system); antigen-IgG antibody complexes may cause tissue injury.	Serum, amniotic fluid
IgM	μ	κ or λ	~950	Similar to IgG; can also serve as antigen receptor on B-lymphocytes.	Serum
IgA	α	κ or λ	160 340 (dimer)	Toxin-neutralizing, agglutinating, opsonizing, secretory immunoglobulin	Serum, secretions, colostrum, saliva, tears, GI tract
IgD	δ	κ or λ	170	Antigen receptor on B-lymphocytes; Marker for mature B cells	Serum, B cell surface
IgE	ε	κ or λ	190	Mediates changes in vascular permeability due to allergy, hypersensitivity, or anaphylactic reactions	Serum

entiated immune cells are derived from hematopoietic stem cells that are localized mostly in bone marrow.

Two classes of lymphocytes, central to immunity against pathogenic challenge, are T- and B-lymphocytes. Some of the key functions of T-lymphocytes are regulation of the development of certain types of immune responses, such as autoimmune response, graft-versus-host reactions, facilitation of antibody production, and enhancement of the microbicidal activity of macrophages, and lysis of virus-infected cells and certain cancer cells. The B-lymphocytes are precursors of antibody-secreting cells and may be involved in antigen presentation to T-lymphocytes. Mature B cells, called "plasma cells," secrete immunoglobulins, antibody molecules that exhibit antitoxin activity and thereby neutralize pathogens. Immuno-globulins are found at high concentrations in blood or plasma.

Antibodies produced by B cells can be further divided into several classes, as listed in Table 10.1. Each immunoglobulin (Ig) contains two antigen binding sites, or epitopes. Immunoglobulin class A (IgA) may exist in a dimeric form, and IgM always exists as a pentameric form of immunoglob-ulin. Regardless of the class of antibody, each subunit of immunoglobulin contains two light (L) chains and two heavy (H) chains held together by four disulfide bridges. The variable domains of all anti-bodies contain amino-acid sequences located in both the H and L chains. They determine the antigen-binding selectivity and the antibody's affinity for antigens (Figure 10.1). The type of H chain deter-mines the class of antibody produced by the plasma cells. Heavy chain γ, μ, α, δ, and ε pro-

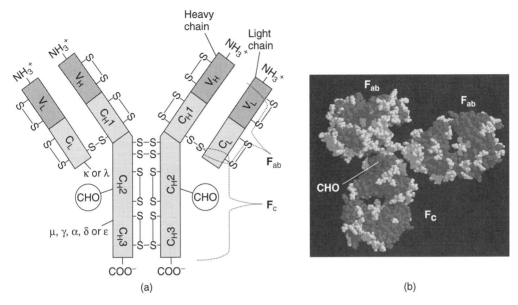

Figure 10.1. Schematic presentation of an immunoglobulin molecule. (**A**) Each IgG molecule contains two heavy and two light chains held together by disulfide (s–s) bridges between them. Conserved regions of heavy and light chains are designated as C_H and C_L while variable regions are designated as V_H and V_L. The hypervariable regions, located within V_H and V_L, are involved in antigen binding and selectivity determinants. The amino-acid sequence of the heavy chain, γ, μ, α, δ or ε distinguish the class of antibody. All classes of antibody can be derived from the κ or λ light chain. Treatment of immunoglobulin with papain, which recognizes and cleaves the heavy chains right above the two disulfide bridges linking the two together, produces one F_c and two F_{ab} fragments. Treatment with pepsin cleaves sequences between the two disulfide bridges producing one F_{ab2} and one F_c fragment. The molecule is glycosylated (designated as CHO) in the F_c region, which is thought to mediate biological functions. The overall dynamic size of immunoglobulin G is estimated to be about 50 nm in diameter. (**B**) A space-filled three-dimensional model of mouse IgG generated with Rasmol modeling software, based on data compiled by L. J. Harris et al. [3]; 1IGY. CHO, carbohydrate attached to IgG heavy chains at the glycosylation sites.

duces IgG, IgM, IgA, IgD, and IgE, respectively. In addition there are regions in the *H* and *L* chains that are constant and do not vary in amino-acid sequence between individuals within the same species. The constant regions of the heavy chains, linked together by two disulfide bridges, form the sites that mediate complement-induced cytolysis and antibody-dependent (Fc receptor-mediated) cellular functions.

In 1959 Porter, using rabbit immunoglobulin, largely consisting of IgG, showed that the molecule can be cleaved into two major fragments approximately equal in size (MW ≃ 50 kDa) by papain, an enzyme found in papaya (Figure 10.1). One of the fragments retains one antigen-binding site and is termed the fragment of antigen binding, or F_{ab}. The other fragment does not have antigen-binding function but is easily crystallizable. This fragment is called F_c. Another enzyme, pepsin, was later shown to cleave IgG below the hinge region distal to the disulfide bridges, generating two different-size fragments. One fragment, containing two antigen binding sites, is called F_{ab_2}, and the other, like that of papain product, is also called F_c [6].

Molecular studies over the years, including sequence analysis of immunoglobulins and their genes, revealed that amino-acid sequences in constant regions of the *H* chains found in F_c fragments are the same in individual subjects within the same species but different among species. The sequence differences between species are

■TABLE 10.2. Physiological properties and disposition of antibodies and serum albumin in humans

Immunoglobulin	MW (kDa)	$t_{1/2}$ (days)	Rate of Synthesis	Blood Concentration	Total Pool (/kg)	Rate of Metabolism (% per day)
IgG (IgG$_1$, IgG$_2$, IgG$_4$)	150	21	32 mg/kg/d (2.2 g/d)	12 mg/ml	1.06 g	6.3%
IgG$_3$	150	7–9	NA	NA	NA	NA
HSA (human serum albumin)	60	15–20	10 g/d	~50 mg/ml	3–4 g	5%
IgM	950	9.6	6.9 mg/kg/d	~1 mg/ml	37 mg	11%
IgA	160	5–6	30 mg/kg/d	~2 mg/ml	~220 mg	25%
IgD	175	3	0.4 mg/kg/d	0.02 mg/ml	1.5 mg	37%
IgE	190	2.5	0.016 mg/kg/d	0.3 mg/ml	~20 ng	NA
Mouse IgG$_2$	150	5–10 & <1[a]	NR	NR	NR	NR

Source: Data partly derived from Waldmann and Strober [44].

Note: NA, Not available.

NR, Not relevant.

[a]Half-life, $t_{1/2}$ for mouse IgG$_2$ is 5 to 10 days for the first dose and this value is reduced to less than 1 day after repeated dosing.

now recognized to cause host-against-mouse allergic, or HAMA, reactions in patients that receive multiple injections of mouse monoclonal antibodies containing mouse immunoglobulin sequences as a part of therapy. These reactions may significantly reduce the circulation time of the antibodies (Table 10.2).

In 1900 Paul Ehrlich proposed that the antigen-antibody interaction could be exploited to produce antibodies that target and neutralize specific antigens. The well-known "magic bullet" theory suggested the potential of using antibodies to treat an array of diseases. However, the implementation of this concept into practice would face many challenges [7].

Animal immunization experiments clearly showed that antibodies produced in a host challenged by an antigen are heterogeneous in specificity and affinity, and in magnitude (Figure 10.2). Furthermore antibody production is a time-dependent event. The life span of antibody-producing B cells is no more than 30 days. Therefore it is not possible to use these B cells as a continuous source of antibody production. Consequently the development of an antibody-producing cell line, derived from select antibody-secreting B cells, as a way to manufacture a unique antibody for therapeutic application was theoretically attractive but technically unattainable.

The therapeutic application of antibodies was limited to heterogeneous mixtures of immunoglobulin molecules found in the serum of immunized animals, but not all of these molecules were target specific. Nevertheless, mixtures of immunoglobulins derived from immunization of animals or humans continue to be used to treat certain diseases, including rabies, hepatitis, and cytomegalovirus infection. Today large quantities of purified immunoglobulins are obtained because of improved methods of antibody production and isolation.

The availability of purified antibodies permitted the detailed characterization of the different classes of immunoglobulins with respect to their biosynthesis, metabolism, and composition in plasma (Table 10.2). The most abundant immunoglobulin molecule in serum is IgG. IgG classes 1, 2,

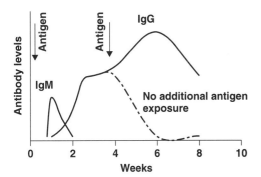

Figure 10.2. Time course of antibody production in the serum of animals following exposure to antigen. Within the first week of immunization with or exposure to an antigen, the initial typical response will be in the form of IgM, which subsides within two weeks. The IgM antibody is replaced by an IgG response that can be enhanced by additional doses of antigen. Typically serum antibody levels will subside within 2 to 4 weeks when no additional antigen remains in the system.

and 4 are metabolized at a slower rate than other immunoglobulins, as reflected by their 21-day half-life. IgG class 3 has a half-life of only 7 to 9 days (see Table 10.2). The half-life of exogenously administered IgG, specifically classes 1 and 2, is comparable to that of serum albumin, which exhibits a half-life of 15 to 20 days. Because of their persistence in the circulation, IgG antibodies used in clinical applications are IgG_1 or IgG_2 but not IgG_3.

Antibodies, with half-lives measured in days, are among the most stable natural proteins, far more stable than larger proteins (MW $\geq$ 150 kDa), which have half-lives measured in minutes or seconds. Indeed, the stability of IgG is comparable to the best-conserved major plasma protein, serum albumin (Table 10.2).

The understanding of how immune responses against pathogens are elicited in animals has allowed production and use of hyperimmune globulin, composed of significant fractions of antibodies that bind and neutralize cellular, viral, and microbial antigens involved in human disease (Table 10.3). Hyperimmune globulins are produced by immunization with antigens often coupled to large proteins or adjuvants, called an antigen complex. The antigen conjugate or antigen complex elicits

■TABLE 10.3. Select hyperimmune globulin or antiserum and therapeutic use

Product	Antigen Target	Therapeutic Use
CytoGram	Cytomegalovirus	Prohylaxis of CMV disease in organ transplant
Nabi-HB Bay HepB	Hepatitis B virus	Acute and perinatal exposure to Hepatitis B
BayRab Imogam Rabies-HT	Rabies virus	Exposure to rabies
BayTet	Tetanus	Prophylaxis against tetanus
Sanstat	Human thymocytes	Treatment of acute rejection in renal transplant
BayRho-D WinRho SDF RhoGam Rho	Rh antigen on red blood cells	Prevention of Rh hemolytic disease in newborn

diverse antibodies, produced by different B cells, which recognize multiple binding domains or epitopes on the target antigen.

Target antigens spanning several hundred amino acids may contain multiple binding epitopes. Each epitope consists of six to eight amino acids. Based on this information, short immunogenic peptides containing a single epitope can be used to generate polyclonal antibodies that recognize and bind only to that epitope. These antibodies are called "monospecific polyclonal" antibodies. However, continuous production of a monospecific antibody from B cells still requires the cells to grow continuously for multiple cycles. This immortal quality of B-cell clones is essential to produce monoclonal antibodies in a continuous, reproducible, and sufficient quantity for therapeutic applications.

10.1.1. Chimeric and Humanized Monoclonal Antibodies

The development of monoclonal antibody technology was made possible in part by the discovery of cancerous B cells in patients with myeloma. As early as 1845, researchers recognized that patients who died from multiple myeloma had bone marrow that unexplainably contained plasma cells [8]. These unique cells were later shown to be immortal B-cell clones that excrete nonspecific antibodies (Bence Jones proteins) that are easily detected in urine. Scientists reasoned that if one could combine the antigen-recognition properties of plasma B cells (obtained from immunized animals) with the immortal characteristic of a myeloma cell, the resultant immortal hybrid B cells should produce target-specific antibodies. The successful generation in 1975 of antibody-secreting hybrid cells (hybridomas) by Georges Kohler and Cesar Milstein [9] reduced this concept to practice.

By fusing an immortal mouse myeloma B-cell line and an immune B-lymphoblast containing a gene that expressed a specific antibody, the resulting hybridoma was demonstrated to continue to secrete target-specific antibody (Figure 10.3). This process is called "hybridoma technology"; it has allowed large-scale production of mouse monoclonal antibodies for human use. Among the first mouse monoclonal antibody (Mab) therapies was anti-CD3 or *Orthoclone OKT3* (muromonab). In 1986, about 10 years after the conception of monoclonal antibody technology, the FDA approved *Orthoclone OKT3* for use in patients with acute rejection of a transplanted kidney. Unfortunately, mouse antibodies are immunogenic to humans. Patients receiving multiple infusions of mouse Mabs exhibit a human–anti-mouse antibody response or human against mouse allergic (HAMA) reactions that manifest with joint swelling, rashes, and kidney failure.

The attempt to generate human Mabs using hybridoma technology was challenging due to a lack of suitable human myeloma cells and the instability of the resultant human hybridoma cells. Over the past 25 years cellular and molecular strategies have been applied to overcome barriers to humanizing mouse Mabs, making mouse antibodies "more human." A better understanding of immunoglobulin gene regulation and expression, together with improved strategies in recombinant DNA technology, allowed scientists to produce functional target-specific recombinant IgG molecules. However, the task of engineering the coding sequences of the heavy and light chains of immunoglobulin genes in a suitable recombinant expression system remains formidable. The fraction of functional antibody recovered from such expression systems is less than 3% of the total expressed protein [10]; however, the field of antibody engineering continues to evolve in improving the yield and quality of the recombinant immunoglobulin molecules.

To reduce the immunogenicity of mouse Mabs, researchers capitalized on knowl-

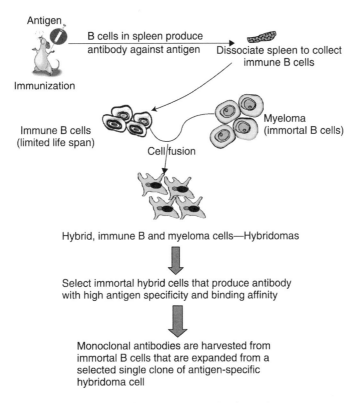

Figure 10.3. Schematic representation of monoclonal antibody production using immortalized hybrid cells that secrete antibodies selective for the target antigen. The mortal, immune B cells isolated from mice immunized with a target antigen are fused with myeloma, immortal B cells that express defective antibodies. The selecting of antigen-specific, immortal hybrid cells (hybridomas) results in identification of unique clones of cells that express antibodies with high specificity and affinity (monoclonal antibodies). These cells are cloned and expanded for large-scale monoclonal antibody preparations.

edge of the immunogenicity determinant that resides predominantly in the constant C_H and C_L regions of IgG [11]. By systematically replacing amino-acid sequences on the mouse Mab, which are not essential for antigen binding, with human sequences, a chimeric protein can be constructed that contains both mouse and human sequences and retains antigen-binding sequences in the F_v fragment, the minimum binding fragment. The F_v fragment is about half the size of the F_{ab} fragment. These antibodies are known as "mouse/human chimeric antibodies." The chimeric antibody retains antigen selectivity and affinity similar to the parent mouse Mab, but the tendency to induce HAMA responses in humans is greatly reduced.

To further reduce the immunogenicity of a chimeric antibody, researchers designed humanized antibodies that replace all mouse amino acid sequences except the hypervariable CDR domains of Ig that represent only a small fraction of the F_v fragment. Some humanized antibodies are now approved for clinical use and many more are being tested for an array of therapeutic indications (Tables 10.4 and 10.5).

As noted, one of the remaining challenges in obtaining human hybridoma cells for generating human monoclonal antibodies is the lack of suitable human myeloma cell lines to generate stable hybrid cells that can be cloned and expanded indefinitely in culture. Many human hybridoma

■TABLE 10.4. Some antibodies and derivatives approved for human use—organized according to the approval date

Antibody	Target	Product Type	Indication	FDA Approval (year)
Orthoclone/OKT3 (Muromonab-CD3)[a]	CD3	Mouse IgG$_2$	Allograft rejection	1986
ReoPro (Abciximab)	Glycoprotein IIb/IIIa receptor	Chimeric F$_{ab}$	Complication of coronary angioplasty	1994
Rituxan (Rituximab)	CD20	Chimeric IgG$_1$	Non-Hodgkin lymphoma	1997
Zenapax (Daclizumab)	CD25	Humanized IgG$_1$	Allograft rejection	1997
Simulect (Basiliximab)	CD25	Chimeric IgG$_1$	Allograft rejection	1998
Remicade (Infliximab)	TNF-α	Chimeric IgG$_1$	Crohn's disease	1998
Herceptin (Trastuzumab)	HER2/neu	Humanized IgG$_1$	Breast cancer	1998
Synagis (Palivizumab)	F protein	Humanized IgG$_1$	RSV virus infection	1998
Mylotarg (Gemtuzumab ozogamicin)	CD32	Humanized IgG$_4$	Relapsed acute myeloid leukemia	2000
Campath (Alemtuzumab)	CD52	Humanized IgG	B-cell chronic lymphocytic leukemia	2001

[a]Nonproprietary name of the antibody. For naming convention and rules, refer to Appendix III.

cells generated over the years either failed after a few replication cycles or reverted to secretion of nonspecific antibodies [12]. Recently a stable human myeloma that may prove useful for generating human monoclonal antibodies has been described. Using this cell line as a starting point, traditional hybridoma technology has been applied to produce stable human Mabs against several antigens [13]. The eventual use of human Mabs produced by this strategy for pharmaceutical purposes is in sight.

Whether humanization satisfactorily reduces the immunogenicity of a murine Mab ultimately requires a direct comparison of the humanized and mouse antibodies in the clinic. Embarking on such an experiment, however, may raise ethical concerns with respect to exposing human subjects to the expected side effects of mouse Mab, which in some cases can be life threatening. Despite the lack of direct evidence that humanization reduces immunogenicity, there is a trend toward producing such products (Table 10.4). Historical comparisons suggest a lower incidence of untoward effects associated with HAMA reactions on administering humanized antibodies rather than mouse antibodies. The reduced incidence of HAMA reactions observed in clinical trials with humanized antibody products suggests a higher degree of safety.

■**TABLE 10.5.** Monoclonal antibodies in clinical development—a partial list

Therapeutic Area	Antigen Target	Monoclonal Antibody		Sponsor Company	Clinical Status
		Name	Characteristics		
Cancer					
Sarcoma	αVβ3 Integrin	Vitaxin	Humanized	Applied Molecular Evolution/Medimmune	II
AML	CD33	Smart M195	Humanized IgG	Protein Design Lab/ Kanebo	III
NHL	CD22	Lymphocide	Humanized IgG	Immunomedics	I/II
	HLA	Smart ID10	Humanized	Protein Design Lab	I
	HLA DR	Lym-1	Murine	Techniclone	II/III
Head and Neck	EGFR	IMC-C225	Chimeric IgG$_1$	Imclone Sys	III
Lung	GD3	BEC2	Murine IgG	ImClone Sys	III
Ovarian	CA 125	OvaRex	Murine	Altarex	II/III
General	VEGF	anti-VEGF	Humanized IgG$_1$	Genentech	III
Infectious diseases					
HIV	gp120	PRO542	CD4-F$_c$ fusion	Progenics/Genzyme	II
CMV	CMV	Protovir	Humanized IgG1	Protein Design Lab/ Novartis	III
Toxic shock	TNFα	MAK-195	Murine F$_{ab_2}$	Knoll Pharma/BASF	III
	CD14	IC14		ICOS	I
Heart					
Myocardial infarction	CD18	anti-CD18	Humanized F$_{ab_2}$	Genentech	II
Stroke	β2-Integrin	LDP 01	Humanized IgG	LeukoSite	I/II
Coronary angioplasty complication	PDGFβ-receptor	CDp860	Humanized F$_{ab_2}$	Celltech	II
Multiple sclerosis					
	VLA-4	Antegren	Humanized IgG	Elan	III
	CD40L	IDEC-131	Humanized	IDEC/Eisai	II
Rheumatoid arthritis					
	TNFα	CDP870	Humanized F$_{ab}$	Celltech	II
	TNFα	D2E7	Human	CAT/BASF	III
	Complement	5G1.1	Humanized	Alexion	II
	CD4	IDEC-151	Primatized IgG$_1$	IDEC/SKB	II
	CD4	MDX-CD4	Human IgG	Medarex/Eisai/Genmab	I

■TABLE 10.5. *Continued.*

Therapeutic Area	Antigen Target	Monoclonal Antibody		Sponsor Company	Clinical Status
		Name	Characteristics		
Autoimmune disease					
	CD3	Smart anti-CD3	Humanized	Protein Design Lab	I/II
	CD4	OKT4A	Humanized IgG	Ortho Biotech	II
Transplantation					
Allograft rejection	CD3	SMART anti CD3	Humanized	Protein Design Lab	I/II
	CD4	OKT4A	Humanized igG	Ortho Biotech	II
	CD147	ABX-CBL	Murine IgM	Abenix	III
	CD18	Anti LFA-1	Murine F_{ab2}	Pasteur-Merieux/Immunotech	III
	β2-Integrin	LDP-01	Humanized IgG	LeukoSite	I/II
GVHD	CD2	BTI-322	Rat IgG	Medimmune/Bio Transplant	II
	CD2	MEDI-507	Humanized	Medimmune	I/II
	CD147	ABX-CBL	Murine IgM	Abenix	III
Ulcerative colitis	α4β7	LDP-02	Humanized	LeukoSite/Genentech	II
Crohn's disease	TNFα	CDP-571	Humanized IgG$_4$	Celltech	II

10.1.2. Antibody Derivatives

Although many therapeutic applications require the prolonged presence of antibody in blood, some treatment situations such as coronary angioplasty procedures to establish reperfusion may not need the extended presence of antibody. Indeed, persistence of antibody may result in severe bleeding. In such situations a smaller antibody fragment such as F_{ab} can be useful. For example, abciximab (*ReoPro*) is a F_{ab} fragment that binds to glycoprotein IIb/IIIa receptors on platelets used to prevent restenosis after revascularization procedures (Table 10.6). The more compact F_{ab} fragment permeates tissues more effectively than the entire antibody and has a higher rate of blood clearance ($t_{1/2} = 1$–2 days) than IgG ($t_{1/2} = 21$ days).

Antibody derivatives may also be used as a means to reduce immunogenicity. One strategy is to recombinantly clone and express F_v fragments, which consist only of the variable regions (V_H and V_L) of the heavy and light chains. V_H and V_L must interact to form F_v fragments capable of antigen binding (i.e., the antigen binding site). This interaction, however, is stabilized only by electrostatic interactions of V_H and V_L and not by covalent bonds.

To compensate for that weak interaction, researchers have directly linked V_H and V_L by adding peptide sequences that covalently hold them together. The product is called a "single-chain F_v protein" or "sF_v

antibody derivative." Employing this strategy, scientists can create recombinant antibody molecules with only the elements essential for antigen binding. However, the compact molecules, F_v and sF_v, often exhibit lower antigen-binding affinity than parent IgG. But a compact size (MW = 30 kDa) allows simpler expression systems that can be adapted for large-scale production. We anticipate that even more compact binding sites will be used for therapeutic purposes in the future. The engineering aspects of producing antibody derivatives is discussed further in Part III.

10.1.3. Disposition of Monoclonal Antibodies

The antigen binding site or epitope recognized by a Mab or a monospecific polyclonal antibody could exist in any body tissue. These epitopes may be one of a myriad of cell-surface proteins found in healthy or cancerous tissues. *In vitro* tests are designed to select antibodies exhibiting target-antigen binding, cell-specific binding, and inhibition or promotion of cell growth. Estimations of therapeutic dose and concentration from these data in whole animals require understanding of the extent and time-dependent distribution of the antibody to target tissues. The degree of antibody penetration and its extent of distribution may play a role in determining the efficiency of an antibody to produce an intended biological function.

In the ideal situation, a Mab would distribute only to high-affinity binding sites. However, antibodies given intravenously (or by other routes, if feasible) also distribute into nontarget tissues. The accumulation of an antibody in nontarget tissue as a result of nonspecific binding can be considerable. In identifying tumor-specific antigens, scientists have found that normal cells located in various tissues in the body also express low levels of these antigens. Therefore the term "tumor-associated" rather than "tumor-specific" antigens may

be more appropriate. Mab distribution to normal tissue through nonspecific binding could pose a major limitation to the use of an antibody and is a particular challenge for antibodies conjugated with radioactive or cytotoxic moieties intended to enhance the function.

Tables 10.1, 10.2, and 10.4 show that IgG molecules that exhibit high serum stability and low plasma clearance (prolonged half-life) are widely used for therapeutic purposes. IgG's long plasma half-life, compared with other proteins, is the result of the interactions between another protein, found in plasma, called the "F_c receptor" ("F_cRN or Brambell receptor") and the F_c regions of IgG. This interaction stabilizes IgG and prevents its degradation in plasma. Animals lacking F_cRN exhibit much faster plasma clearance of IgG. F_cRN recognizes the amino-acid sequences in the F_c fragment of IgG's heavy chains (Figure 10.1), and the recognition is exquisitely specific and sensitive to changes in amino-acid sequences that span CH_2 and CH_3 domains within F_c [14–17]. The interactions between F_cRN and IgG F_c have been localized to amino acid residues 288, 307, 309, 311, 433, 435, and 436. In other words, binding determinants are interspersed within the CH_2 and CH_3 regions of IgG F_c [18,19]. These sequences are highly conserved within the same species but vary between species.

Understanding the role of F_c and F_cRN has allowed engineering of hybrid molecules that take advantage of F_c-dependent plasma half-life extension in designing fusion proteins intended for therapeutic use. An example is the construct of the TNF receptor-F_c fusion protein, which results in a significant increase in the half-life of the TNF receptor [20–22]. *Enbrel*, a recombinant TNF-α-receptor-F_c fusion protein, has provided effective treatment for rheumatoid arthritis patients.

In addition to amino-acid sequences, the specific carbohydrate moiety or glycosyl group attached to IgG may play a role in antibody function, stability, distribution, and clearance (see Chapter 5, Figures 5.5 and 5.6). The carbohydrate portion is attached to the IgG molecule in the constant CH_2 domain within F_c fragments via an N-linked glycan. The N-linkage involves asparagine amino acids located within CH_2 regions of the heavy chains of the IgG. The glycan is a polysaccharide containing N-acetyl glucosamine, fructose, mannose, and terminal sialated galactose. The glycosylation process occurs during post-translational modifications of IgG synthesis.

The degree of glycosylation on IgG influences tissue distribution by means of cell surface recognition and binding. For example, native IgG contains terminal sialated galactose—attached through the N-linked glycan—which prevents galactose receptor-mediated clearance of IgG in hepatocytes. Sialation is an important factor in the prolonged half-life of IgG. Any modification of the glycan will lead to a reduced half-life. Because scavenger phagocytic cells in blood and other tissues express receptors that recognize mannose residues, partial fragmentation of IgG exposes mannose residues and results in clearance through mannose-receptor-mediated, phagocytic pathways.

On internalization by scavenger cells, IgG and its derivatives are degraded inside endosomes and lysosomes by proteolysis and other intracellular processes. The intracellular degradation processes, including those mediated through galactose receptors on hepatocytes as well as mannose receptors on macrophages and Kufper cells (liver macrophages), may work in concert to inactivate IgG molecules as a part of physiological IgG degradation or turnover.

10.1.4. Extravascular Tissue Penetration

Very often antibodies are required to bind to cells and tissues located outside of blood

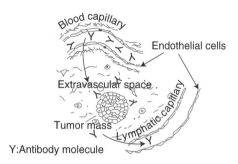

Figure 10.4. Schematic representation of antibody distribution into extravascular space across endothelial cells lining blood capillaries and into the tumor mass. Some of these antibody molecules may be further distributed into lymphatic capillaries.

and lymphatic systems. In such scenarios, movement of immunoglobulin out of the capillary lumen and into tissue space is essential for biologic and pharmacologic activities [23]. The movement across the capillary endothelium, however, is limited by the relatively large size of the IgG molecule (Figure 10.4). Despite this limitation it is possible that IgG could move across the endothelial cells lining the blood and lymphatic capillary wall by several other mechanisms, including perfusion through extracellular matrices that form between cells or tight junctions (via convection or diffusion), nonspecific transcytosis, and receptor-mediated transcytosis.

IgG molecules found outside the extravascular space can accumulate and penetrate into a tumor mass within a tissue by several mechanisms. The rate of IgG accumulation in tissue and subsequent penetration into the tumor mass depends on the affinity and avidity of the antibody for a tumor-associated antigen. According to one model, based on computer simulations, penetration of the tumor mass is influenced by diffusion and convection, while binding affinity determines residence time of the antibody molecule bound to antigen expressed on tumor cells [23–25].

A tumor mass, however, could contain different types of cells with diverse tumor-

■TABLE 10.6. Effects of antibody size on relative plasma clearance

Antibody Molecule	MW	Relative Plasma Clearance (expressed as $t_{1/2}$)	
Native intact human IgG	150,000	Slowest *CL*	($t_{1/2} \sim 21$ days)
Chimeric intact mouse/human IgG	150,000		($t_{1/2} \sim 10$–12 days)
Intact mouse IgG	150,000	↓	($t_{1/2} \sim 4$–6 days)
F_{ab_2}	100,000		($t_{1/2} \sim 1$–2 days)
F_{ab}	50,000	Highest *CL*	($t_{1/2} \sim$ hours)

antigen expression, assembled as multiple layers. In this context an antibody with high affinity to tumor antigen will favor its accumulation at outer layers of the tumor and reduce penetration into the tumor cells located in the central core. Therefore, the development of a very high-affinity antitumor antibody may not be the best strategy to yield the most effective drug. Such an antibody might be trapped on the surface of the tumor and not realize its potential. An antibody with a more moderate affinity for the tumor antigen might be more effective.

Needless to say, intratumor penetration by antibodies remains a challenge. Even highly potent cytotoxic radionuclide-labeled antibodies designed to destroy cells neighboring those bound to the antibody do not completely eliminate abnormal cells in the core of a solid tumor.

One strategy to increase the elimination of tumor cells inside the mass is to increase antibody penetration. This can be achieved by reducing the size of the antibody. Size reduction can be realized by treating IgG with papain or pepsin to produce F_{ab} or F_{ab_2} fragments, respectively. F_{ab} and F_{ab_2} fragments range from one-third to two-thirds the size of IgG, while retaining antibody-antigen recognition. The smaller and more compact antibody-binding domains provide better penetration into the central core of a tumor mass several millimeters in diameter (Table 10.6). The removal of F_c domains, however, also increases the plasma clearance of IgG.

Therefore size reduction may not always provide the intended advantage. Nevertheless, enhanced distribution into the central core of the tumor is likely to result in better overall control of tumor growth and disease progression. While the benefit of improved tissue penetration is debated, it is clear that modification of antibody molecules by reducing their size through enzyme digestion will invariably increase the plasma clearance.

10.1.5. Antibody Conjugates

Most of the antibodies designed for therapeutic use derive from the IgG class of immunoglobulins. They provide antigen neutralization and subsequent cell-mediated immune responses, thereby clearing pathogens, microbes, virus-infected cells, and abnormal cells. In many cases, however, the "cytotoxic potency" provided by IgG may not provide sufficient therapeutic benefit. To increase its cell-killing ability, potent bacterial toxins or radioactive compounds have been conjugated to the antibody molecules. These modified molecules are called "antibody conjugates."

Highly potent bacterial toxins such as ricin and diphtheria can completely inhibit cellular protein synthesis at very low levels [26]. The bacterial toxin exerts cytotoxicity through enzymatic inactivation of factors essential for protein synthesis (e.g., ribosomal RNA, elongation factor 2 or EF2). Inactivation of these proteins, which are limited in quantity, leads to cell death. In theory, only one to five toxin molecules are needed to completely inhibit protein synthesis. Typically the binding activity of these toxins is localized in one of two domains that can be separated from the domain that promotes cytotoxicity. The binding domain in the B chain of ricin, for example, can be removed to prevent antigen-independent binding, and replaced with IgG. Thereby, the antibody-toxin conjugate exhibits selective destruction of cells recognized by the antibody. Antibody-toxin conjugates are also called "immunotoxins."

Therapeutic applications of antibody-toxin conjugates may be limited because (1) antibody-toxin conjugates distribute nonspecifically to organs such as the liver and cause severe toxicity, (2) the bacterial toxin is immunogenic to humans, (3) tumor-associated antigens are often found in normal tissue (although levels are low), and (4) premature release of toxin from the antibody conjugate leads to systemic toxicity.

The minimum antibody binding domain, F_v, has also been linked to toxin to produce a more compact, recombinant immunotoxin. The recombinant F_v-toxin appears to be more stable and exhibits less liver accumulation [27,28]. The immunogenicity of F_v-toxin as a consequence of the highly antigenic determinants of bacterial toxin remains a barrier to development.

Nevertheless, immunotoxins may be useful for purging cancer cells in bone marrow aspirate isolated from patients undergoing autologous bone marrow transplantation [29,30]. When no suitable bone marrow donor is available, the patients' own marrow cells are collected for repopulation of immune cells after radiation and chemotherapy. The small fraction of leukemic cells in autologous bone marrow cells can be removed with immunotoxins before they are reintroduced to the patient for restoring their immune system. Because this procedure

calls for cancer-cell removal *ex-vivo*, immunogenicity of the antibody-toxin conjugate does not pose a problem.

Radiolabeled antibody is another kind of antibody conjugate designed to enhance antibody-mediated cytotoxicity [31,32]. The limited ability of antibodies to penetrate the tumor cell mass is compensated by the ability of radioisotopes to act some distance away from the site of antibody binding. Radiolabeled antibodies are also called "radioimmunoconjugates." The effective killing radius achieved with radioimmunoconjugates is much greater than with antibody-toxin conjugates. However, as a result of enhanced cytotoxic potency, radioimmunoconjugates increase damage to healthy tissue if the radioisotope dissociates prematurely and accumulates in liver, spleen, and bone marrow. Stable coupling procedures to reduce premature release of radionuclide from radioimmunoconjugates have been described.

Improved radioimmunoconjugate technology has allowed the use of these biologics as contrast-enhancing agents for the diagnosis of cancer [33,34]. For such applications the radioisotopes attached to the antibodies usually exhibit low energy and short half-lives. Integration of radioimmunoconjugates with tomography [i.e., positron emission tomography (PET) and single-photon emission tomography (SPECT)] extends the capability to detect cancers and other abnormal tissues. Radioimmunoconjugates used for tumor and myocardial imaging are ProstaScint (In^{111}), MyosScint (In^{111}), and Verluma (Tc^{99}).

10.1.6. Antibody and Derivatives in Development

Humanization of Antibodies

A better understanding of antibody disposition and technical developments in antibody engineering have provided essential knowledge for transforming mouse monoclonal antibodies into chimeric and humanized forms with improved safety and efficacy. Nearly all monoclonal antibody products introduced after 1997 (constituting 80% of all marketed Mab products) are either chimeric or humanized [11]. Because of the success biotechnology and pharmaceutical companies have had in developing monoclonal antibody products, IgG molecules are among the most-favored class of macromolecules for the treatment of disease. Some of the antibodies and derivatives that have progressed to clinical testing in various therapeutic areas are listed in Table 10.5. Many more are in preclinical development. A report demonstrating successful isolation of stable human hybridomas has raised the hope of achieving human monoclonal antibodies for therapeutic use [13]

High-Efficiency Screening Procedures to Optimize Binding Affinity

As noted, F_v molecules derived from the *CDR* domain of IgG have a lower binding affinity than their IgG counterpart with identical *CDR* sequences. Therefore binding-affinity optimization is needed to improve the biologic activity of F_v molecules. The detailed structural elucidation of antibody-antigen binding sites and advances in recombinant DNA technology have allowed development of phage (molecular display systems) to efficiently identify an F_v containing unique *CDR* sequences that produce a high-binding affinity [35,36]. Phage are rapidly replicating bacterial viruses capable of carrying short DNA sequences. By random mutation of *CDR* sequences and expression of F_v on the coat of the phage, researchers can identify and isolate phage with maximum binding affinity to the target antigen [37–39].

The affinity of the antibody binding domain, F_v, displayed by phage can be directly measured by determining the affinity of the phage particle for the antigen. There is no need to purify the expressed protein. This efficient screening procedure greatly improves identification of recombinant phage containing *CDR* sequences with high-affinity binding characteristics. The high-affinity phage clone is then expanded, and *CDR* DNA sequences corresponding to the high-affinity binding to target antigen are identified. These sequences are subsequently used to construct recombinant expression systems for large-scale production.

Variations in *CDR* sequences corresponding to binding affinity for any antigen can be systematically and efficiently determined by molecular display technologies. Pertinent *CDR* sequences and their corresponding antigens, together with binding-affinity profiles, are compiled in binding-site libraries. These libraries permit researchers to engineer an antibody with unique binding characteristics.

Binding-site libraries may allow ready identification of antibody sequences for specific target antigens without having to collect and identify *CDR* sequences from immune B cells of immunized animals. Continued developments in this area may allow additional enhancements in the binding affinities of therapeutic antibodies. This high throughput strategy may also be useful to construct and improve the production or function of existing Mab's.

Improving Antibody Function Using Bispecific Antibodies

Bispecific antibodies (BsAbs) have been demonstrated to be a way to expand the biologic functions of antibodies [40]. Typically antibodies recognize a single antigen and mediate biologic functions through F_c domains. BsAbs, however, recognize two antigens. In addition to their capability to bind to an intended target (e.g., tumor-associated antigen), BsAbs also recognize antigens that induce some effector cell functions to either evoke cell-mediated

■TABLE 10.7. Some examples of bispecific antibodies (BsAb) in clinical trials

BsAb's Antigen Recognition

Target 1	Target 2	Name	Indication
CD3	Glioma		Glioma [45,46]
CD3	CD19	SHR-1	B-cell malignancies [47]
CD3	Folate-R	OC-TR	Ovarian [48]
CD3	EGP-2	BIS-1	Renal [49]
CD16	CD30	HRS-3/A9	Hodgkins lymphoma
CD16	HER2	2B1	HER2-positive tumor [50]
CD64	HER2	MDX-H210	Breast cancer [51]
CD64	EGF-R	MDX-447	Solid tumors [52,53]

killing or reduce undesirable side effects of antibodies [41–43]. BsAb molecules can be produced by chemical, genetic, and cellular techniques. The insight gained from recent studies about pathophysiology of antibody-mediated immune responses allows scientists to identify useful targets that may serve as a second epitope to enhance the primary antigen target in the treatment of cancer. Some BsAbs have progressed to early clinical trials (Table 10.7), but their therapeutic potential remains to be seen. The targets listed in the table include a wide array of antigens designed to induce cancer cell differentiation, apoptosis, and killing.

10.1.7. Summary

Monoclonal antibodies and Mab derivatives are among the fastest growing class of molecules being developed for a range of human diseases. Sophisticated, high-efficiency screening technologies developed recently may expedite the identification of antibodies with high potency and enhanced pharmaceutical properties. Monoclonal antibodies, regardless of how they are made and what antigens they recognize, exhibit similar disposition profiles and therefore produce predictable pharmacokinetic properties. The development of new antibodies and antibody derivatives for therapeutic use is greatly simplified

by using techniques for identification of sequence determinants that control antigen recognition. These determinants can be introduced in recombinant systems for large-scale production. In theory, the number of therapeutic antibodies that can be developed into drugs is limited only by the ability to identify antigens that hold the key to the therapeutic target. Such antigens could come from a range of pharmacologically and pathologically important receptors, cytokines, interferons, and other effector molecules.

REFERENCES

1. Jenner, E., *The origin of the vaccine inoculation.* 1801, London: D.N. Shury. 12.
2. Pasteur, L., and J. Lister, *Germ theory and its applications to medicine and on the antiseptic principle of the practice of surgery.* Great minds series. 1996, Amherst, NY: Prometheus Books. 144.
3. Harris, L.J., S.B. Larson, K.W. Hasel, J. Day, A. Greenwood, and A. McPherson, *The three-dimensional structure of an intact monoclonal antibody for canine lymphoma.* Nature, 1992. **360**(6402): 369–72.
4. Kantha, S.S., *A centennial review: the 1890 tetanus antitoxin paper of von Behring and Kitasato and the related developments.* Keio J Med, 1991. **40**(1): 35–9.

5. Paul, W.E., *Fundamental immunology*. 3rd ed. 1993, New York: Raven Press. xvii, 1490.

6. Porter, R.R., *The structure of the heavy chain of immunoglobulin and its relevance to the nature of the antibody-combining site: the Second CIBA Medal Lecture*. Biochem J, 1967. **105**(2): 417–26.

7. Mann, J., *The elusive magic bullet: the search for the perfect drug*. 1999, New York: Oxford University Press. x, 209.

8. Kyle, R.A., *Multiple Myeloma: diagnostic challenges and standard therapy*. Semin Hematol, 2001. **38**(2 Suppl 3): 11–4.

9. Kohler, G., and C. Milstein, *Continuous cultures of fused cells secreting antibody of predefined specificity*. Nature, 1975. **256**(5517): 495–7.

10. Field, H., G.T. Yarranton, and A.R. Rees, *Expression of mouse immunoglobulin light and heavy chain variable regions in Escherichia coli and reconstitution of antigen-binding activity*. Protein Eng, 1990. **3**(7): 641–7.

11. Clark, M., *Antibody humanization: a case of the "Emperor's new clothes"?* Immunol Today, 2000. **21**(8): 397–402.

12. Koropatnick, J., J. Pearson, and J.F. Harris, *Extensive loss of human DNA accompanies loss of antibody production in heteromyeloma hybridoma cells*. Mol Biol Med, 1988. **5**(2): 69–83.

13. Karpas, A., A. Dremucheva, and B.H. Czepulkowski, *A human myeloma cell line suitable for the generation of human monoclonal antibodies*. Proc Natl Acad Sci USA, 2001. **98**(4): 1799–804.

14. Hemmings, W.A., and E.W. Williams, *The attachment of IgG to cell components: a reconsideration of Brambell's receptor hypothesis of protein transmission*. Proc R Soc Lond B Biol Sci, 1974. **187**(1087): 209–19.

15. Telleman, P., and R.P. Junghans, *The role of the Brambell receptor (FcRB) in liver: protection of endocytosed immunoglobulin G (IgG) from catabolism in hepatocytes rather than transport of IgG to bile*. Immunology, 2000. **100**(2): 245–51.

16. Junghans, R.P., *Finally! The Brambell receptor (FcRB). Mediator of transmission of immunity and protection from catabolism for IgG*. Immunol Res, 1997. **16**(1): 29–57.

17. Holland, P.W., A.M. Hacker, and N.A. Williams, *A molecular analysis of the phylogenetic affinities of Saccoglossus cambrensis Brambell & Cole (Hemichordata)*. Philos Trans R Soc Lond B Biol Sci, 1991. **332**(1264): 185–9.

18. Kim, J.K., M. Firan, C.G. Radu, C.H. Kim, V. Ghetie, and E.S. Ward, *Mapping the site on human IgG for binding of the MHC class I-related receptor, FcRn*. Eur J Immunol, 1999. **29**(9): 2819–25.

19. Shields, R.L., A.K. Namenuk, K. Hong, Y.G. Meng, J. Rae, J. Briggs, D. Xie, J. Lai, A. Stadlen, B. Li, J.A. Fox, and L.G. Presta, *High resolution mapping of the binding site on human IgG1 for Fc gamma RI, Fc gamma RII, Fc gamma RIII, and FcRn and design of IgG1 variants with improved binding to the Fc gamma R*. J Biol Chem, 2001. **276**(9): 6591–604.

20. DeLa Cadena, R.A., A. Majluf-Cruz, A. Stadnicki, M. Tropea, D. Reda, J.M. Agosti, R.W. Colman, and A.F. Suffredini, *Recombinant tumor necrosis factor receptor p75 fusion protein (TNFR:Fc) alters endotoxin-induced activation of the kinin, fibrinolytic, and coagulation systems in normal humans*. Thromb Haemost, 1998. **80**(1): 114–8.

21. Murray, K.M., and S.L. Dahl, *Recombinant human tumor necrosis factor receptor (p75) Fc fusion protein (TNFR:Fc) in rheumatoid arthritis*. Ann Pharmacother, 1997. **31**(11): 1335–8.

22. Eason, J.D., S.L. Wee, T. Kawai, H.Z. Hong, J. Powelson, M. Widmer, and A.B. Cosimi, *Recombinant human dimeric tumor necrosis factor receptor (TNFR:Fc) as an immunosuppressive agent in renal allograft recipients*. Transplant Proc, 1995. **27**(1): 554.

23. Fujimori, K., D.G. Covell, J.E. Fletcher, and J.N. Weinstein, *A modeling analysis of monoclonal antibody percolation through tumors: a binding-site barrier*. J Nucl Med, 1990. **31**(7): 1191–8.

24. Fujimori, K., D.G. Covell, J.E. Fletcher, and J.N. Weinstein, *Modeling analysis of the global and microscopic distribution of immunoglobulin G, F(ab')2, and Fab in tumors*. Cancer Res, 1989. **49**(20): 5656–63.

25. van Osdol, W., K. Fujimori, and J.N. Weinstein, *An analysis of monoclonal antibody distribution in microscopic tumor*

nodules: consequences of a "binding site barrier". Cancer Res, 1991. **51**(18): 4776–84.

26. Hall, W.A., *Immunotoxin therapy*. Neurosurg Clin N Am, 1996. **7**(3): 537–46.

27. Haggerty, H.G., W.A. Warner, C.R. Comereski, W.M. Peden, L.E. Mezza, B.D. Damle, C.B. Siegall, and T.J. Davidson, *BR96 sFv-PE40 immunotoxin: nonclinical safety assessment*. Toxicol Pathol, 1999. **27**(1): 87–94.

28. Brinkmann, U., *Recombinant antibody fragments and immunotoxin fusions for cancer therapy*. In Vivo, 2000. **14**(1): 21–7.

29. Laurent, G., D. Maraninchi, E. Gluckman, J.P. Vernant, J.M. Derocq, M.H. Gaspard, B. Rio, M. Michalet, J. Reiffers, F. Dreyfus, et al., *Donor bone marrow treatment with T101 Fab fragment-ricin A-chain immunotoxin prevents graft-versus-host disease*. Bone Marrow Transplant, 1989. **4**(4): 367–71.

30. Preijers, F.W., T. De Witte, J.M. Wessels, J.P. Meyerink, C. Haanen, and P.J. Capel, *Cytotoxic potential of anti-CD7 immunotoxin (WT1-ricin A) to purge ex vivo malignant T cells in bone marrow*. Br J Haematol, 1989. **71**(2): 195–201.

31. Goldenberg, D.M., *Targeted therapy of cancer with radiolabeled antibodies*. J Nucl Med, 2002. **43**(5): 693–713.

32. Burke, J.M., J.G. Jurcic, and D.A. Scheinberg, *Radioimmunotherapy for acute leukemia*. Cancer Control, 2002. **9**(2): 106–13.

33. Hadley, S.W., and D.S. Wilbur, *Evaluation of iodovinyl antibody conjugates: comparison with a p-iodobenzoyl conjugate and direct radioiodination*. Bioconjug Chem, 1990. **1**(2): 154–61.

34. Waibel, R., R. Alberto, J. Willuda, R. Finnern, R. Schibli, A. Stichelberger, A. Egli, U. Abram, J.P. Mach, A. Pluckthun, and P.A. Schubiger, *Stable one-step technetium-99m labeling of His-tagged recombinant proteins with a novel Tc(I)-carbonyl complex*. Nat Biotechnol, 1999. **17**(9): 897–901.

35. Hoogenboom, H.R., and P. Chames, *Natural and designer binding sites made by phage display technology*. Immunol Today, 2000. **21**(8): 371–8.

36. Hoogenboom, H.R., *Overview of antibody phage-display technology and its applications*. Methods Mol Biol, 2002. **178**: 1–37.

37. Thompson, J., T. Pope, J.S. Tung, C. Chan, G. Hollis, G. Mark, and K.S. Johnson, *Affinity maturation of a high-affinity human monoclonal antibody against the third hypervariable loop of human immunodeficiency virus: use of phage display to improve affinity and broaden strain reactivity*. J Mol Biol, 1996. **256**(1): 77–88.

38. De Ciechi, P.A., C.S. Devine, S.C. Lee, S.C. Howard, P.O. Olins, and M.H. Caparon, *Utilization of multiple phage display libraries for the identification of dissimilar peptide motifs that bind to a B7-1 monoclonal antibody*. Mol Divers, 1996. **1**(2): 79–86.

39. Griffin, H.M., and W.H. Ouwehand, *A human monoclonal antibody specific for the leucine-33 (P1A1, HPA-1a) form of platelet glycoprotein IIIa from a V gene phage display library*. Blood, 1995. **86**(12): 4430–6.

40. Staerz, U.D., and M.J. Bevan, *Hybrid hybridoma producing a bispecific monoclonal antibody that can focus effector T-cell activity*. Proc Natl Acad Sci USA, 1986. **83**(5): 1453–7.

41. Mezzanzanica, D., S. Canevari, S. Menard, S.M. Pupa, E. Tagliabue, A. Lanzavecchia, and M.I. Colnaghi, *Human ovarian carcinoma lysis by cytotoxic T cells targeted by bispecific monoclonal antibodies: analysis of the antibody components*. Int J Cancer, 1988. **41**(4): 609–15.

42. Gilliland, D.G., Z. Steplewski, R.J. Collier, K.F. Mitchell, T.H. Chang, and H. Koprowski, *Antibody-directed cytotoxic agents: use of monoclonal antibody to direct the action of toxin A chains to colorectal carcinoma cells*. Proc Natl Acad Sci USA, 1980. **77**(8): 4539–43.

43. Gilliland, L.K., M.R. Clark, and H. Waldmann, *Universal bispecific antibody for targeting tumor cells for destruction by cytotoxic T cells*. Proc Natl Acad Sci USA, 1988. **85**(20): 7719–23.

44. Waldmann, T.A., and W. Strober, *Metabolism of immunoglobulins*. Prog Allergy, 1969. **13**: 1–110.

45. Nitta, T., K. Sato, H. Yagita, K. Okumura, and S. Ishii, *Preliminary trial of specific targeting therapy against malignant glioma*. Lancet, 1990. **335**(8686): 368–71.

46. Nitta, T., K. Sato, K. Okumura, and S. Ishii, *Induction of cytotoxicity in human T cells coated with anti-glioma x anti-CD3 bispecific antibody against human glioma cells*. J Neurosurg, 1990. **72**(3): 476–81.

47. De Gast, G.C., A.A. Van Houten, I.A. Haagen, S. Klein, R.A. De Weger, A. Van Dijk, J. Phillips, M. Clark, and B.J. Bast, *Clinical experience with CD3 × CD19 bispecific antibodies in patients with B cell malignancies.* J Hematother, 1995. **4**(5): 433–7.

48. Canevari, S., D. Mezzanzanica, A. Mazzoni, D.R. Negri, V. Ramakrishna, R.L. Bolhuis, M.I. Colnaghi, and G. Bolis, *Bispecific antibody targeted T cell therapy of ovarian cancer: clinical results and future directions.* J Hematother, 1995. **4**(5): 423–7.

49. Kroesen, B.J., A. ter Haar, H. Spakman, P. Willemse, D.T. Sleijfer, E.G. de Vries, N.H. Mulder, H.H. Berendsen, P.C. Limburg, T.H. The, et al., *Local antitumour treatment in carcinoma patients with bispecific-monoclonal-antibody-redirected T cells.* Cancer Immunol Immunother, 1993. **37**(6): 400–7.

50. Weiner, L.M., J.I. Clark, M. Davey, W.S. Li, I. Garcia de Palazzo, D.B. Ring, and R.K. Alpaugh, *Phase I trial of 2B1, a bispecific monoclonal antibody targeting c-erbB-2 and Fc gamma RIII.* Cancer Res, 1995. **55**(20): 4586–93.

51. van Ojik, H.H., R. Repp, G. Groenewegen, T. Valerius, and J.G. van de Winkel, *Clinical evaluation of the bispecific antibody MDX-H210 (anti-Fc gamma RI × anti-HER-2/neu) in combination with granulocyte-colony-stimulating factor (filgrastim) for treatment of advanced breast cancer.* Cancer Immunol Immunother, 1997. **45**(3–4): 207–9.

52. Curnow, R.T., *Clinical experience with CD64-directed immunotherapy. An overview.* Cancer Immunol Immunother, 1997. **45**(3–4): 210–5.

53. Wallace, P.K., J.L. Romet-Lemonne, M. Chokri, L.H. Kasper, M.W. Fanger, and C.E. Fadul, *Production of macrophage-activated killer cells for targeting of glioblastoma cells with bispecific antibody to FcgammaRI and the epidermal growth factor receptor.* Cancer Immunol Immunother, 2000. **49**(9): 493–503.

■ SECTION TWO ■

■ 10.2. MONOGRAPHS

MUROMONAB-CD3

Trade name: ***Orthoclone OKT3***

Manufacturer: OrthoBiotech Products, Raritan, NJ

Indications: Treatment of acute allograft rejection in renal transplant patients and the treatment of steroid-resistant acute allograft rejection in cardiac and hepatic transplant patients

Approval date: 1 June 1986

Type of submission: Product license application

A. General description Muromonab-CD3 is a murine monoclonal antibody (MW: 150kDa) to the CD3 antigen of human T cells. It functions as an immunosuppressant. The antibody is a biochemically purified IgG$_{2a}$ immunoglobulin with a heavy chain of approximately 50kDa and a light chain of approximately 25kDa. Muromonab CD3 is directed to a glycoprotein with a molecular weight of 20kDa in the human T-cell surface that is essential for T-cell functions.

B. Indications and use *Orthoclone OKT3* is indicated for the treatment of acute allograft rejection in renal transplant patients and for the treatment of steroid-resistant acute allograft rejection in cardiac and hepatic transplant patients.

C. Administration
Dosage form Each 5ml ampule of *Orthoclone OKT3* contains muromonab-CD3 5mg in a sterile solution.
Route of administration *Orthoclone OKT3* is administered as an intravenous bolus over less than one minute. The preparation should not be administered by infusion or in conjunction with other drug solutions.

Recommended dosage and monitoring requirements For adults and children weighing greater than 30 kg, the recommended intravenous dose of *Orthoclone OKT3* is 5 mg per day for 10 to 14 days following initial signs and symptoms of rejection. The initial recommended dose is 2.5 mg per day in pediatric patients weighing less than or equal to 30 kg. Concomitant intravenous methylprednisolone is highly recommended. Acetaminophen and antihistamine, given with the first dose, may also help reduce some early reactions. The dosage of other immunosuppressive agents used in conjunction with *Orthoclone OKT3* should be reduced to the lowest level compatible with an effective therapeutic response.

D. Pharmacology and pharmaceutics

Clinical pharmacology According to Micromedex, allograft recipients previously unexposed to donor antigens reject the graft largely through the T-cell system. If the recipients have previously been exposed to donor antigens by blood transfusion or a previous transplant, they undergo humoral rejection in which B-cell mechanisms predominate. Muromonab-CD3 exerts its activity by binding to CD3 antigen on the surface of T-lymphocytes. Binding to the CD3 antigen inactivates the adjacent T-cell receptor portion of the T-lymphocyte cell membrane and prevents activation of the T-lymphocyte. CD3 lymphocyte levels fall precipitously within a few hours of administration of muromonab-CD3.

Pharmacokinetics A rapid decrease in T-lymphocytes is observed within a few minutes after administration of *Orthoclone OKT3*. The volume of distribution of muromonab CD3 is approximately 6.5 liters; its half-life is 18 hours. Clinical experience has demonstrated that serum levels greater than or equal to 0.8 µg/ml of muromonab CD3 blocks the function of cytotoxic T cells *in vitro* and *in vivo*. Reduced T-cell clearance or low plasma muromonab CD3 levels provide a basis for adjusting *Orthoclone OKT3* dosage or for discontinuing therapy.

Disposition No information available.

Drug interactions When using combinations of immunosuppressive agents, the dose of each agent, including *Orthoclone OKT3*, should be reduced to the lowest level compatible with an effective therapeutic response so as to reduce the potential for and severity of infections and malignant transformations.

E. Therapeutic response In a comparison of *Orthoclone OKT3* with conventional high-dose steroid therapy in reversing acute renal allograft rejection, the one-year Kaplan-Meier (actuarial) estimates of graft survival rates were 62% and 45% for *Orthoclone OKT3* and steroid-treated patients, respectively. At two years the rates were 56% and 42%. *Orthoclone OKT3* was studied for use in reversing acute cardiac and hepatic allograft rejection in patients who were unresponsive to high doses of steroids. The rate of reversal was 90% in acute cardiac allograft rejection and 83% in hepatic allograft rejection.

F. Role in therapy According to Micromedex, *Orthoclone OKT3* is effective in reversing acute renal, hepatic, cardiac, and kidney/pancreas transplant rejection episodes that are resistant to conventional treatment (i.e., high-dose steroids and antithymocyte globulin) and offers a vital therapeutic alternative to retransplantation. *Orthoclone OKT3* offers no advantage over other immunosuppressive agents for prevention of transplant rejection, and may increase the incidence of side effects and infections.

G. Other applications Other possible uses of *Orthoclone OKT3* are for prophylaxis and treatment of acute graft-versus-host disease (as an adjunct with allogeneic bone marrow transplantation); treatment of rejection after lung transplantation; prophylaxis and treatment of rejection after pancreatic allograft transplantation; and psoriasis.

DACLIZUMAB

Trade name: *Zenapax*

Manufacturer: Hoffman-La Roche Inc., Nutley, NJ

Indications: Prophylaxis of acute organ rejection in patients receiving renal transplants, as part of an immunosuppressive regimen that includes cyclosporine and corticosteroids

Approval date: 110 December 1997

Type of submission: Biologics license application

A. General description Daclizumab is an immunosuppressive, humanized IgG$_1$ monoclonal antibody produced by recombinant DNA technology. It binds specifically to the alpha subunit (p55 alpha, CD25, or Tac subunit) of the human high-affinity interleukin-2 (IL-2) receptor that is expressed on the surface of activated lymphocytes, where it acts as an interleukin-2 (IL-2) receptor antagonist. Daclizumab is a composite of human (90%) and murine (10%) antibody sequences. Its estimated molecular weight is 144 kDa. Daclizumab is derived from a murine anti-Tac monoclonal antibody by grafting of the murine antigen-binding site to a prototype human IgG$_1$ constant domain and the Eu myeloma antibody variable region framework.

B. Indications and use *Zenapax* has demonstrated efficacy in and is approved for the prophylaxis of acute organ rejection in adult renal transplant recipients. It is used as part of an immunosuppressive regimen that includes cyclosporine and corticosteroids.

C. Administration

Dosage form *Zenapax* 25 mg/5 ml is supplied as a sterile concentrate for further dilution and intravenous administration.

Route of administration *Zenapax* is administered via a peripheral or central vein over a 15-minute period.

Recommended dosage and monitoring requirements *Zenapax* is used as part of an immunosuppressive regimen that includes cyclosporine and corticosteroids. The recommended dose of *Zenapax* for the prophylaxis of rejection episodes in patients undergoing renal transplantation is 1 mg/kg intravenously over 15 minutes, no more than 24 hours before the transplant, repeated every 14 days for a total of five doses. No dosage adjustment is necessary for patients with severe renal impairment. No dosage adjustments based on other identified covariates (age, gender, proteinuria, and race) are required for renal allograft patients. Anaphylactic reactions following administration of proteins can occur and medications for the treatment of severe hypersensitivity reactions should be available for immediate use.

D. Pharmacology and pharmaceutics

Clinical pharmacology Acute episodes of renal allograft rejection occur in 20% to 50% of all recipients within the first 6 months after transplantation. One approach to immunomodulation blocks IL-2 signaling by inhibition of IL-2 receptor engagement. Daclizumab functions as an IL-2 receptor antagonist that binds with high-affinity to the Tac subunit of the high-affinity IL-2 receptor complex and inhibits IL-2 binding. Daclizumab binding is highly specific for Tac, which is expressed on activated but not resting lymphocytes. Daclizumab has a slightly lower affinity than parent murine monoclonal anti-Tac antibody, but is less immunogenic and more immunosuppressive, and has a longer half-life. Administration of *Zenapax* inhibits IL-2-mediated activation of lymphocytes, a critical pathway in the cellular immune response involved in allograft rejection.

Pharmacokinetics In clinical trials, peak serum concentration (mean ± SD) of daclizumab rose between the first dose (21

± 14 µg/ml) and fifth dose (32 ± 22 µg/ml). *In vivo* data suggest that serum levels of 5 to 10 µg/ml are necessary for saturation of the Tac subunit of the IL-2 receptors to block the responses of activated T-lymphocytes. Population pharmacokinetic analysis gave the following values for a reference patient (45-year-old male Caucasian patient with a body weight of 80 kg and no proteinuria): systemic clearance = 15 ml/h, volume of distribution = 6 liters. The estimated terminal elimination half-life for the reference patient was 20 days (480 hours), which is similar to the terminal elimination half-life for human IgG (18–23 days). Estimates of terminal elimination half-life ranged from 11 to 38 days for the 123 patients included in the population analysis. The influence of body weight on systemic clearance supports the dosing of *Zenapax* on a milligram per kilogram (mg/kg) basis.

Disposition   No information available.

Drug interactions   In a randomized, double-blind study, *Zenapax* or placebo was added to an immunosuppressive regimen of cyclosporine, mycophenolate mofetil, and steroids to assess tolerability, pharmacokinetics, and drug interactions. The addition of *Zenapax* did not result in an increased incidence of adverse events or a change in the types of adverse events reported. The following medications have been administered in clinical trials with *Zenapax* with no incremental increase in adverse reactions: cyclosporine, mycophenolate mofetil, ganciclovir, acyclovir, azathioprine, and corticosteroids.

E. Therapeutic response   The primary efficacy endpoint of two randomized, double-blind, placebo-controlled, multi-center trials was the proportion of patients who developed a biopsy-proven acute rejection episode within the first 6 months following transplantation. These trials compared a dose of 1.0 mg/kg of *Zenapax* with placebo when each was administered as part of standard immunosuppressive regimens containing either cyclosporine and corticosteroids (double-therapy trial) or cyclosporine, corticosteroids, and azathioprine (triple-therapy trial) to prevent acute renal allograft rejection. The incidence of acute rejection was significantly lower in the *Zenapax*-treated group in both the double-therapy and triple-therapy trials.

F. Role in therapy   In the setting of renal transplantation, *Zenapax*, administered prophylactically may improve graft survival and mortality when added to standard immunosuppressive regimens.

G. Other applications   In limited studies, *Zenapax* has produced responses in virus-associated adult T-cell leukemia. *Zenapax* is also effective as treatment for acute graft-versus-host disease (GVHD) but is ineffective as GVHD prophylaxis. A recent study in seven patients with type 1 diabetes and a history of severe hypoglycemia and metabolic instability demonstrated that pancreatic islet transplantation can result in insulin independence with excellent metabolic control when glucocorticoid-free immunosuppression, with a regimen consisting of sirolimus, tacrolimus, and daclizumab, is combined with the infusion of an adequate islet mass. Another study showed that daclizumab, added to cyclosporine, mycophenolate mofetil, and prednisone, safety reduced the frequency and severity of cardiac-allograft rejection during the induction period.

H. Other considerations   Daclizumab has been designated an orphan product for use in the prevention of acute renal allograft rejection.

BASILIXIMAB

Trade name: *Simulect*

Manufacturer: Novartis Pharmaceutical Corp., East Hanover, NJ

Indications: Prophylaxis of acute organ rejection in patients receiving renal transplantation when used as part of an immunosuppressive regimen that includes cyclosporine and corticosteroids

Approval date: 12 May 1998

Type of submission: Biologics license application

A. General description Basiliximab is a chimeric (murine/human) monoclonal antibody (IgG_{1k}), produced by recombinant DNA technology. It specifically binds to and blocks the interleukin-2R receptor (alpha)-chain (CD25), which is expressed on the surface of activated T-lymphocytes in response to antigenic challenge. Based on the amino-acid sequence, the calculated molecular weight of the protein is 144 kDa. Basiliximab is a glycoprotein obtained from fermentation of an established mouse myeloma cell line genetically engineered to express plasmids containing the human heavy and light chain constant region genes and mouse heavy and light chain variable region genes encoding the antibody that binds selectively to primate IL-2R (alpha).

B. Indications and use Basiliximab is indicated for the prophylaxis of acute organ rejection in patients receiving renal transplants. It is administered with an immunosuppressive regimen that includes cyclosporine and corticosteroids.

C. Administration
    Dosage form Simulect is a sterile lyophilized powder, packaged in glass vials. Each vial contains basiliximab 20 mg, to be reconstituted in sterile water for injection.
    Route of administration Simulect is intended for central or peripheral intravenous administration only.
    Recommended dosage and monitoring requirements In adult patients, the recommended regimen of Simulect is two doses of 20 mg each, administered as an intravenous infusion over 20 to 30 minutes. The first dose should be given within 2 hours prior to transplantation surgery. The second dose should be given 4 days after transplantation. The second dose should be withheld if complications occur. For children and adolescents from 2 up to 15 years of age, the recommended regimen is two doses of $12 mg/m^2$ each, up to a maximum of 20 mg/dose. Conventional dual immunosuppression with cyclosporine and corticosteroids is administered concurrently. Three-drug immunosuppression (cyclosporine, steroids, and azathioprine) is not recommended due to a potentially increased risk of post-transplant lymphoproliferative disease.

D. Pharmacology and pharmaceutics
    Clinical pharmacology Basiliximab is a chimeric (mouse/human) interleukin-2 receptor antagonist. It is directed against the interleukin-2 receptor-alpha chain (CD25) on activated T-lymphocytes, and is a potent inhibitor of interleukin-2-mediated activation of lymphocytes, a critical pathway in the cellular immune response involved in allograft rejection. Another anti-CD25 monoclonal antibody, daclizumab (*Zenapax*), has the same indication as basiliximab.
    Pharmacokinetics Complete and consistent blocking of the IL-2 receptor is maintained as long as serum basiliximab levels exceed 0.2 μg/ml. As concentrations fall below this level, expression of IL-2 receptor returns to pre-therapy values within 1–2 weeks. Basiliximab has a steady-state volume of distribution of 9 liters and total body clearance of 41 ml/h. Its elimination half-life ranges from 7 to 10 days. There is a dose-proportional increase in peak concentration (C_{max}) and area under the curve (AUC) up to the highest tested single dose of 60 mg. No clinically relevant influence of body weight or gender on distribution volume or clearance has been observed in adult patients. Elimination

half-life was not influenced by age (20–69 years), gender or, race. A study of the pharmacokinetics of *Simulect* in pediatric renal transplantation patients, children (2–11 years of age), and adolescents (12–15 years of age), suggests that in children the volume of distribution of basiliximab at steady state is 5.2 ± 2.8 liter, its half-life is 11.5 ± 6.3 days, and its clearance is 17 ± 6 ml/h. Distribution volume and clearance in children are reduced by about 50% compared with adult renal transplantation patients. Disposition in adolescents was similar to that in adult renal transplantation patients.

Disposition   No information available.

Drug interactions   No formal drug–drug interaction studies have been conducted. The following medications have been administered in clinical trials with *Simulect* with no incremental increase in adverse reactions: azathioprine, corticosteroids, cyclosporine, mycophenolate mofetil, and muromonab-CD3.

E. Therapeutic response   The safety and efficacy of *Simulect*, when added to a standard immunosuppressive regimen comprised of cyclosporine and corticosteroids, were assessed in two placebo-controlled trials. The primary end point in both studies was the incidence of death, graft loss, or an episode of acute rejection during the first 6 months post-transplantation. Patients receiving *Simulect* experienced a significantly lower incidence of biopsy-confirmed rejection episodes at both 6 and 12 months after transplantation, but there was no difference in the rate of delayed graft function, patient survival, or graft survival between *Simulect*-treated patients and placebo-treated patients in either study.

F. Role in therapy   Either *Simulect* or *Zenapax* (daclizumab), which has the same mechanism of action, should be considered as add-on rejection prophylaxis (with cyclosporine and corticosteroids) in first renal transplant recipients. Although direct comparisons are lacking, these agents appear comparable in efficacy. An advantage of *Simulect* is a shorter, more simplified dose regimen that does not require return of the patient for outpatient administration; a potential advantage of *Zenapax* is a longer duration of sustained immunosuppression (up to 3 months).

G. Other applications   *Simulect* may have application in the prevention of liver allograft rejection and in the treatment of psoriasis.

ETANERCEPT

Trade name: ***Enbrel***

Manufacturer: Amgen Inc., Thousand Oak, CA

Indications: Reduction in signs and symptoms of moderately to severely active rheumatoid arthritis in patients who have had an inadequate response to one or more disease-modifying antirheumatic drugs

Approval date: 2 November 1998

Type of submission: Biologics license application

Supplement approval date: 27 May 1999

New indication: Polyarticular-course juvenile rheumatoid arthritis

Supplement approval date: 6 June 2000

New indication: Reducing the signs and symptoms and delaying structural damage in patients with moderately to severely active rheumatoid arthritis, including those who have not previously failed treatment with disease-modifying drugs

A. General description   Etanercept is a dimeric fusion protein consisting of the extracellular ligand-binding portion of the human 75 kDa (p75) tumor necrosis factor receptor (TNFR) linked to the F_c portion of human IgG$_1$. The F$_c$ component of etanercept contains the CH2 domain, the CH3 domain and hinge region, but not the CH1 domain of IgG1. Etanercept is produced by recombinant DNA technology in a Chinese hamster ovary (CHO) mammalian cell expression system. It consists of 934 amino acids and has an apparent molecular weight of approximately 150 kDa.

B. Indications and use *Enbrel* is indicated for reducing signs and symptoms and delaying structural damage in patients with moderately to severely active rheumatoid arthritis. The drug can be used in combination with methotrexate in patients who do not respond adequately to methotrexate alone. Enbrel is also indicated for reducing signs and symptoms of moderately to severely active polyarticular-course juvenile rheumatoid arthritis in patients who have had an inadequate response to one or more disease-modifying drugs.

C. Administration

Dosage form *Enbrel* is supplied as a sterile, lyophilized powder for parenteral administration after reconstitution with sterile bacteriostatic water for injection. Each single-use vial of *Enbrel* contains etanercept 25 mg.

Route of administration *Enbrel* is intended for subcutaneous administration. Patients may self-inject only if their physician determines that it is appropriate, after proper training in injection technique.

Recommended dosage and monitoring requirements The recommended dose of *Enbrel* for adult patients with rheumatoid arthritis is 25 mg given twice weekly as a subcutaneous injection. The recommended dose in children with polyarticular-course juvenile rheumatoid arthritis, ages 4 to 17 years, is 0.4 mg/kg (maximum 25 mg per dose), subcutaneously twice weekly. The twice-weekly doses should be 72 to 96 hours apart. Methotrexate, glucocorticoids, salicylates, nonsteroidal anti-inflammatory drugs (NSAIDs), or analgesics may be

continued during treatment with *Enbrel*. *Enbrel* therapy should be discontinued in a patient who develops a severe infection or sepsis.

D. Pharmacology and pharmaceutics

Clinical pharmacology Tumor necrosis factor (TNF) is a polypeptide cytokine produced by activated macrophages and T cells; after cleavage from the cell surface by proteolysis. The soluble form of the transmembrane protein can circulate in the serum. TNF is an essential component of the inflammatory process and is associated with the inflammation and joint destruction of rheumatoid arthritis (RA). Concentrations of TNF are elevated in the synovial fluid of patients with RA. Two distinct TNF receptors have been discovered (the 75 kDa, or p75 receptor; and the 55 kDa, or p55 receptor); they exist in cell-surface and soluble forms, and both bind TNF. Monomeric fragments from the extracellular portion of the receptor can leave the cell surface and become soluble TNF receptors (sTNFRs). Elevated concentrations of sTNFRs have been found in the circulation of RA patients. Chronic RA may result from an imbalance between cytokines and their natural inhibitors. Etanercept is a soluble, dimeric, recombinant human p75 TNFR, fused to the F$_c$ fragment of human immunoglobulin G1 (IgG$_1$), developed for neutralization of TNF-alpha. Etanercept binds *in vivo* to TNF and lymphotoxin alpha, reducing TNF biological activities.

Pharmacokinetics After administration of *Enbrel* 25 mg by a single subcutaneous injection to patients with RA, a median half-life of 115 hours (range: 98–300 hours) was observed with a clearance of 89 ml/h. Based on available data, individual patients may undergo a two- to five-fold increase in serum levels of etanercept with repeated dosing. Pharmacokinetic parameters were not different between men and women and did not vary with age in adult patients. No formal pharmacokinetic studies have been conducted to examine the effects of renal or hepatic impairment. The clearance of etanercept is reduced slightly in children ages 4 to 8 years, compared with adults. Children under 4 years of age have not been studied.

Disposition No information available.

Drug interactions Specific drug interaction studies have not been conducted with *Enbrel*. Although no data are available regarding the effects of vaccination in patients receiving etanercept therapy, vaccination with live vaccines is not recommended. The possibility exists for etanercept to affect host defenses against infections since the cellular immune response may be altered.

E. Therapeutic response The safety and efficacy of *Enbrel* were assessed in three controlled studies. The results of all three trials were expressed in percentage of patients with improvement in RA using American College of Rheumatology (ACR) response criteria. In one study *Enbrel* reduced arthritic signs and symptoms more rapidly and provided significantly greater protection against new joint erosion and existing erosion progression in patients with early, active RA than did methotrexate. *Enbrel* significantly reduced development of new joint erosions and prevented progression of existing joint erosions over one year, in early, active RA. *Enbrel* was markedly superior to placebo in the treatment of juvenile RA, in a controlled clinical study of 69 patients who were either refractory to, or intolerant of methotrexate.

F. Role in therapy *Enbrel* helps reduce the symptoms of moderate to severe, active RA in patients who have not responded well to other treatments. It can also be used in combination with methotrexate if patients do not benefit enough from use of methotrexate alone. Although many patients with RA respond well to currently available treatments, many are also disabled and suffer severe pain from the disease. In clinical trials, approximately 59% of patients treated with *Enbrel*, compared to 11% of the untreated groups experienced a significant reduction in symptoms such as tender, swollen, and painful joints after 6 months of treatment. *Enbrel* has also been studied in children aged 4 to 17 years old with moderate to severe juvenile RA with results similar to those of the adult studies.

G. Other applications Intravenous *Enbrel* has been used to attenuate OKT3 (muromonab-CD3)-induced acute clinical syndrome in renal transplant patients. Early clinical trials have also revealed etanercept to be useful in the treatment of psoriatic arthritis. In a 12-week study, 87% of *Enbrel*-treated patients met the ACR criteria for 20% improvement, compared with 23% of placebo-controlled patients.

H. Other considerations *Enbrel* has been designated an orphan product for use in the treatment of juvenile rheumatoid arthritis and Wegener's granulomatosis.

INFLIXIMAB

Trade name: *Remicade*

Manufacturer: Centocor, Inc., Malvern, PA

Indications: Treatment of moderately to severely active Crohn's disease for the reduction of the signs and symptoms in patients who have an inadequate response to conventional therapies; and treatment of patients with fistulizing Crohn's disease for the reduction in the number of draining enterocutaneous fistula(s)

Approval date: 24 August 1998

Type of submission: Biologics license application

Supplement approval date: 10 November 1999

New indication: Reduction in signs and symptoms of rheumatoid arthritis (RA) in patients who have had an inadequate response to methotrexate.

A. General description Infliximab is a glycosylated human-murine chimeric IgG1 monoclonal antibody specific for human and chimpanzee TNFα. Like all IgGl molecules it has a molecular weight of about 149 kDa and is comprised of two identical heavy and light chains that associate by disulfide bonds and noncovalent interactions. Infliximab specifically binds both soluble TNFα homotrimer and the membrane-bound precursor form of TNFα. The antibody is produced by cultured cells transfected with a DNA expression construct containing murine immunoglobulin variable and human constant region genes.

B. Indications and use *Remicade* is indicated for the reduction of the symptoms of moderate to severe Crohn's disease in patients who have not responded well to traditional treatments, including corticosteroids and other immunosuppressants, and antibiotics, and to close enterocutaneous fistulas. *Remicade*, with concomitant methotrexate, is indicated for the reduction in signs and symptoms, and inhibiting the progression of structural damage due to RA in patients who have had an inadequate response to methotrexate alone.

C. Administration
   Dosage form *Remicade* is supplied as a sterile, lyophilized powder for intravenous infusion, following reconstitution with sterile water for injection. Each single-use vial contains infliximab 100 mg.
   Route of administration *Remicade* is administered by intravenous infusion over 2 hours.
   Recommended dosage and monitoring requirements The recommended dose of *Remicade* for reducing signs and symptoms of moderate to severe Crohn's disease

activity is 5 mg/kg. In patients responding to an initial infusion of *Remicade*, up to four infusions given at 8-week intervals may be administered to sustain clinical benefit. The recommended dose for closure of enterocutaneous fistulae is also 5 mg/kg; two additional 5 mg/kg doses should be administered at 2 and 6 weeks following the first infusion. The recommended initial dose of *Remicade* for the treatment of RA is 3 mg/kg body weight, given intravenously over a period of no less than two hours. Additional 3 mg/kg infusions are recommended at 2 and 6 weeks after the first dose, followed by additional 3 mg/kg doses every 8 weeks. *Remicade* should be given in combination with methotrexate.

D. Pharmacology and pharmaceutics
   Clinical pharmacology Clinical, animal, and genetic evidence indicates that TNFα plays a role in the inflammatory process seen in patients with Crohn's disease. Infliximab neutralizes the biological activity of TNFα by binding with high affinity to the soluble and transmembrane forms of the inflammatory cytokine and inhibits binding of TNFα with its receptors. Biological activities attributed to TNFα include induction of pro-inflammatory cytokines, enhancement of leukocyte migration, activation of neutrophil and eosinophil functional activity, induction of acute phase reactants and other liver proteins, as well as tissue degrading enzymes. Anti-TNFα antibodies reduce disease activity in a colitis animal model, and decrease synovitis and joint erosions in a murine model of collagen-induced arthritis. Infliximab prevents disease in transgenic mice that develop polyarthritis as a result of constitutive expression of human TNFα, and, when administered after disease onset, allows eroded joints to heal.
   Pharmacokinetics Following single doses of *Remicade* up to 20 mg/kg, maximum drug concentration and area under the curve (AUC) were proportional

to dose, and no dose effect was observed on either total body clearance or volume of distribution at steady state. Infliximab was distributed primarily within the vascular compartment. No pharmacokinetic differences were found between *Remicade* responders and nonresponders. A terminal half-life of 8.0 to 9.5 days has been reported after a single infusion of *Remicade* 5 mg/kg. No important differences in clearance or volume of distribution were observed in patient subgroups defined by age or weight. It is not known if there are differences in clearance or volume of distribution between gender subgroups or in patients with marked impairment of hepatic or renal function.

Disposition   No information available.

Drug interactions   Although no data are available on the response to vaccination or on the secondary transmission of infection by live vaccines in patients receiving *Remicade* therapy, it is recommended that live vaccines not be given concurrently. *Remicade's* label includes the following; "Specific drug interaction studies, including interactions with methotrexate, have not been conducted. The majority of patients in rheumatoid arthritis or Crohn's disease clinical trials received one or more concomitant medications. In rheumatoid arthritis, concomitant medications besides methotrexate were nonsteroidal anti-inflammatory agents, folic acid, corticosteroids and/or narcotics.

E. Therapeutic response   The safety and efficacy of a single intravenous dose of *Remicade* were assessed in a randomized, double-blind, placebo-controlled study of patients with moderate to severe active Crohn's disease who had failed standard therapy. The primary end point was the proportion of patients who experienced a clinical response, defined as a minimum decrease in the Crohn's Disease Activity Index from baseline at the 4-week evaluation and without an increase in Crohn's

disease medications or surgery for Crohn's disease. Patients who responded at week 4 were followed to week 12. At week 4, 16% of the placebo patients achieved a clinical response compared with 82% of the patients receiving *Remicade* 5 mg/kg. The proportion of patients responding gradually diminished over the 12 weeks of the evaluation period. In a controlled study of patients with fistulizing Crohn's disease, a clinical response was observed in 26% of patients in the placebo arm and in 68% of patients in the *Remicade* 5 mg/kg arm.

The safety and efficacy of *Remicade* when given in conjunction with methotrexate (MTX) were assessed in a multicenter, randomized, double-blind, placebo-controlled study of 428 patients with active rheumatoid arthritis despite treatment with MTX. All patients were to have received MTX for ≥ 6 months and be on a stable dose ≥ 12.5 mg/week for 4 weeks prior to study. All *Remicade* and placebo groups continued their stable dose of MTX and folic acid. In addition to MTX, patients received placebo or *Remicade* by intravenous infusion at weeks 0, 2, and 6 followed by additional infusions every 4 or 8 weeks thereafter. The primary end point was the proportion of patients at week 30 who attained an improvement in signs and symptoms as measured by the American College of Rheumatology criteria (ACR 20). An ACR 20 response is defined as at least a 20% improvement in both tender and swollen joint counts and in 3 of 5 clinical criteria. At week 30, 43/86 (50%) of patients treated every 8 weeks with 3 mg/kg of *Remicade* plus MTX attained an ACR 20 compared with 18/88 (20%) of patients treated with placebo plus MTX ($p < 0.001$).

F. Role in therapy   Medical management of Crohn's disease is largely empirical and is designed to reduce inflammation. Therapies mainly consist of corticosteroids, antibiotics, aminosalicylates, and immunomodulatory agents. *Remicade* is the first approved treatment for Crohn's disease. Although not a cure, in the short-term, *Remicade* can have a dramatic impact on the quality of life of patients with severe forms of Crohn's disease. Improvement in patients who have taken *Remicade* is measured in terms of the number of liquid or soft stools, number and severity of abdominal cramps, and the overall sense of well-being. *Remicade* treatment also reduced the number of draining fistulas that occur in some cases of Crohn's disease.

In patients with rheumatoid arthritis, treatment with infliximab plus methotrexate was significantly more effective than placebo plus methotrexate in reducing signs and symptoms, as well as inhibiting the progression of structural (joint) damage; efficacy was sustained for up to 54 weeks. *Remicade* represents an important advance in the treatment of rheumatoid arthritis, with tolerability concerns raised by early studies having been eased somewhat by more recent data in larger patient numbers.

G. Other applications Infliximab was well tolerated and may provide clinical benefit for some patients with steroid-refractory ulcerative colitis.

H. Other considerations Infliximab has been designated an orphan product for use in the treatment of Crohn's disease.

ALEMTUZUMAB

Trade name: *Campath*

Manufacturer: Millennium and ILEX Partners, LP, Cambridge, MA

Distributed by: Berlex Laboratories, Richmond, CA

Indications: Treatment of patients with B-cell chronic lymphocytic leukemia who have been treated with alkylating agents and who have failed fludarabine therapy

Approval date: 7 May 2001

Type of submission: Biologics license application

A. General description Alemtuzumab is a recombinant DNA-derived humanized monoclonal antibody (Campath-1H) that is directed against the 21 to 28 kDa cell-surface glycoprotein, CD52. CD52 is expressed on the surface of normal and malignant B- and T-lymphocytes, NK cells, monocytes, macrophages, and tissues of the male reproductive system. The Campath-1H antibody is an IgG_1 kappa with human variable framework and constant regions, and complementarity-determining regions from a murine (rat) monoclonal antibody (Campath-1G). The Campath-1H antibody has an approximate molecular weight of 150 kDa. *Campath* is produced in mammalian cell (Chinese hamster ovary) suspension culture.

B. Indications and use *Campath* is indicated for the treatment of B-cell chronic lymphocytic leukemia (B-CLL) in patients who have been treated with alkylating agents and who have failed fludarabine therapy. Determination of the effectiveness of Campath is based on overall response rates. Comparative randomized trials demonstrating increased survival or clinical benefits such as improvement in disease-related symptoms have not yet been conducted.

C. Administration

Dosage form *Campath* is a sterile isotonic solution for injection. Each single use ampoule of Campath contains alemtuzumab 30 mg.

Route of administration *Campath* is administered by means of an intravenous infusion.

Recommended dosage and monitoring requirements *Campath* therapy should be initiated at a dose of 3 mg administered as a 2-hour intravenous infusion daily. When the *Campath* 3 mg daily dose is tolerated, the daily dose should be escalated to 10 mg and continued. When the 10 mg dose is tolerated, the maintenance dose of *Campath* 30 mg may be initiated. The maintenance dose of *Campath* is 30 mg/day administered three times per week on alternate days (i.e., Monday, Wednesday, and Friday) for up to 12 weeks. In most patients, escalation to 30 mg can be accomplished in 3 to 7 days. Premedication should be given prior to the first dose, at dose escalations, and as clinically indicated. The premedication used in clinical studies was diphenhydramine 50 mg and acetaminophen 650 mg administered 30 minutes prior to *Campath* infusion. Patients should receive anti-infective prophylaxis to minimize the risks of serious opportunistic infections.

D. Pharmacology and pharmaceutics

Clinical pharmacology Alemtuzumab binds to CD52, a nonmodulating antigen that is present on the surface of essentially all B- and T-lymphocytes, a majority of monocytes, macrophages, and NK cells, and a subpopulation of granulocytes. There does not appear to be CD52 expression on erythrocytes or hematopoietic stem cells. Alemtuzumab's proposed mechanism of action is antibody-dependent lysis of leukemic cells following cell surface binding.

Pharmacokinetics The pharmacokinetic profile of alemtuzumab was studied in a rising-dose trial in non-Hodgkin's lymphoma (NHL) and chronic lymphocytic leukemia (CLL). Campath was administered once weekly for a maximum of 12 weeks. Following intravenous infusions over a range of doses, the maximum serum concentration (C_{max}) and the area under the curve (AUC) showed relative dose pro-

portionality. The overall average half-life over the dosing interval was about 12 days. The pharmacokinetic profile of *Campath* administered as a 30 mg intravenous infusion three times per week was evaluated in CLL patients. Peak and trough levels of *Campath* rose during the first few weeks of treatment, and appeared to approach steady state by approximately week 6, although there was marked interpatient variability. The rise in serum *Campath* concentration corresponded with the reduction in malignant lymphocytosis.

Disposition No information available.

Drug interactions No information available.

E. Therapeutic response
The safety and efficacy of *Campath* were evaluated in an open-label study of 93 patients with B-cell chronic lymphocytic leukemia (B-CLL). A partial response was observed in 31% of patients and a complete response in 2%. The median duration of response was 7 months.

F. Role in therapy
No alternative therapy is available.

GEMTUZUMAB OZOGAMICIN

Trade name: *Mylotarg*

Manufacturer: Wyeth-Ayerst Laboratories, Philadelphia, PA

Indications: Treatment of patients with CD33 positive acute myeloid leukemia (AML)

Approval date: 1 June 2000

Type of submission: New drug application

A. General description
Gemtuzumab ozogamicin is composed of a recombinant humanized IgG$_4$, kappa antibody conjugated with a cytotoxic antitumor antibiotic, calicheamicin. The anti-CD33 antibody is produced by mammalian cell suspension

culture using a myeloma NSO cell line and is purified under conditions that remove or inactivate viruses. The constant region and framework regions contain human sequences while the complementarity-determining regions are derived from a murine antibody that binds CD33. This antibody is linked to N-acetyl-gamma calicheamicin via a bifunctional linker. Gemtuzumab ozogamicin has approximately 50% of the antibody loaded with 4 to 6 moles calicheamicin per mole of antibody. The remaining 50% of the antibody is not linked to the calicheamicin derivative. Gemtuzumab ozogamicin has a molecular weight of 151 to 153 kDa.

B. Indications and use *Mylotarg* is indicated for the treatment of patients with CD33 positive AML in first relapse who are 60 years of age or older and who are not considered candidates for cytotoxic chemotherapy.

C. Administration

Dosage form *Mylotarg* is a sterile, preservative-free lyophilized powder containing 5 mg of drug conjugate in a 20 ml vial. The contents of the vial are reconstituted with sterile water for injection.

Route of administration On reconstitution, *Mylotarg* is administered as an intravenous infusion.

Recommended dosage and monitoring requirements In patients with AML in first untreated relapse, two 2-hour intravenous doses of 9 mg/m^2, separated by a 14-day interval, are recommended. Patients should receive diphenhydramine and acetaminophen one hour before *Mylotarg* administration.

D. Pharmacology and pharmaceutics

Clinical pharmacology The antibody portion of *Mylotarg* binds specifically to the CD33 antigen, a sialic acid-dependent adhesion protein found on the surface of leukemic blasts in more than 80% of patients with AML, as well as on the surface of immature normal cells of myelomonocytic lineage, but not on normal hematopoietic stem cells. Binding of the anti-CD33 antibody portion of *Mylotarg* with the CD33 antigen results in the formation of a complex that is internalized. Upon internalization, the calicheamicin derivative is released inside the lysosomes of the myeloid cell. The released calicheamicin derivative binds to DNA in the minor groove resulting in DNA double strand breaks and cell death. *In vitro*, gemtuzumab ozogamicin demonstrated specific cytotoxicity against CD33+ leukemia cells in colony-forming assays.

Pharmacokinetics According to the product label, after administration of the first recommended dose of gemtuzumab ozogamicin, the elimination half-lives of total and unconjugated calicheamicin were about 45 and 100 hours, respectively. After the second dose, the half-life of total calicheamicin was increased to about 60 hours and the area under the curve (AUC) was about twice that following the first dose. The pharmacokinetics of unconjugated calicheamicin did not appear to change from dose one to dose two.

Disposition Metabolic studies indicate hydrolytic release of the calicheamicin derivative from gemtuzumab ozogamicin. Many metabolites of this derivative were found after *in vitro* incubation of gemtuzumab ozogamicin in human liver microsomes and cytosol, and in HL-60 promyelocytic leukemia cells. Metabolic studies characterizing the possible isozymes involved in the metabolic pathway of *Mylotarg* have not been performed.

Drug interactions There have been no formal drug-interaction studies performed with *Mylotarg*.

E. Therapeutic response The effectiveness of *Mylotarg* is based on overall response rates. There are no controlled trials demonstrating a clinical benefit, such as improvement in disease-related symptoms or increased survival, compared with any other treatment. The overall response rate for three pooled monotherapy studies in patients with CD33 positive AML in first relapse was 29%, consisting of 16% of patients with complete remission (CR), and 13% of patients with complete remission with the exception of platelet recovery ≥100,000/μl (CRp). The median time to remission was 60 days for both CR and CRp.

F. Role in therapy According to Micromedex, single-agent therapy with gemtuzumab ozogamicin has produced remission in some patients with relapsed or refractory acute myelogenous leukemia. The selective ablation of leukemic cells with this agent is an advantage over conventional chemotherapy.

G. Other considerations Gemtuzumab ozogamicin has been designated an orphan product for use in the treatment of CD33-positive AML.

RITUXIMAB

Trade name: **_Rituxan_**

Manufacturer: Genentech, Inc., South San Francisco, CA

Indications: Treatment of patients with relapsed or refractory low-grade or follicular, B-cell non-Hodgkin's lymphoma (NHL)

Approval date: 26 November 1997

Type of submission: Biologics license application

A. General description Rituximab is a genetically engineered murine/human chimeric monoclonal antibody directed against the CD20 antigen found on the surface of normal and malignant B-lymphocytes. The antibody is an IgG 1 kappa immunoglobulin containing murine light- and heavy-chain variable region sequences (F_{ab} domain) and human constant region sequences (F_c domain). Rituximab is composed of two heavy chains of 451 amino acids and two light chains of 213 amino acids and has a molecular weight of about 145 kDa. The chimeric anti-CD20 antibody is produced by mammalian cell (Chinese hamster ovary) suspension culture.

B. Indications and use _Rituxan_ is the first FDA-approved monoclonal antibody for the treatment of cancer. Several clinical trials have demonstrated efficacy of this chimeric antibody against NHL. _Rituxan_ is indicated for the treatment of patients with relapsed or refractory low-grade or follicular B-cell non-Hodgkin's lymphoma (indolent lymphoma).

C. Administration

Dosage form _Rituxan_ is a sterile, preservative-free liquid concentrate for injection. It is supplied at a concentration of 10 mg/ml in either 100 mg (10 ml) or 500 mg (50 ml) single-use vials.

Route of administration _Rituxan_ is administered as an intravenous infusion.

Recommended dosage and monitoring requirements The recommended dose of _Rituxan_ is 375 mg/m^2 given as an intravenous infusion once weekly for four doses (days 1, 8, 15, and 22). _Rituxan_ may be administered in an outpatient setting. It should not be administered as an intravenous push or bolus. The rituximab solution for infusion should be administered intravenously at an initial rate of 50 mg/h. If hypersensitivity or infusion-related events do not occur, escalate the infusion rate in 50 mg/h increments every 30 minutes, to a maximum of 400 mg/h. To minimize toxicity and wastage of medicine, it has been recommended to initiate treatment with a small dose on day one with prophylactic acetaminophen, diphenhydramine, allopurinol, and hydration. After treatment, patients should be monitored for thrombocytopenia and evidence of rapid tumor lysis. High pretreatment levels of human antimurine antibody (HAMA) may indicate risk of more severe infusion reactions; pretreatment with antihistamines may be useful. Patients who develop clinically significant cardiopulmonary events should have *Rituxan* discontinued and receive medical treatment.

D. Pharmacology and pharmaceutics

Clinical pharmacology Rituximab binds specifically to the antigen CD20, a hydrophobic transmembrane protein with a molecular weight of approximately 35 kDa located on pre-B- and mature B-lymphocytes. The antigen is also expressed on >90% of B-cell non-Hodgkin's lymphomas, but is not found on hematopoietic stem cells, pro-B cells, normal plasma cells, or other normal tissues. CD20 regulates an early step(s) in the activation process for cell cycle initiation and differentiation. *In vitro*, the F_{ab} domain of rituximab binds to the CD20 antigen on B-lymphocytes, and the F_c domain recruits immune effector functions to mediate B-cell lysis. Possible mechanisms of cell lysis include complement-dependent cytotoxicity and antibody-dependent cell mediated cytotoxicity.

Pharmacokinetics The half-life of rituximab has been reported to be about 60 hours (range: 11–105 hours) after the first infusion of *Rituxan* and 174 hours (range: 26–442 hours) after the fourth infusion. After administration, rituximab is taken up by B-lymphocytes and then degraded throughout the body by proteolysis. There is no appreciable excretion of this polypeptide. The wide range of half-lives may reflect the variable tumor burden among patients and the changes in CD20 positive (normal and malignant) B-cell populations upon repeated administrations. When *Rituxan* was administered at weekly intervals for four doses to NHL patients, the peak and trough serum levels of rituximab were inversely correlated with baseline values for the number of circulating CD20 positive B cells and measures of disease burden. Median steady-state serum levels were higher for responders compared to nonresponders. Rituximab was detectable in the serum of patients 3 to 6 months after completion of treatment.

Disposition Rituximab is degraded throughout the body by proteolysis.

Drug interactions Live virus and live bacteria vaccines should not be administered to a patient receiving an immunosuppressive chemotherapeutic agent. At least three months should elapse between the discontinuation of chemotherapy and vaccination with a live vaccine.

E. Therapeutic response A multicenter, open-label, single-arm study was conducted in patients with relapsed or refractory low-grade or follicular B-cell NHL who received 375 mg/m² of *Rituxan* given as an intravenous infusion, weekly for four doses. The overall response rate (ORR) was 48% with a 6% complete response and a 42% partial response rate. In a second multicenter, multiple-dose study in patients with relapsed or refractory B-cell NHL, the ORR was 46%.

F. Role in therapy *Rituxan* is recommended for low-grade or follicular B-cell lymphoma in patients relapsing after or refractory to conventional chemotherapy. Complete responses are infrequent but partial responses appear durable.

G. Other applications According to Micromedex, further potential roles for *Rituxan* include combination with chemotherapy in low-grade or intermediate-grade disease, first-line therapy of B-cell lymphoma, and use in other malignancies with CD20-antigen expression (e.g., chronic lymphocytic leukemia). Treatment with rituximab for post-transplant lymphoproliferative disease associated with Epstein-Barr virus has been effective and well tolerated.

H. Other considerations *Rituxan* has been granted orphan product status for use in the treatment of non-Hodgkin's B-cell lymphoma.

TRASTUZUMAB

Trade name: *Herceptin*

Manufacturer: Genentech, Inc., South San Francisco, CA

Indications: For treatment of patients with metastatic breast cancer whose tumors overexpress the HER2 protein and who have received one or more chemotherapy regimens for their metastatic disease. Trastuzumab in combination with paclitaxel is also indicated for treatment of patients with metastatic breast cancer whose tumors overexpress HER2 protein and who have not received chemotherapy for their metastatic disease

Approval date: 25 September 1998

Type of submission: Biologics license application

A. General description Trastuzumab is a recombinant DNA-derived humanized monoclonal antibody that selectively targets HER2/neu, the extracellular domain of the human epidermal growth factor (EGF) receptor 2 protein. It is engineered by grafting the complementarity-determining regions (CDR) of the parent murine antibody into the consensus framework of a human IgGl. Trastuzumab is produced from Chinese hamster ovary cells maintained in cell-culture systems at Genentech.

B. Indications and use *Herceptin* as a single agent is indicated for the treatment

of patients with metastatic breast cancer whose tumors overexpress the HER2 protein and who have received one or more chemotherapy regimens for their metastatic disease. *Herceptin* in combination with paclitaxel (*Taxol*) is indicated for treatment of patients with metastatic breast cancer whose tumors overexpress the HER2 protein and who have not received chemotherapy for their metastatic disease. *Herceptin* should only be used in patients whose tumors have HER2 protein overexpression.

C. Administration

Dosage form *Herceptin* is a sterile, lyophilized powder for injection after reconstitution with bacteriostatic water for injection. Each vial of *Herceptin* contains trastuzumab 440 mg.

Route of administration Trastuzumab is administered intravenously.

Recommended dosage and monitoring requirements In patients with HER2-overexpressing metastatic breast carcinoma, an initial loading dose of 4 mg/kg, given intravenously over 90 minutes, followed by weekly infusions of 2 mg/kg over 30 minutes, has been administered either alone or combined with chemotherapy. *Herceptin* should not be given as an intravenous push or bolus. Left ventricular function should be evaluated in all patients prior to and during treatment with *Herceptin*.

D. Pharmacology and pharmaceutics

Clinical pharmacology Approximately 25% to 30% of breast and ovarian cancers overexpress HER2/neu. There is ample preclinical evidence to indicate that HER2/neu is oncogenic. HER2/neu overexpression has been correlated with poor clinical outcome: shorter disease-free survival, shorter overall survival, more rapid disease progression (higher incidence of metastasis), and resistance to chemotherapy. Trastuzumab, as a mediator of antibody-dependent cellular cytotoxicity has been shown, in both *in vitro* assays and in animals, to inhibit the proliferation of human tumor cells that overexpress HER2.

Pharmacokinetics From data used to perform a pharmacokinetics simulation analysis, a weekly dose of 100 mg was selected for study and found to provide trough levels within the serum levels thought to be efficacious (10–20 pg/ml), based on preclinical studies. A loading dose of 250 mg was added to the dosing regimen to attain the target levels more quickly. Experience in early clinical development suggested that rather than administer a fixed dose of 250 and 100 mg, a body weight adjusted dose of 4 mg/kg as a loading dose and 2 mg/kg as a maintenance dose would improve the consistency of response. Furthermore a minimum target trough level of 20 pg/ml was selected as the lower limit of serum levels to be maintained upon repeated dosing. Various factors were found to modify the pharmacokinetics of trastuzumab including dose and shed antigen. Patients with higher baseline shed antigen levels were more likely to have lower serum trough concentrations. However, with weekly dosing, most patients with elevated shed antigen levels achieved target serum concentrations of trastuzumab by week 6. Early studies demonstrated that trastuzumab clearance decreased with increasing dose. Concomitantly, half-life ($t_{1/2}$) increased with increasing dose. The half-life averaged 1.7 and 12 days at dose levels of 10 and 500 mg, respectively. The volume of distribution remained basically unchanged, about 44 ml/kg, with increases in dose. In studies using a loading dose of 4 mg/kg followed by a weekly maintenance dose of 2 mg/kg, a mean half-life of 5.8 days (range: 1 to 32 days) was observed. Data suggest that the disposition of trastuzumab is not altered based on age or serum creatinine (up to 2.0 mg/deci liter).

Disposition No information available.

Drug interactions No formal clinical drug–drug interaction studies were conducted to investigate co-administration of *Herceptin* with cisplatin, doxorubicin or epirubicin plus cyclophosphamide, or paclitaxel. A comparison of serum levels of trastuzumab given in combination with various chemotherapeutic agents did not suggest the possibility of any pharmacokinetic interactions except in combination with paclitaxel. Although not statistically significant, mean serum trough concentrations of trastuzumab were consistently elevated, about 1.5-fold, when *Herceptin* was administered in combination with paclitaxel. However, trastuzumab and paclitaxel were used concurrently in clinical trials with positive outcome results. The concurrent administration of anthracyclines, cyclophosphamide, and trastuzumab increased the incidence and severity of cardiac dysfunction during clinical trials.

E. Therapeutic response Compared with patients randomized to chemotherapy alone, patients with metastatic breast cancer who had not been previously treated with chemotherapy for metastatic disease randomized to *Herceptin* and chemotherapy experienced a significantly longer time to disease progression. They also experienced a higher overall response rate (ORR), a longer median duration of response, and a higher one-year survival rate. Treatment effects were observed both in patients who received *Herceptin* plus paclitaxel and in those who received *Herceptin* with anthracycline plus cyclophosphamide; however, the magnitude of the effects was greater in the paclitaxel subgroup. In patients with HER2-overexpressing metastatic breast cancer who had relapsed following one or two prior chemotherapy regimens for metastatic disease, *Herceptin* as a single agent provided an ORR (complete plus partial response) of 14%.

F. Role in therapy In untreated HER2-overexpressing metastatic breast cancer, the addition of *Herceptin* to paclitaxel increases response rates and time to progression; it should be considered in selected patients. The degree of HER2 overexpression appears to be a predictor of treatment effect. A higher response rate in untreated patients is also seen when the antibody is added to an anthracycline plus cyclophosphamide regimen, although benefits over the anthracycline regimen alone are only modest, and the risk of cardiotoxicity is significant. *Herceptin* alone may be a useful alternative in pretreated patients unable to tolerate further chemotherapy; typical toxicities of chemotherapy (e.g., neutropenia and stomatitis) have not been observed; however, cardiac dysfunction can occur even with single-agent therapy.

G. Other applications *Herceptin* has been combined with cisplatin in the treatment of heavily pretreated metastatic breast cancer. Treatment of patients with ovarian cancer is under investigation. A recent study demonstrated that *Herceptin* increased the clinical benefits of first-line chemotherapy—doxorubicin (or epirubicin) and cyclophosphamide or paclitaxel—in metastatic breast cancer that overexpressed HER2.

PALIVIZUMAB

Trade name: *Synagis*

Manufacturer: MedImmune, Inc., Gaithersburg, MD

Co-Marketed by: Ross Products Division, Abbott Laboratories, Inc., Columbus, Ohio

Indications: Prophylaxis of serious lower respiratory tract disease, caused by respiratory syncytial virus (RSV), in pediatric patients at high risk of RSV disease

Approval date: 19 June 1998

Type of submission: Biologics license application

A. General description According to the label, palivizumab is a humanized monoclonal antibody (IgG₁) produced by recombinant DNA technology, directed to an epitope in the A antigenic site of the F protein of respiratory syncytial virus (RSV). Palivizumab is a composite of human (95%) and murine (5%) antibody sequences. The human heavy chain sequence was derived from the constant domains of human IgG_1 and the variable framework regions of the *V H* genes *Cor* and *Cess*. The human light chain sequence was derived from the constant domain of *C[kgr]* and the variable framework regions of the *V L* gene *K104* with *J[kgr]–4*. The murine sequences were derived from a murine monoclonal antibody, in a process that involved the grafting of the murine complementarity-determining regions into the human antibody framework. Palivizumab is composed of two heavy chains and two light chains and has a molecular weight of approximately 148 kDa.

B. Indications and use *Synagis* is indicated for the prevention of serious lower respiratory tract disease caused by RSV in infants and children with bronchopulmonary dysplasia (BPD) or a history of premature birth (< 35 weeks gestation) who are under 24 months of age at the time of first administration.

C. Administration

Dosage form *Synagis* is supplied in single-use vials as lyophilized powder to deliver either 50 or 100 mg when reconstituted with sterile water for injection.

Route of administration Reconstituted *Synagis* is to be administered by intramuscular injection only.

Recommended dosage and monitoring requirements The recommended dose of *Synagis* is 15 mg/kg of body weight intramuscularly. Patients, including those who develop an RSV infection, should receive monthly doses throughout the RSV season.

The first dose should be administered prior to commencement of the season. In the Northern Hemisphere the RSV season typically begins in November and lasts through April, but it may begin earlier or persist later in certain communities.

D. Pharmacology and pharmaceutics

Clinical pharmacology RSV, an enveloped RNA virus of the paramyxovirus family, is a common human respiratory pathogen and is the major cause of respiratory tract illness in children. Infants at highest risk for severe RSV infections include those with premature birth, BPD, and congenital heart disease. *Synagis* exhibits neutralizing and fusion-inhibitory activity against RSV. A study of the *in vivo* neutralizing activity of palivizumab determined that the monoclonal antibody reduced the quantity of RSV in the lower respiratory tract.

Pharmacokinetics Serum levels considered therapeutic ($40 \mu g/ml$ or greater) are achieved in most high-risk infants on intramuscular administration of the recommended dose. In studies in adult volunteers, palivizumab had a pharmacokinetic profile similar to a human IgG_1 antibody in regard to the volume of distribution and the half-life (mean 18 days). In pediatric patients less than 24 months of age, the mean half-life of palivizumab was 20 days.

Disposition No information available.

Drug interactions No formal drug–drug interaction studies have been conducted with *Synagis*. In a large clinical trial the proportions of patients in the placebo and *Synagis* groups who received routine childhood vaccines, influenza vaccine, bronchodilators, or corticosteroids were similar, and no difference in adverse reactions was observed between the treatment and placebo groups among patients receiving these agents.

E. Therapeutic response
The safety and efficacy of *Synagis* were assessed in a placebo-controlled study of RSV disease prophylaxis among high-risk pediatric patients. RSV hospitalizations occurred among 10.6% of patients in the placebo group and 4.8% of patients in the *Synagis* group, a 55% reduction. The reduction of RSV hospitalization was observed both in patients enrolled with a diagnosis of BPD and patients enrolled with a diagnosis of prematurity without BPD. The reduction of RSV hospitalization was observed throughout the course of the RSV season.

F. Role in therapy
According to Micromedex, *Synagis* is the second product licensed for RSV disease. The first one, *RespiGam* (RSV immune globulin) is made from human plasma. Both are approved for use in high-risk infants less than 2 years old with lung problems or prematurity. Both products are effective in reducing the number of hospitalizations due to RSV. Although both products must be given in five monthly doses, palivizumab is given intramuscularly, rather than by intravenous infusion over a period of hours. In addition palivizumab is more concentrated than *RespiGam*, an advantage since infants with certain pulmonary diseases may retain excess fluids. Ribavirin aerosol remains the agent of choice by many pediatricians for acute treatment of children, although the efficacy and role of ribavirin in infants with RSV disease remains controversial.

G. Other applications
Intravenous palivizumab has been used for the acute treatment of RSV-infected patients, but treatment in this setting is not yet a labeled indication.

ABCIXIMAB

Trade name: ***ReoPro***

Manufacturer: Centocor B.V., Leiden, The Netherlands

Distributed by: Eli Lilly and Company, Indianapolis, IN

Indications: Adjunct to percutaneous coronary intervention for the prevention of cardiac ischemic complications

Approval date: 1 February 1998

Type of submission: Biologic license application

A. General description
Abciximab is the F_{ab} fragment of the chimeric human-murine monoclonal antibody 7E3. It binds to the glycoprotein (gp) IIb/IIIa receptor of human platelets and inhibits platelet aggregation. The chimeric antibody is produced by continuous perfusion in mammalian cell-culture. The 48 kDa F_{ab} fragment is purified from cell-culture supernatant.

B. Indications and use
ReoPro is indicated as an adjunct to percutaneous coronary intervention for the prevention of cardiac ischemic complications in patients undergoing percutaneous coronary intervention or patients with unstable angina not responding to conventional medical therapy when percutaneous coronary intervention is planned within 24 hours.

C. Administration
Dosage form *ReoPro* is a sterile, preservative-free, buffered solution for intravenous use. Each single-use vial contains abciximab 2 mg/ml in water for injection.

Route of administration *ReoPro* is administered as an intravenous bolus, followed by a continuous intravenous infusion.

Recommended dosage and monitoring requirements The recommended dosage of *ReoPro* is an intravenous bolus of 0.25 mg/kg administered 10 to 60 minutes before the start of percutaneous transluminal coronary angioplasty (PTCA), followed by a continuous infusion of 0.125 µg/kg per

minute (to a maximum of 10 µg/min) for 12 hours. Patients with unstable angina may be treated with a *ReoPro* 0.25 mg/kg intravenous bolus followed by an 18 to 24 hour intravenous infusion of 10 µg/min, concluding one hour after the percutaneous coronary intervention. *ReoPro* is intended for use with aspirin and heparin and has been studied only in that setting.

D. Pharmacology and pharmaceutics
Clinical pharmacology Abciximab binds to the intact platelet gp IIb/IIIa receptor, which is a member of the integrin family of adhesion receptors and the major platelet surface receptor involved in platelet aggregation. Abciximab inhibits platelet aggregation by preventing the binding of fibrinogen, von Willebrand factor, and other adhesive molecules to gp IIb/IIIa receptor sites on activated platelets. Abciximab binds with similar affinity to the vitronectin receptor, which mediates the procoagulant properties of platelets and the proliferative properties of vascular endothelial and smooth muscle cells.

Pharmacokinetics Abciximab has an initial half-life of about 10 minutes and a second phase half-life of about 30 minutes, probably related to rapid binding to the platelet gp IIb/IIIa receptors. Although abciximab remains in the circulation for up to 10 days in a platelet-bound state, platelet function generally recovers over the course of 48 hours.

Disposition No information available.

Drug interactions Concurrent administration of *ReoPro* with oral anticoagulants is contraindicated. Limited experience with the concurrent use of *ReoPro* and low molecular weight dextran during a PTCA procedure has also shown an increased risk of a major bleeding episode. Concurrent administration of *ReoPro* and heparin may also increase the risk of bleeding, but *ReoPro* is intended for use with aspirin and heparin and has only been studied in that setting. *ReoPro* has been administered safely to patients with ischemic heart disease treated concomitantly with a broad range of medications used in the treatment of angina, myocardial infarction, and hypertension. These medications have included beta-blockers, calcium antagonists, angiotensin converting enzyme inhibitors, and intravenous and oral nitrates.

E. Therapeutic response *ReoPro* has been studied in placebo-controlled trials that evaluated its effects in patients undergoing percutaneous coronary intervention. The primary end point was the occurrence of any of the following events within 30 days of percutaneous coronary intervention: death, myocardial infarction (MI), or the need for urgent intervention for recurrent ischemia. A lower incidence of the primary end point was observed in the *ReoPro* (bolus plus infusion) arm in all three studies.

F. Role in therapy *ReoPro* has demonstrated unequivocal efficacy in reducing ischemic complications associated with high-risk PTCA. However, mortality rates have not been changed, and use of *ReoPro* may not reduce the need for emergency coronary artery bypass grafting in these patients. In addition, bleeding complications and transfusion requirements are increased with the usual regimen of *ReoPro*. Despite these disadvantages, the potential of reducing myocardial infarction and repeat angioplasty often favors use of abciximab in this setting.

G. Other applications In patients with acute myocardial infarction, coronary stenting plus *ReoPro* leads to a greater degree of myocardial salvage and a better clinical outcome than does fibrinolysis with a tissue plasminogen activator. As compared with placebo, early administration of *ReoPro* in patients with acute myocardial infarction improves coronary patency before stenting, the success rate of the

stenting procedure, the rate of coronary patency at 6 months, left ventricular function, and clinical outcomes.

IBRITUMOMAB TIUXETAN

Trade name: *Zevalin*

Manufacturer: IDEC Pharmaceuticals Corp., San Diego, CA

Indications: Treatment of patients with B-cell non-Hodgkin's lymphoma

Approval date: 19 February 2002

Type of submission: Biologics license application

A. General description Ibritumomab tiuxetan is the immunoconjugate resulting from a stable thiourea covalent bond between the monoclonal antibody ibritumomab and the link-chelator tiuxetan. This link-chelator provides a high-affinity, conformationally restricted chelation site for indium-111 or yttrium-90. Ibritumomab, the monoclonal antibody moiety, is a murine kappa IgG$_1$ monoclonal antibody directed against the CD20 antigen, which is found on the surface of normal and malignant B-lymphocytes. Ibritumomab is produced in Chinese hamster ovary cells and is composed of two murine gamma 1 heavy chains of 445 amino acids each and two kappa light chains of 213 amino acids each.

B. Indications and use *Zevalin*, as part of the *Zevalin* therapeutic regimen, is indicated for the treatment of patients with relapsed or refractory low-grade, follicular, or transformed B-cell non-Hodgkin's lymphoma, including patients with rituximab refractory follicular non-Hodgkin's lymphoma.

C. Administration
Dosage form *Zevalin* kits provide for the radiolabeling of ibritumomab tiuxetan with In-111 and with Y-90. Each of the kits

for the preparation of a single dose of In-111 Zevalin and a single dose of Y-90 Zevalin includes four vials: one containing ibritumomab tiuxetan 3.2 mg in normal saline, one containing a sodium acetate solution, one containing formulation buffer, and one empty reaction vial. The contents of all vials are sterile. The indium-111 chloride sterile solution must be ordered separately.

Route of administration In-111 Zevalin and Y-90 Zevalin are administered by means of intravenous injections.

Recommended dosage and monitoring requirements The Zevalin therapeutic regimen is administered in two steps. Step 1 includes a single infusion of rituximab preceding a fixed dose of In-111 Zevalin administered as an intravenous push. Step 2 follows step 1 by 7 to 9 days and consists of a second infusion of rituximab prior to a fixed dose of Y-90 Zevalin administered as an intravenous push. The dose of rituximab is lower when used as part of the Zevalin therapeutic regimen, as compared with the dose of rituximab when used as a single agent. The dose of Y-90 Zevalin should be reduced in patients with mild thrombocytopenia. Y-90 Zevalin should not be given to patients with altered biodistribution of In-111 Zevalin. Patients who develop severe infusion reactions should have therapy discontinued immediately and receive medical treatment. Careful monitoring for and management of cytopenias and their complications for up to 3 months after treatment with the Zevalin therapeutic regimen are necessary. The Zevalin therapeutic regimen is intended as a single-course treatment.

D. Pharmacology and pharmaceutics

Clinical pharmacology Ibritumomab tiuxetan binds specifically to the CD20 antigen with an apparent affinity (K_d) of 14 to 18 nM. The antigen is expressed on pre-B- and mature B-lymphocytes and on greater than 90% of B-cell non-Hodgkin's lymphomas. The CD20 antigen is not shed from the cell surface and does not internalize upon antibody binding. The complementarity-determining regions of ibritumomab bind to the antigen on B-lymphocytes and induce apoptosis in CD20+ B-cell lines in vitro. The chelate, tiuxetan, which binds In-111 or Y-90, is covalently linked to the amino group of exposed lysines and arginines contained within the antibody. The beta emission from Y-90 induces cellular damage by the formation of free radicals in the target and neighboring cells. Ibritumomab tiuxetan binding occurs on lymphoid cells throughout the body but not on nonlymphoid tissues or gonadal tissues.

Pharmacokinetics In patients receiving the Zevalin therapeutic regimen, the mean effective half-life of Y-90 activity in blood was 30 hours.

Disposition No information available.

Drug interactions No formal drug interaction studies have been performed with Zevalin.

E. Therapeutic response The safety and efficacy of the Zevalin therapeutic regimen were evaluated in two trials enrolling a total of 197 subjects. Study 1 was a single-arm study of 54 patients with relapsed follicular lymphoma refractory to treatment with rituximab. The primary efficacy end point, the overall response rate, was 74%, and the complete response rate was 15%. In a secondary analysis comparing response to the Zevalin therapeutic regimen with that observed in the most recent treatment with rituximab, the overall response rate, the complete response rate, and the median duration of response were greater for patients treated with Zevalin. Study 2 was a well-controlled trial comparing the Zevalin therapeutic regimen with rituximab in 143 patients with relapsed or refractory non-Hodgkin's lymphoma. The overall response rate was significantly greater (80% vs. 56%) for patients treated with *Zevalin*. In clinical studies, administration of *Zevalin* therapeutic regimen resulted in sustained depletion of circulating B cells, and at 4 weeks the median number of circulating B cells was zero. B-cell recovery began at approximately 12 weeks following treatment, and the median level of B-cells was within the normal range within 9 months after treatment.

F. Role in therapy Radioimmunotherapy is a promising new area of cancer treatment that combines the targeting power of monoclonal antibodies with the cell-damaging ability of localized radiation. When infused into a patient, these radiation-carrying antibodies circulate in the body until they locate and bind to the surface of specific cells, and then deliver their cytotoxic radiation directly to malignant cells. *Zevalin* is the first treatment for cancer that includes a monoclonal antibody that is combined with a radioactive chemical. It is indicated for the treatment of a type of non-Hodgkin's lymphoma (NHL). *Zevalin* must be used along with rituximab, an already approved biotechnology product for low-grade B-cell NHL. Both drugs are monoclonal antibodies that target white blood cells (B cells) including the malignant B cells involved in NHL. The *Zevalin* treatment regimen is more toxic than treatment with rituximab. More than half of the patients receiving *Zevalin* in clinical trials experienced serious reductions in white blood cells and platelets persisting for 3 to 4 weeks. Patients with low-grade or follicular NHL may remain in remission for years, but eventually they have relapses that occur more frequently over the course of the disease. As patients are treated for subsequent relapses, their responses to treatment diminish. *Zevalin* will provide another treatment option for these patients.

11

VACCINES

Shiu-Lok Hu & Rodney J.Y.Ho

■ 11.1. WHY VACCINES?

With the possible exception of clean water, vaccines were the single most important contributing factor to improvements in global health in the twentieth century. This was especially evident after the introduction of improved tissue culture techniques in the late 1940s, which enabled mass production of vaccines against various infectious diseases [1]. Epidemiological data illustrated a clear correlation between the introduction of mass vaccination programs and reduced incidence of once-common childhood diseases, such as diphtheria, pertussis, measles, mumps, and rubella (Figure 11.1). Smallpox, an endemic, highly contagious, and often fatal disease in many human populations for thousands of years, was eradicated in the late 1970s through the use of vaccines first developed by Edward Jenner two centuries ago [2]. At the dawn of the twenty-first century, we are poised to witness the global eradication of another human disease, poliomyelitis [3], thanks to the vaccines developed in the 1950s by Jonas Salk and Albert Sabin [4,5]. These achievements attest not only to the effectiveness of these particular vaccines but also to the importance of vaccines in general as part of the armamentarium of modern biopharmaceuticals.

Biotechnology and Biopharmaceuticals, by Rodney J. Y. Ho and Milo Gibaldi
ISBN 0-471-20690-3 Copyright © 2003 by John Wiley & Sons, Inc.

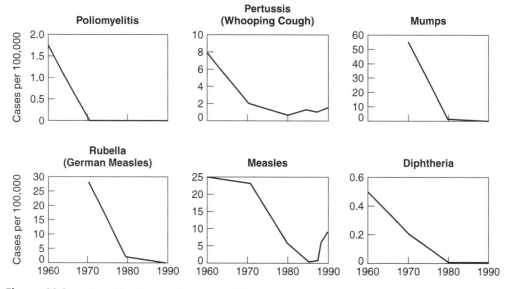

Figure 11.1. Reduced incidence of common childhood diseases due to the introduction of vaccines. (Adapted from Ann NY Acad Sci 1994, 729; 8–18).

■ 11.2. HOW DO VACCINES WORK?

It is not the focus of this chapter to discuss detailed mechanisms underlying vaccine protection. However, to appreciate the rationale and the strengths and weaknesses of different vaccination approaches, it is helpful to have a rudimentary understanding of the immune system and how it responds to infection and vaccination.

The first line of defense, launched within minutes after exposure to a pathogen or foreign antigen, generally involves relatively nonspecific immune responses. Pro-inflammatory cytokines (e.g., IL-1, TNF-α, and IFN-γ) and proteins of the complement-mediated cascade mediate these responses. A number of cell types, including natural killer cells, macrophages, and neutrophils, work in concert to achieve a partial clearing of the pathogen or pathogen-infected cells. This initial, non-antigen-specific, immune response is termed innate immunity.

An important consequence of this early immune response is the recruitment of

antigen-presenting cells (APC), such as dendritic cells, macrophages, and lymphocytes, to the site of infection. These cells play central roles in mounting antigen-specific or adaptive responses against the invading pathogen. Two types of immune response can be generated, one mediated by antibodies (humoral response) and one mediated by T-lymphocytes (cell-mediated response). Antibody responses are initiated by B-lymphocytes, which recognize specific epitopes on foreign antigens through interaction with immunoglobulin receptors on their cell surface.

Internalization, processing, and presentation of antigens through class II major histocompatibility complex (MHC II) to helper T cells leads to the activation and differentiation of B-lymphocytes, which, upon maturation, become plasma cells that secrete antigen-specific antibodies (Figure 11.2). These antibodies can recognize and inactivate the pathogen itself, or mediate lysis of pathogen-infected cells through the action of the complement cascades. Additional details on mechanisms of time-dependent antibody responses are discussed in Chapter 10, section 10.1 on antibodies and derivatives.

APC can also process (or degrade) endogenously expressed antigens, and present them through the class I histocompatibility complex (MHC I) to T-lymphocytes, resulting in the activation and proliferation of antigen-specific cytotoxic T-lymphocytes (CTLs), which in turn can recognize and lyse pathogen-infected cells (Figure 11.2).

The onset of antigen-specific T- and B-cell responses usually coincides with the initial clearance of the infecting agent. After this acute phase, infection subsides and the circulating antigens are cleared (usually within 3–6 weeks); most of the effector lymphocytes die off or revert to memory phenotype. Once established, immunologic memory may last a long time and form the basis for protection against future infection by the same pathogen.

Vaccines mimic natural infection by establishing pathogen-specific immunologic memory in vaccinated individuals. Upon re-exposure to the pathogen, individuals with prior exposures, or those who have been vaccinated, are able to mount a more rapid and robust response than unvaccinated individuals because of existing memory T and B cells that recognize the pathogen. The balance between cellular and humoral immunity elicited by the vaccine can differ depending on the nature of the vaccine and how it is delivered (e.g., the dose, route, regimen, and adjuvant used). Effective vaccines often establish a broad spectrum of pathogen-specific immune responses. Activation and expansion of pathogen-specific T helper cells, cytotoxic T-lymphocytes, and antibodies produced by mature B cells leads to the inactivation of the pathogen itself or the clearance of pathogen-infected cells, resulting in the prevention or amelioration of disease.

In addition to the quality, specificity, and magnitude of the immunologic responses elicited, non-immunologic factors may also contribute to the success or failure of vaccines against a given pathogen. These include some intrinsic properties of the pathogen, such as antigenic variability, ability to establish cellular latency, and the existence of natural reservoirs. Table 11.1 summarizes some of the major factors that influence the feasibility of successful vaccine development.

■ 11.3. TRADITIONAL VACCINE APPROACHES

Traditional vaccines make use of the incapacitated whole pathogen. This approach is based on the use of biological, chemical, or physical means to weaken or eliminate the disease-causing (pathogenic) properties of the organism, while retaining its ability to induce an effective immune response. In other words, a good vaccine retains immunogenic activity without eliciting host pathogenic responses. The majority of vaccines currently approved for clinical use can be divided into live attenuated and whole-killed vaccines.

Live attenuated vaccines retain their ability to infect and replicate in the vaccinated host. However, infection is usually self-limiting, resulting in an absence or a milder form of the disease normally caused by the pathogen. These vaccines are usually made by repeated passage ("adapting") of the pathogen in nonnatural hosts, or in tissue cultures, and selection for variants that show reduced pathogenicity (attenuation).

The most notable example of live attenuated vaccines is the smallpox vaccine, first developed by Edward Jenner, although the origin of the vaccine (vaccinia virus) remains obscure. More recent examples of live attenuated vaccines include most of the viral vaccines currently in use, such as measles/mumps/rubella (MMR) and varicella zoster (VZV) vaccines, and some

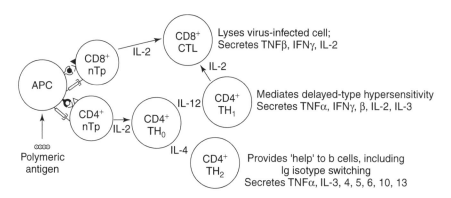

APC, antigen-presenting cell; ⱱ and ⱱ , class I and II major histocompatibility complex (MHC) antigens;

o and ●, peptides from degraded antigen bound to MHC molecules;

⊾, T cell receptor; CD4, △, and CD8, ▲, T-cell differentiation antigens;

|, costimulator molecules on APC;

⊣, ligand on T cell recognizing costimulator molecules;

nTp, naive precursor T cells;

TH$_0$, early activated CD4$^+$ T cell; IL, interleukin;

TH$_1$, TH$_2$, and cytotoxic T lymphocytes (CTLs), regulatory or effector T cells.

(A)

Figure 11.2. Antigen presentation and activation of T- and B-cell-mediated immune responses. (Adapted from Ada [57]).

of the bacterial vaccines, such as *Salmonella typhi* Ty21, a vaccine against typhoid fever, and Bacillus Calmette-Guerin (BCG), a vaccine against tuberculosis (Table 11.2). Because of the replicating nature of live attenuated vaccines, the effective dose must be sufficiently low to ensure safety. Despite this precaution, secondary transmission of the live attenuated vaccine could occur through accidental contact, or in some cases, through aerosols (e.g., sneezing), leading to accidental "immunization of neighbors." In a few reported cases, such "herd immunization" produced a broader protection in accidentally vaccinated individuals [6,7].

Mimicking natural infection, live attenuated vaccines often generate long-lasting immunity against the pathogen. Because antigens are presented as replication-competent bacteria or viruses, both humoral and cell-mediated immune responses, including CTLs, are generated. On the other hand, potential disadvantages of live attenuated vaccines include insufficient attenuation and reversion (leading to some limited pathogenicity) [8]. In some cases (e.g., rotavirus vaccine) altered pathogenicity has also been noted [9].

Another approach is the whole-killed or inactivated vaccine. These vaccines are derived from the pathogen itself, rather than a weakened form, as are live attenuated vaccines. Chemical (e.g., formalin, ether, and beta-propiolactone) or physical (e.g., heat, ultraviolet, and gamma irradiation) means, or combinations thereof, are used to inactivate the pathogen. Notable examples of this type of viral vaccine include the inactivated polio vaccine

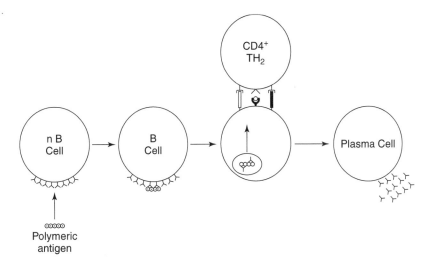

n B cell, immunocompetent, naive B cell; ∞, polymeric antigen;

⅄, Ig receptor; ⅄, class II MHC antigen receptor;

○, peptide from degraded foreign antigen; ⋉, T cell receptor;

▯, costimulator molecule on the B cell and the corresponding ligand, CD28.

on the T cell; ▮, CD40 differentiation antigen on the B cell and the

corresponding ligand on the T cell; Y, secreted Ig molecules.

(B)

Figure 11.2. *Continued*

(IPV), hepatitis A, influenza, and Japanese encephalitis virus vaccines, and bacterial vaccines against pertussis, plague, cholera, and typhoid (Table 11.2). This approach has an obvious advantage in that it does not require attenuated strains of the pathogen, which could be difficult to generate for many pathogens. Manufacturing processes are usually quantifiable and well defined.

Nevertheless, the inactivated vaccine approach suffers from several shortcomings. The process could inactivate not only the organism but also destroy the antigen conformation required to elicit protective immunity. Incomplete inactivation (e.g., virion aggregation resulting in poor exposure to the inactivating agent) could result in disease transmission or severe complications [10]. Because of the nonreplicating nature of inactivated vaccines, multiple immunizations with relatively high doses are required.

Very often whole-killed vaccines are formulated with adjuvants, which are designed to enhance vaccine persistence and induction of immune responses. However, the only adjuvant currently approved by FDA for clinical use is alum, in the form of vaccines complexed with aluminum hydroxide or aluminum sulfate. Even with the help of alum adjuvants, inactivated vaccine antigens are presented to APC extracellularly, as opposed to intracellularly, leading to a bias toward antibody-mediated responses. Little or no cell-mediated response to whole-killed vaccines with alum adjuvant renders some vaccines ineffective.

■TABLE 11.1. Biological properties of a pathogen that influence the likelihood of successful vaccine development

Properties	Favoring	Opposing
Antigenic variation	Few	Many
Latency	No	Yes (e.g., integration of viral genome)
Transmission	Only cell-free transmission	Cell-free and cell-to-cell transmission
Site of infection/latency	Accessible to immune surveillance	Infect cells in immunoprivileged sites
Target cells	Does not target or compromise immune cells	Target crucial cells of the immune system
Immune enhancement	No	Yes (e.g., infection of macrophage/monocytes)
Immunity from natural exposure	Protects against future infection	Does not protect
Antigenic target(s)	Few, simple	Multiple, complex
Suitable animal model	Yes	No
Natural reservoir	No	Yes

■ 11.4. SUBCELLULAR AND RECOMBINANT SUBUNIT VACCINES

Rather than using the whole organism, scientists can also make vaccines from specific subcellular components of a pathogen. A distinction can be made between subcellular and subunit vaccines in that subcellular vaccines refer to those made from cellular organisms, whereas subunit vaccines refer to those made from components of noncellular pathogens, such as viruses. The subcellular approach is especially appropriate (and indeed preferred) when the disease-causing component or fraction of the pathogen is known. This is the case for a number of bacterial infections where the toxin(s) produced by the bacteria, rather than the organism itself, causes the disease. The first subcellular vaccines developed in the 1920s were chemically inactivated diphtheria and tetanus toxins. Chemically inactivated toxins are called toxoids.

Subcellular vaccines approved by the FDA for human use include those against pneumococcus, meningococcus, *H. influen-zae*, *B. burgdorferi* (Lyme disease), and *B. pertussis* (acellular pertussis vaccine) (Table 11.2). The immunogens in these vaccines are composed of bacterial proteins, polysaccharides, or conjugates of both. Conjugate vaccines are especially important for immunizing children younger than age two. Infants of this age do not have immune systems that recognize T-cell-independent immunogens, such as polysaccharides [11,12], but they do recognize T-cell-dependent immunogens, such as proteins. Immunization with polysaccharides conjugated to carrier proteins (e.g., detoxified diphtheria toxin, *Neisseria* outer membrane protein, or tetanus toxoid) allows these infants to generate protective antibodies and immune memory against organisms, such as *H. influenzae* type B, which cause invasive diseases predominantly in children under two years.

With the advent of recombinant DNA technology, it is feasible to produce subcellular or subunit vaccines using heterologous expression systems, rather than partially purified components from the native source

■TABLE 11.2. Summary of currently licensed vaccines

Pathogen	Types of Vaccines Available for Each Pathogen		
	Attenuated	Whole Killed	Subcellular/Subunit
Viral			
	Smallpox		
	Polio	Polio	
	Measles/mumps/rubella		
	Varicella zoster		
	Influenza (cold-adapted)	Influenza	
	Adenovirus		
	Yellow fever		
		Rabies	
		Japanese encephalitis	
		Hepatitis A	
			Hepatitis B
Bacterial			
	Tuberculosis (BCG)		
	Salmonella typhi Ty21a	*Salmonella typhi*	Vi capsular polysaccharide
		Cholera	
		Plague	
		Lyme disease	
		Perstussis	Diptheria/tetanus/pertussis (acellular)
			Hemophilus influenza b
			Streptococcus pneumoniae
			Meningococcal

Note: See Appendiexes I and II for detailed descriptions of licensed vaccines.

(i.e., the pathogen itself). An example is the recombinant hepatitis B virus vaccine based on the viral surface antigen produced in yeast (e.g., *Recombivax HB*) or mammalian cell cultures (e.g., *Engerix*-B). Recombinant DNA technology has also enabled the introduction of specific alterations in vaccines produced from both native and recombinant sources to enhance their safety and efficacy. Acellular pertussis vaccines [13] and vaccines against Lyme disease [14,15] are examples of such vaccines currently approved for human use.

Improved safety is the primary advantage of subcellular and subunit vaccines over whole-killed vaccines. Because the pathogen itself is not used as the source material, vaccines manufactured by recombinant DNA technologies provide a greater margin of safety. Production of recombinant proteins in bacterial or mammalian hosts also allows for the possibility of greater yield and purity, better control of quality, and lower overall costs. On the other hand, to prepare subcellular or subunit vaccines, one must have previously identified the key antigenic target(s) to include in the vaccine. Even if the key component is known, immune responses against a single target antigen may be insufficient in protecting against a wide spectrum of subspecies [16,17] or may be ineffective in protecting populations with diverse genotypes [18,19].

By their very nature, subcellular and subunit vaccines are nonreplicative and often share drawbacks similar to inactivated vaccines in their need for repeated dosing and help with adjuvants (or conjugation to proteins) to enhance their responses. Despite these drawbacks safety considerations are prompting extensive efforts to develop or improve vaccines against a multitude of pathogens through recombinant subunit vaccine approaches (Box 11.1). These efforts attest to the tremendous promise and potential advantages of this approach.

BOX 11.1. HOW A VACCINE PRODUCT IS APPROVED FOR HUMAN USE BY THE FOOD AND DRUG ADMINISTRATION IN THE UNITED STATES

The FDA's Center for Biologics Evaluation and Research, CBER, is responsible for regulating vaccines in the United States. Current authority for the regulation of vaccines resides primarily in Section 351 of the Public Health Service Act and specific sections of the Federal Food, Drug and Cosmetic Act.

Clinical vaccine development follows the same general pathway as for drugs and other biologics. A sponsor who wishes to begin clinical trials with a vaccine must submit an Investigational New Drug (IND) application to the FDA. The IND describes the vaccine, its method of manufacture, and quality control tests for release. Also included is information about the vaccine's safety and ability to elicit a protective immune response (immunogenicity) in animal testing, as well as the proposed clinical protocol for studies in humans.

Clinical testing of vaccines in human subjects is typically done in three phases, as is the case for drugs or biologics. Initial human studies, referred to as phase I, are safety and immunogenicity studies performed in a small number of closely monitored subjects. Phase II studies are dose-ranging studies and may enroll hundreds of subjects. Finally, phase III trials typically enroll thousands of individuals and provide the critical documentation of effectiveness and important additional safety data required for licensing. At any stage of the clinical or animal studies, if data raise significant concerns about either safety or effectiveness, the FDA may request additional information or studies, or may halt ongoing clinical studies.

The successful completion of all three phases of clinical development is followed by the submission of a Biologics License Application (BLA). The license application must include the efficacy and safety information necessary for FDA reviewers to make a risk–benefit judgment. Also during this stage the proposed manufacturing facility undergoes a pre-approval inspection, during which production of the vaccine is examined as it is in progress.

Following FDA's review of a license application for a new indication, the sponsor and the FDA may present their findings to the FDA's Vaccines and Related Biological Products Advisory Committee. This independent expert committee (scientists, physicians, biostatisticians, and a consumer representative) provides advice to the Agency regarding the safety and efficacy of the vaccine for the proposed indication.

Vaccine approval also requires the provision of adequate product labeling to allow health care providers to understand the vaccine's proper use, including its potential benefits and risks, to communicate with patients and parents, and to safely deliver the vaccine to the public.

To ensure safety, the FDA continues to oversee the production of vaccines after the vaccine and the manufacturing processes are approved. After licensure, monitoring of the product and of production activities, including periodic facility inspections, must continue as long as the manufacturer holds a license for the product. If requested by the FDA, manufacturers are required to submit the results

(Continued on next page)

of their own tests for potency, safety, and purity for each vaccine lot. They may also be required to submit samples of each vaccine lot to the FDA for testing. However, if the sponsor describes an alternative procedure that provides continued assurance of safety, purity and potency, CBER may determine that routine submission of lot release protocols (showing results of applicable tests) and samples is not necessary.

Before a vaccine is given to the general population, all potential adverse events cannot be anticipated. Thus, many vaccines undergo phase IV studies—formal studies on a vaccine once it is on the market. Also the government relies on the Vaccine Adverse Event Reporting System to identify problems after marketing begins. (*Adapted from an FDA document*)

◼ 11.5. FUTURE DIRECTIONS

In addition to the classical approaches described above, scientific and technological advances in recent years have engendered a number of experimental approaches. Some of these are likely to have major impacts on the future direction of vaccine development.

Our concept of what constitutes a vaccine has undergone major modifications. Rather than some weakened forms of the pathogen or portions of it, DNA molecules encoding the relevant antigens have been used as vaccines [20–25]. Such DNA vaccines direct the expression of the relevant antigens in the vaccinated host to elicit the desired immune responses. Approaches devised to deliver these vaccines include "naked" DNA, DNA adsorbed onto gold particles for delivery through skin by means of a gene gun [26], and bacterial (e.g., *Vibrio, Salmonella, Listeria, and Mycobacteria*) and viral (poxviruses, adenoviruses, adeno-associated viruses, and Venezuelan encephalitis and other alpha viruses) vectors [27–36]. By their very nature as infectious agents, bacteria and viruses are uniquely suitable as vectors for efficient delivery and expression of foreign (vaccine) antigens to relevant target cells or tissues.

With better understanding of the precise nature of the antigen-binding epitopes for protective immune responses, researchers are exploring peptides and peptidomimetics as vaccines [37–39]. Peptidomimetic vaccines contain antigenic fragments that mimic the antibody-binding site. In cases where the relevant epitope can be defined by a monoclonal antibody, anti-idiotype antibodies (which may mimic the structure of the antigen itself) could be used as vaccine immunogens [40,41].

There have also been major advances in how vaccines are formulated and delivered. Currently the only adjuvant approved for general use in human vaccines is alum (aluminum salts). It is moderately effective in potentiating humoral immunity but does not generally elicit $CD8^+$ T-cell-mediated responses. In recent years a number of adjuvant and delivery systems have been developed that are able to not only augment the magnitude but also to influence the specificity and quality of vaccine responses [42]. Some of these adjuvants are designed specifically to enhance mucosal immunity, others to allow delivery of vaccines via skin patches (see Table 11.3). Mucosal immunity involves antibody- and cell-mediated responses at secretory sites, such as the nasal membrane, gastrointestinal tract, lungs, and genitalia. Most contact-mediated transmissions of pathogens occur through these sites. Consequently mucosal immunity is the first line of defense against infection.

■TABLE 11.3. Novel adjuvant systems and vaccine delivery technologies

Vaccine Enhancement Techniques	Example	References
Adjuvants		
Emulsions	SAF (Syntex)	[65–68]
	MF59 (Chiron)	
	Provax (IDEC)	
	TiterMax (Vaxcel)	
Copolymers	CRL 1005;L121	[68]
Liposomes	MTP-PE liposomes; IL-7 liposomes; cholera toxin-B subunit expressed liposomes; IL-2 liposome	[69–73]
Saponins	QS-21; Quil A; ISCOM	[74,75]
Bacterial DNA	CpG oligonucleotides	[76]
Cytokines	GM-CSF; IL-2; IL-12 IFN-α; IFN-γ, FTL-3	[43,44]
Co-stimulatory molecules	CD40-L; CD80/CD86	[43,77]
Microbial-derived material and other immunomodulator	MLP-A; DETOX (Corixa) α-Galactosylceramide (Kirin)	[78,79]
Delivery systems		
Biolistic delivery	Gene gun; biojector	[26,33,80]
Biodegradable microparticles	PLG microparticles	[59,60]
Immuno-targeting	Immunoglobulin complex Complement	[81–83]
Toxin-mediated delivery	Pertussis exotoxin Anthrax exotoxin	[84]
Dendritic cells (DC)	Antigen-pulsed DC	[45–47]
Edible vaccines	Genetically modified plants	[85–87]

Source: Adapted in part from Sheikh et al. [42].

■TABLE 11.4. Novel vaccine approaches

Vaccine Approaches	Examples	References
DNA	Naked	[21–24]
	Gene Gun	[26]
	Lipid-conjugated or liposomes	[58]
	Microparticles	[59,60]
Viral vectors	Poxvirus	[31,32]
	Adenovirus	[33,34]
	Alphavirus	[36,61]
	Adeno-associated virus	[35,62]
	Poliovirus	[63]
	Yellow fever virus	[64]
Bacterial vectors	*Salmonella*	[28]
	Listeria monocytogenes	[29]
	Vibrio cholera	[27]
	BCG	[30]
Peptide and mimetics		[37,38]
Anti-idiotype antibodies		[40,41]

Recent advances in understanding how immune responses are generated allows a particularly interesting approach using cytokines, co-stimulatory molecules, and professional antigen-presenting cells (e.g., dendritic cells) to enhance or otherwise modify vaccine responses [43–45]. It is likely that one or more of these adjuvant and delivery systems will be incorporated in the design of future vaccines.

Scientists expect that disease targets and fields of application for future vaccines will be greatly expanded. Besides continuing efforts to improve existing vaccines, there is a need for vaccines against newly emerging (e.g., HIV, ebola, and West Nile viruses) and re-emerging (e.g., tuberculosis and malaria) diseases (Table 11.4). The threat of bioterrorism also serves to emphasize the need for vaccines against hitherto relatively obscure agents, such as smallpox, anthrax, and tularemia. Finally, the use of vaccines will not be restricted to the pre-

vention of infectious diseases. In a broader sense, vaccines are agents that potentiate and modify the body's immune system to achieve therapeutic effects. Examples of these "therapeutic vaccines" include cancer vaccines in advanced-stage clinical trials [46–51], as well as contraceptive vaccines [52], and vaccines against heart disease, diabetes, autoimmune diseases, and alcohol and drug addiction [53–56] (Table 11.5). Therefore the concept of vaccination and the importance of vaccines as therapeutic agents are likely to expand significantly in the future.

■ 11.6. SUMMARY

During the past century our society has seen significant progress in public health. A child born in the United States one hundred years ago had a life expectancy of 47 years. Today the same child would be

■TABLE 11.5. New and re-emerging disease targets for future generations of vaccines

Disease Area	Examples	References
Infectious		
Emerging diseases	AIDS	[88–90]
	Ebola	[91,92]
	West Nile	[64]
	Dengue	[93]
	Multi-drug	[94]
	Resistant bacteria	
Re-emerging diseases	Tuberculosis	[95]
	Malaria	[96,97]
	Plague	[98]
	Influenza	[99]
Agents for biodefense	Anthrax	[100]
	Tularemia	[101]
	Rift Valley fever	[102]
Noninfectious	Cancer	[48–51,103]
	Contraceptive	[52]
	Drug abuse	[55,56]
	Diabetes	[54,104]
	Auto-immunity	[53,105,106]

expected to live 30 years longer. Together with clean water, the availability of safe and efficacious vaccines has been the most significant factor contributing to the increase in life expectancy and to the quality of life.

The eradication of smallpox in the 1970s stands as one of the crowning achievements of global public health efforts in the twentieth century. At the beginning of the twenty-first century, polio is on the verge of global eradication. Once-common childhood diseases, such as mumps, measles, and rubella, have been brought under control in most parts of the world. The majority of these victories have been achieved with only a handful of "conventional" approaches, including live attenuated vaccines, whole-killed vaccines, and subcellular or subunit vaccines. Recent advances in molecular biology, immunology, genomics and proteomics, and drug delivery technologies are already having major impacts on experimental vaccines, both in terms of how they are designed and delivered and how they are used. Beyond their traditional role as immunoprophylactic agents against infectious diseases, vaccines of the future are likely to find use as immunotherapeutic agents against cancer and other noninfectious diseases.

REFERENCES

1. Enders, J., T. Weller, and F. Robbins, *Cultivation of the Lansing strain of poliomyelitis virus in cultures of various human embryonic tissues.* Science, 1949. **109**: 85–7.

2. Jenner, E., *An inquiry into the causes and effects of the variolae vaccinae, a disease discovered in some of the western counties of England, particularly Gloucestershire, and known by the name of the cow pox.* 1798, London: S. Low.

3. *From the Centers for Disease Control and Prevention. Progress toward global eradication of poliomyelitis, 2001.* Jama, 2002. **287**(15): 1931–2.

4. Salk, J., U. Krech, and J. Younger, *Formaldehyde and safety testing of experimental poliomyelitis vaccines.* Am J Publ Health, 1954. **44**: 563–70.

5. Sabin, A.B., *Oral poliovirus vaccine. History of its development and prospects for eradication of poliomyelitis.* Jama, 1965. **194**(8): 872–6.

6. Anderson, R.M., and R.M. May, *Immunisation and herd immunity.* Lancet, 1990. **335**(8690): 641–5.

7. Yoshida, H., H. Horie, K. Matsuura, T. Kitamura, S. Hashizume, and T. Miyamura, *Prevalence of vaccine-derived polioviruses in the environment.* J Gen Virol, 2002. **83**(Pt 5): 1107–11.

8. Macadam, A.J., C. Arnold, J. Howlett, A. John, S. Marsden, F. Taffs, P. Reeve, N. Hamada, K. Wareham, J. Almond, et al., *Reversion of the attenuated and temperature-sensitive phenotypes of the Sabin type 3 strain of poliovirus in vaccinees.* Virology, 1989. **172**(2): 408–14.

9. Murphy, T.V., P.M. Gargiullo, M.S. Massoudi, D.B. Nelson, A.O. Jumaan, C.A. Okoro, L.R. Zanardi, S. Setia, E. Fair, C.W. LeBaron, M. Wharton, J.R. Livengood, and J.R. Livingood, *Intussusception among infants given an oral rotavirus vaccine.* N Engl J Med, 2001. **344**(8): 564–72.

10. Brown, F., *Review of accidents caused by incomplete inactivation of viruses.* Dev Biol Stand, 1993. **81**: 103–7.

11. Smith, D.H., G. Peter, D.L. Ingram, A.L. Harding, and P. Anderson, *Responses of children immunized with the capsular polysaccharide of Hemophilus influenzae, type b.* Pediatrics, 1973. **52**(5): 637–44.

12. Kayhty, H., V. Karanko, H. Peltola, and P.H. Makela, *Serum antibodies after vaccination with Haemophilus influenzae type b capsular polysaccharide and responses to reimmunization: no evidence of immunologic tolerance or memory.* Pediatrics, 1984. **74**(5): 857–65.

13. Pizza, M., A. Covacci, A. Bartoloni, M. Perugini, L. Nencioni, M.T. De Magistris, L. Villa, D. Nucci, R. Manetti, M. Bugnoli, et al., *Mutants of pertussis toxin suitable for vaccine development.* Science, 1989. **246**(4929): 497–500.

14. Steere, A.C., V.K. Sikand, F. Meurice, D.L. Parenti, E. Fikrig, R.T. Schoen, J. Nowakowski, C.H. Schmid, S. Laukamp, C. Buscarino, and D.S. Krause, *Vaccination against Lyme disease with recombinant Borrelia burgdorferi outer-surface lipoprotein A with adjuvant. Lyme Disease Vaccine Study Group.* N Engl J Med, 1998. **339**(4): 209–15.

15. Sigal, L.H., J.M. Zahradnik, P. Lavin, et al., *A vaccine consisting of recombinant Borrelia burgdorferi outer-surface protein A to prevent Lyme disease. Recombinant Outer-Surface Protein A Lyme Disease Vaccine Study Consortium.* N Engl J Med, 1998. **339**(4): 216–22.

16. Lovrich, S.D., S.M. Callister, L.C. Lim, B.K. DuChateau, and R.F. Schell, *Seroprotective groups of Lyme borreliosis spirochetes from North America and Europe.* J Infect Dis, 1994. **170**(1): 115–21.

17. Polacino, P., V. Stallard, J.E. Klaniecki, D.C. Montefiori, A.J. Langlois, B.A. Richardson, J. Overbaugh, W.R. Morton, R.E. Benveniste, and S.L. Hu, *Limited breadth of the protective immunity elicited by simian immunodeficiency virus SIVmne gp160 vaccines in a combination immunization regimen.* J Virol, 1999. **73**(1): 618–30.

18. Alper, C.A., M.S. Kruskall, D. Marcus-Bagley, D.E. Craven, A.J. Katz, S.J. Brink, J.L. Dienstag, Z. Awdeh, and E.J. Yunis, *Genetic prediction of nonresponse to hepatitis B vaccine.* N Engl J Med, 1989. **321**(11): 708–12.

19. Marescot, M.R., A. Budkowska, J. Pillot, and P. Debre, *HLA linked immune response to S and pre-S2 gene products in hepatitis B vaccination.* Tissue Antigens, 1989. **33**(5): 495–500.

20. Wolff, J.A., R.W. Malone, P. Williams, W. Chong, G. Acsadi, A. Jani, and P.L. Felgner, *Direct gene transfer into mouse muscle in vivo.* Science, 1990. **247**(4949 Pt 1): 1465–8.

21. Tang, D.C., M. DeVit, and S.A. Johnston, *Genetic immunization is a simple method for eliciting an immune response.* Nature, 1992. **356**(6365): 152–4.

22. Ulmer, J.B., J.J. Donnelly, S.E. Parker, G.H. Rhodes, P.L. Felgner, V.J. Dwarki, S.H. Gromkowski, R.R. Deck, C.M. DeWitt, A. Friedman, et al., *Heterologous protection against influenza by injection of DNA encoding a viral protein.* Science, 1993. **259**(5102): 1745–9.

23. Wang, B., K.E. Ugen, V. Srikantan, M.G. Agadjanyan, K. Dang, Y. Refaeli, A.I. Sato, J. Boyer, W.V. Williams, and D.B. Weiner, *Gene inoculation generates immune responses against human immunodeficiency virus type 1.* Proc Natl Acad Sci USA, 1993. **90**(9): 4156–60.

24. Robinson, H.L., L.A. Hunt, and R.G. Webster, *Protection against a lethal influenza virus challenge by immunization with a haemagglutinin-expressing plasmid DNA.* Vaccine, 1993. **11**(9): 957–60.

25. Ulmer, J.B., *An update on the state of the art of DNA vaccines.* Curr Opin Drug Discov Devel, 2001. **4**(2): 192–7.

26. Fynan, E.F., R.G. Webster, D.H. Fuller, J.R. Haynes, J.C. Santoro, and H.L. Robinson, *DNA vaccines: protective immunizations by parenteral, mucosal, and gene- gun inoculations.* Proc Natl Acad Sci USA, 1993. **90**(24): 11478–82.

27. Killeen, K., D. Spriggs, and J. Mekalanos, *Bacterial mucosal vaccines: Vibrio cholerae as a live attenuated vaccine/vector paradigm.* Curr Top Microbiol Immunol, 1999. **236**: 237–54.

28. Russmann, H., H. Shams, F. Poblete, Y. Fu, J.E. Galan, and R.O. Donis, *Delivery of epitopes by the Salmonella type III secretion system for vaccine development.* Science, 1998. **281**(5376): 565–8.

29. Lieberman, J., and F.R. Frankel, *Engineered Listeria monocytogenes as an AIDS vaccine.* Vaccine, 2002. **20**(15): 2007–10.

30. Barletta, R.G., B. Snapper, J.D. Cirillo, N.D. Connell, D.D. Kim, W.R. Jacobs, and B.R. Bloom, *Recombinant BCG as a candidate oral vaccine vector.* Res Microbiol, 1990. **141**(7–8): 931–9.

31. Paoletti, E., *Applications of pox virus vectors to vaccination: an update.* Proc Natl Acad Sci USA, 1996. **93**(21): 11349–53.

32. Moss, B., *Genetically engineered poxviruses for recombinant gene expression, vaccination, and safety.* Proc Natl Acad Sci USA, 1996. **93**(21): 11341–8.

33. Davis, A.R., B. Kostek, B.B. Mason, C.L. Hsiao, J. Morin, S.K. Dheer, and P.P. Hung,

Expression of hepatitis B surface antigen with a recombinant adenovirus. Proc Natl Acad Sci USA, 1985. **82**(22): 7560–4.

34. Xiang, Z.Q., Y. Yang, J.M. Wilson, and H.C. Ertl, *A replication-defective human adenovirus recombinant serves as a highly efficacious vaccine carrier.* Virology, 1996. **219**(1): 220–7.

35. Clark, K.R., and P.R. Johnson, *Gene delivery of vaccines for infectious disease.* Curr Opin Mol Ther, 2001. **3**(4): 375–84.

36. Tubulekas, I., P. Berglund, M. Fleeton, and P. Liljestrom, *Alphavirus expression vectors and their use as recombinant vaccines: a minireview.* Gene, 1997. **190**(1): 191–5.

37. Moe, G.R. and D.M. Granoff, *Molecular mimetics of Neisseria meningitidis serogroup B polysaccharide.* Int Rev Immunol, 2001. **20**(2): 201–20.

38. Cunto-Amesty, G., T.K. Dam, P. Luo, B. Monzavi-Karbassi, C.F. Brewer, T.C. Van Cott, and T. Kieber-Emmons, *Directing the immune response to carbohydrate antigens.* J Biol Chem, 2001. **276**(32): 30490–8.

39. Nakouzi, A., P. Valadon, J. Nosanchuk, N. Green, and A. Casadevall, *Molecular basis for immunoglobulin M specificity to epitopes in Cryptococcus neoformans polysaccharide that elicit protective and nonprotective antibodies.* Infect Immun, 2001. **69**(5): 3398–409.

40. Grzych, J.M., M. Capron, P.H. Lambert, C. Dissous, S. Torres, and A. Capron, *An anti-idiotype vaccine against experimental schistosomiasis.* Nature, 1985. **316**(6023): 74–6.

41. Dalgleish, A.G., and R.C. Kennedy, *Anti-idiotypic antibodies as immunogens: idiotype-based vaccines.* Vaccine, 1988. **6**(3): 215–20.

42. Sheikh, N.A., M. al-Shamisi, and W.J. Morrow, *Delivery systems for molecular vaccination.* Curr Opin Mol Ther, 2000. **2**(1): 37–54.

43. Berzofsky, J.A., J.D. Ahlers, and I.M. Belyakov, *Strategies for designing and optimizing new generation vaccines.* Nat Rev Immunol, 2001. **1**(3): 209–19.

44. Scheerlinck, J.Y., *Genetic adjuvants for DNA vaccines.* Vaccine, 2001. **19**(17–19): 2647–56.

45. Steinman, R.M., and M. Pope, *Exploiting dendritic cells to improve vaccine efficacy.* J Clin Invest, 2002. **109**(12): 1519–26.

46. Hsu, F.J., C. Benike, F. Fagnoni, T.M. Liles, D. Czerwinski, B. Taidi, E.G. Engleman, and R. Levy, *Vaccination of patients with B-cell lymphoma using autologous antigen-pulsed dendritic cells.* Nat Med, 1996. **2**(1): 52–8.

47. Nestle, F.O., S. Alijagic, M. Gilliet, Y. Sun, S. Grabbe, R. Dummer, G. Burg, and D. Schadendorf, *Vaccination of melanoma patients with peptide- or tumor lysate-pulsed dendritic cells.* Nat Med, 1998. **4**(3): 328–32.

48. Thurner, B., I. Haendle, C. Roder, et al., *Vaccination with mage-3A1 peptide-pulsed mature, monocyte-derived dendritic cells expands specific cytotoxic T cells and induces regression of some metastases in advanced stage IV melanoma.* J Exp Med, 1999. **190**(11): 1669–78.

49. Wolchok, J.D., and P.O. Livingston, *Vaccines for melanoma: translating basic immunology into new therapies.* Lancet Oncol, 2001. **2**(4): 205–11.

50. Foy, T.M., G.R. Fanger, S. Hand, C. Gerard, C. Bruck, and M.A. Cheever, *Designing HER2 vaccines.* Semin Oncol, 2002. **29**(3 Suppl 11): 53–61.

51. Bronte, V., *Genetic vaccination for the active immunotherapy of cancer.* Curr Gene Ther, 2001. **1**(1): 53–100.

52. Delves, P.J., T. Lund, and I.M. Roitt, *Antifertility vaccines.* Trends Immunol, 2002. **23**(4): 213–19.

53. Garren, H. and L. Steinman, *DNA vaccination in the treatment of autoimmune disease.* Curr Dir Autoimmun, 2000. **2**: 203–16.

54. Jaeckel, E., M. Manns, and M. Von Herrath, *Viruses and diabetes.* Ann NY Acad Sci, 2002. **958**: 7–25.

55. Carrera, M.R., J.A. Ashley, L.H. Parsons, P. Wirsching, G.F. Koob, and K.D. Janda, *Suppression of psychoactive effects of cocaine by active immunization.* Nature, 1995. **378**(6558): 727–30.

56. Fox, B.S., K.M. Kantak, M.A. Edwards, K.M. Black, B.K. Bollinger, A.J. Botka, T.L. French, T.L. Thompson, V.C. Schad, J.L.

Greenstein, M.L. Gefter, M.A. Exley, P.A. Swain, and T.J. Briner, *Efficacy of a therapeutic cocaine vaccine in rodent models.* Nat Med, 1996. **2**(10): 1129–32.

57. Plotkin, S.A., and W.A. Orenstein, *Vaccines.* 3rd ed. 1999, Philadelphia: Saunders. xix, 1230.

58. Gregoriadis, G., *DNA vaccines: a role for liposomes.* Curr Opin Mol Ther, 1999. **1**(1): 39–42.

59. Jones, D.H., J.C. Clegg, and G.H. Farrar, *Oral delivery of micro-encapsulated DNA vaccines.* Dev Biol Stand, 1998. **92**: 149–55.

60. Vajdy, M., and D.T. O'Hagan, *Microparticles for intranasal immunization.* Adv Drug Deliv Rev, 2001. **51**(1–3): 127–41.

61. Lundstrom, K., *Alphavirus vectors for gene therapy applications.* Curr Gene Ther, 2001. **1**(1): 19–29.

62. Manning, W.C., X. Paliard, S. Zhou, M. Pat Bland, A.Y. Lee, K. Hong, C.M. Walker, J.A. Escobedo, and V. Dwarki, *Genetic immunization with adeno-associated virus vectors expressing herpes simplex virus type 2 glycoproteins B and D.* J Virol, 1997. **71**(10): 7960–2.

63. Crotty, S., C.J. Miller, B.L. Lohman, M.R. Neagu, L. Compton, D. Lu, F.X. Lu, L. Fritts, J.D. Lifson, and R. Andino, *Protection against simian immunodeficiency virus vaginal challenge by using Sabin poliovirus vectors.* J Virol, 2001. **75**(16): 7435–52.

64. Monath, T.P., *Prospects for development of a vaccine against the West Nile virus.* Ann NY Acad Sci, 2001. **951**: 1–12.

65. Byars, N.E., and A.C. Allison, *Adjuvant formulation for use in vaccines to elicit both cell-mediated and humoral immunity.* Vaccine, 1987. **5**(3): 223–8.

66. Ott, G., G.L. Barchfeld, D. Chernoff, R. Radhakrishnan, P. van Hoogevest, and G. Van Nest, *MF59. Design and evaluation of a safe and potent adjuvant for human vaccines.* Pharm Biotechnol, 1995. **6**: 277–96.

67. Hanna, N., and K. Hariharan, *Development and application of PROVAX adjuvant formulation for subunit cancer vaccines.* Adv Drug Deliv Rev, 1998. **32**(3): 187–97.

68. Hunter, R.L., M.R. Kidd, M.R. Olsen, P.S. Patterson, and A.A. Lal, *Induction of long-lasting immunity to Plasmodium yoelii malaria with whole blood-stage antigens and copolymer adjuvants.* J Immunol, 1995. **154**(4): 1762–9.

69. Ho, R.J., R.L. Burke, and T.C. Merigan, *Disposition of antigen-presenting liposomes in vivo: effect on presentation of herpes simplex virus antigen rgD.* Vaccine, 1994. **12**(3): 235–42.

70. Bui, T., T. Dykers, S.L. Hu, C.R. Faltynek, and R.J. Ho, *Effect of MTP-PE liposomes and interleukin-7 on induction of antibody and cell-mediated immune responses to a recombinant HIV-envelope protein.* J Acquir Immune Defic Syndr, 1994. **7**(8): 799–806.

71. Dong, P., C. Brunn, and R.J. Ho, *Cytokines as vaccine adjuvants: current status and potential applications.* Pharm Biotechnol, 1995. **6**: 625–43.

72. Lian, T., T. Bui, and R.J. Ho, *Formulation of HIV-envelope protein with lipid vesicles expressing ganglioside GM1 associated to cholera toxin B enhances mucosal immune responses.* Vaccine, 1999. **18**(7–8): 604–11.

73. Lian, T., and R.J. Ho, *Trends and developments in liposome drug delivery systems.* J Pharm Sci, 2001. **90**(6): 667–80.

74. Kensil, C.R., J.Y. Wu, C.A. Anderson, D.A. Wheeler, and J. Amsden, *QS-21 and QS-7: purified saponin adjuvants.* Dev Biol Stand, 1998. **92**: 41–7.

75. Morein, B., M. Villacres-Eriksson, and K. Lovgren-Bengtsson, *Iscom, a delivery system for parenteral and mucosal vaccination.* Dev Biol Stand, 1998. **92**: 33–9.

76. Krieg, A.M., and H.L. Davis, *Enhancing vaccines with immune stimulatory CpG DNA.* Curr Opin Mol Ther, 2001. **3**(1): 15–24.

77. Liebowitz, D.N., K.P. Lee, and C.H. June, *Costimulatory approaches to adoptive immunotherapy.* Curr Opin Oncol, 1998. **10**(6): 533–41.

78. Rudbach, J., D. Johnson, and J. Ulrich, *Ribi adjuvants: chemistry, biology and utility in vaccines for human and veterinary medicine,* in *The Theory and Practical Application of Adjuvants,* S.-T. Des, ed. 1995, New York: Wiley. 287–13.

79. Gonzalez-Aseguinolaza, G., L. Van Kaer, C.C. Bergmann, J.M. Wilson, J. Schmieg, M. Kronenberg, T. Nakayama, M. Taniguchi, Y. Koezuka, and M. Tsuji, *Natural killer T cell ligand alpha-galactosylceramide enhances protective immunity induced by malaria vaccines.* J Exp Med, 2002. **195**(5): 617–24.

80. Epstein, J.E., E.J. Gorak, Y. Charoenvit, et al., *Safety, tolerability, and lack of antibody responses after administration of a PfCSP DNA malaria vaccine via needle or needle-free jet injection, and comparison of intramuscular and combination intramuscular/intradermal routes.* Hum Gene Ther, 2002. **13**(13): 1551–60.

81. Bot, A.I., D.J. Smith, S. Bot, L. Dellamary, T.E. Tarara, S. Harders, W. Phillips, J.G. Weers, and C.M. Woods, *Receptor-mediated targeting of spray-dried lipid particles coformulated with immunoglobulin and loaded with a prototype vaccine.* Pharm Res, 2001. **18**(7): 971–9.

82. Barber, B.H., *The immunotargeting approach to adjuvant-independent subunit vaccine design.* Semin Immunol, 1997. **9**(5): 293–301.

83. Guyre, P.M., R.F. Graziano, J. Goldstein, P.K. Wallace, P.M. Morganelli, K. Wardwell, and A.L. Howell, *Increased potency of Fc-receptor-targeted antigens.* Cancer Immunol Immunother, 1997. **45**(3–4): 146–8.

84. Glenn, G.M., D.N. Taylor, X. Li, S. Frankel, A. Montemarano, and C.R. Alving, *Transcutaneous immunization: a human vaccine delivery strategy using a patch.* Nat Med, 2000. **6**(12): 1403–6.

85. Kong, Q., L. Richter, Y.F. Yang, C.J. Arntzen, H.S. Mason, and Y. Thanavala, *Oral immunization with hepatitis B surface antigen expressed in transgenic plants.* Proc Natl Acad Sci USA, 2001. **98**(20): 11539–44.

86. Lauterslager, T.G., D.E. Florack, T.J. van der Wal, J.W. Molthoff, J.P. Langeveld, D. Bosch, W.J. Boersma, and L.A. Hilgers, *Oral immunisation of naive and primed animals with transgenic potato tubers expressing LT-B.* Vaccine, 2001. **19**(17–19): 2749–55.

87. Mason, H.S., H. Warzecha, T. Mor, and C.J. Arntzen, *Edible plant vaccines: applications for prophylactic and therapeutic mol-ecular medicine.* Trends Mol Med, 2002. **8**(7): 324–9.

88. Letvin, N.L., D.H. Barouch, and D.C. Montefiori, *Prospects for vaccine protection against HIV-1 infection and AIDS.* Annu Rev Immunol, 2002. **20**: 73–99.

89. Gupta, K., M. Hudgens, L. Corey, et al., *Safety and immunogenicity of a high-titered canarypox vaccine in combination with rgp120 in a diverse population of HIV-1-uninfected adults: AIDS Vaccine Evaluation Group Protocol 022A.* J Acquir Immune Defic Syndr, 2002. **29**(3): 254–61.

90. Schultz, A.M. and J.A. Bradac, *The HIV vaccine pipeline, from preclinical to phase III.* Aids, 2001. **15**(Suppl 5): S147–58.

91. Sullivan, N.J., A. Sanchez, P.E. Rollin, Z.Y. Yang, and G.J. Nabel, *Development of a preventive vaccine for Ebola virus infection in primates.* Nature, 2000. **408**(6812): 605–9.

92. Geisbert, T.W., P. Pushko, K. Anderson, J. Smith, K.J. Davis, and P.B. Jahrling, *Evaluation in nonhuman primates of vaccines against Ebola virus.* Emerg Infect Dis, 2002. **8**(5): 503–7.

93. Rigau-Perez, J.G., G.G. Clark, D.J. Gubler, P. Reiter, E.J. Sanders, and A.V. Vorndam, *Dengue and dengue haemorrhagic fever.* Lancet, 1998. **352**(9132): 971–7.

94. Appelbaum, P.C., *Resistance among Streptococcus pneumoniae: implications for drug selection.* Clin Infect Dis, 2002. **34**(12): 1613–20.

95. Cohn, D.L., *Use of the bacille Calmette-Guerin vaccination for the prevention of tuberculosis: renewed interest in an old vaccine.* Am J Med Sci, 1997. **313**(6): 372–6.

96. Doolan, D.L., and S.L. Hoffman, *Nucleic acid vaccines against malaria.* Chem Immunol, 2002. **80**: 308–21.

97. Richie, T.L. and A. Saul, *Progress and challenges for malaria vaccines.* Nature, 2002. **415**(6872): 694–701.

98. Titball, R.W., and E.D. Williamson, *Vaccination against bubonic and pneumonic plague.* Vaccine, 2001. **19**(30): 4175–84.

99. Palese, P., and A. Garcia-Sastre, *Influenza vaccines: present and future.* J Clin Invest, 2002. **110**(1): 9–13.

100. Baillie, L., *The development of new vaccines against Bacillus anthracis.* J Appl Microbiol, 2001. **91**(4): 609–13.

101. Ellis, J., P.C. Oyston, M. Green, and R.W. Titball, *Tularemia*. Clin Microbiol Rev, 2002. **15**(4): 631–46.

102. Pittman, P.R., C.T. Liu, T.L. Cannon, R.S. Makuch, J.A. Mangiafico, P.H. Gibbs, and C.J. Peters, *Immunogenicity of an inactivated Rift Valley fever vaccine in humans: a 12-year experience.* Vaccine, 1999. **18**(1–2): 181–9.

103. Jonasch, E., *Melanoma vaccination: state-of-the-art and experimental approaches.* Expert Rev Anticancer Ther, 2001. **1**(3): 427–40.

104. von Herrath, M.G., and J.L. Whitton, *DNA vaccination to treat autoimmune diabetes.* Ann Med, 2000. **32**(5): 285–92.

105. Ramshaw, I.A., S.A. Fordham, C.C. Bernard, D. Maguire, W.B. Cowden, and D.O. Willenborg, *DNA vaccines for the treatment of autoimmune disease.* Immunol Cell Biol, 1997. **75**(4): 409–13.

106. Schwartz, M., *Protective autoimmunity as a T-cell response to central nervous system trauma: prospects for therapeutic vaccines.* Prog Neurobiol, 2001. **65**(5): 489–96.

12

OTHER PRODUCTS: MONOGRAPHS

BECAPLERMIN

Trade name: *Regranex Gel*

Manufacturer: OMJ Pharmaceuticals, Inc., Manati, Puerto Rico

Distributed by: McNeil Pharmaceutical, Raritan, NJ

Indications: Treatment of lower extremity diabetic neuropathic ulcers that extend into the subcutaneous tissue or beyond, and have adequate blood supply

Approval date: 16 December 1997

Type of submission: Biologics license application

A. General description Becaplermin, a recombinant human platelet-derived growth factor (rhPDGF-BB), is produced by recombinant DNA technology by insertion of the gene for the B chain of PDGF into the yeast *Saccharomyces cerevisiae.*

Becaplermin has a molecular weight of about 25 kDa and is a homodimer composed of two identical polypeptide chains bound together by disulfide bonds.

B. Indications and use *Regranex Gel* is indicated for the topical treatment of deep diabetic foot and leg ulcers that have an adequate blood supply. Patients with diabetes who develop chronic ulcers of the foot and leg are at higher risk for local and systemic infections and amputation.

C. Administration

Dosage form *Regranex Gel* is a non-sterile, preserved, sodium carboxymethyl-cellulose-based topical preparation containing becaplermin. Each gram of *Regranex Gel* contains becaplermin 100 µg.

Route of administration A measured quantity of *Regranex Gel* should be spread evenly over the ulcerated area once daily to yield a thin continuous layer. The site(s)

Biotechnology and Biopharmaceuticals, by Rodney J. Y. Ho and Milo Gibaldi
ISBN 0-471-20690-3   Copyright © 2003 by John Wiley & Sons, Inc.

of application should then be covered by a saline moistened dressing and left in place for about 12 hours. The dressing should then be removed and the ulcer rinsed with saline to remove residual gel and covered with a second moist dressing for the remainder of the day.

Recommended dosage and monitoring requirements The amount of *Regranex Gel* to be applied will vary depending on the size of the ulcer area. The physician should recalculate the amount of gel to be applied at weekly or biweekly intervals depending on the rate of change in ulcer area.

D. Pharmacology and pharmaceutics

Clinical pharmacology Native PDGF is a growth factor derived from blood platelets that mediates tissue repair. Becaplermin has biological activity similar to that of endogenous PDGF; it promotes the chemotactic recruitment and proliferation of cells involved in wound repair—fibroblasts, smooth muscle cells, monocytes, and neutrophils—and enhances the formation of granulation tissue.

Pharmacokinetics No information available.

Disposition No information available.

Drug interactions Whether becaplermin interacts with other topical medication applied to the ulcer site is not known.

E. Therapeutic response
The primary end point assessed in four randomized controlled trials was the incidence of complete ulcer closure within 20 weeks. In one study, the incidence of complete ulcer closure for *Regranex Gel* 0.003% was 48% versus 25% for placebo gel. In another study, the incidence of complete ulcer closure for *Regranex Gel* 0.01% was 50% versus 36% for *Regranex Gel* 0.003%. One study failed to detect a difference between *Regranex Gel* 0.01% and good ulcer care alone.

F. Place in therapy *Regranex Gel* is indicated in treatment of chronic and nonhealing lower-extremity diabetic ulcers, but only as adjunct to surgical debridement. When combined with standard care—removal of infected and dead tissue, daily cleanings, and pressure relief—becaplermin can improve the likelihood of complete healing. Becaplermin works by recruiting cells and promoting cell growth, which help to heal the ulcer.

G. Other applications Whether *Regranex Gel* provides significant benefit in the treatment of superficial diabetic neuropathic ulcers or nondiabetic ulcers has not been determined.

FOMIVIRSEN

Trade name: ***Vitravene***

Manufacturer: Abbott Laboratories, McPherson, KS

Distributed by: CIBA Vision, Duluth, GA

Indications: Local treatment of cytomegalovirus (CMV) retinitis in patients with acquired immunodeficiency syndrome

Approval date: 1 September 1998

Type of submission:

A. General description Fomivirsen sodium (molecular weight about 7 kDa) is a phosphorothioate oligonucleotide, 21 nucleotides in length, with the following sequence: 5'-GCG TTT GCT CTT CTT CTT GCG-3'.

B. Indications and use Intravitreal *Vitravene* is indicated for local treatment of CMV retinitis in AIDS patients who are refractory to or intolerant of other treatment or when other treatments are contraindicated.

C. Administration

Dosage form *Vitravene* is a sterile, aqueous, preservative-free, buffered solution supplied in 0.25 ml single-use vials for intravitreal injection. Each ml of *Vitravene* solution contains fomivirsen sodium 6.6 mg.

Route of administration Fomivirsen is administered intravitreally with topical and/or local anesthesia.

Recommended dosage and monitoring requirements For the treatment of CMV retinitis in AIDS patients, the recommended dose of *Vitravene* is 330 µg (0.05 ml) by intravitreal injection; for induction, one dose should be injected every other week for two doses; for maintenance, after induction, one dose should be given once every 4 weeks after induction.

D. Pharmacology and pharmaceutics

Clinical pharmacology Fomivirsen is a synthetic phosphorothioate oligonucleotide that inhibits human CMV replication through an antisense mechanism. It is the first and, so far, only antisense molecule to receive FDA approval. The compound is complementary to messenger RNA of the immediate–early (IE2) transcriptional unit of CMV; fomivirsen binding to this messenger RNA results in selective inhibition of IE2 proteins that are needed for CMV replication. *In vitro*, fomivirsen was significantly more potent than ganciclovir against CMV (at least 30-fold). Additive antiviral activity was seen when fomivirsen was combined with ganciclovir or foscarnet.

Pharmacokinetics No information on human pharmacokinetics is available. In the rabbit model, half-life in the vitreous humor was 62 hours and 79 hours in the retina, while the retinal half-life in monkeys was 78 hours. Systemic exposure to fomivirsen following single or repeated intravitreal injection in monkeys was not quantifiable.

Disposition Metabolites of fomivirsen are detected in the retina and vitreous humor in animals. Fomivirsen sodium is metabolized by exonucleases in a process that sequentially removes residues from the terminal ends of the oligonucleotide yielding shortened oligonucleotides and mononucleotide metabolites.

Drug interactions *Vitravene* is not recommended for use in patients who have recently (2–4 weeks) been treated with either intravenous or intravitreal cidofovir because of the risk of exaggerated ocular inflammation. Results from *in vitro* tests demonstrated no inhibition of anti-CMV activity of fomivirsen by zidovudine or zalcitabine.

E. Therapeutic response Intravitreal *Vitravene* therapy is effective in delaying disease progression in AIDS patients with refractory or newly diagnosed CMV retinitis; depending on the patient population, the median time to progression has ranged from 2 to 11 weeks.

F. Role in therapy *Vitravene* may be recommended in AIDS patients with sight-threatening CMV retinitis unresponsive to usual agents of choice (i.e., intravenous ganciclovir, ganciclovir intraocular implant, intravenous cidofovir, and intravenous foscarnet). *Vitravene* may also be recommended in both advanced and newly diagnosed patients who have relative contraindications to (e.g., renal insufficiency) or develop complications on current regimens of choice (i.e., renal failure or neutropenia with foscarnet or cidofovir; dose-limiting neutropenia or thrombocytopenia with ganciclovir). *Vitravene* offers an advantage over standard therapy by virtue of a more convenient dose schedule and apparent lack of systemic toxicity.

G. Other applications The combination of *Vitravene* and ganciclovir is under investigation for AIDS-related cytomegalovirus retinitis in previously treated and untreated patients.

ALPHA₁-PROTEINASE INHIBITOR

Trade name: *Prolastin*

Manufacturer: Bayer Corporation, Pharmaceutical Division, Elkhart, IN

Indications: Replacement therapy in patients with congenital alpha-1 antitrypsin deficiency who have panacinar emphysema

Approval date: 2 December 1987

Type of submission: Product license application

A. General description *Prolastin* is prepared from pooled human plasma of normal donors; it has a molecular weight of 52 kDa. To reduce the potential risk of transmission of infectious agents, *Prolastin* has been heat-treated in solution at $60 \pm 0.5°C$ for not less than 10 hours. However, no procedure has been found to be totally effective in removing viral infectivity from plasma fractionation products.

B. Indications and use *Prolastin* is indicated for chronic replacement therapy in patients with congenital alpha-1 antitrypsin deficiency who have panacinar emphysema, generalized obstructive emphysema affecting all lung segments, with atrophy and dilatation of the alveoli and destruction of the vascular bed. As some individuals with alpha-1 antitrypsin deficiency will not go on to develop panacinar emphysema, only those with early evidence of such disease should be considered for chronic replacement therapy with *Prolastin*. Patients with the PiMZ or PiMS phenotypes of alpha-1 antitrypsin deficiency should not be considered for treatment, as they appear to be at small risk for panacinar emphysema. *Prolastin* is not indicated for use in patients other than those with PiZZ, PiZ(null), or Pi(null)(null) phenotypes.

C. Administration

Dosage form *Prolastin* is supplied in single-use vials with the total alpha₁-proteinase inhibitor functional activity, in milligrams, 500 or 1000, stated on the label of each vial. A suitable volume of sterile water for injection is provided. When the product is reconstituted it contains *Prolastin* 25 mg/ml.

Route of administration *Prolastin* must be administered by the intravenous route.

Recommended dosage and monitoring requirements The recommended dosage of *Prolastin* is 60 mg/kg body weight administered once weekly. This dose is intended to increase and maintain a level of functional alpha₁-proteinase inhibitor in the epithelial lining of the lower respiratory tract, providing adequate anti-elastase activity in the lungs of individuals with alpha-1 antitrypsin deficiency. The recommended dosage of 60 mg/kg takes approximately 30 minutes to infuse.

D. Pharmacology and pharmaceutics

Clinical pharmacology Alpha-1 antitrypsin deficiency is a chronic, hereditary, usually fatal, autosomal recessive disorder in which a low concentration of alpha₁-proteinase inhibitor is associated with slowly progressive, severe, panacinar emphysema that most often manifests itself in the third to fourth decades of life. The pathogenesis of development of emphysema in alpha-1 antitrypsin deficiency is believed to be due to a chronic biochemical imbalance between elastase and alpha₁-proteinase inhibitor (the principal inhibitor of neutrophil elastase), which is deficient in alpha-1 antitrypsin disease. As a result it is believed that alveolar structures are unprotected from chronic exposure to elastase released from a chronic low-level burden of neutrophils in the lower respiratory tract, resulting in progressive degradation of elastin tissues. The eventual outcome is the development of emphysema. A large number of phenotypic variants of alpha-1 antitrypsin deficiency exists. The most severely affected individuals are those with the PiZZ variant, typically characterized by alpha$_1$-proteinase inhibitor serum levels <35% normal.

Pharmacokinetics The threshold serum level of alpha$_1$-proteinase inhibitor that is needed for adequate anti-elastase activity in the lung is 80 mg/dl. Clinical trials have demonstrated alpha$_1$-proteinase inhibitor peak serum levels of greater than 300 mg/dl following intravenous weekly doses of 60 mg/kg over 6 months. The average serum level with one month of therapy was 138 to 163 mg/dl, with a trough of greater than 80 mg/dl. The mean elimination half-life of alpha$_1$-proteinase inhibitor is about 4 days.

Disposition Endogenous alpha$_1$-proteinase inhibitor undergoes catabolism in the intravascular space. The exact process of catabolism is not well defined; desialyzation of the terminal sugars on the oligosaccharide side chain may occur in the initial process.

Drug interactions No information available

E. Therapeutic response Within a few weeks of commencing replacement therapy, bronchoalveolar lavage studies de-

monstrated significantly increased levels of alpha$_1$-proteinase inhibitor and functional antineutrophil elastase capacity in the epithelial lining fluid of the lower respiratory tract of the lungs, as compared to levels prior to initiating the program of chronic replacement therapy with *Prolastin*. Long-term controlled clinical trials to evaluate the effect of chronic replacement therapy with *Prolastin* on the development of or progression of emphysema in patients with congenital alpha-1 antitrypsin deficiency have not been performed.

F. Role in therapy According to Micromedex, *Prolastin* will effectively cause elevation of serum and lung fluid levels of alpha$_1$-proteinase inhibitor. The drug is used as replacement therapy for patients with congenital deficiency who demonstrate clinical signs of emphysema. It is anticipated that replacement therapy will slow the progression of emphysema-related changes for these patients, but limited clinical trials have not demonstrated that emphysema is effectively treated with this drug.

G. Other consideration Alpha$_1$-proteinase inhibitor has been designated an orphan product for use as supplementary or replacement therapy for alpha-1 antitrypsin deficiency.

Part III

FUTURE DIRECTIONS

13

ADVANCED DRUG DELIVERY

■ 13.1. RATIONALE AND BASIC PRINCIPLES

The general goal of drug delivery is to maximize the fraction of a drug dose delivered to target tissues and cells directly responsible for therapeutic effects and to limit exposure of other tissues to the drug, as this exposure may elicit untoward effects.

As discussed in Chapter 5, most biotechnology products (i.e., proteins and peptides) are formulated in sterile solution or suspensions designed for injection. A survey of the biotechnology products available in the United States reveals that over 90% of protein and peptide drugs are designed for parenteral administration. Biopharmaceuticals relying on other routes of administration constitute a very small fraction of the total number of products (Figure 13.1).

While some parenteral injections, such as intravenous administration, provide rapid and predictable access to the circulation and tissues, therapeutic proteins are rapidly cleared from the system, and thus such administrations may result in very short durations of action. Regardless of route of administration, therapeutic proteins may exhibit limited distribution outside of endothelial cells lining blood vessels. This may be advantageous for thrombolytic agents, such as tissue plasminogen activator, which is used for rapid fibrinolytic actions at

Biotechnology and Biopharmaceuticals, by Rodney J. Y. Ho and Milo Gibaldi
ISBN 0-471-20690-3   Copyright © 2003 by John Wiley & Sons, Inc.

the site of a blood clot. Significant lytic activity outside of the vasculature may have severe consequences. Therefore these agents should be cleared rapidly from the circulation when the clot is resolved. Most biopharmaceuticals designed for chronic therapies, however, require sustained presence of drugs in target tissues and cells that are some distance away from blood capillaries and outside of blood vessels.

In mediating therapeutic response, protein pharmaceuticals are exposed to competing pathways of distribution in blood and highly perfused tissue that could sometimes produce untoward responses. Only a small fraction of drug is distributed to the sites of action. The schematic presented in Figure 13.2 depicts some of the

potential pathways, including distribution of significant fractions of drug to nontarget tissues and excretory fluids. In addition therapeutic proteins may be subjected to proteases in plasma and cells that degrade and convert the protein to its inactive form.

To modulate the rate and extent of therapeutic protein exposure in target tissues, scientists have considered other routes of administration. These alternative routes of drug delivery will be discussed below. All available data suggest that protein bioavailability is expected to be low for nonparenteral routes of administration (i.e., less than 10%).

13.1.1. Routes of Therapeutic Protein Administration

Parenteral

Most therapeutic proteins and peptides are formulated as solution or suspensions for injection (i.e., intravenous, subcutaneous, intramuscular, intraperitoneal, intravitreal, or intrathecal). These routes of administration are described in Table 5.5. Parenteral administration provides a more predictable therapeutic profile than oral administra-

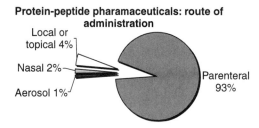

Figure 13.1. Routes of administration of biopharmaceuticals.

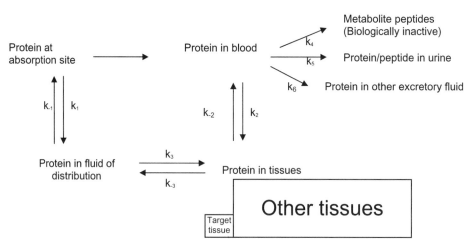

Figure 13.2. Disposition of therapeutic proteins after injection. In order to reach target tissue from the injection site, administered dose of protein is exposed to competing pathways. Target tissue is exposed to only a small fraction of the total dose.

tion but frequently incurs additional costs due to the requirement of trained health care professionals to implement the therapy. In addition the high degree of sterility and purity standards required for such products, coupled with the rapid elimination of proteins, contributes to the high cost of using biopharmaceuticals compared with the use of small-molecule drugs that do not require injection.

Oral

Oral drug administration is the preferred and most widely used route for small molecule drugs because of the ease of formulation, patient convenience and compliance, and usually good absorption in the intestine. Orally administered therapeutic proteins, however, are exposed to high levels of proteases, such as pepsin, in the stomach that can readily inactivate the drug. Even if proteins are protected by enteric coating to prevent drug release in the stomach, pancreatic proteases—trypsin, carboxypeptidase A, chymotrypsin, and esterases—found in the small intestine could degrade the protein. According to one estimate, it only takes about 10 minutes for a therapeutic protein to degrade when it is taken orally [1]. One of the strategies to minimize protease-catalyzed protein degradation is by modifying the L-amino acid, the naturally occurring form of amino acids that are protease substrates, to the D-amino-acid isomer at strategic peptide sequences. Converting native L-ala$_2$-glucagon-like peptide 1 to D-ala$_2$-glucagon-like peptide-1 results in a peptide that is resistant to breakdown by dipeptidylpeptidase IV, which specifically cleaves the L-ala$_2$ form. D-ala$_2$-glucagon-like peptide-1 encased in microspheres has been shown to increase biological activity of glucagon after oral administration, as reflected by reduced glycemic response to glucose challenge in normal and diabetic subjects (by 41 ± 12% and 27 ± 5%, respectively) [2].

Consequently only a small fraction of

■TABLE 13.1. Factors influencing protein and peptide absorption

Physical-chemical characteristics of therapeutic protein

Molecular weight and structure
Partition coefficient
Stability
PK
Electrostatic and net charge

Physiological

Degradative enzymes
Resistance of mucosal barrier
Membrane transporters
Metabolism at mucosa
pH changes
Disease conditions

protein surviving proteolytic cleavage is absorbed into the blood, and that small fraction is further subject to metabolism as it passes through the liver as part of first-pass drug metabolism. Therefore, in all likelihood, only a negligible fraction of an oral dose of protein is available for therapeutic action. Some of the factors influencing the rate and degree of protein and peptide absorption are listed in Table 13.1.

Table 13.2 lists the molecular weight and bioavailability from the gastrointestinal tract of proteins, peptides, and other molecules. The data collected to date suggest that oral absorption of molecules that are greater than 2000 daltons generally is less than 2%. Some barriers to protein absorption may be overcome by drug delivery strategies that will be described later, but an oral dosage form of a therapeutic protein will continue to be a challenge for the foreseeable future.

Buccal

The mucous membranes of the oral cavity are highly vascular and relatively permeable. Drugs designed to dissolve under the tongue or in the cheek pouch but not to be

■TABLE 13.2. Molecular weight and bioavailability of proteins, drugs, and peptides absorbed from gastrointestinal tract

Low (<5% absorption)	MW	Moderate (20–40% absorption)	MW	High (>50% absorption)	MW
Saquinavir	671	Ampicillin	349	Aminocephalosporins	~350
1-deamino-9-D-Arg vasopressin	1,007	Di- or tri-peptides	250–350	Enalapril	377
Leuprolide	1,208	Lisinopril	405	Oseltamivir phosphate	410
Empedopeptin	1,250	Tacrolimus	822	Talampicillin	482
Insulin	5,700	Cyclosporine	1,203		
Serum albumin	50,000				

Note: MW, molecular weight, expressed in daltons.

swallowed are examples of buccal dosage forms. Buccal administration is convenient and thereby promotes compliance. The mucosa in the oral cavity exhibits a lower protease activity than intestinal mucosa. The membrane barrier is similar to that of skin in that it is assembled as stratified squamous epithelium. As a result it is permeable to peptides with about 5 to 10 amino-acid residues. Permeation enhancers designed to transiently interrupt the mucosal barrier may be needed to promote absorption of larger peptides and proteins.

Rectal and Vaginal

Both the rectal and vaginal cavities provide sufficient volume to retain dosage forms and allow a reasonable degree of absorption of small molecule drugs. Rectal and vaginal administration of peptides and proteins are attractive because of the cavities' large mucosal surface area (200–400 cm^2) and richly vascularized nature (with lymphatic and blood vessels). Because only about one-third of the drug absorbed into rectal veins is shunted to the liver, the remaining two-thirds avoids first-pass metabolism.

Comparison of oral and rectal routes of administration of lidocaine showed that

bioavailability was 33% after oral administration and 67% after rectal administration [3], a difference most likely explained by partial avoidance of first-pass metabolism after rectal administration. However, the ability to retain rectal and vaginal dosage forms may be important to produce adequate and consistent drug absorption. Study results with aspirin suppositories indicate that bioavailability of aspirin is 54% to 64% for individuals with retention times of less than 5 hours but greater than 80% for individuals with retention times of 10 or more hours [4].

To achieve a reasonable degree of absorption, peptide and protein pharmaceuticals may need permeation enhancers to promote passage across mucosal cells. Another concern is that studies of insulin and enkephalin in animals suggest that protease activity may be high, especially in the rectal cavity [5,6]. On the other hand, the density of lymphatic vessels and drainage therein at these sites may be advantageous compared with other routes of administration.

Ocular

Typically less than 3% of topically applied drugs in ocular dosage forms—solutions,

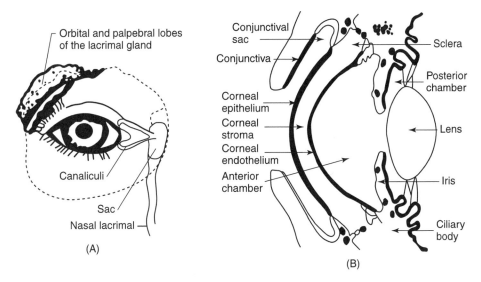

Figure 13.3. Schematic representation of the eye. (A) lacrimal drainage; (B) cross section of the eye. Physiological factors diminishing the fraction of drug delivery across the corneal epithelium to the aqueous humor include tears that drain into the lacrimal duct and limited permeation of protein and peptide across the corneal epithelium.

suspensions, or ointments—permeate the corneal epithelium, reach the aqueous humor, and then make their way into the systemic circulation. As soon as ophthalmic solutions of proteins and peptides are applied, physiologic protective mechanisms such as tear dilution, lacrimal drainage, protein binding, and enzymatic degradation are provoked, all working in concert to lower the effective concentration of the drug and limit passage across the corneal epithelium, the major barrier and rate-limiting step of intraocular delivery of proteins and peptides (Figure 13.3). According to one study, only a very small fraction (<0.1%) of leucine enkephalin, and even a protease-resistant D-ala$_2$-met-enkepalin derivative of this anti-inflammatory peptide, made their way into aqueous humor as intact peptides [7]. If delivery efficiency could be improved, the ocular route might be used to provide safe and effective local delivery of proteins and peptides of therapeutic interest in ophthalmology. Some candidate proteins and peptides for ophthalmic therapies are listed in Table 13.3.

■**TABLE 13.3.** Candidate proteins and peptides for ophthalmic therapies

Therapeutic Area	Candidates
Inflammatory	Enkephalin and derivatives
	Substance P
Immunologic response	Cyclosporin
	Interferons
Wound healing	Epidermal growth factor
	Fibronectin
	Insulin-like growth factor
Local effects in acqueous humor	Neurotensin
	Vasopressin
	VI peptide
	LH-RH and derivatives

Nasal

Therapeutic proteins designed for nasal delivery must cross mucosal epithelial cells of about 3 mm thickness, coated with degradative enzymes. The total surface of the nasal mucosa is about 200 cm^2, and the nasal cavity can accommodate about 1.5 ml of drug solution. Because the nasal cavity is rich in vasculature and lymphatic drainage, and passage through the nasal mucosa avoids first-pass metabolism by liver enzymes, it may be an attractive route of delivery for small proteins and peptides. Adequate protein absorption, however, will probably require an absorption enhancer. Intranasal delivery of small peptides that are less than 10 amino acids in length have been shown to exhibit fair to good bioavailability [e.g., 75% for somatostatin (6 aa), 10% for desmopressin (9 aa), and 100% for metkephamid (5 aa)] [8]. Peptides of greater than 20 amino-acid length such as glucagon (29 aa) and calcitonin (32 aa) produce less than 1% bioavailability.

Pulmonary

The upper respiratory tract, like the nasal cavity, consists of epithelium with columnar structure (30–40 μm thick) that is relatively impermeable to large molecules (Figure 13.4). On the other hand, the deep lung consists of alveolar (air sac) epithelium that is only 0.1 to 0.2 μm thick and substantially more permeable to peptides and proteins. When insulin (51 aa) and growth hormone (192 aa) are instilled into the deep lung, about 10% to 15% of these therapeutic agents are bioavailable. Because the alveoli surface is large, an estimated

75 m² for a 70 kg person, significant drug absorption should be achieved through deep-lung delivery of aerosolized pharmaceuticals. This route of drug delivery avoids first-pass liver metabolism, and therefore is an attractive route for protein administration. The rich blood supply in alveoli, designed for the blood–gas exchange process, also facilitates the rate of protein drug permeation across alveolar epithelium.

Aerosolized DNase (dornase) is a therapeutic protein designed for alveoli delivery to achieve local effects in the deep lung. Aerosolized DNase is formulated as a pulmonary dosage form, targeted for deep-lung delivery to reduce opportunistic infections due to the increased viscosity of mucus in the lung that affects respiratory function in patients with cystic fibrosis.

Transdermal and Topical

While uncontrolled topical delivery of proteins and peptides for local application such as open wound healing is easily achieved, it is much more challenging to use transdermal delivery systems to deliver protein across intact skin. In theory, transdermal delivery could provide constant drug release for days, avoids first-pass metabolism, and could allow drug effects to be rapidly terminated by simply removing

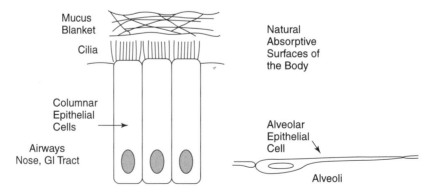

Figure 13.4. Comparison of epithelial cells lining the nose, the gastrointestinal tract, and the alveoli.

the patch. However, absorption across the skin for molecules larger than 1000 daltons has proved to be difficult, even with the addition of permeation enhancers.

In developing protein and peptide transdermal delivery systems, one must be mindful of the high interindividual variation in drug absorption across the skin. Large variations in bioavailability have been demonstrated with fentanyl patches, initially intended for postoperative pain relief but later abandoned due to unacceptable variability among individuals receiving the same dose [9]. Response in postoperative patients to the application of a fentanyl patch ranged from ineffective pain relief to severe respiratory depression, and effects were correlated with variations in plasma fentanyl levels [9].

The amount of drug that can be delivered thought the skin is limited by the size of the dosage form. As demonstrated with $2.5\,cm^2$ scopolamine skin patches, less than 1.5 mg is delivered over 3 days or 0.5 mg/day. Therefore a drug requiring delivery of 200 mg/day, for example, with the same permeability as scopolamine, would need a barely imaginable patch with a surface area of $1000\,cm^2$. Nevertheless, recent developments that employ ion-current or iontophoresis to enhance skin penetration may overcome this problem. Iontophoresis used an electrical current that mediates bulk water flow through endocrine glands in the skin and across the epidermis. While it was originally thought that iontophoresis would only enhance the transdermal absorption of polar and ionic drugs, this strategy may even promote absorption of nonionic and neutral proteins and peptides across the skin [10,11].

■ 13.2. PHYSIOLOGIC AND MECHANISTIC APPROACHES

While parenteral administration allows direct access of therapeutic proteins into blood and may provide a rapid response, injectable protein formulations often require frequent administration that sometimes incurs alternating periods of drug insufficiency and high drug concentrations in patients. As Figure 13.5 shows, switching from intravenous administration to subcutaneous administration often increases the proportion of a dosing interval in which drug concentration stays within the thera-

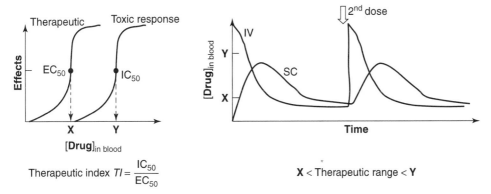

Interactions between therapeutic index and time course of plasma drug concentration

Therapeutic index $TI = \dfrac{IC_{50}}{EC_{50}}$

X < Therapeutic range < Y

Figure 13.5. Relationship between therapeutic index, derived from concentration-response curves (panel A), and time course of plasma drug concentration after intravenous (IV) and subcutaneous (SC) administration. Note that IV dosing of a drug exposes patients to excessive and fleeting plasma drug concentrations; the problem is much less pronounced with SC dosing.

peutic range. A survey of protein therapeutics from several therapeutic areas indicates that (1) they are administered mainly by intramuscular, intravenous, or subcutaneous injection, (2) their dose range is relatively small and often described in activity units (i.e., international units or IU) rather than mass (e.g., mg), (3) their half-life tends to be short, and (4) most are eliminated from the body by renal and hepatic mechanisms (Table 13.4). Some of the larger proteins and protein aggregates in blood will be eliminated through nonspecific phagocytic uptake of scavenger cells mainly of reticuloendothelial origin.

Proteins and peptides are subjected to various physiological processes that may render them inactive, as schematically presented in Figure 13.2. The degree and mechanism of inactivation has been described in detail in Chapter 5, and these mechanisms, in relation to molecular size, are summarized in Figure 13.6. Hydrolysis of polypeptides by carboxy peptidases and proteases play a significant role in metabolizing small proteins (less than 1200 daltons). Larger proteins, up to 6 kDa, are cleared mainly by fluid phase endocytosis and biliary secretion, and proteins up to 50 kDa are largely cleared by renal filtration. Six to 250 kDa proteins are often involved in receptor-mediated endocytosis

processes and are metabolized by these intracellular processes. Partial degradation and denaturation of proteins lead to exposure of hydrophobic domains that give rise to protein aggregation. Protein aggregates, typically 300 to 400 kDa in molecular weight, are recognized as particulate matter and readily phagocytized by reticuloendothelial system (RES) cells. While most therapeutic proteins are less than 250 kDa, some multimeric enzymes may have molecular weights greater than 250 kDa, and they may be cleared by a combination of mechanisms. The basic understanding of protein clearance mechanisms has allowed development of novel delivery strategies, including attachment of polyethylene glycol to reduce phagocytosis of proteins by RES cells and to increase half-life.

Most endothelial cell membranes are ordinarily impermeable to proteins. Transport across these barriers occurs only with the aid of receptor-mediated or other transport processes. However, many active sites (receptors) are located on cell surfaces and there is no need to permeate the cell. To achieve an adequate intracellular concentration, relatively large amounts of protein must be administered. Proteins administered by nonparenteral means and intended for systemic effects, such as intranasally and directly into the lungs,

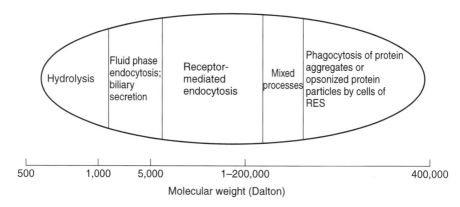

Figure 13.6. Schematic representation of size-dependent protein disposition mechanisms. Depending on molecular weight, a protein or peptide predominantly undergoes selective inactivation.

■TABLE 13.4. Pharmacokinetic and disposition profile of some protein drugs

Protein Drug	MW (kDa)	Route of Administration	Dose Range	Half-life α	Half-life β	Route of Elimination
Oncology and transplantation						
Interferon-alfa	18–23	IM/SC	2–10 × 10⁶ IU/m²	90–120 min	6–7 h	Renal and hepatic
Interferon-beta	23	IM/SC		13 min	6–7 h	
Interferon-gamma	20	IM/SC		25–35 min	1–2 h	
OKT3 (muronomab)	~150	IV	5 mg/day	60 min	6–10 d	RES and hepatic
Epoetin	30.4	IV	50–500 IU/kg		6–13 h	Renal and hepatic[a]
Granulocyte macrophage colony-stimulating factor (GM-CSF)	15–30	SC/IV	250 μg/m²	5 min	5 h	Renal and hepatic[a]
Granulocyte colony-stimulating factor (G-CSF)	18.8	IV/SC	5 μg/kg/d	30 min	3–4 h	Renal and hepatic[a]
Interleukin-2	15.5	IM/SC	2–6 × 10⁶ IU/m²	6–10 min	60–90 min	Renal and hepatic
Endocrine and metabolism						
Somatotropin	~192	IM/SC	0.05 mg/kg	20 min		Hepatic
Adenosoine deaminase	90 (+170)	IM	~15 IU/kg	~3 d (variable)		Hepatic[a]
α1-antitrypsin	52	IV	60 mg/kg	~4 d		Hepatic[a]
DNase	20	Inhalation	2.5 mg	limited systemic levels		Pulmonary and renal
Cardiovascular						
Alteplase (r-tPA)	68	IV	1–2 mg/kg (3 parts)	4 min		Hepatic and renal
Streptokinase	46	IV	0.25–1.5 × 10⁶ IU	23 min		Hepatic and renal
Urokinase	34	IV	4400 IU/kg	20 min		Hepatic and renal
Abciximab (Reopro; Fab)	47.6	IV	0.25 mg/kg	10 min	30 min	Hepatic and RES
Antimicrobial						
Human monoclonal antibody against lipid A	~150	IV		15–30 h	21–26 d	Hepatic and RES
Hyper immune globulin against cytomegalovirus (human)	~150	IV	250 mg	15–30 h	21–26 d	Hepatic and RES

[a]predicted major route(s) of elimination; RES, reticulo-endothelical system; IM, intramuscular; SC, subcutaneous; IV, intravenous

have an additional barrier in reaching the circulation—the epithelial lining.

Pharmaceutical scientists assess and express drug permeation across membrane barriers in terms of flux. Flux measures the molar unit of a drug that permeates a resistant barrier (e.g., skin or gastrointestinal epithelial cells) per unit time and surface area (Box 13.1). Permeation enhancers, such as alcohols and surfactants, increase flux by modulating resistance factors that counteract drug diffusion across barriers at the site of administration.

Fundamental understanding of protein disposition and permeation in and out of the systemic circulation has led to invention of several novel approaches to overcome some of the limitations encountered in using proteins as pharmaceuticals. These drug delivery strategies provide an increase in safety (or reduced dose-dependent toxicities), an improvement in drug efficacy (or potency), or both. These novel drug delivery strategies can generally be categorized as (1) controlled release, (2) permeation enhancement, (3) modulation of clearance, and (4) localization or targeting to the site of actions.

13.2.1. Controlled and Sustained Release

Controlled and sustained release strategies aim to maintain plasma drug levels within the therapeutic concentration range for longer periods than are possible with immediate release dosage forms, thereby minimizing periods of excessive or ineffective drug concentrations, and to reducing the frequency of administration (Figure 13.5). Controlled release allows less frequent administration with tighter control of peak and trough concentrations to maintain intended responses than is possible with a bolus injection. It also allows safe targeting of higher plasma drug levels for drugs with short half-lives. In some cases this strategy indirectly improves efficacy. Some of the benefits of controlled delivery

of biopharmaceuticals are improved patient compliance, fewer injections, and potentially fewer adverse effects.

Controlled release can be achieved using infusion devices in hospital and clinical settings. Advances in miniaturization of these devices is beginning to provide portability. However, these devices are not ideal for most applications due to their cost and complex training requirements.

Biodegradable drug carriers composed of biopolymers or lipid membrane vesicles (liposomes) can be used to formulate protein drugs in colloidal or suspension dosage forms. These biodegradable carriers can release incorporated protein in a controlled and sustained manner.

Polymer-Based Drug Carriers

A biopolymer-based drug carrier designed for protein delivery must meet the following requirements. In addition to controlling the release of drug, (1) the carrier must be biocompatible and degraded products must be nontoxic, (2) the carrier must incorporate the protein in a sufficiently gentle manner to retain bioactive conformation, and (3) the carrier must be able to incorporate the protein in pharmaceutical scale [12].

More than a dozen biocompatible and biodegradable polymers have been described and studied for their potential use as carriers for therapeutic proteins (Table 13.5). However, some of the monomer building blocks such as acrylamide and its derivatives are neurotoxic. Incomplete polymerization or breakdown of the polymer may result in toxic monomer. Among the biopolymers, polylactide cofabricated with glycolide (PLG) is one of the most well studied and has been demonstrated to be both biocompatible and biodegradable [12]. PLG polymers are hydrolyzed *in vivo* and revert to the monomeric forms of glycolic and lactic acids, which are intermediates in the citric acid metabolic pathway.

Large-scale preparation procedures

BOX 13.1. ESTIMATING THE RATE OF A DRUG MOVING ACROSS A BARRIER

Drug delivery scientists consider that the overall process of drug delivery from a site of administration to the site of action can be described by the term flux (J) (Figure 13.7). The flow of drug from site A to B is restricted by the "barrier," which is in the form of cells such as those of the gastrointestinal tract (for oral drug absorption), the blood brain barrier (BBB) (for drug crossing into brain tissue), or specific cellular membranes (for intracellular delivery). The movement of drugs from site A to site B requires input of electro- and physiochemical energy (e.g., a concentration gradient), biological energy (e.g., transporters), or mechanical energy (e.g., modifying the permeability of endothelial cell membranes). The rate of flux, in theory, can be modulated by regulating these energy sources and can be mathematically expressed.

Flux is expressed as

$$J = \frac{D}{h}(C_A - C_B)$$

where D is the diffusion coefficient (in cm^2/h), h is the thickness (in cm) of the barrier, and flux J is the flow rate of a drug in compartment A across a barrier of h thickness and into compartment B. At the drug absorption site A, the rate of change can be expressed as

$$-\frac{dC_A}{dt} = \frac{D}{h}(C_A - C_B)$$

from which permeability of a drug across the barrier; D/h can be estimated by measuring time-dependent change in drug concentration at sites A and B.

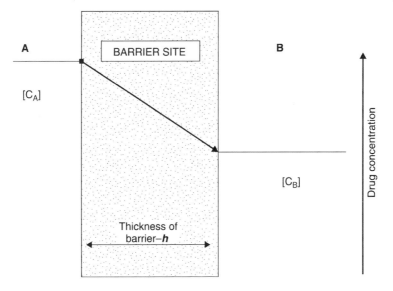

Figure 13.7. Schematic representation of drug flux across a barrier (e.g., cell or synthetic membranes).

■TABLE 13.5. Examples of biodegradable polymers tested for sustained protein release

Polymeric Matrix	Monomer	References
Polylactide-co-glycolide co-polymers	Lactic acid	[15]
	Glycolic acid	
Poly-(vinyl alcohol)-lactide copolymer	Vinyl alcohol	[147]
	Lactic acid	
Branched polyoxyethylene-lactide	Oxyethylene	[146]
	Lactic acid	
Poly-(orthoesters) terpolymers	Orthoesters	[145]
(poly(N-isopropyiacrylamide-co-	Isoproyiacrylamide	[143]
butylmethacrylate-co-acrylic acid))	Acrylic acid	
	Butylmethacrylate	
Poly-(acrylic acid)	Acrylic acid	[144]

have been established for formulations of low-molecular-weight drugs in PLG microspheres employing harsh conditions such as phase separation, solvent evaporation, and spray drying that require the use of heat and organic solvents. These processes, however, denature and degrade proteins, and thereby reduce their biological activity. Recently, an improved microsphere fabrication process, ProLease, was developed to overcome some of the hurdles (see Bartus et al. [12] for additional details), permitting production of bioactive human growth hormone (hGH) with high encapsulation efficiency (>95%) in pharmaceutical scale. In the process, protein encapsulation into PLG occurs at a low temperature (–40°C in liquid nitrogen), and organic solvents for emulsion stabilization and encapsulation are avoided [12].

The biopolymer-protein complex must be optimized to achieve the desired controlled release rate of protein. This is accomplished by means of both empirical and systematic modifications of protein stabilization agents and polymer composition and fabrication conditions. Stable formulations of hGH and α-interferon have been designed to produce sustained release profiles suitable for therapeutic application. Early efforts to achieve long-acting protein products relied simply on the *in vitro* release rate of protein in an aqueous medium. These *in vitro* measures, however, do not translate directly to the *in vivo* situation because protein release in the body is controlled by a combination of factors, including polymer hydration, protein disassociation, and absorption into the circulation.

Despite the complexity in vitro protein release from a biopolymer-protein complex has guided the development of PLG microspheres containing hGH stabilized with Zn^{++} that produce sustained therapeutic levels of hGH in humans for up to 28 days after a single subcutaneous dose (Figure 13.8). The improved pharmacokinetic profile achieved with PLG-formulated hGH is reflected in the results of clinical efficacy studies. Subcutaneous injections of PLG-hGH complex (*Nutropin Depot*) either at 0.75 mg/kg twice monthly or 1.5 mg/kg once monthly are effective in increasing the growth rate of pediatric patients with growth hormone deficiency [13]. PLG formulation of hGH provides sustained plasma levels over four weeks, allowing a reduction in the frequency of hGH injection from daily or weekly to monthly [12,13]. *Nutropin Depot* has been used since 1999.

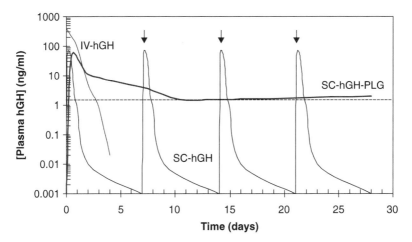

Figure 13.8. Effects of route and sustained release formulation on the time course of human growth hormone concentration in plasma. Shown is the average time course of human growth hormone (hGH) in plasma after intravenous (0.02 mg/kg) and subcutaneous (0.1 mg/kg) administration in humans. Arrows indicate weekly subcutaneous dosing of hGH in solution. A single dose of the same protein formulated in polylactide-co-glycolide (PLG) microspheres (0.75 mg/kg) given subcutaneously sustains human growth hormone levels in plasma for at least one month.

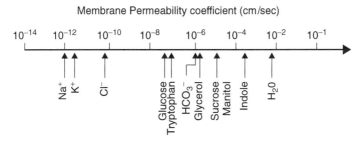

Figure 13.9. Membrane permeability coefficience of solutes. Solute permeabilities across typical lipid bilayers of liposomes or lipid vesicles are presented as their respective coefficients in cm/s. In the absence of other transport processes, it would require 10^{12} s to move Na^+ across 1 cm distance. When there is a concentration difference across a membrane, multiplying the concentration difference (mole/ml; equivalent to mole/cm^3) by the permeability coefficient (cm/s) allows estimation of flow rate (mole/s-cm^2). For example, a concentration difference of 10^{-6} mole/cm^3 Na^+ (or 1×10^{-3} M Na^+) would provide a flow of 10^{-6} mole/cm$^3 \times 10^{-12}$ cm/s $= 10^{-18}$ mole/s through 1 cm^2, or 0.006 mole/s through 1 μm^2 of a membrane bilayer.

Lipid-Membrane-Based Drug Carriers

Bilayered lipid vesicles or liposomes resemble cell membranes. They are constructed with synthetic or semisynthetic phospholipids (Table 13.6) and provide selective passage of solutes—ionic or non-ionic—at a controlled rate. Rates of passage, as measured by permeability coefficients, are summarized in Figure 13.9. While water can permeate across the lipid membrane with relative ease, passage of sodium and potassium ions is slower by about 10 orders of magnitude.

Permeability of proteins and peptides across lipid membranes is expected to be

■TABLE 13.6. Some of the commonly used phospholipids and the attributes of head and fatty acyl (tail) groups

Domain	Effect on Liposome Membrane	Functional Attribute on Lipid Bilayer
Tail group–Fatty acyl chains: R_1 and R_2 (C14–18 in length)		
Increase degree of saturation	Increase rigidity; decrease fluidity	Elevate $T_c{}^a$
Increase chain length of R_1 and R_2	Increase thickness of bilayer	Elevate T_c
Varying degree of saturation and chain length on R_1 and R_2	Decrease order of membrane packing	Lower T_c (compared to phospholipid with two identical fatty acyl tails)
Head group: R_3		
Choline: $-CH_2-CH_2-N(CH_3)_3{}^+$	Some surface hydration	Neutral charge
Ethanolamine: $-CH_2-CH_2-NH_3{}^+$	Minimum degree of surface hydration	Neutral charge
Serine: $-CH_2-CH(COO^-)-NH_3{}^+$	Some surface hydration	Negative charge
Glycerol: $-CH_2-C(OH)CH_2OH$	Some surface hydration	Negative charge
PEG (ethanolamine): $-CH_2-CH_2-NH-PEG$	Enhanced surface hydration and steric effect	Negative charge

^{a}Tc; Phase transition temperature of phospholipid bilayer.

even slower than sodium or potassium. Therefore liposomes could provide sustained release of encapsulated proteins and peptides. While *in vitro* release of a protein or peptide from a liposome is controlled by membrane permeability, *in vivo* release is complicated by degradation of the liposomal lipid head and tail groups (fatty acyl chains) by enzymes (e.g., phospholipases C, A1, and A2). By taking account of all of these factors, one can develop a protein product encapsulated in a liposome formulation that provides slow but complete release of the biopharmaceutical. The phospholipid breakdown products of a spent liposome could be recycled *in vivo*

for membrane biosynthesis; they generally exhibit low or no toxicity. Commonly used phospholipids used to construct liposomal drug carriers and the attributes of head and tail groups are described in Table 13.6.

Liposomes are colloidal particles that can be prepared with (phospho)-lipid molecules derived from either natural sources or chemical synthesis (recently reviewed by Lian and Ho [14]). The potential application of liposomes as biodegradable or biocompatible drug carriers to enhance the potency and reduce the toxicity of therapeutic agents was recognized in 1960. In the 1960s and 1970s various methods for liposome preparation were developed as models of biological membranes. By 1970 liposomes were proposed as drug carriers to modify a drug's therapeutic index by reducing toxicity or increasing efficacy (or both) of the parent drug.

Early research in liposomal drug preparation was beset with problems, including insufficient understanding of liposome disposition and clearance, inaccurate extrapolation of *in vitro* liposome-cell interactions or liposome targeting data, and insufficient stability and circulation time of liposome-based drugs. Advances in the late 1980s and early 1990s, including a detailed understanding of lipid polymorphisms, physiological mechanisms of *in vivo* liposome disposition, and lipid–drug and lipid–protein interactions, overcame many of the early disappointments. The result was liposome designs with increased stability both *in vitro* and *in vivo*, improved biodistribution, and optimized residence time of liposomes in the systemic circulation. The goal of using liposomes as drug carriers in pharmaceutical applications was realized in the mid-1990s with several FDA-approved liposome-formulated drugs, mainly antifungal and anticancer therapies.

Cytokines such as interleukin (IL)-1, IL-2, IL-6, IL-7, interferon-gamma, and granulocyte macrophage colony-stimulating factor (GM-CSF) have been formulated in liposomes and evaluated in laboratory animals [15–17]. The data to date indicate that (1) procedures required for encapsulating proteins in liposomes are sufficiently gentle to do no harm to the protein, (2) most liposome encapsulated proteins are biologically active, both *in vitro* and *in vivo*, and (3) the route of administration may greatly influence the disposition of liposomes, thereby modifying the overall half-life and residence time of the liposome-encapsulated proteins. In one study with liposome-encapsulated IL-7 given subcutaneously to guinea pigs, the apparent plasma half-life of IL-7 was increased by as much as 26-fold with about equivalent overall bioavailability compared with that of the soluble form of IL-7 (Table 13.7) [16]. These liposomes are relatively large, in the range of 1 to 5 µm diameter, and diffuse slowly from the injection site. Subcutaneous administration of a mixture of 40:60, free and liposome-encapsulated IL-7 exhibit a mean residence time of 7.9 days, compared with a residence time of four hours with soluble IL-7 given subcutaneously [16]. These results indicate that proteins encapsulated in relatively large liposomes can be administered subcutaneously to provide sustained and controlled release to improve the therapeutic index of proteins.

Because the size and surface net charge of liposomes can also influence the rate and extent of tissue distribution, these characteristics could be used for target-specific delivery of liposome-encapsulated proteins and drugs. These issues will be discussed later in the context of "Site-specific delivery strategies."

13.2.2. Permeation Enhancement

The major limitation to developing therapeutic proteins in nonparenteral dosage forms is the poor permeability of these water-soluble and hygroscopic macromolecules across the tissue layers at the site of drug administration. These tissue layers include the epithelium of the gastrointestinal tract, the stratum corneum of the skin, and the epithelium lining of the alveoli and nasal cavity.

Poor absorption of small-molecule drugs across such external barriers has been presumed to be due to the drugs' physiochemical properties, which can be systematically modified by medicinal chemists, and the nature of the physiological barrier, which can sometimes be transiently altered. To overcome physiological barriers, scientists have studied a series of synthetic and natural compounds exhibiting absorption enhancement properties. These molecules, included in the dosage formulation to enhance absorption, are

■TABLE 13.7. Select pharmacokinetic parameters of subcutaneous IL-7 in liposome and soluble form

Treatment	Route	Dose (μg)	Peak Blood Concentration (ng/ml)[a]	Half-life β (min)	AUC_{0-t}/dose (min/ml)[b]	Au (% Injected Dose)	Bioavailability (% of IV)	
							AUC_{0-t}[c]	Au^d
Soluble IL-7	IV	42	2402 ± 331	116 ± 10	15.6 ± 0.7	93.9 ± 10.2		
Soluble IL-7	SC	39	49.7 ± 8.5	115 ± 12	9.9 ± 0.3	60.6 ± 4.2	68.6 ± 1.7	72.2 ± 5
Mixed IL-7 Liposomes	SC	39	20.7 ± 3.1	1516 ± 366	9.2 ± 0.2	54.6 ± 6.1	63.6 ± 13	65.1 ± 7.3
Liposome-encapsulated IL-7	SC	39	2.6 ± 0.4	3018 ± 897	9.4 ± 3.9	57.3 ± 0.4	64.4 ± 24.9	73.6 ± 0.5

Note: Guinea pigs were injected with ^{125}I-IL-7 at the indicated dose and route. Blood samples were collected, and the pharmacokinetic parameters were determined based on the time course of blood IL-7 profile.

[a]Mean ± SD of the peak blood IL-7 data was normalized by the injected dose of IL-7.

[b]Blood AUC was determined from 0–7200 min time-blood concentration curve, corrected for the dose.

[c]Bioavailability determined from blood AUC, corrected for dose.

[d]Bioavailability determined for cumulative amount excreted in urine (Au) after correcting for dose.

generally known as permeation enhancers. Significant improvement in bioavailability of proteins delivered by nonparenteral means could provide formulations that can be self-administered, and thereby improve safety and potentially reduce the overall cost of therapy.

Since 1964 organic solvents such as dimethyl sulfoxide (DMSO), and *N,N*-dimethyl formamide (DMF) have been used to enhance skin penetration of peptide derivatives including antibiotics and vasopressin [18]. While DMSO enhanced skin penetration, local irritation and systemic effects, including changes in the lens of the eye, prompted a search for safer permeation enhancers. Researchers identified potentially useful natural and synthetic enhancers. Many of these compounds formed mixed micelles that exhibited concentration-dependent effects on membranes such as the intestinal mucosa. These enhancers contain both hydrophobic and hydrophilic side chains and are known as amphoteric compounds (Figure 13.10). At high concentrations in aqueous solutions, they form micelles that momentarily permeate and disrupt cell membranes, allowing the passage of proteins and peptides across absorption barriers. As the concentration of permeation enhancers are reduced, they form soluble monomers, instead of polymeric micelles, that are

membrane inactive, thereby allowing the cell membrane to return to its normal state and resume its role as a physiologic barrier.

The accumulated data from cellular and animal studies on permeation enhancers indicate that these agents can be categorized into four classes, according to their effectiveness and safety (Table 13.8) [18,19]. While water-miscible organic solvents (class IV agents) and surfactants (class III agents) provide comparable permeation enhancement, the higher tendency to produce local and systemic side effects render them less attractive. The natural and synthetic compounds of classes I and II, such as fatty acids and bile salt derivatives, have moderate-to-strong action on membrane permeability but allow the membrane to recover quickly. They are considered to be better choices.

Both interferon-α and -β have been shown to be absorbed across the rectal mucosa when given with the fatty acid linolenic acid as mixed micelles. In the absence of a permeation enhancer, rectally administered interferons (165 aa) do not achieve detectable levels in blood nor lymph. With 0.56% linolenic acid, a significant degree of lymphatic absorption of interferon-α and -β was detected [20]. The rectally absorbed interferon first distributes into the lymphatics, and interferon concentration is much higher in lymph than

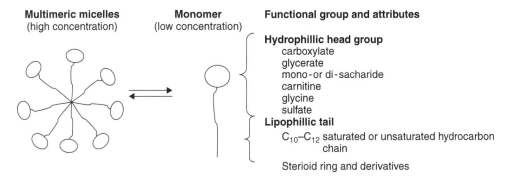

Figure 13.10. A schematic representation and summary of major permeation enhancers and their functional attributes.

■TABLE 13.8. General classification of permeation enhancers

Class	Effectiveness	Safety and Recovery	Examples
I	Strong and rapid action	Relatively safe due to fast recovery rate	Acylcarnitine Alkylsaccharides such as OG, DM, and LM Azone Fatty acids such as oleic, capric, linoleic acids, and their monoglycerides
II	Moderate and fast activity	Relatively safe due to fast recovery rate	Bile salts such as cholate, taurocholate and derivatives, UDCA, CDCA, SCG, STDHF Homovanilate Salicylates such as 3- or 5-methoxy-salicylate and salicylate
III	Strong to moderate activity	May exhibit tissue disturbance because these molecules tend to remained associated with cells	Surfactants such as SDS, Triton X-100, BL 9-EX, Brij, polyoxy-23 lauryl ether Chelating agents such as EDTA, EGTA, citric acid, phytic acid and enamine derivatives
IV	Moderate activity and remained as water miscible solvents	Comparatively safe but may produce systemic side effects at high concentrations	Ethanol, DMSO, DMF, and DMAC

Source: Adapted from Muranishi [109] and Augst [110].

Abbreviations: OG, *n*-octyl-β-D-glucopyranoside; DM, dodecylmaltoside; LM, *n*-lauryl-β-D-glucopyranoside; Azone, 1-dodecyl azacyclohepatan-2-one, SCG, sodium glycocholate; STDHF, sodium tauro-24,25 dihydrofusidate; UCDA, ursodeoxycholate; CDCA, chenodeoxycholate; SDS, sodium dodecylsulfate; DMSO, dimethyl sulfoxide; DMF, *N,N*-dimethyl formamide; DMAC, *N,N*-dimethyl acetamide.

in blood [18]. Interferon administered by intravenous injection cannot preferentially distribute in the lymph, and blood concentrations are higher than lymph concentrations. Therefore increased distribution of interferons to lymph nodes might be made possible by rectal administration with permeation enhancers, thereby providing a unique opportunity for interferon use in cancer therapies.

Parenteral calcitonin (32 aa) formulations are used to regulate calcemia. Intranasal delivery of calcitonin, a potentially convenient alternative, produces low bioavailability and is not useful as such. Permeation enhancers, including mixed micelle formulations composed of either *n*-lauryl beta-D-maltopyranoside, sodium glycocholate, and linolenic acid, or a steroid derivative, sodium tauro-24,25 dihydrofusidate (STDHF), have been evaluated for use as permeation enhancers for intranasal calcitonin. While both formulations provided improved bioavailability of calcitonin, STDHF appeared to be a safer agent [8,21]. Calcitonin bioavailablity after intranasal administration with STDHF is about 8% but not measurable in its absence [8].

STDHF has also been shown to enhance intranasal delivery of insulin. The bioavailability of insulin (51 aa) formulated as a nasal drop or spray containing STDHF was 16% to 47% in healthy volunteers [22]. The degree of permeation enhancement was dependent on the molar ratio of STDHF to insulin. Insulin bioavailability is negligible in the absence of the enhancer. A similar degree of permeation enhancement was achieved with a powder formulation [22].

In addition to improving drug absorption, permeation enhancers that form mixed micelles with peptides or proteins may also provide protection against metabolic processes, including protease-dependent inactivation, efflux transport, and protein denaturation, all of which could lead to reduced bioavailability. A permeation enhancer that may also inhibit some competing metabolic processes could greatly improve the poor and inconsistent bioavailability of some short and cyclic polypeptides such as cyclosporine (11 aa cyclic peptide) (Figure 13.11).

Efflux transporters, such as multidrug resistant protein, known as P-glycoprotein (Pgp), and related multidrug-resistant associated protein isoforms (MRPs), have been shown to be important for structurally diverse lipophilic compounds. Pgp and several MRPs, which were first identified in multidrug-resistant tumor cells, were shown to be important in reducing the bioavailability of even well-absorbed drugs, such as verapamil, quinidine, and digoxin, and hydrophilic compounds, such as peptides and peptidomimetics. As much as a 10-fold increase in oral bioavailability of Pgp substrates such as paclitaxel was demonstrated by inhibiting the Pgp-dependent secretory transport with a Pgp-specific inhibitor, PSC833. The role of Pgp transport in bioavailability has also been shown to be important for cyclosporine. Because a number of permeation enhancers, such as polysorbate, polyoxyl 40 hydrogenated castor oil or POE-35 castor oil (Cremophor EL), and tocophorol derivatives have been shown to interfere with Pgp activity, these agents could be used to overcome resistance to antitumor agents and also improve the bioavailability of orally administered polypeptides such as cyclosporine. The emulsion dosage form of cyclosporine (*Neoral*) is an example where the inclusion of permeation enhancers that interfere with secretory export of drug gives rise to about twofold higher bioavailability of cyclosporine [23]. The use of this permeation-enhancer combination produces both safe and improved therapeutic outcomes of patients on cyclosporine [24–27].

Today a number of permeation enhancers with multifunctional effects are being developed and evaluated for inhalational, buccal, ocular, intranasal, and intravaginal delivery of peptides and proteins. Some of these dosage forms may also be in a powder form such as those designed for dry-powder inhalation devices for protein and peptide delivery through lung alveoli. These devices will be discussed in the context of physical drug delivery strategies (see below).

13.2.3. Modulation of Drug Clearance

Most proteins currently approved for therapeutic application are cleared by the

Cyclosporine A

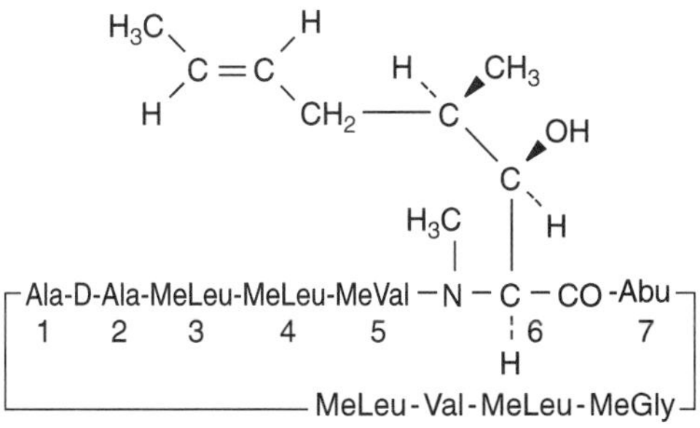

Figure 13.11. Structure of cyclosporine A, an 11 amino-acid cyclic peptide. Cyclosporine is chemically designated as [R-[R*, R*-(E)]]-cyclic(L-alanyl-D-alanyl-N-methyl-L-leucyl-N-methyl-L-leucyl-N-methyl-L-valyl-3-hydroxy-N,4-dimethyl-L-2-amino-6-octenoyl-L-(alpha)-amino-butyryl-N-methylglycyl-N-methyl-L-leucyl-L-valyl-N-methyl-L-leucyl).

kidneys and liver (Table 13.4). Mechanistic studies have shown that small to intermediate size proteins (i.e., from a few thousand daltons to 50,000 daltons) are often eliminated renally if no ligand-specific reabsorption exists. Classic size-limitation studies on glomerular filtration have clearly demonstrated that macromolecules less than 50 kDa are cleared from blood as a part of renal elimination. Because most therapeutic proteins are less than 50 kDa, renal elimination is an important route of drug clearance (Table 13.4).

On the other hand, large macromolecules are subjected to nonspecific phagocytic uptake. This can occur in liver and blood by means of scavenger cells (i.e., reticuloendothelial system [RES] cells). Large proteins such as adenosine deaminase (170 kDa) and somatotropin (192 kDa) accumulate in the liver and are cleared by RES cells (Table 13.4).

A drug delivery strategy that reduces both renal and hepatic clearance would reduce the total dose of protein or peptide needed for effectiveness, decrease the frequency of dosing, and possibly improve the safety profile. A proven approach for reducing hepatic and renal clearance is surface modification of proteins and peptides to increase hydrophilicity and, in the process, increase the overall hydrodynamic size of the protein, thereby reducing renal filtration. The increased hydration of surface-modified proteins also camouflages the molecule and thereby shields it from recognition by RES cells (phagocytic uptake).

Polyethylene glycol (PEG), a hygroscopic polymer, was the first molecule to be considered for modifying the surface of proteins. It has proved to be very successful. PEG, available in molecular weights ranging from a few hundred to several thousand daltons, have been used as parenteral excipients for decades with proven clinical safety.

Adenosine deaminase (ADA) was the first therapeutic enzyme coupled to PEG with the aim of reducing clearance and thereby overcoming the short half-life of ADA. Patients deficient in ADA are unable to regulate purine metabolism. As a result purine metabolites (e.g., adenosine monophosphate) accumulate to cytotoxic levels in B-lymphocytes and lead to severe B-cell depletion that presents clinically as severe combined immunodeficiency syndrome (SCIDS). While intramuscular injection of unmodified ADA provides some relief, antibodies develop rapidly against the protein and prevent it from being useful as replacement therapy. Even in the absence of antibodies, unmodified ADA's plasma half-life is only a few minutes.

Surface modification of ADA by covalent attachment or grafting of PEG to the protein has been shown to improve the plasma half-life of ADA in mice to about 28 hours after intravenous injection and reduce antibody formation by about fourfold. Fourteen weekly injections of PEG-ADA are required to elicit antibody formation in mice; antibody formation occurs after a single injection of unmodified ADA [28]. Administration of PEG-modified ADA (pegylated ADA) in humans also shows a significant increase in plasma half-life ($t_{1/2}$ = 72 hours) compared with unmodified ADA, permitting a weekly dosing schedule to produce sufficient therapeutic effect (i.e., notable improvement of immune function) [29]. Currently available PEG-ADA (*Adagen*) contains 11 to 17 PEG molecules (5 kDa each) on each molecule of ADA. The molecular weight of PEG-ADA is 260 kDa, compared with a molecular weight of 170 kDa for unmodified ADA.

The success of PEG-ADA led to other applications of this surface-modification strategy. In particular, the ability of PEG to increase the hydrodynamic size of IL-2

■TABLE 13.9. Effective size of IL-2 conjugated to polyethylene glycol (PEG) and IL-2 clearance in rats

IL-2 Modified with	PEG size (kDa)	PEG: IL-2 (mole ratio)	Effective Size (kDa)	Clearance (liters/h)	Volume (ml/kg)	Half-life $t_{1/2\beta}$ (min)
None	—	—	19.5	19.2	100	44
PEG	0.35	5	21	16.1	89	57
PEG	4	1	40	10.0	93	44
PEG	4	2	66	3.8	95	162
PEG	10	1	103	2.0	80	263
PEG	4	3	104	2.0	86	292
PEG	5	4–5	208	0.8	60	370
PEG	5	>5	280	0.8	106	409
PEG	20	2	326	1.8	91	256

Source: Adapted from Knauf et al. [111].

Note: Soluble recombinant interleukin-2 (IL-2) was modified with various lengths of polyethylene glycol (PEG) polymers with average molecular weight listed. The final PEG to IL-2 ratio is listed as molar ratio, and their respective hydrodynamic size was determined based on their sendimentation coefficient. Hence the molecular weight listed in Effective Size is not the sum of the PEG and IL-2 with consideration of molar ratio of the PEG-IL-2 conjugates.

and reduce renal filtration has been well studied [30,31]. Unmodified soluble IL-2 is rapidly cleared from plasma with a half-life of about 44 minutes. Surface modification with a varying number of PEG molecules for each IL-2 molecule and with PEGs of varying chain length (molecular weight) has increased IL-2's half-life by as much as 20-fold (Table 13.9). A rapid decrease in clearance of IL-2 was observed as the effective molecular size of PEG-IL-2 increased from 19.5 to 70 kDa; additional increases in molecular size produced less dramatic effects on systemic clearance [30]. The biphasic size-dependent reduction in clearance was later demonstrated to be due to renal filtration, specifically due to the permeability threshold of the glomerular membrane. It is well-documented that the glomerular membrane retains proteins larger than albumin (50 kDa).

Collectively the surface-modification studies of ADA and IL-2 with PEG poly-mers indicate that increasing surface hydration or hydrophilicity can reduce phagocytic-dependent protein clearance by RES cells, while overall increases in effective molecular size may reduce renal clearance, leading to overall reduction in systemic clearance and increase in plasma half-life of therapeutic proteins. As a result, smaller doses of the protein may be administered less frequently and may improve safety and efficacy.

Other hygroscopic polymers such as polyoxyethylated glycerol have been tested with varying degrees of success. Some of the therapeutic proteins successfully modified with PEG are interferon (Peginterferon alfa-2b or *PEG-Intron*), granulocyte colony-stimulating factor (Pegfilgrastim or *Neulasta*), and L-asparaginase (*Pegaspargase* or *Oncaspar*). A similar surface modification can be achieved by other molecular approaches, and some examples are discussed in Section 13.4.

13.2.4. Localization or Targeting to Sites of Actions

Developing targeting strategies to preferentially localize therapeutic proteins and other drugs at key sites to produce maximum pharmacologic effects has long been the ultimate goal of pharmaceutical scientists. In theory, preferential localization of drugs to target tissues and cells could be achieved using drug carriers and target recognition molecules such as ligands, receptors, or antibodies. In practice, however, identification of target tissues and cells of interest could be complex.

Effector cells and tissues often are either not well defined or distributed widely throughout the body. In addition specific carriers or targeting molecules capable of providing differential distribution to target sites are not always available. Despite the challenges, targeting is a worthwhile goal. Not only can targeting improve effectiveness, but by keeping the drug away from sites associated with toxicity, targeting improves safety. Regardless of the strategy used to improve safety and efficacy, successful targeting of drug to tissues and cells could provide a significant increase in the therapeutic index of protein therapeutics.

The selected cell-specific drug delivery system, identified and optimized with *in vitro* cell- and tissue-based evaluation systems, must provide a significant degree of target-tissue accumulation *in vivo* to be clinically useful. Tissue distribution must be considered early in the course of drug development. Liposomes carrying specific ligands were designed and shown to produce target-selective delivery to cancer cells *in vitro*, but the results did not directly correlate with the degree of tumor growth suppression in extravascular space because of the poor permeability of liposomes across the vasculature. Although the liposome had an avid affinity for the target, it could not reach the target, and target-specific delivery to tumor cells was not realized [14]. This limitation was only recognized late in the development of liposome delivery systems and represented a very costly oversight.

Size-Dependent Accumulation of Particles in Phagotocytic Cells and Lymphoid Tissues

Most proteins in solution assume an overall hydrodynamic size of only a few nanometers (nm) in diameter. Upon hydration, immunoglobulin G (IgG), for example, exhibits a diameter of about 5 nm. In the absence of receptor-mediated uptake, protein and peptide particles are too small to be recognized by RES cells. While avoidance of phagocytosis is useful for increasing plasma residence time of proteins, for a protein required to accumulate in phagocytic target cells, an increase in the apparent size of the molecule is needed for efficient delivery. Encapsulation of proteins and peptides in liposome carriers, 100 nm or larger, have been shown to greatly improve protein localization in phagocytically active RES cells, leading to therapeutic protein accumulation in lymphoid organs, such as the spleen and lymph nodes.

Vaccines based on live-attenuated or inactivated HIV or herpes simplex virus (HSV) are not acceptable because of the risk of infection. On the other hand, vaccines constructed with recombinant antigens of HIV and HSV, which avoid the risk of infection, are poorly immunogenic and elicit insufficient cell-mediated protection against infection and reactivation. The inability to elicit strong cell-mediated immune responses is believed to be due to the inability to localize viral antigens to antigen-presenting cells, many of which are phagocytically active. By incorporating soluble recombinant HSV or HIV antigens in liposomes, researchers have demonstrated preferential delivery to macrophages in spleen, a secondary immune organ [32]. The enhanced delivery to antigen-presenting cells resulted in the induction of cell-mediated immune responses, detectable in the form of selective elimination of virus-infected cells [33,34]. The immune enhancement of HSV surface antigen, glycoprotein-D, by this approach was shown to reduce recurrent outbreaks in guinea pigs infected with genital herpes (HSV-2) [33].

The mechanism by which liposomes enhance antigen-specific immune response is not fully understood. *In vivo* liposome disposition studies indicate that large liposomes are taken up efficiently by macrophages in blood and tissue, including the liver and spleen. Since macrophages are thought to be predominantly responsible for processing and presenting liposome-associated or encapsulated antigens, the liposome formulation provides an excellent way to enhance antigen delivery and presentation for both humoral and cellular immune responses to vaccines. Mechanisms of liposome-mediated immune enhancement have been reviewed [35]. Antigens taken up by macrophages are presented in the context of MHC class II antigens. Antigens delivered to other phagocytic endothelial cells, such as Langerhan's and dendritic cells may also enhance antigen presentation in the context of MHC class I antigens [36].

The influence of physicochemical properties of liposomes, such as charge density, membrane fluidity, and epitope density, on the immune response elicited by antigens has been extensively studied [37]. In addition to antigens, other immune stimulators that are amphoteric muramyl peptides or lipid-soluble compounds, such as monophosphoryl lipid A or muramyl tripeptidyl-phosphatidylethanolamine, can also be incorporated into liposomes to increase their adjuvant effect in eliciting immune responses [34].

Recently a new liposome-based hepatitis A vaccine (*Epaxal*), developed by the Swiss Serum and Vaccine Institute, has been tested in humans [38]. This vaccine contains formalin-inactivated hepatitis A virus particles attached to liposomes together with influenza virus hemagglutinin. The advantage of the vaccine is that it not only induces antibodies to hepatitis A antigen but could also elicit an immune response against influenza virus protein on the liposome surface [39]. The inclusion of influenza hemagglutinin in the preparation also enhances the binding and fusion of hepatitis A liposomes to target cells.

Particle size-dependent targeting of drugs to spleen and lymph nodes can also be achieved with other lipid–drug complexes. An example is amphotericin B bound to phospholipids and cholesterol. The amphotericin B-lipid complex accumulates preferentially in RES cells *in vivo*, thereby improving delivery of amphotericin B to fungal-infected phagocytes in spleen. As a consequence of this distribution, the kidney is exposed to less free amphotericin B, and nephrotoxicity is reduced, improving the therapeutic index of the parent drug [40,41].

Enzyme-Mediated, Tissue-Selective Targeting of Peptide Prodrugs

Neurologically active peptides such as enkephalin and thyrotropin-releasing hormone (TRH) analogues are easily synthesized in pharmaceutical scale. However, their pharmacologic activity after systemic administration is poor due mainly to the low levels of accumulation in brain tissues. The results of permeation studies to date indicate that the blood-brain barrier is poorly permeable to hydrophilic drugs but easily breached by hydrophobic, usually uncharged, drugs. However, hydrophobic drugs, once in the brain, easily diffuse back to the systemic circulation. Hydrophilic drugs, on the other hand, if made to overcome the blood-brain barrier may be locked into brain tissue (Figure 13.12).

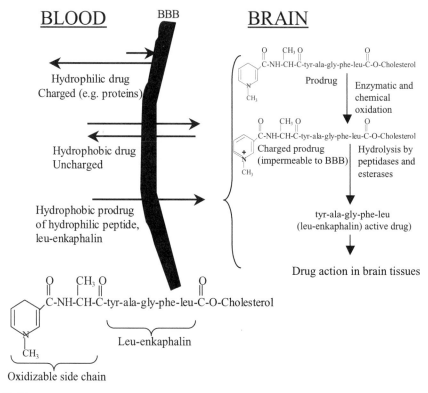

Figure 13.12. Schematic representation of enzyme-mediated targeting of enkephalin using a prodrug strategy.

Researchers have reasoned that if the water-soluble peptide leu-enkephalin could be conjugated with an oxidizable hydrophobic chain, and further shielded with a hydrophobic domain provided by cholesterol via an ester linkage, the modified leu-enkephalin could become sufficiently hydrophobic to breach the blood-brain barrier. The oxidation of one of the engineered domains in the brain would produce an ionic form that cannot redistribute back into blood, and is essentially locked in. The ester linkages could then be hydrolyzed by esterases in brain tissues to release biologically active leu-enkephalin (Figure 13.12).

This approach, known as the chemical-based delivery system (CDS), has been demonstrated to enhance the analgesic effects of leu-enkephalin in the rat tail flick latency test [42]. Similar derivatives of other pharmacologically active peptides, including TRH agonists, have been synthesized and demonstrated to enhance brain accumulation [43]. However, other tissues, including the liver and gut wall, also express oxidative metabolic enzymes and esterases, so that premature release of active drug from the product may occur and produce untoward effects. Successful engineering of chemical side chains and linkages that can be modified by enzymes preferentially found in the brain would greatly improve this tissue-targeted drug delivery strategy.

Ligand-Receptor-Based Targeting for Tissue- and Cell-Specific Drug Delivery

Target-tissue identification is challenging because for most therapeutic applications, the target tissues that mediate pharmacologic and toxicologic responses are not fully characterized. Another level of complexity comes from incomplete understanding of the target-specific ligand-receptor interaction and the degree of receptor expression on target cells versus bystander cells. However, exclusive or predominant expression of these receptors on target cells and tissues rarely occurs. The same thinking can be applied to antigen-antibody interactions in that either a ligand or a receptor could be used as an antigen to produce a selective monoclonal antibody for targeting. Generation of monoclonal antibodies is discussed in Chapter 10. In theory, ligand-receptor-based targeting to tissues and cells is only limited by our ability to identify target-specific antigens or ligands and develop antibodies against them. These antibody or targeting molecules can be conjugated to cytotoxic agents such as toxins or drugs, or expressed on drug carriers loaded with therapeutic agents.

Based on these principles, cell-associated ligands, receptors (i.e., galactose receptors on hepatocytes or folate receptors in actively growing cells and tumors), or antigens, such as carcino-embryonic-antigen (CEA), could be used for targeting. Tumor antigens identified so far are also expressed on nontarget, normal cells, albeit in low density. In fact *tumor-specific antigens*—the term used to describe antigens isolated from certain tumor cells and tissues (e.g., prostate antigen)—are now referred to as *tumor-associated antigens*, which are expressed at much higher levels in tumor cells than in normal cells. Therefore, knowledge of the receptor's density in target tissues compared to nontarget sites,

and the degree of blood perfusion of target tissues, allows one to refine the design of targeted drug delivery systems.

Directing Toxins to Tumor Cells with Antibody Conjugates

While antibody molecules, particularly IgG, exhibit intrinsic biologic activities such as neutralizing pathogen, binding to receptors, and receptor-mediated consequences, these effects may not be sufficiently potent for therapeutic applications. To increase antibody potency against tumor cells, highly potent bacterial toxins are conjugated to antibody. In this case the antibody provides binding selectivity to the toxin molecules. Antibody-toxin conjugates, which can be produced chemically or by a recombinant approach as a fusion protein, are generally known as *immunotoxins*.

Most of the toxins used to construct immunotoxins interfere with protein synthesis by either catalytic inactivation of elongation factor through ADP-ribosylation or by directly inactivating ribosomes [44]. It has been estimated that only one to five toxin molecules per cell are required to completely abrogate intracellular protein synthesis, and elicit cytotoxicity. Very often bacterial toxins contain two domains. Typically the A subunit contains cytotoxic sequences, and the B subunit contains binding activity. The two subunits could be separated to retain only the cytotoxic component, thereby eliminating toxin-directed nonspecific target recognition (Table 13.10). Toxins studied in immunotoxin development include diphtheria toxin (derived from *Corynebacterium diphtheriae*), ricin, abrin, mistletoe lectin, and modeccin. Some toxins that are expressed as a single-chain protein do not contain binding domains (e.g., saporin and gelonin), Pseudomonas exotoxin A (derived from *Pseudomonas aeruginosa*) contains a binding domain, but it can be

■TABLE 13.10. Properties of toxins often used in construction of immunotoxins

Toxin	Source	Polypeptides[a]	Inhibition Mechanism of Protein Synthesis
Diphtheria toxin	Bacterium	A–B	ADP ribosylation of elongation factor-2
Pseudomonas exotoxin A	Bacterium	One	ADP ribosylation of elongation factor-2
Ricin	Plant	A–B	
Abrin	Plant	A–B	
Mistletoe lectin	Plant	A–B	Inactivate 60s ribosomes by preventing the assembly of elongation factor 1 and elongation factor 2
Modeccin	Plant	A–B	
Pokeweed antiviral protein (PAP)	Plant	A	
Saporin	Plant	A	
Gelonin	Plant	A	

[a]Toxins can be expressed in either one or two polypeptides. The binding activity of toxin resides in B domain while enzymatic activity that produce intracellular toxicity through inhibition of protein synthesis resides in A domain. Toxin with only A domain do not exhibit binding activity toward human cells, while "one" domain exhibit both binding and enzymatic activity.

muted to prevent binding to intrinsic targets. These cytotoxic subunits are then chemically linked to an antibody selective for a target antigen (e.g., Erb-2 breast cancer-associated antigen).

A newer class of immunotoxins, recombinant immunotoxins, contains the variable minimum antigen binding domain F_v of an antibody fused with a toxin gene and expressed by recombinant DNA technology. These recombinant toxins, like conventional immunotoxins, can be directed to target antigens or receptors. Anti-Tac-PE38 (targeted to CD25 antigen and designated as LMB-2) and RFB4-PE38 (targeted to CD22 antigen and designated as BL22) are recombinant immunotoxins containing truncated Pseudomonas exotoxin (PE) and have been shown to exhibit clinical activity in patients with hematologic malignancies. BL22 and LMB-2 also show a lower degree of vascular leak syndrome and are thought to be less immunogenic than conventional immunotoxin conjugates. Several other recombinant immunotoxins are in clinical testing, and major responses have been reported, particularly in hematological malignancies [45]. Recombinant immunotoxins based on the

F_v domain have a molecular weight of 60 to 65 kDa, compared with a molecular weight of 200 kDa for conventional immunotoxins.

A close relative of immunotoxins is the growth factor fusion toxin, in which antibody is replaced with a ligand or growth factor to provide selectively of the toxin domain in immunotoxin. One such molecule, containing human interleukin-2 (IL-2) fused to truncated diphtheria toxin (denileukin diftitox or *Ontak*), was approved by the FDA in 1999.

The progress made in the identification and characterization of target receptors and the more compact size of recombinant immunotoxin molecules suggest that the immunotoxin field is ready to make a major impact on the treatment of cancers. A more compact protein should allow a higher degree of penetration into solid tumor masses, while the increased differentiation power due to improvement in tumor-associated antigen recognition should yield improved safety. If the smaller and more compact immunotoxins are proven to be less immunogenic molecules, so that immune-related symptoms could be better managed, repeated dosing could be considered to further improve their clinical effects.

Active Targeting of Liposomes Based on Ligand-Receptor Interactions

Most active liposome target delivery systems use chemically coupled ligands expressed on liposome membranes. By means of this strategy, a variety of ligands or receptors, such as antibodies, growth factors, cytokines, hormones, and toxins, have been anchored and expressed on liposome surfaces so that drugs, proteins, and nucleic acids may be introduced into target cells (Table 13.11). *In vitro*, liposomes coated with monoclonal antibodies (immunoliposomes) can provide target-specific binding to cells [46]. However, *in vivo*, key issues must be addressed before active targeting of liposomes using ligand-receptor interactions can be realized: (1) Liposomes expressing specific targeting molecules must circulate in blood long enough to localize in and perfuse target organs or tissues, and eventually interact with and bind to the target cells. Without sufficient residence time in the systemic circulation, most of the liposomes will be cleared without making contact with cell targets. (2) The ligand or receptor on the target cells must provide sufficient specificity. (3) The targeting molecule(s) expressed on the liposome surface must be sufficiently stable *in vivo* and exhibit minimum potential of removal by serum proteins.

The elucidation of liposome disposition mechanisms and the development of long-circulating liposomes (i.e., pegylated liposomes) have brought us closer to the goal of target-specific liposome drug delivery systems. Initial targeting experiments using antibody expressed on PEG-coated liposomes demonstrated that PEG may interfere with target cell binding [47,48]. Further studies suggested that PEG length is critical for target binding by immuno-

liposomes. To overcome the PEG barrier, attachment of antibody to a PEG molecule has provided specificity for liposome binding *in vitro* to cells that express surface antigen or receptor [49]. Similar methods have been used with success to bind liposomes to tissue in the lung [50] and brain [51] *in vivo*. Pegylated liposomes with antibodies have long survival times in the circulation and demonstrate target recognition in vivo [52,53]. Methods for coupling monoclonal antibodies at PEG termini have been developed [49,54]. However, PEG liposomes with antibody seem to lose their advantage in treating advanced solid tumors, due to limited perfusion of the tumor interior [55].

In addition to antibodies, other ligands, such as apo-E (apolipoprotein E), a high-affinity glycoprotein ligand for the LDL receptor, can be incorporated into small liposomes to target tumor cells that express a high density of LDL receptors [56]. Because folic acid receptors are overexpressed in human cancer cells [57], folic acid anchored on liposomes increases uptake and internalization of liposomes by such cells [58,59]. The discovery that actively growing tumor cells express a high density of transferrin receptors (required for ion uptake in tumors) led to the development of strategies to express transferrin on liposomes and the enhanced delivery of anthracyclines (e.g., doxorubicin and daunorubicin) and methotrexate to tumor cell lines *in vitro* [60]. Whether significant improvement in controlling tumor growth *in vivo* parallels improved target delivery *in vitro* remains to be seen.

While anchoring targeting molecules such as antibodies or other ligands on the liposome surface provides target selectivity *in vitro* and *in vivo*, the cost and reproducibility of these constructs with sufficient purity and in sufficient quantity for pharmaceutical application continue to be barriers to their development. The lipopeptide approach (i.e., attaching an acylated

■TABLE 13.11. Select list of ligands and receptors tested as liposome targeting agents

Molecule	Ligand/Receptor	Target Cell/Tissue	References
Antibodies			
H1817	E-selectin	Human umbilical vein Endothelial cells	[77]
CD19	CD19	B-cell lymphoma	[78]
CD35	CD35	Hematopoietic progenitors	[79]
N-12A5	erbB-2 (Her/2)	Tumor cells	[80]
Anti-HLA-DR	HLA-DR (MHC-II)	Lymph nodes	[81]
Anti-Selectin		Inflammatory sites	[82]
G-22	Glioma-associated antigen	Glioma cells	[83]
Transferrin	Transferrin receptor rat glioma cells		
Proteins or peptides			
HIV-gp 120	CD4	CD4 cells	[84]
HIV-gp 120 peptides	CD4	HIV-infected cells	[85]
Cytokines			
IL-2	IL-2 receptor	T cells	[86]
Growth Factors			
Transferrin	Transferrin receptor	Tumor cells/tumor, and crossing of blood-brain barrier	[60,87,88]
Beta nerve growth factor (NGF)		NGF receptor	[89]
Hexose and derivatives			
Galactoside	Asialoglycoprotein receptor	Hepatocyte	[90]
Galactosylated histone	Asialoglycoprotein receptor	Hepatocyte	[91,92]
Asialofetuin (AF)	Asialoglycoprotein receptor	Hepatocyte	[92]
Other			
Apo E/Apo B	LDL receptor	Tumor cells	[56]
Folate	Folate receptor	Tumor cells	[93]
Fibrinogen	Fibrin/thrombi	Fibrous atheroma thrombi	[94]
ICAM-1	ICAM-1	Atherosclerotic	[94]
Cholera toxin B subunit	GM-1	Mouse epithelial cells and tissues	[95,96]

peptide to the liposome surface) may produce more efficient and reproducible outcomes [61]. However, large-scale synthesis of lipopolypeptides may be a costly venture. Recently a single-chain, minimum binding domain of an antibody molecule (scF$_v$, MW: 25–27 kDa) was cloned and successfully produced in pharmaceutical scale in a cost-effective manner [62]. The prospect of using such recombinant monoclonal antibody derivatives, anchored on the liposome surface, to target a wide range of tissues and cells for therapeutic application is in sight.

In theory, amino acids subject to acylation by posttranslational modification in a bacterial host can be inserted in the DNA sequences of F$_v$ molecules to produce recombinant acylated F$_v$ products. Acylated F$_v$ molecules can be incorporated into liposomes with the ease and efficiency required for successful pharmaceutical scale preparation. F$_v$ molecules with specificity to a wide range of target antigens, including a number of tumor-associated antigens, vesicular endothelial cell growth factors (e.g., VEGF, endostatin) [63], and T-cell receptors (CD3, CD34) can be engineered by high-throughput screening systems, including phage display expression technology [64]. Incorporation of acylated antibodies and peptides has been well characterized and proven to be an efficient process [46,65].

Although target delivery of liposomes has shown promising results *in vitro*, *in vivo* and clinical data are needed to support these observations. In coupling ligands to liposomes, one must consider the final size of the liposome, coupling methods, and binding characteristics. Ligand receptor-specific liposome delivery can be achieved *in vivo* only when a significant fraction of liposomes perfuses the target tissue, allowing the interaction of targeted liposomes with their respective ligands or receptors on target cells [14,66]. Only then will the efficiency of liposome-mediated delivery become dependent on ligand-receptor affinity. The practicality for a large-scale pharmaceutical preparation of targeted liposomes must be considered early in the drug design and development process if we are to further advance the clinical use of liposome drug delivery systems to improve the safety and efficacy of highly potent therapeutic compounds.

■ 13.3. APPROACHES USING DEVICES

Drugs administered by intravenous injection and those injected at a site resulting in rapid absorption leads to high initial blood levels (Figure 13.5). Intravenous infusion is one way to eliminate high peak plasma drug levels as well as to deliver drugs for medical conditions that require constant and prolonged blood levels. Steady-state plasma drug concentrations are maintained by controlling the rate of intravenous infusion, after accounting for the drug's elimination rate, with an infusion device or pump that is powered by electricity or battery. The energy source for driving the device can also be provided by elastic rubber bellows, and the release rate is controlled by a transducer that regulates the flow of drug solution from the bellows (Figure 13.13). These devices maintain constant drug levels in plasma, but their routine long-term use for intravenous infusion is costly and likely to result in high infection rates. Subcutaneous infusion, on the other hand, while costly, has proved practical and is used by patients with diabetes who require insulin.

Controlled-rate infusion devices, as well as feedback-regulated drug delivery devices, have the potential to improve drug safety and efficacy. With advancement in computer technology and miniaturization of precision motors, highly compact computer-controlled infusion pumps are now available for insulin delivery. These pumps, weighing about 3 to 4 ounces, can be

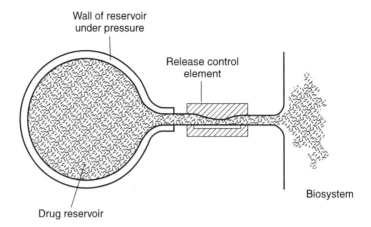

Figure 13.13. Schematic representation of a physically driven based portable infusion device.

programmed to provide bolus, multistep bolus, continuous, continuous complex or bolus-delay rates of drug delivery. These devices, produced by several manufacturers under the names Anima, Disetronic, Minimed, and Dana Dibecare, are approved by the FDA for insulin delivery. They are likely to be useful also for the delivery of other drugs. Additional enhancements of such devices, in late-stage development, include further miniaturization— Debiotech SA has produced a device weighing only 2 ounces—and replacement of infusion catheters with less intrusive, microneedles, less than $\frac{1}{8}$ inch in length, and fabricated on a matrix that serves as a skin patch. An example is SimpleChoice patch, in development by SpectRx.

Initial attempts to use implanted drug-concentration sensors to control the delivery of drugs based on concentration in plasma found that the drug-sensing device lost effectiveness after a few months. Ineffectiveness was often attributed to tissue buildup on the semipermeable membrane essential for sampling of drug in biological fluid. In theory, this problem could be overcome by means of a external drug-sensing device such as the ionotophoretic drug sensor, Biographer, developed to measure plasma glucose levels. Such a sensor device

could be used to directly regulate delivery rate to provide precise blood levels of insulin needed at any given time according to constantly changing glucose levels in diabetic patients.

Implanted devices to provide controlled drug delivery have also been developed. For example, simple bellows with an injection port and catheter can be placed under the skin, and the catheter directed to a target organ or tissue such as the intrathecal space, brain tissue (Ommaya reservoir), or the cephalic vein draining to the heart. Because this strategy provides direct access to the target site, distribution to nonspecific sites is reduced, and may lead to improvements in safety and efficacy. The self-sealing septum (injection port) of the implanted bellows, residing just below the skin, allows self-injection by patients and carries a significantly reduced risk of injection-related infections.

Polymeric membrane-based implantable drug delivery systems have been developed to deliver growth hormone, hormonal contraception, and leuprolide for the treatment of prostate cancer. Implanted sialic tubes loaded with levonorgestrel (*Norplant* system) is used outside the United States to provide five years of sufficient sustained release of the progestin to prevent pregnancy. In addition to sialic tubes, other implantable materials such as polylactide and ethylene-vinyl acetate copolymers of various shapes and sizes are being evaluated to sustain drug delivery for up to 100 weeks.

Besides injections and implants, proteins and peptides can be delivered across the skin by iontophoresis. Iontophoreis relies on the flow of current to increase the permeability of the epidermal layer of skin, the principal barrier to drug delivery. At one time, the current was thought to increase the flux of charged molecules but not that of neutral molecules. More recent observations, however, suggest that enhanced drug permeation is the result of increased bulk flow of water across the skin, carrying the drug in solution along with it. It was also determined that the battery producing the current in original designs of iontophoretic devices could be replaced by chemically derived power sources. While promising, iontophoresis is limited by the amount of current that can be applied before eliciting skin reactions. The iontophoretic process can be used also in reverse, to sample interstitial fluid for monitoring levels of markers such as glucose so that the appropriate dose of insulin can be selected.

Several portable inhalation devices have been developed and are being tested to determine whether they improve protein and peptide delivery via the airways. Aerosolized DNase has been shown in patients with cystic fibrosis to significantly reduce the buildup of mucus in the lung and the incidence of infections. Devices for delivery of therapeutic proteins to deep-lung alveoli to achieve systemic effects are also in development. These products are formulated so that the device aerosolizes the protein in a defined particle size range that cannot be easily achieved by means of conventional metered dose inhalers.

Systematic studies reveal a relationship between aerosol size and fraction of aerosolized drug deposited in the deep lung. The aerosol particle size for optimal alveoli drug delivery is 2 to 5 µm in diameter. Particles smaller than 2 µm penetrate into alveoli but exhibit poor accumulation, while particles larger than 5 µm tend to deposit in the upper airway. Because metered-dose inhalers produce only a very small fraction of particles 1 to 3 µm in diameter, and proteins are often subjected to denaturation by the current configuration of existing devices, they are not suitable for delivery of proteins into deep lung. Nebulizers that could be made to produce optimal particle size range are too bulky and impractical.

Inhalation devices for systemic delivery of peptides and proteins include dry powder inhalers, being studied by Inhale Therapeutic Systems (San Carlos, CA) and Elán (formally Dura Pharmaceuticals, San Diego, CA), and inhalers based on suspensions, being evaluated by Aradigm (Hayward, CA) and Aerogen (Santa Clara, CA) [67]. These devices are designed to (1) protect more delicate macromolecules in the aerosolization process, (2) provide reproducibility that cannot be achieved with current metered-dose inhaler systems, and (3) allow dispensing of doses that are in the mg range instead of the µg range, which is the limit of most metered-dose inhalers. Many of these devices have been tested in humans and produce reasonable systemic bioavailability (15–28%) of proteins such as insulin, calcitonin, and interleukin-4 receptor (Table 13.12).

■ 13.4. MOLECULAR APPROACHES

The progress in recombinant technology and molecular design and has allowed development of molecular instead of chemical modification to improve delivery of biopharmaceuticals. Chemical modifications require less effort and time than molecular approaches, which require sequence

■TABLE 13.12. Aerosolized proteins and peptides tested in humans with portable devices for pulmonary delivery

Peptide or Protein (Amino-acid Length)	Dosage Configuration	Device	Bioavailability (Relative to IM or SC Route)	References
Calcitonin (32 aa)	Dry powder	Duramed (Elán)	28%	[97]
Insulin (52 aa)	Dry powder	Inhale (Inhance)	20–25%	[67,98,99]
	Liquid suspension	Aerogen (Aerodose)	16%	[100]
Interleukin-4 receptor (800 aa)	Liquid suspension	Aradigm (AERx)	—	[101]

modification by recombinant DNA techniques to generate molecular variants. However, the production of a chemically modified drug may be more costly. Furthermore, most chemically conjugated proteins are heterogeneous products. For example, the product of antibody conjugated to diphtheria toxin (A chain) with a 1:1 average mole ratio also contains antibody molecules with no toxin and with two toxins in the mixture. On the other hand, cloned and recombinantly expressed antibody-toxin fusion conjugates produce a product wherein each antibody is linked faithfully to one toxin molecule.

In an ideal situation, chemical modification of a protein could be used to collect proof of principle for improvement in drug delivery, with the ultimate goal of replacing the initial lead compound with a more reproducible and scalable molecular product. However, this strategy may not always be feasible when amino-acid sequence modification would also reduce or eliminate biologic activity. In this scenario, advanced protein-modeling techniques (proteomics), which are discussed in the context of the integration of drug discovery and development (see Chapter 16), could assist selection of sequence modification with the highest probability of success.

Native human insulin self-associates to hexameric unit, which limits its absorption rate from injection sites (e.g., subcutaneous or intramuscular space). Based on the knowledge gained from therapeutic use of insulin in humans since the 1920s, molecular modifications have been made to improve its properties. These products, several of which have been approved by the FDA in the last few years, are the result of improved drug design based on the understanding of protein structure coupled with advances in recombinant DNA technology. Insulin variants are designed either to overcome the slow onset of pharmacologic effects that results from injection of native insulin or to prolong the duration of glucose control (Table 13.13). Additionally modifications on insulin A and B chains are being made to reduce inactivation—due to deamidation, dimerization, and disulfide exchange in storage (see Chapter 5)—and metabolism [68,69].

Chemical conjugation of polyethylene glycol is the most frequently used technique to reduce the renal and hepatic clearance of therapeutic proteins. An alternative strategy is to reduce RES cell uptake and renal filtration by strategically adding glycosyl (carbohydrate polymer) groups. This could be accomplished by creating additional N or O glycosylation sites and allowing mammalian cells to add glycosyl groups with terminal sialic acid, as a part of the post-translation modification process. Additional glycosyl groups increase apparent molecular weight and degree of protein hydration, and reduce liver and RES cell uptake. Indeed, this molecular modification

■TABLE 13.13. Molecular derivatives of insulin and their pharmacologic attributes

Insulin	A Chain 21	B Chain 20 & 21 Insertion	28	29	30	Functional Attributes[a]
	Amino-acid Position					
Insulin-human	Asn	None	Pro	Lys	Thr	Peak at 4–5 hours and dissipated by 14–15 hours; $t_{1/2}$ = 1.5 hours
Insulin-Lispro (Humalog)	Asn	None	*Lys*	*Pro*	Thr	Rapid acting and peak at 1 hour, instead of 3 hours; effect dissipated by 6–8 hours; $t_{1/2}$ = 1 hour
Insulin-asp (Novolog)	Asn	None	*Asp*	Lys	Thr	Rapid acting and peak at 1 hour and sustained for another 2 hours; $t_{1/2}$ = 1.3 hours
Insulin-glargine	*Gly*	*Arg & Arg*	Pro	Lys	Thr	Sustained release over 24 hours without peak response

[a]Based on glucose utilization rate after subcutaneous administration.

strategy was used to develop a long-acting human recombinant erythropoietin.

About 40% of human erythropoietin mass is contributed by the carbohydrate moieties of this heavily glycosylated protein, and it is believed that the degree of glycosylation modulates receptor binding activity. The negatively charged sialic acids at the terminal ends of glycosyl groups have been shown to be important in reducing protein clearance from blood by RES cells and liver hepatocytes (Chapter 5). Naturally produced as well as recombinant forms of human erythropoietin (produced in mammalian cells) exhibit small but measurable heterogeneity in the degree of hydroxylation. A systematic study revealed that increasing glycosylation increases plasma half-life and biologic activity, while receptor-binding affinity is somewhat compromised (Table 13.14).

With this key information, researchers at Amgen have genetically engineered additional glycosylation sites to increase the mass of carbohydrate on each molecule of erythropoietin while retaining biologic activity. After searching for variants meeting these criteria, one, with five N-linked groups and one O-linked glycosyl group, was shown to produce an optimum pharmacokinetic and pharmacodynamic profile. The new erythropoietin variant, termed *novel erythropoiesis stimulating protein* or NESP (*Aranesp*: darbepoetin alfa) has been evaluated in humans and shown to increase plasma half-life of the parent protein by about 3.5-fold (from 8.5 to 25.3 hours). This improvement has allowed weekly, instead of three-times-weekly dosing to achieve similar levels of erythropoiesis [70]. The FDA approved *Aranesp* in 2001. Such a strategy may be useful for optimizing other glycoproteins to improve their pharmacokinetic and therapeutic profiles.

Other molecular designs to improve drug delivery rely on engineering of proteins with a drug-targeting domain from one molecule and an effector domain from another. Fusion proteins have been successfully designed to demonstrate target selectivity of effector molecules that would otherwise distribute nonspecifically to nontarget tissues and lead to untoward effects. Initially, fusion proteins were designed using an antibody with a

■TABLE 13.14. Molecular and biological characteristics of erythropoietin molecules with varying degree of glycosylation

Description	rhuEPO	4 Chain Analogue	NESP-5 Chain Analogue
Glycosyl chain number	3 N-linked and 1 O-linked	4 N-linked and 1 O-linked	5 N-linked and 1 O-linked
Amino-acid length	165 aa	165 aa	165 aa
Apparent molecular weight	30.4 kDa	33.8 kDa	37.1 kDa
Molecular weight contribution by glycosyl group	40%	46%	51%
Sialic-acid residues per molecule	~14	~18	~22
Receptor binding affininty			
Serum half-life			
Biologic activity			

Source: Egrie and Browne [150].

Abbreviations: rhuEPO, recombinant human erythropoietin; NESP, novel erythropoiesis stimulating protein derived from genetic modification of rhuEPO to create two additional N-linked gylcosylation sites.

minimum binding domain (F_v) and a cytotoxic effector such as diphtheria toxin A. This strategy has the same rationale as that of antibody-toxin conjugates. The fusion product with a minimum antibody-binding domain is now optimized to be more compact (55–60 kDa), and the smaller recombinant fusion product could be produced more consistently in pharmaceutical scale.

Advances in cellular and molecular research have led to the discovery of ligands, antigens, and receptors on the surface of target cells (e.g., tumors). Development of simple and effective cloning techniques, sequence characterization, and expression procedures have provided markers—growth factors, cytokines, and T-cell antigens—that are often overexpressed by abnormal cell processes, such as carcinogenesis and other pathological processes. These developments have allowed generation of fusion proteins containing the target recognition domains, such as interleukins, growth factors, or antibody derivatives, and effector domains, including TNF-α, Pseudomonas endotoxin or diph-

theria toxin (Table 13.15). One of these fusion proteins (*Ontak*) was approved for the treatment of cancer in 1999.

Instead of linking a cytotoxic domain to a target recognition domain, amino-acid sequences of the F_c fragment of IgG can be linked to increase the half-life of ligands and receptors with pharmacologic properties. As discussed in Chapter 10, the uniquely long circulation half-life of IgG molecules, about 30 days instead of a few minutes or a few hours for proteins of comparable molecular weight, is due to the binding of F_cRN (Brambell receptor) to sequences of amino acids found in the Fc domain of IgG, which hinders clearance. Pharmacologically active molecules such as CD4 (the receptor for HIV envelope protein gp120) or the TNF receptor (capable of blocking TNF-mediated inflammatory responses) are molecularly linked to F_c to generate a recombinant fusion protein exhibiting both target specificity and prolonged plasma half-life (Table 13.5). An example is the construct of a TNF receptor-F_c fusion protein. Etanercept (*Enbrel*), a recombinant TNF-α-receptor-

■**TABLE 13.15.** Examples of fusion proteins engineered to contain a target recognition and highly potent effector domain

Fusion Protein	Targeting Domain		Effector Domain		Comments
	Description	Attributes	Description	Attributes	
TNFr-II: F_c (*Etanercept* or *Enbrel*)	Tumor necrosis factor (TNF) receptor II (p75)	Binds to TNF and neutralizes its biologic activities	IgG$_1$ F$_c$ sequences	Increases half-life of the final product due to F_c recognition sequences	Provides improved TNFr half-life (from 20 minutes to 73 hours) (data from BLA of *Enbrel*) Approved for treatment of rheumatoid arthritis
TNFr-I: F_c	TNF receptor I (p55)	Binds to TNF and neutralizes its biologic activities	IgG$_1$ F$_c$ sequences	Increases half-life of the final product due to F_c recognition sequences	Provides improved TNFr half-life (from 20 minutes to 82 hours) in baboon [103]
IL-2: DAB$_{389}$ (*Ontak*)	Interleukin-2 (IL-2)	Binds to IL-2 receptors expressed on cancer cells and T-cell lymphoma	DAB$_{389}$ is a binding-deficient variant of diphtheria toxin	Cytotoxicity	Approved for treatment of T-cell lymphoma
IL-4: PE$_{38}$[KDEL]	Interleukin-4	Binds to IL-4 receptor overexpressed on myeloma and glioma cells	PE$_{38}$[KDEL] is a binding-deficient variant of Pseudomonas exotoxin (PE)	Cytotoxicity	Under investigation as a potential intratumor treatment for patients with glioma [104]
IL-6: PE$_{4E}$	Interleukin-6	Binds to IL-6 receptors overexpressed on cancer and myeloma cells	PE$_{4E}$ is a binding-deficient variant of PE	Cytotoxicity	In preclinical development for multiple myeloma [105]
TGFα-PE$_{40}$ (TP40)	Tissue growth factor-alpha (TGF-α)	Binds to TGF receptors overexpressed on solid tumors	PE$_{40}$ is a binding-deficient variant of PE	Cytotoxicity	Evaluated in phase I testing for bladder cancer [106]
MFE23 (F$_v$): TNF-α	MFE23 contains F$_v$ amino-acid sequences of antibody binding domain	Binds to carcinoembryonic antigen (CEA) expressed on many cancer cells	TNF-α	Tumoricidal activity and associated inflammatory responses	The F$_v$ domain of MFE23 antibody fragment increases the target selectivity of TNF in mice [107,108]

F_c fusion protein, approved in 1998, has provided effective treatment for rheumatoid arthritis patients.

In theory, the number of fusion proteins with a target sensor and effector domain is limited mainly by our ability to identify and characterize target molecules that are overexpressed. The insight gained from the use of therapeutic proteins in humans and the understanding of common mechanisms of protein disposition can be put to use to create genetically engineered macromolecules with optimum distribution and disposition properties. Advanced computation techniques that provide efficient simulation of the molecular and structural variables that are important for binding recognition, effector functions, and disposition characteristics promise to improve our ability to further refine the design of fusion proteins.

■ 13.5. SUMMARY

Advances in protein and peptide research with regard to distinctive disposition characteristics, molecular understanding of sequence determinants, detailed elucidation of the intricate barriers to protein absorption, and study of the mechanisms of distribution and elimination have led to development of novel drug delivery strategies. Advanced drug delivery research has yielded products with improved therapeutic index and other improvements in safety and efficacy, such as reducing the frequency needed to provide similar or even better therapeutic responses. Drug delivery strategies have also benefited from the miniaturization of microprocessor-controlled programmable devices to fine-tune the therapeutic levels of biopharmaceuticals for chronic therapies, especially for drugs with short half-lives. Research to discover sites of drug action through proteomic modeling and high-throughput screening has yielded targets that have been used to

construct fusion proteins with added effector function to enhance pharmacologic activity. As the efficiency of identifying drug targets has improved and cloning technology has matured, pharmaceutical formulation and delivery have become a major limiting step in bringing new molecular entities onto the market.

REFERENCES

1. Woodley, J.F., *Enzymatic barriers for GI peptide and protein delivery.* Crit Rev Ther Drug Carrier Syst, 1994. **11**(2–3): 61–95.
2. Joseph, J.W., J. Kalitsky, S. St-Pierre, and P.L. Brubaker, *Oral delivery of glucagon-like peptide-1 in a modified polymer preparation normalizes basal glycaemia in diabetic db/db mice.* Diabetologia, 2000. **43**(10): 1319–28.
3. de Boer, A.G., D.D. Breimer, H. Mattie, J. Pronk, and J.M. Gubbens-Stibbe, *Rectal bioavailability of lidocaine in man: partial avoidance of "first-pass" metabolism.* Clin Pharmacol Ther, 1979. **26**(6): 701–9.
4. Nowak, M.M., B. Brundhofer, and M. Gibaldi, *Rectal absorption from aspirin suppositories in children and adults.* Pediatrics, 1974. **54**(1): 23–6.
5. Sayani, A.P., I.K. Chun, and Y.W. Chien, *Transmucosal delivery of leucine enkephalin: stabilization in rabbit enzyme extracts and enhancement of permeation through mucosae.* J Pharm Sci, 1993. **82**(11): 1179–85.
6. Ritschel, W.A., and G.B. Ritschel, *Rectal administration of insulin.* Methods Find Exp Clin Pharmacol, 1984. **6**(9): 513–29.
7. Lee, V.H., L.W. Carson, S.D. Kashi, and R.E. Stratford, Jr., *Metabolic and permeation barriers to the ocular absorption of topically applied enkephalins in albino rabbits.* J Ocul Pharmacol, 1986. **2**(4): 345–52.
8. Lee, W.A., R.D. Ennis, J.P. Longenecker, and P. Bengtsson, *The bioavailability of intranasal salmon calcitonin in healthy volunteers with and without a permeation enhancer.* Pharm Res, 1994. **11**(5): 747–50.

9. Fiset, P., C. Cohane, S. Browne, S.C. Brand, and S.L. Shafer, *Biopharmaceutics of a new transdermal fentanyl device.* Anesthesiology, 1995. **83**(3): 459–69.

10. Chien, Y.W., O. Siddiqui, W.M. Shi, P. Lelawongs, and J.C. Liu, *Direct current iontophoretic transdermal delivery of peptide and protein drugs.* J Pharm Sci, 1989. **78**(5): 376–83.

11. Pillai, O., V. Nair, R. Poduri, and R. Panchagnula, *Transdermal iontophoresis. Part II: Peptide and protein delivery.* Methods Find Exp Clin Pharmacol, 1999. **21**(3): 229–40.

12. Bartus, R.T., M.A. Tracy, D.F. Emerich, and S.E. Zale, *DRUG DELIVERY:Sustained Delivery of Proteins for Novel Therapeutic Products.* Science, 1998. **281**(5380): 1161–2.

13. Reiter, E.O., K.M. Attie, T. Moshang, Jr., B.L. Silverman, S.F. Kemp, R.B. Neuwirth, K.M. Ford, and P. Saenger, *A Multicenter Study of the Efficacy and Safety of Sustained Release GH in the Treatment of Naive Pediatric Patients with GH Deficiency.* J Clin Endocrinol Metab, 2001. **86**(10): 4700–6.

14. Lian, T., and R.J. Ho, *Trends and developments in liposome drug delivery systems.* J Pharm Sci, 2001. **90**(6): 667–80.

15. Anderson, P.M., D.C. Hanson, D.E. Hasz, M.R. Halet, B.R. Blazar, and A.C. Ochoa, *Cytokines in liposomes: preliminary studies with IL-1, IL-2, IL-6, GM-CSF and interferon-gamma.* Cytokine, 1994. **6**(1): 92–101.

16. Bui, T., C. Faltynek, and R.J. Ho, *Differential disposition of soluble and liposome-formulated human recombinant interleukin-7: effects on blood lymphocyte population in guinea pigs.* Pharm Res, 1994. **11**(5): 633–41.

17. Kedar, E., H. Gur, I. Babai, S. Samira, S. Even-Chen, and Y. Barenholz, *Delivery of cytokines by liposomes: hematopoietic and immunomodulatory activity of interleukin-2 encapsulated in conventional liposomes and in long-circulating liposomes.* J Immunother, 2000. **23**(1): 131–45.

18. Muranishi, S., *Absorption enhancers.* Crit Rev Ther Drug Carrier Syst, 1990. **7**(1): 1–33.

19. Aungst, B.J., *Intestinal permeation enhancers.* J Pharm Sci, 2000. **89**(4): 429–42.

20. Yoshikawa, H., K. Takada, S. Muranishi, Y. Satoh, and N. Naruse, *A method to potentiate enteral absorption of interferon and selective delivery into lymphatics.* J Pharmacobiodyn, 1984. **7**(1): 59–62.

21. Yamamoto, A., S. Okumura, Y. Fukuda, M. Fukui, K. Takahashi, and S. Muranishi, *Improvement of the pulmonary absorption of (Asu1,7)-eel calcitonin by various absorption enhancers and their pulmonary toxicity in rats.* J Pharm Sci, 1997. **86**(10): 1144–7.

22. Lee, W.A., B.A. Narog, T.W. Patapoff, and Y.J. Wang, *Intranasal bioavailability of insulin powder formulations: effect of permeation enhancer-to-protein ratio.* J Pharm Sci, 1991. **80**(8): 725–9.

23. Barone, G., C.T. Chang, M.G. Choc, Jr., et al., *The pharmacokinetics of a microemulsion formulation of cyclosporine in primary renal allograft recipients. The Neoral Study Group.* Transplantation, 1996. **61**(6): 875–80.

24. Levy, G.A., *Neoral/cyclosporine-based immunosuppression.* Liver Transpl Surg, 1999. **5**(4 Suppl 1): S37–47.

25. Keown, P., and D. Niese, *Cyclosporine microemulsion increases drug exposure and reduces acute rejection without incremental toxicity in de novo renal transplantation. International Sandimmun Neoral Study Group.* Kidney Int, 1998. **54**(3): 938–44.

26. Senel, F.M., S. Yildirim, H. Karakayali, G. Moray, and M. Haberal, *Comparison of Neoral and Sandimmun for induction and maintenance immunosuppression after kidney transplantation.* Transpl Int, 1997. **10**(5): 357–61.

27. Barone, G., C.M. Bunke, M.G. Choc, Jr., et al., *The safety and tolerability of cyclosporine emulsion versus cyclosporine in a randomized, double-blind comparison in primary renal allograft recipients. The Neoral Study Group.* Transplantation, 1996. **61**(6): 968–70.

28. Davis, S., A. Abuchowski, Y.K. Park, and F.F. Davis, *Alteration of the circulating life and antigenic properties of bovine adeno-*

sine deaminase in mice by attachment of polyethylene glycol. Clin Exp Immunol, 1981. **46**(3): 649–52.

29. Hershfield, M.S., R.H. Buckley, M.L. Greenberg, A.L. Melton, R. Schiff, C. Hatem, J. Kurtzberg, M.L. Markert, R.H. Kobayashi, A.L. Kobayashi, et al., *Treatment of adenosine deaminase deficiency with polyethylene glycol-modified adenosine deaminase.* N Engl J Med, 1987. **316**(10): 589–96.

30. Knauf, M.J., D.P. Bell, P. Hirtzer, Z.P. Luo, J.D. Young, and N.V. Katre, *Relationship of effective molecular size to systemic clearance in rats of recombinant interleukin-2 chemically modified with water-soluble polymers.* J Biol Chem, 1988. **263**(29): 15064–70.

31. Katre, N.V., M.J. Knauf, and W.J. Laird, *Chemical modification of recombinant interleukin 2 by polyethylene glycol increases its potency in the murine Meth A sarcoma model.* Proc Natl Acad Sci USA, 1987. **84**(6): 1487–91.

32. Ho, R.J., R.L. Burke, and T.C. Merigan, *Disposition of antigen-presenting liposomes in vivo: effect on presentation of herpes simplex virus antigen rgD.* Vaccine, 1994. **12**(3): 235–42.

33. Ho, R.J., R.L. Burke, and T.C. Merigan, *Antigen-presenting liposomes are effective in treatment of recurrent herpes simplex virus genitalis in guinea pigs.* J Virol, 1989. **63**(7): 2951–8.

34. Bui, T., T. Dykers, S.L. Hu, C.R. Faltynek, and R.J. Ho, *Effect of MTP-PE liposomes and interleukin-7 on induction of antibody and cell-mediated immune responses to a recombinant HIV-envelope protein.* J Acquir Immune Defic Syndr, 1994. **7**(8): 799–806.

35. Rao, M., and C.R. Alving, *Delivery of lipids and liposomal proteins to the cytoplasm and Golgi of antigen-presenting cells.* mangala.rao@na.amedd.army.mil. Adv Drug Deliv Rev, 2000. **41**(2): 171–88.

36. Lian, T., T. Bui, and R.J. Ho, *Formulation of HIV-envelope protein with lipid vesicles expressing ganglioside GM1 associated to cholera toxin B enhances mucosal immune responses.* Vaccine, 1999. **18**(7–8): 604–11.

37. Kersten, G.F., and D.J. Crommelin, *Liposomes and ISCOMS as vaccine formulations.* Biochim Biophys Acta, 1995. **1241**(2): 117–38.

38. Ambrosch, F., G. Wiedermann, S. Jonas, B. Althaus, B. Finkel, R. Gluck, and C. Herzog, *Immunogenicity and protectivity of a new liposomal hepatitis A vaccine.* Vaccine, 1997. **15**(11): 1209–13.

39. Gluck, R., *Liposomal presentation of antigens for human vaccines.* Pharm Biotechnol, 1995. **6**: 325–45.

40. Adedoyin, A., C.E. Swenson, L.E. Bolcsak, A. Hellmann, D. Radowska, G. Horwith, A.S. Janoff, and R.A. Branch, *A pharmacokinetic study of amphotericin B lipid complex injection (Abelcet) in patients with definite or probable systemic fungal infections.* Antimicrob Agents Chemother, 2000. **44**(10): 2900–2.

41. Chopra, R., *AmBisome in the treatment of fungal infections: the UK experience.* J Antimicrob Chemother, 2002. **49**(Suppl 1): 43–7.

42. Bodor, N., L. Prokai, W.M. Wu, H. Farag, S. Jonalagadda, M. Kawamura, and J. Simpkins, *A strategy for delivering peptides into the central nervous system by sequential metabolism.* Science, 1992. **257**(5077): 1698–700.

43. Prokai, L., K. Prokai-Tatrai, and N. Bodor, *Targeting drugs to the brain by redox chemical delivery systems.* Med Res Rev, 2000. **20**(5): 367–416.

44. Pastan, I.I., and R.J. Kreitman, *Immunotoxins for targeted cancer therapy.* Adv Drug Deliv Rev, 1998. **31**(1–2): 53–88.

45. Kreitman, R.J., *Toxin-labeled monoclonal antibodies.* Curr Pharm Biotechnol, 2001. **2**(4): 313–25.

46. Huang, A., S.J. Kennel, and L. Huang, *Interactions of immunoliposomes with target cells.* J Biol Chem, 1983. **258**(22): 14034–40.

47. Torchilin, V.P., *Immunoliposomes and PEGylated immunoliposomes: possible use for targeted delivery of imaging agents.* Immunomethods, 1994. **4**(3): 244–58.

48. Klibanov, A.L., K. Maruyama, A.M. Beckerleg, V.P. Torchilin, and L. Huang, *Activity of amphipathic poly(ethylene glycol) 5000 to prolong the circulation time of liposomes depends on the liposome size*

and is unfavorable for immunoliposome binding to target. Biochim Biophys Acta, 1991. **1062**(2): 142–8.

49. Hansen, C.B., G.Y. Kao, E.H. Moase, S. Zalipsky, and T.M. Allen, *Attachment of antibodies to sterically stabilized liposomes: evaluation, comparison and optimization of coupling procedures.* Biochim Biophys Acta, 1995. **1239**(2): 133–44.

50. Maruyama, K., T. Takizawa, T. Yuda, S.J. Kennel, L. Huang, and M. Iwatsuru, *Targetability of novel immunoliposomes modified with amphipathic poly(ethylene glycol)s conjugated at their distal terminals to monoclonal antibodies.* Biochim Biophys Acta, 1995. **1234**(1): 74–80.

51. Huwyler, J., D. Wu, and W.M. Pardridge, *Brain drug delivery of small molecules using immunoliposomes.* Proc Natl Acad Sci USA, 1996. **93**(24): 14164–9.

52. Blume, G., and G. Cevc, *Molecular mechanism of the lipid vesicle longevity in vivo.* Biochim Biophys Acta, 1993. **1146**(2): 157–68.

53. Maruyama, K., N. Takahashi, T. Tagawa, K. Nagaike, and M. Iwatsuru, *Immunoliposomes bearing polyethyleneglycol-coupled Fab' fragment show prolonged circulation time and high extravasation into targeted solid tumors in vivo.* FEBS Lett, 1997. **413**(1): 177–80.

54. Mercadal, M., J.C. Domingo, J. Petriz, J. Garcia, and M.A. de Madariaga, *A novel strategy affords high-yield coupling of antibody to extremities of liposomal surface-grafted PEG chains.* Biochim Biophys Acta, 1999. **1418**(1): 232–8.

55. Yuan, F., M. Leunig, S.K. Huang, D.A. Berk, D. Papahadjopoulos, and R.K. Jain, *Microvascular permeability and interstitial penetration of sterically stabilized (stealth) liposomes in a human tumor xenograft.* Cancer Res, 1994. **54**(13): 3352–6.

56. Versluis, A.J., P.C. Rensen, E.T. Rump, T.J. Van Berkel, and M.K. Bijsterbosch, *Low-density lipoprotein receptor-mediated delivery of a lipophilic daunorubicin derivative to B16 tumours in mice using apolipoprotein E-enriched liposomes.* Br J Cancer, 1998. **78**(12): 1607–14.

57. Weitman, S.D., R.H. Lark, L.R. Coney, D.W. Fort, V. Frasca, V.R. Zurawski, Jr., and B.A. Kamen, *Distribution of the folate receptor GP38 in normal and malignant cell lines and tissues.* Cancer Res, 1992. **52**(12): 3396–401.

58. Lee, R.J., and P.S. Low, *Delivery of liposomes into cultured KB cells via folate receptor-mediated endocytosis.* J Biol Chem, 1994. **269**(5): 3198–204.

59. Gabizon, A., A.T. Horowitz, D. Goren, D. Tzemach, F. Mandelbaum-Shavit, M.M. Qazen, and S. Zalipsky, *Targeting folate receptor with folate linked to extremities of poly(ethylene glycol)-grafted liposomes: in vitro studies.* Bioconjug Chem, 1999. **10**(2): 289–98.

60. Singh, M., *Transferrin As A targeting ligand for liposomes and anticancer drugs.* Curr Pharm Des, 1999. **5**(6): 443–51.

61. Ogawa, Y., H. Kawahara, N. Yagi, M. Kodaka, T. Tomohiro, T. Okada, T. Konakahara, and H. Okuno, *Synthesis of a novel lipopeptide with alpha-melanocyte-stimulating hormone peptide ligand and its effect on liposome stability* [published erratum appears in Lipids, 1999; **34**(6): 643]. Lipids, 1999. **34**(4): 387–94.

62. Neri, D., H. Petrul, and G. Roncucci, *Engineering recombinant antibodies for immunotherapy.* Cell Biophys, 1995. **27**(1): 47–61.

63. Ferrara, N., and K. Alitalo, *Clinical applications of angiogenic growth factors and their inhibitors.* Nat Med, 1999. **5**(12): 1359–64.

64. Krebber, A., S. Bornhauser, J. Burmester, A. Honegger, J. Willuda, H.R. Bosshard, and A. Pluckthun, *Reliable cloning of functional antibody variable domains from hybridomas and spleen cell repertoires employing a reengineered phage display system.* J Immunol Methods, 1997. **201**(1): 35–55.

65. Ho, R.J., B.T. Rouse, and L. Huang, *Target-sensitive immunoliposomes: preparation and characterization.* Biochemistry, 1986. **25**(19): 5500–6.

66. Lasic, D.D., *Novel applications of liposomes.* Trends Biotechnol, 1998. **16**(7): 307–21.

67. Patton, J., *Breathing life into protein drugs.* Nat Biotechnol, 1998. **16**(2): 141–3.

68. Brange, J., *The new era of biotech insulin analogues.* Diabetologia, 1997. **40**(Suppl 2): S48–53.

69. Nielsen, L., S. Frokjaer, J. Brange, V.N. Uversky, and A.L. Fink, *Probing the mechanism of insulin fibril formation with insulin mutants.* Biochemistry, 2001. **40**(28): 8397–409.

70. Heatherington, A.C., J. Schuller, and A.J. Mercer, *Pharmacokinetics of novel erythropoiesis stimulating protein (NESP) in cancer patients: preliminary report.* Br J Cancer, 2001. **84**(Suppl 1): 11–16.

71. Cleland, J.L., A. Daugherty, and R. Mrsny, *Emerging protein delivery methods.* Curr Opin Biotechnol, 2001. **12**(2): 212–19.

72. Jung, T., A. Breitenbach, and T. Kissel, *Sulfobutylated poly(vinyl alcohol)-graft-poly(lactide-co-glycolide)s facilitate the preparation of small negatively charged biodegradable nanospheres.* J Control Release, 2000. **67**(2–3): 157–69.

73. Breitenbach, A., Y.X. Li, and T. Kissel, *Branched biodegradable polyesters for parenteral drug delivery systems.* J Control Release, 2000. **64**(1–3): 167–78.

74. Heller, J., J. Barr, S.Y. Ng, H.R. Shen, K. Schwach-Abdellaoui, S. Einmahl, A. Rothen-Weinhold, R. Gurny, and S. Emmahl, *Poly(ortho esters)—their development and some recent applications.* Eur J Pharm Biopharm, 2000. **50**(1): 121–8.

75. Serres, A., M. Baudys, and S.W. Kim, *Temperature and pH-sensitive polymers for human calcitonin delivery.* Pharmaceutical Research, 1996. **13**(2): 196–201.

76. Lyampert, I.M., E.N. Semenova, V. Sanina, N.G. Puchkova, and A.V. Nekrasov, *Immunogenicity of a polypeptide fragment of the streptococcal group A M protein conjugated with an acrylic acid N-vinylpyrrolidone copolymer.* Biomed Sci, 1991. **2**(4): 410–14.

77. Spragg, D.D., D.R. Alford, R. Greferath, C.E. Larsen, K.D. Lee, G.C. Gurtner, M.I. Cybulsky, P.F. Tosi, C. Nicolau, and M.A. Gimbrone, Jr., *Immunotargeting of liposomes to activated vascular endothelial cells: a strategy for site-selective delivery in the cardiovascular system.* Proc Natl Acad Sci USA, 1997. **94**(16): 8795–800.

78. Lopes de Menezes, D.E., L.M. Pilarski, and T.M. Allen, *In vitro and in vivo targeting of immunoliposomal doxorubicin to human B-cell lymphoma.* Cancer Res, 1998. **58**(15): 3320–30.

79. Mercadal, M., C. Carrion, J.C. Domingo, J. Petriz, J. Garcia, and M.A. de Madariaga, *Preparation of immunoliposomes directed against CD34 antigen as target.* Biochim Biophys Acta, 1998. **1371**(1): 17–23.

80. Goren, D., A.T. Horowitz, S. Zalipsky, M.C. Woodle, Y. Yarden, and A. Gabizon, *Targeting of stealth liposomes to erbB-2 (Her/2) receptor: in vitro and in vivo studies.* Br J Cancer, 1996. **74**(11): 1749–56.

81. Dufresne, I., A. Desormeaux, J. Bestman-Smith, P. Gourde, M.J. Tremblay, and M.G. Bergeron, *Targeting lymph nodes with liposomes bearing anti-HLA-DR Fab' fragments.* Biochim Biophys Acta, 1999. **1421**(2): 284–94.

82. Bendas, G., A. Krause, R. Schmidt, J. Vogel, and U. Rothe, *Selectins as new targets for immunoliposome-mediated drug delivery. A potential way of anti-inflammatory therapy.* Pharm Acta Helv, 1998. **73**(1): 19–26.

83. Kito, A., J. Yoshida, N. Kageyama, N. Kojima, and K. Yagi, *Liposomes coupled with monoclonal antibodies against glioma-associated antigen for targeting chemotherapy of glioma.* J Neurosurg, 1989. **71**(3): 382–7.

84. Schreier, H., P. Moran, and I.W. Caras, *Targeting of liposomes to cells expressing CD4 using glycosylphosphatidylinositol-anchored gp120. Influence of liposome composition on intracellular trafficking.* J Biol Chem, 1994. **269**(12): 9090–8.

85. Slepushkin, V.A., Salem, II, S.M. Andreev, P. Dazin, and N. Duzgunes, *Targeting of liposomes to HIV-1-infected cells by peptides derived from the CD4 receptor.* Biochem Biophys Res Commun, 1996. **227**(3): 827–33.

86. Konigsberg, P.J., R. Godtel, T. Kissel, and L.L. Richer, *The development of IL-2 conjugated liposomes for therapeutic purposes.* Biochim Biophys Acta, 1998. **1370**(2): 243–51.

87. Shi, N., Y. Zhang, C. Zhu, R.J. Boado, and W.M. Pardridge, *Brain-specific expression*

of an exogenous gene after i.v. administration. Proc Natl Acad Sci USA, 2001. **98**(22): 12754–9.

88. Shi, N., and W.M. Pardridge, *Noninvasive gene targeting to the brain.* Proc Natl Acad Sci USA, 2000. **97**(13): 7567–72.

89. Rosenberg, M.B., X.O. Breakefield, and E. Hawrot, *Targeting of liposomes to cells bearing nerve growth factor receptors mediated by biotinylated nerve growth factor.* J Neurochem, 1987. **48**(3): 865–75.

90. Sliedregt, L.A., P.C. Rensen, E.T. Rump, P.J. van Santbrink, M.K. Bijsterbosch, A.R. Valentijn, G.A. van der Marel, J.H. van Boom, T.J. van Berkel, and E.A. Biessen, *Design and synthesis of novel amphiphilic dendritic galactosides for selective targeting of liposomes to the hepatic asialoglycoprotein receptor.* J Med Chem, 1999. **42**(4): 609–18.

91. Junbo, H., Q. Li, W. Zaide, and H. Yunde, *Receptor-mediated interleukin-2 gene transfer into human hepatoma cells.* Int J Mol Med, 1999. **3**(6): 601–8.

92. Wu, J., P. Liu, J.L. Zhu, S. Maddukuri, and M.A. Zern, *Increased liver uptake of liposomes and improved targeting efficacy by labeling with asialofetuin in rodents.* Hepatology, 1998. **27**(3): 772–8.

93. Lundberg, B., K. Hong, and D. Papahadjopoulos, *Conjugation of apolipoprotein B with liposomes and targeting to cells in culture.* Biochim Biophys Acta, 1993. **1149**(2): 305–12.

94. Demos, S.M., H. Alkan-Onyuksel, B.J. Kane, K. Ramani, A. Nagaraj, R. Greene, M. Klegerman, and D.D. McPherson, *In vivo targeting of acoustically reflective liposomes for intravascular and transvascular ultrasonic enhancement.* J Am Coll Cardiol, 1999. **33**(3): 867–75.

95. Lian, T., and R.J. Ho, *Cholera toxin B-mediated targeting of lipid vesicles containing ganglioside GM1 to mucosal epithelial cells.* Pharm Res, 1997. **14**(10): 1309–15.

96. Harokopakis, E., N.K. Childers, S.M. Michalek, S.S. Zhang, and M. Tomasi, *Conjugation of cholera toxin or its B subunit to liposomes for targeted delivery of antigens.* J Immunol Methods, 1995. **185**(1): 31–42.

97. Deftos, L.J., J.J. Nolan, B.L. Seely, P.L. Clopton, G.J. Cote, C.L. Whitham, L.J. Florek, T.A. Christensen, and M.R. Hill, *Intrapulmonary drug delivery of salmon calcitonin.* Calcif Tissue Int, 1997. **61**(4): 345–7.

98. Skyler, J.S., W.T. Cefalu, I.A. Kourides, W.H. Landschulz, C.C. Balagtas, S.L. Cheng, and R.A. Gelfand, *Efficacy of inhaled human insulin in type 1 diabetes mellitus: a randomised proof-of-concept study.* Lancet, 2001. **357**(9253): 331–5.

99. Cefalu, W.T., J.S. Skyler, I.A. Kourides, W.H. Landschulz, C.C. Balagtas, S. Cheng, and R.A. Gelfand, *Inhaled human insulin treatment in patients with type 2 diabetes mellitus.* Ann Intern Med, 2001. **134**(3): 203–7.

100. Fishman, R.S., D. Guinta, F. Chambers, R. Quintana, and D.A. Shapiro, *Insulin administration via the AeroDose (TM) inhaler: Comparison to subcutaneously injected insulin.* Diabetes, 2000. **49**: A9–A10.

101. Sangwan, S., J.M. Agosti, L.A. Bauer, B.A. Otulana, R.J. Morishige, D.C. Cipolla, J.D. Blanchard, and G.C. Smaldone, *Aerosolized protein delivery in asthma: gamma camera analysis of regional deposition and perfusion.* J Aerosol Med, 2001. **14**(2): 185–95.

102. Egrie, J.C., and J.K. Browne, *Development and characterization of novel erythropoiesis stimulating protein (NESP).* Nephrol. Dial. Transplant., 2001. **16**(90003): 3–13.

103. Rosenberg, J.J., S.W. Martin, J.E. Seely, et al., *Development of a novel, nonimmunogenic, soluble human TNF receptor type I (sTNFR-I) construct in the baboon.* J Appl Physiol, 2001. **91**(5): 2213–23.

104. Puri, R.K., *Development of a recombinant interleukin-4-Pseudomonas exotoxin for therapy of glioblastoma.* Toxicol Pathol, 1999. **27**(1): 53–7.

105. Rozemuller, H., W.J. Rombouts, I.P. Touw, D.J. FitzGerald, R.J. Kreitman, I. Pastan, A. Hagenbeek, and A.C. Martens, *Treatment of acute myelocytic leukemia with interleukin-6 Pseudomonas exotoxin fusion protein in a rat leukemia model.* Leukemia, 1996. **10**(11): 1796–803.

106. Goldberg, M.R., D.C. Heimbrook, P. Russo, M.F. Sarosdy, R.E. Greenberg, B.J.

Giantonio, W.M. Linehan, M. Walther, H.A. Fisher, E. Messing, et al., *Phase I clinical study of the recombinant oncotoxin TP40 in superficial bladder cancer.* Clin Cancer Res, 1995. **1**(1): 57–61.

107. Cooke, S.P., R.B. Pedley, R. Boden, R.H. Begent, and K.A. Chester, *In vivo tumor delivery of a recombinant single-chain Fv::tumor necrosis factor: a fusion protein.* Bioconjug Chem, 2002. **13**(1): 7–15.

108. Cooke, S.P., R.B. Pedley, R. Boden, R.H. Begent, and K.A. Chester, *In vivo tumor delivery of a recombinant single chain Fv:: tumor necrosis factor-alpha fusion protein.* Bioconjug Chem, 2002. **13**(2): 385.

109. Muranishi, S., *Absorption enhancers.* Crit Rev Ther Drug Carrier Syst, 1990. **7**(1): 1–33.

110. Aungst, B.J., *Intestinal permeation enhancers.* J Pharm Sci, 2000. **89**(4): 429–42.

111. Knauf, M.J., D.P. Bell, P. Hirtzer, Z.P. Luo, J.D. Young, and N.V. Katre, *Relationship of effective molecular size to systemic clearance in rats of recombinant interleukin-2 chemically modified with water-soluble polymers.* J Biol Chem, 1988. **263**(29): 15064–70.

INDIVIDUALIZATION OF DRUG REGIMENS: INTEGRATION OF PHARMACOKINETIC AND PHARMACOGENETIC PRINCIPLES IN DRUG THERAPY

■ 14.1. OVERVIEW OF FACTORS GOVERNING INTERINDIVIDUAL VARIATIONS

Patients on drug therapy manifest different responses to the same regimen. A drug may have little or no therapeutic effects in some patients but elicit adverse effects at low doses in others. Interindividual variation in drug efficacy and safety has resulted in the failure of drug candidates in clinical trials and the withdrawal of other drugs from the market.

Intersubject variation in drug response is thought to cost lives and dramatically increase health care costs [1,2]. While

Biotechnology and Biopharmaceuticals, by Rodney J. Y. Ho and Milo Gibaldi
ISBN 0-471-20690-3   Copyright © 2003 by John Wiley & Sons, Inc.

matching the patient with the right drug and dosing regimen tailored to his or her therapeutic need would overcome these problems, efforts to use therapeutic drug concentration monitoring to individualize dosing have had limited clinical impact. Therapeutic drug monitoring, albeit cumbersome and expensive, is sometimes used as part of drug safety and efficacy management for a small number of drugs with low therapeutic indexes, such as anticonvulsants, digoxin, some chemotherapeutic agents, and some immunosuppressants.

With advances in medical genetics and the study of the genetic basis of variation in drug disposition and biologic activity (pharmacogenetics), some interindividual variability can be predicted. In such cases the overall response to a drug is determined by variations in key proteins regulating disposition and activity. These advances have also led to development of medicinal agents designed to interfere with proteins or processes involved in medical conditions and perhaps benefit only defined individuals. In principal, genetic information—genotyping and phenotyping—can account for differences in drug response among individuals and identify individuals with disease-causing mutations that might be countered with targeted therapies.

While an individual's phenotype can directly assess a clinical measureable biologic function of interest (e.g., whether an individual is a fast or slow metabolizer of a certain drug), phenotyping is often expensive and not readily adapted for routine clinical practice. Analysis of genetic information—genotyping—might prove to be more efficient and better able to be adapted to the individualization of drug therapy.

■ 14.2. HISTORICAL PERSPECTIVE ON PHARMACOGENETICS

We have learned a great deal about the role of genetics in determining the safety and efficacy of drug therapy and the molecular basis of disease during the past decade. The study of the association between genetics and response to drug therapy is called *pharmacogenetics*. Pharmacogenetics is distinctly different from pharmacogenomics in that it is only one area of pharmacogenomic research. According to a leading pharmaceutical researcher's definition, pharmacogenetics refers to people, including gene identification and selecting the "right medicine for the right patient." Pharmacogenomics, on the other hand, includes other applications of genetic information related to drug response, including gene expression, alteration of protein function, and pathological consequences [3]. Pharmacogenetic investigations started more than 150 years ago.

By 1910 biological chemists in Europe discovered that humans were capable of transforming ingested drugs and chemicals into other products before excreting them. This observation, which ignited interest in linkage to Mendel's discovery, 44 years earlier, of the fundamental laws of heredity, invited the inference of receptors by scientists in Germany and France in the 1870s, and led to the suggestion that genetic material was crucial in directing chemical transformations in humans and other animals. These pioneering investigators envisioned enzymes as detoxifiers of drugs and other chemicals. In this role, enzymes enabled people to use drugs effectively and then excrete them harmlessly. These visionary scientists stressed that because of the failure of enzymes to detoxify drugs, some people might experience a clinical response very much out of proportion to that seen in an average person.

In the first part of the twentieth century, reports appeared on differences among Whites, Blacks, and Chinese in response to drugs and chemicals. During the 1920s and 1930s, individual differences in sensory perception of chemicals were studied. Human family studies showed that the failure to perceive bitterness after tasting phenylthiourea was transmitted from parents to children as an autosomal recessive trait. In the late 1940s another major advance emerged from studies of sickle cell anemia and the sickle cell trait. These studies provided the first irrefutable proof that the response of individuals to their environment was linked directly to the proteins they synthesize. The studies proved the autosomal transmission of this disorder from parents to offspring and demonstrated that a change in the globin molecule involving the replacement of a single amino acid by another was the basis of the disease.

Wendell W. Weber, who has authored two books on pharmacogenetics, noted [4]: "In 1953, the molecular basis of heredity, the double helix of DNA, was described. In 1956, human chromosomes had been visualized, enumerated, and soon thereafter one form of cancer (chronic myeloid leukemia) had been associated with an aberrant chromosome (the Philadelphia chromosome). . . . The development of electrophoresis and advances in chromatography permitted complex proteins and smaller molecules to be separated and analyzed." Consequently, he noted, "Protein polymorphism, initially associated with enzymes that occurred in multiple forms, was soon recognized as a phenomenon of much broader biological significance. For the first time, new relationships between the metabolic fate of exogenous substances in humans and the genetic control of human drug response were coming into view."

For many in the pharmaceutical sciences, the most important application of pharmacogenetics is as a predictor of drug response. CYP2D6, one of many cytochrome P-450 drug-metabolizing enzymes, typifies the pharmacogenetic diversity that has been identified in humans (Table 14.1) [5]. CYP2D6 is involved in the metabolism of at least 30 drugs, many of which are widely used. At least 48 nucleotide variations that create 53 CYP2D6 alleles have been identified in the *CYP2D6* gene. Of these, one is found in all human populations, some occur in several, while others are limited to a single population. Some of these variations lead to multiple copies of the enzyme, while others lead to the absence of the enzyme. Consequently a drug dose that produces the desired response in the average person can be therapeutically ineffective in some people or unsafe in others [5]. New methodology for gene analysis reflective of CYP2D6 activity allows clinical decisions to be made quickly [6].

Genetic polymorphisms in drug metabolism are certainly one reason for the differences in how patients respond to drugs. Another reason, of looming importance, is mutations in genes coding for a receptor or another protein that controls drug response; such mutations are being identified with increasing frequency. The genes for more than a dozen inherited traits or polymorphisms—among them cystic fibrosis, insulin receptor resistance, thrombophilia, estrogen resistance, and HIV resistance—have been identified. In these cases pharmacogenetics may serve as a guide to the development and application of drug therapy. In thrombophilia, for example, variation in the structure of a blood-clotting protein predicts susceptibility to deep-vein thrombosis. Molecular techniques have shown that the susceptible phenotype is associated with a single change in factor V—the Leiden mutation—that substitutes a glutamine for an arginine in the molecule. Not everyone who carries the variant protein experiences deep-vein thrombosis, but all carriers are lifelong candidates for oral anticoagulation, which requires close monitoring to maintain efficacy and avoid serious bleeding.

The promise of pharmacogenetics is that by determining a patient's genotype, physicians will be able to make better prescribing decisions. Determining genotype prior to a prescribing decision may improve care by increasing the proportion of patients for whom the drug is beneficial or by decreasing the risk of adverse events. In such circumstances the key question is: Which gene(s) can predict response to a specific drug?

■TABLE 14.1. Some polymorphically expressed metabolic enzymes and transporters that may be amenable to genotype analysis for individualization of drug regimens

Metabolic Enzyme[a]	Genetic Variation (Alleles)	Frequency of Genetic Variant [% and (Race[b])]	Functional Consequence [% Frequency (Race[b]) in Population]	Drug Substrate Implicated in Adverse Drug Reactions
CYP•2C9	CYP2C9*2	8–20 (W)	Reduced metabolism in 2–6% (W) population	Fluoxetine, Ibuprofen, Imipramine, Isoniazid, Naproxen, Phenytoin, Piroxicam, Rifampin, Verapamil, Warfarin, NSAIDs
	CYP2C9*3	6–9 (W)		
CYP•2C19	CYP2C19*2A	13 (W); 29 (C); 25 (AA); 21(K); 14 (E)	Reduced metabolism in 2–6% (W); 15–17% (C); 18–23% (J) population	Fluoxetine, Imipramine, Isoniazid, Nortriptyline, Phenytoin, Rifampin, Warfarin, Omeprazole, Diazepam
	CYP2C19*3A	0.3 (W); 12 (J) or (K)		
	CYP2C19*4	0.6 (W)		
CYP•2D6	CYP2D6*2A	28–30 (W); 20 (C); 12 (J)	Reduced metabolism in 3–10% (W); <2% (C); <2% (J); <2% (AA) population	Diltiazem, Fluoxetine, Imipramine, Metoprolol, [†]Nortriptyline, Theophylline, Haloperidol, Perphenazine, antiarrythmic drugs, Phenformin, SSRI antidepressants
	CYP2D6*3A	21 (W)		
	CYP2D6*3B	2 (W)		
	CYP2D6*4A,B	20–23 (W); 7–9(AA); 9 (Af)		
	CYP2D6*5	2–5 (W); 10–13 (J)		
	CYP2D6*6A	2 (W)		
	CYP2D6*7	<1–2 (W)		
	CYP2D6*8	<1 (W)		
	CYP2D6*9	2 (W)		
	CYP2D6*10A,B	2–5 (W); 43–51 (C); 33–60 (J)		
	CYP2D6*11 or *12	<1 (W)		
	CYP2D6*17	0 (W); 26 (AA); 9–34 (A); 19 (K)		
	CYP2D6*36	9 (K); 31(C) and (J)		

■ TABLE 14.1. *Continued*

Metabolic Enzyme[a]	Genetic Variation (Alleles)	Frequency of Genetic Variant [% and (Race[b])]	Functional Consequence [% Frequency (Race[b]) in Population]	Drug Substrate Implicated in Adverse Drug Reactions
CYP•2E1	CYP2E1*2	0 (W); 1 (As)	Reduced metabolism in 0% (W); 1% (As) population	Fluoxetine, Isoniazid, Theophylline, Verapamil
CYP•3A4	CYP3A4*2	0 (W); 1 (As)	Lower substrate affinity in 3% (W); 0% (As) population	Cyclosporine A, Midazolam, opioid analgesics
UGT•2	UGT2B7	IPD	Reduced metabolism in 10–15% (W) population	Ibuprofen, Naproxen, Irinotecan
NAT•2	NAT2*5A NAT2*5B NAT2*5C NAT2*5A,5B,5C NAT2*6A NAT2*6A,6B NAT2*7A NAT2*7A,7B NAT2*7B NAT2*13 NAT2*14A NAT2*14A,14B	1–4 (W) 38–45 (W) 1–4 (W) 43–46 (W); 30 (AA) 24–30 (W) 26–31 (W); 23 (AA) 1 (W) 1–2 (W); 5 (AA), 21–24 (AI) 1 (W) 2 (W) <0.6 (W) <1 (W); 8 (AA)	Slow drug acetylators in 50–59% (W); 41% (AA); 20% (C); 8–10* (J); 92% (Egp) population	Isoniazid, Sulfonamides, Procainamide, Hydralazine
PgP	MDR1 (3435 C→T)	24 (W)	Increase plasma drug concentrations in 24% (W) population	Digoxin, Saquinavir, Ritonavir, Indinavir, Nelfinavir

*Data source [25–28], and Web site (http://www.immki.se/CYPalleles).

[a] Abbreviations for metabolic proteins, CYP, cytochrome P450 metabolizing isoenzymes (the gene of specific isoenzyme is listed in italic form, i.e., *CYP2C19*); UTG, UDP-glucuronosyl-transferase; NAT, N-acetyltransferase; Pgp, P-glycoprotein, product of the multidrug resistance transporter *MDR1* gene.

[b] Abbreviations for race, W, White or Caucasian; C, Chinese; AA, African American; K, Korean; E, Ethiopian; J, Japanese; Af, African; A, Asian; AI, American Indian; Egp, Egyptian; As, Asian or oriental; IPD, insufficient population data collected from population study.

Candidate gene association studies have been the approach usually used to establish a genetic basis for individual drug responses. These experiments consist of statistical analysis of the relationship between a drug response trait (phenotype) and variants in a selected candidate gene (genotype). Research scientists can establish a relationship by demonstrating that particular genotypes are more prevalent in individuals who have a specific phenotype than those who do not, or by demonstrating that the mean value of a phenotype scale is dependent on genotype. Candidate genes are selected because they are known to be in a pathway related to the phenotype of interest or because experiments have suggested a mechanism by which variation in the gene could account for variation in clinical outcome.

Genome-wide linkage analysis has been used to establish the genetic causes of many Mendelian diseases, but because it relies on patterns of genetic and phenotypic changes within families, the analysis is rarely applicable for drug response traits. Therefore another strategy has been proposed—that a genome-wide map of thousands of single nucleotide polymorphisms (SNPs) will be useful for pharmacogenetic discovery. SNPs are simply locations along the chromosomes where a single base varies among different people. Where some have a guanine in a given string of nucleotides, for example, others might have a cytosine. The premise of SNP mapping is in the assumption that common diseases such as cancer must be caused in part by common mutations. And the most common mutations in the genome are SNPs, which occur about every 1000 bases.

The author of a recent review on pharmacogenetics observed: "The proposed strategy is to determine SNP marker frequencies in two populations: cases (responders or affected individuals) and controls (nonresponders or unaffected individuals). Since markers are closely spaced throughout the genome, there should be detectable marker frequency differences corresponding to each genetic determinant" [7]. The increased frequency of the joint inheritance of two genes that are closely linked on a chromosome is termed linkage disequilibrium. Association studies based on SNPs can work because of linkage disequilibrium.

Linkage disequilibrium-based association studies using candidate genes from selected genomic regions may be an efficient step toward a genome-wide association study. Within those regions, SNPs are selected at random locations, rather than specifically within the genes. Those SNPs are then used to genotype cases and controls. Using biological intuition, if successful, greatly reduces the time and cost required to determine an association. At today's prices it would cost as much as $50 million to conduct a genome-wide association study [7].

■ 14.3. PHARMACOGENETICS: DRUG METABOLISM AND TRANSPORT

14.3.1. Cytochrome P450 3A Genetics in Drug Metabolism

Various defense mechanisms have evolved to protect humans and other animals from the thousands of chemicals present in food, drinks, and the environment. Of particular interest are cytochrome P450 (CYP) enzymes, which catalyze the final step in the incorporation of oxygen into organic molecules. They frequently convert xenobiotics, including human-made chemicals and drugs, into less toxic products but can also transform nontoxic chemicals into toxic or carcinogenic-reactive species.

There are more than 50 known human P450 genes (Table 14.1). The most highly expressed subfamily is CYP3A, which includes the isoforms CYP3A4, CYP3A5, CYP3A7, and CYP3A43. These isoforms account for as much as 30% of total P450 content in liver and have an important role in the oxidative metabolism of at least 50% of all drugs. The most abundant

CYP isoform expressed in liver and gut is CYP3A4.

Hepatic expression of CYP3A4 is known to vary by more than 50-fold among individuals, and CYP3A4 enzyme function *in vivo* (drug clearance) varies by at least 20-fold. The level of CYP3A4 expression may determine which patients best respond to certain drugs and which patients experience side effects or even toxicity when the same dosage is given. The causes of the variability in constitutive CYP3A4 are unknown, but a better understanding of these mechanisms may allow individual patients to receive tailored drug dosages. Recent analyses suggest that, depending on the drug, 60% to 90% of interpatient variability in CYP3A4 function is caused by genetic factors.

Polymorphisms in the coding regions of *CYP2C9*, *CYP2C19*, and *CYP2D6* underlie the variability in expression of these enzymes and have been associated with distinct phenotypes throughout the population. The few reported mutations in the coding region of *CYP3A4*, however, have not been observed to have a profound effect on enzyme expression or function, but the possibility that other as-yet-unidentified polymorphisms affect its expression cannot be ruled out. The unimodal distribution in drug clearance of CYP3A4 substrates indicates that this is likely a multigenic process.

It has been proposed that the variability in CYP3A4 expression may be regulated at the transcriptional level, either through polymorphisms in upstream regulatory elements or in the genes encoding the transcription factors themselves. Rifampin, an anti-infective agent that induces CYP3A4 transcription, does so by activating a member of the nuclear receptor superfamily PXR (also called SXR or PAR). Rifampin is a ligand of PXR, bound at regulatory regions of the *CYP3A4* gene, allowing it to activate transcription. Targeted disruption of mouse PXR abolishes *CYP3A* induction but does not affect the level of constitutive CYP3A expression [8].

The publication of the map of human genome sequence variation, reported to contain 1.42 million SNPs, will allow a comprehensive search for genetic loci that are likely to regulate expression of CYP3A enzymes. Because DNA elements that control expression of CYP3A4 have been localized to far upstream enhancers, searching for genetic polymorphism in regulatory regions by traditional methods such as sequencing is exhausting. The SNP map will facilitate the search for polymorphisms in distant regions [7].

"Polymorphisms in the *CYP3A4* gene or other genetic loci controlling expression and function of CYP3A4 may explain the person-to-person variations seen in intensity and duration of drug action, as well as in the occurrence of side effects. Understanding the genetic basis of differences in CYP3A function will allow us someday to determine the proper drug dosages for individual patients, and achieve an optimal therapeutic response with minimal side effects" [8].

14.3.2. Cytochrome P450 SNP Could Be the Key to Variations in Drug Response

Researchers report that variations in the cytochrome *P4503A5* gene, largely ignored up until now, with CYP3A4 getting all the attention, determine how individuals respond to drugs metabolized by CYP systems in the liver and the intestine. A common variant leads to some individuals not making a significant amount of the CYP3A5 protein [9].

The investigators used DNA in liver samples from diverse individuals to identify the most common single nucleotide polymorphisms in the CYP3A family of genes. They then looked for associations between the SNPs and differences in drug metabolism and clearance of the anesthetic midazolam by these individuals. They found that one SNP, CYP3A5*3, results in a shorter, less active CYP3A5 protein. People with this SNP metabolize midazolam more slowly than do people with the CYP3A5*1 variant (which encodes a full-length protein). The researchers plan to undertake pharmacokinetic and pharmacodynamic studies of drugs metabolized by the CYP3A family of enzymes to see how genotype affects efficacy. The lead author predicted, ". . . in a reasonable amount of time people will be doing targeted dosing based on genotyping" [10].

14.3.3. Genotypic Differences in CYP2C19 May Modulate Disposition of Drug Therapy for *H. pylori* Infection and Cure Rates

Proton pump inhibitors (PPIs) such as omeprazole and lansoprazole are mainly metabolized by cytochrome P450 2C19 (CYP2C19), and genotype status may modulate the therapeutic effects of these agents. To explore this possibility, medical investigators in Japan examined whether CYP2C19 genotype status was related to eradication of *H. pylori* on treatment with triple therapy—PPI, clarithromycin, and amoxicillin. They also attempted to establish a treatment plan after failure to eradicate *H. pylori*. Eradication of *H. pylori* is the most effective strategy known for the treatment of peptic ulcer [11].

The study population consisted of CYP2C19-genotyped patients infected with *H. pylori* who had completed initial treatment with omeprazole 20 mg or lansoprazole 30 mg twice daily, and clarithromycin 200 mg and amoxicillin 500 mg three times a day for 1 week. Patients in whom the infection was not eradicated after initial treatment were retreated with lansoprazole 30 mg and amoxicillin 500 mg four times a day for 2 weeks.

The patients were classified into the following three genotype groups: homozygous extensive metabolizers ($n = 88$), heterozy-

gous extensive metabolizers ($n = 127$), and poor metabolizers ($n = 46$). Overall, the infection was eliminated in 87% of patients. Eradication rates were 73%, 92%, and 98% in the homozygous extensive, heterozygous extensive, and poor metabolizer groups, respectively. Thirty-four of 35 patients without eradication had an extensive metabolizer genotype of CYP2C19, and 19 of those patients were infected with clarithromycin-resistant strains of *H. pylori*. The researchers found no amoxicillin-resistant strains. On retreatment with high-dose lansoprazole and high-dose amoxicillin, therapy succeeded in 30 of 31 patients with extensive metabolizer genotype of CYP2C19 who had failed to respond to initial treatment.

The investigators concluded: "The results of this study suggest that CYP2C19 genotyping appears to be one of the predictable determinants for a PPI-based *H. pylori* eradication therapy with the aid of bacterial sensitivity testing. . . . If the CYP2C19 genotype status is determined before treatment, an optimal dose of a PPI may be prescribable on the basis of pharmacogenetic or pharmacogenomic status. This predetermined strategy should increase the eradication rates of *H. pylori* achieved by the initial treatment . . ." [11].

14.3.4. Pharmacogenetics and Altered Drug Metabolism of Chemotherapeutic Agents for Cancer

One of the main messages of the 2001 American Association of Cancer Research meeting was this: variability in response to drugs can lead to therapeutic failure or adverse drug reactions in individuals or subpopulations of patients [12]. It is now recognized that all patients do not react in the same way to chemotherapy, but in some cases treatment response can be predicted. Pharmacogenetics provides the opportunity to tailor drug treatments and decrease uncertainties of therapy.

Polymorphisms in cytochrome P450 drug-metabolizing enzymes are associated with both efficacy and adverse drug reactions. One research group used a SNP screening approach to study drug disposition in previously untreated patients with advanced breast cancer. The researchers found 22 unique SNPs in 12 genes involved in drug metabolism. They also found a statistically significant association of three genes with drug metabolism rates, tumor response, and patient survival following chemotherapy with high-dose cyclophosphamide, cisplatin, and carmustine. In particular, about one-quarter of patients with decreased metabolism of cyclophosphamide correlated with a significantly shorter survival. Decreased cyclophosphamide metabolism was linked to a SNP in the promoter region of *CYP3A4* or *CYP3A5*. These findings could lead potentially to a simple pretreatment genotype analysis to determine what dose of drug or even which drug combination would be most effective for each patient.

Oncologists are moving closer to correlating genetic data with phenotypic data and using this information to help forecast a patient's response to chemotherapy and predict toxic effects. A medical center in Alabama is already testing patients with colorectal cancer for polymorphisms in the gene involved in the metabolism of fluorouracil before initiating treatment with the drug. In particular, scientists at the center look for polymorphisms in the gene coding for dihydropyrimidine dehydrogenase.

14.3.5. Genetic Polymorphisms of Alcohol Dehydrogenase Modulates Alcohol Disposition and May Relate to Increased Risk of Myocardial Infarction

Moderate alcohol consumption is consistently associated with a reduced risk of myocardial infarction (MI). Studies of alcohol metabolism have shown that the class I alcohol dehydrogenase (ADH) isoenzymes, encoded by *ADH1*, *ADH2*, and *ADH3*, oxidize ethanol and other small aliphatic alcohols. *ADH2* and *ADH3* have polymorphisms that produce isoenzymes with distinct kinetic properties. Among White populations, however, variant alleles are relatively uncommon at the *ADH2* locus (present in less than 10% of the population) but common at the *ADH3* locus (present in 40–50%). At the *ADH3* locus, the γ_1 allele differs from the γ_2 allele by two amino acids. Pharmacokinetic studies show a 2.5-fold difference in the maximal velocity of ethanol oxidation between the homodimeric γ_1 isoenzyme (associated with a fast rate) and the homodimeric γ_2 isoenzyme (associated with a slow rate).

Effects of moderate alcohol consumption on the risk of MI and the role of *ADH3* genotype have been evaluated. Moderate drinkers with the γ_2 variant were found to have a lower risk of heart attack than those who disposed of alcohol more quickly [13].

The investigators identified 396 patients with newly diagnosed cases of acute MI among men in the Physician's Health Study. The patients were matched with one or two controls. The researchers determined the *ADH3* genotype ($\gamma_1\gamma_1$, $\gamma_1\gamma_2$, or $\gamma_2\gamma_2$) in all subjects.

As compared with men homozygous for the allele associated with the fast rate of ethanol oxidation (γ_1), those homozygous for the allele associated with a slow rate of ethanol oxidation (γ_2) had a significantly reduced risk of acute MI (relative risk, 0.65). Moderate alcohol consumption was associated with a lower risk of a heart attack in all three genotype groups. Among men who were homozygous for the γ_1 allele, those who consumed at least one drink per day had a relative risk of MI of 0.62, as compared with the risk among men who consumed less than one drink per week. Men who consumed at least one drink per day and were homozygous for the γ_2 allele had the greatest reduction in risk (relative risk, 0.14). These men also had the highest plasma HDL levels, which seem to protect against coronary heart disease. The researchers confirmed the interaction among the *ADH3* genotype, the level of alcohol consumption, and the HDL level in an independent study of postmenopausal women. They concluded, "Moderate drinkers who are homozygous for the slow-oxidizing *ADH3* allele have higher HDL levels and substantially decreased risk of myocardial infarction" [13].

14.3.6. Cellular Transporters and Drug Resistance

About three-quarters of cancer patients are intrinsically unresponsive to or develop resistance to anticancer drugs. In epilepsy, the situation is also problematic as the entire arsenal of anticonvulsant drugs fails up to 30% of patients. Fortunately solutions in the area of cancer research are emerging. Scientists have developed tests that can predict whether a patient is likely to respond to the most common chemotherapeutic agents. There is now the potential of developing novel drugs that, given with current chemotherapy, can overcome multidrug resistance. Researchers are hoping that, should newly emerging theories on drug resistance hold true, the progress made in the cancer field may lead to benefits for other drug-resistant diseases.

While there seems to be little in common between intractable epilepsy and cancer, researchers in each field have focused on a common mechanism underlying multidrug resistance: a cellular pump called P-glycoprotein (Pgp). Pgp protects cells from toxic substances by actively excreting the toxic agent. The pumps reside in tissues that are extensively exposed to toxic material: liver, lungs, kidney, intestine, placenta, and blood-brain barrier. How-

ever, the pumps not only protect tissue but also influence the uptake of drugs. Almost 30% of medicinal agents are thought to be substrates of Pgp and thereby challenged in reaching target sites.

A link between Pgp and drug-resistant epilepsy was reported in 1999, when researchers found high levels of the Pgp gene, *MDR1*, in brain lesions of a 4-month-old patient unresponsive to drug therapy. This finding may explain why anticonvulsant drugs such as phenytoin and phenobarbital, both of which are Pgp substrates, may not reach pharmacologically effective concentrations in critical parts of the brain in drug-resistant patients. This phenomenon is well known in oncology.

Applying a pharmacogenetic approach, researchers have identified 15 polymorphisms for *MDR1*, only one of which correlates with poor drug uptake. A test for this polymorphism could allow physicians to predict drug uptake from the outset and eliminate the administration of useless therapy. There is, however, a larger problem in that chemotherapy itself induces Pgp expression. Stifling the Pgp transporter with an inhibitor could resolve both issues. A newly discovered, orally effective molecule, OC144-093, is entering phase II trials. Not only is this agent a potent Pgp inhibitor, so far it also appears to be safe. Tempering enthusiasm, however, is the understanding that Pgp expression is but one of several mechanisms of drug resistance identified by cancer researchers.

Other researchers are studying polymorphisms and ethnic variation. In one study 1280 people from 10 different ethnic groups were evaluated for the C3435T polymorphism in the gene coding for Pgp. The investigators found that while 20% to 47% of White and Asian people were homozygous for the polymorphism leading to underexpression of Pgp and sensitivity to Pgp-dependent drugs, less than 6% of Africans had this genotype. These differences may have important implications for use of Pgp-dependent drugs in cancer patients of African origin [12].

14.3.7. MDR1 Expression Predicts Outcome of Liver Transplantation

Living-donor liver transplantation and immunosuppressive therapy with tacrolimus (*Prograf*) are key elements in the recovery of patients from end-stage liver failure. The optimal tacrolimus dosage regimen, however, has not been established, because in large measure the bioavailability of orally administered tacrolimus is highly variable. In this light, researchers have examined whether barriers to tacrolimus absorption—multidrug resistance protein (MDR1) or cytochrome P450 3A4 (CYP3A4)—are important pharmacokinetic factors and prognostic indicators of liver transplantation outcome.

The investigators measured messenger ribonucleic acid (mRNA) expression levels of MDR1 and CYP3A4 in mucosal cells of the upper jejunum collected during living-donor liver transplantation in 48 recipients. Tacrolimus was initiated at an oral dose of 0.075 mg/kg every 12 hours and adjusted on the basis of trough levels in whole blood.

The mRNA expression level of MDR1, but not the level of CYP3A4, was inversely related to the concentration/dose ratio of tacrolimus. High levels of MDR1 expression were associated with low bioavailability and reductions in survival rates after transplantation. Achieving a tacrolimus trough level of about 11 ng/ml required a daily dose of 0.13 mg/kg in the high-MDR1 group and 0.07 mg/kg in the low-MDR1 group. For the overall study population, the 600-day survival rates in the high- and low-MDR1 groups were 70.8% and 95.6%, respectively. The findings suggest that intestinal MDR1 variability should be taken into consideration for tacrolimus therapy, especially the determination of the initial dosage regimen, as a pharmacokinetic and prognostic factor for live-donor liver transplantation recipients [14].

◾ 14.4. PHARMACOGENETICS: THERAPEUTIC RESPONSE

14.4.1. SNPs Predispose to Cancer and Alter Response to Therapy

Alteration of a single nucleotide in a gene can determine whether the gene will predispose an individual to developing cancer. At the same time these SNPs can alter gene function in ways that affect response to therapy or chemoprophylaxis.

Researchers (at an American Association for Cancer Research conference on Genetic Modification of Cancer Susceptibility, held in 2001) suggested that this is the case for *SRD5A2* gene variants, which encode steroid type 2,5-alpha reductase, an enzyme that catalyzes the conversion of testosterone to dihydrotestosterone (DHT). They proposed that increases in the activity of this enzyme, resulting in higher levels of DHT, increase a man's likelihood of developing benign prostatic hypertrophy (BPH) and prostate cancer. Their experiments revealed that a particular variant of *SRD5A2* was associated with a sevenfold increased risk of prostate cancer in African-American men, a fourfold risk in Hispanic men, and a threefold risk in White men.

The investigators found that the described 10 SNP variants of *SRD5A2* differ in their biologic effects, not only in how much they affect levels of DHT, but also in their response to finasteride, a competitive inhibitor of 2,5-alpha reductase. Finasteride (*Proscar*) is marketed for the treatment of BPH and is under investigation for the chemoprevention of prostate cancer. The investigators found a 200-fold range in activity of the enzymes encoded by these gene variants and as much as a 60-fold difference in the ability of finasteride to inhibit the enzyme's activity.

Information about the pharmacogenetic variation in the *SRD5A2* gene was not available in 1993, when the National Cancer Institute-sponsored Prostate Cancer Prevention Trial began recruiting 18,000 participants to determine whether finasteride can prevent prostate cancer in men 55 years and older. All of the participants in the active treatment arm of the placebo-controlled trial are receiving the same dosage 5 mg daily finasteride. When the trial ends in 2003, the investigators hope to determine whether finasteride has a role in preventing prostate cancer. Absence of information on the participants' *SRD5A2* genotype, however, may make it more difficult to interpret the findings.

Inevitably some of the men given finasteride will develop prostate cancer. Would those who had *SRD5A2* variants producing a weak response to finasteride have fared better if they had been given a higher dosage of the drug? As the Prostate Cancer Prevention Trial draws to a close, it may be fruitful to explore this possibility by genotyping individual participants for *SRD5A2* and assessing whether certain genotypes are linked to a particular response to finasteride [15].

14.4.2. Altered Response to Chemotherapy in Women with BRCA1 Mutation

According to findings presented at the 2001 annual meeting of the American Association for Cancer Research, women who harbor mutations in *BRCA1*, a gene implicated in some types of hereditary breast cancer, may respond differently to chemotherapy than those lacking such mutations. The *BRCA1* gene is believed to function normally as a tumor suppressor and transcriptional (DNA to RNA) regulator. Moreover it encodes a protein involved in the cellular response to DNA damage. *BRCA1* mutations increase the susceptibility to breast, ovarian, and prostate cancer. The frequency of *BRCA1* mutations in the general population is 1 in 833, and in Ashkenazi Jewish women 1 in 107. Women with *BRCA1* mutations face a 50% to 80% lifetime risk of developing breast cancer and develop it at an earlier age than those without such mutations.

Investigators found that human breast cancer cell lines with *BRCA1* mutations showed a twofold to fourfold increase in apoptosis after treatment with ionizing radiation, cisplatin, or doxorubicin, compared with cells free of mutations. They also found that *BRCA1* tumor cell lines were resistant to other agents, such as paclitaxel (*Taxol*) and docetaxel (*Taxotere*), treatments used commonly in ovarian cancer and advanced-stage breast cancers.

Differences in drug sensitivity could be traced to the levels of another protein, Bcl2, which is also implicated in apoptosis. Loss of Bcl2 results in higher levels of programmed cell death after DNA damage. Normal *BRCA1* regulates the expression of Bcl2. Breast tumors expressing *BRCA1* mutations lacked or had reduced levels of the Bcl2 protein and, consistently, were resistant to chemotherapy with taxanes, which induce cell death through a Bcl2 pathway. They remained susceptible to chemotherapy that interacted directly with DNA, independent of Bcl2.

Hence women with *BRCA1* mutations who develop breast or ovarian cancer may not be good candidates for treatment with certain types of chemotherapy agents such as *Taxol*. Doxorubicin, cisplatin, or other newly discovered agents might be a better choice. In the future, genetic profiling may point the way to optimal treatment of cancer.

14.4.3. Pharmacogenetics and Methotrexate Toxicity

Linking pharmacogenetics and adverse drug reactions has proved fruitful in some instances. Methotrexate, an antifolate often used to prevent graft-versus-host disease after bone marrow transplantation (BMT), has several side effects such as oral mucositis, which can impede eating, talking, or even breathing.

Methylenetetrahydrofolate reductase (MTHFR) is a key enzyme in folate metabolism. Researchers have investigated whether the 677 C→T polymorphism in the *MTHFR* gene, which leads to an amino-acid substitution and decreased enzyme activity, modifies toxicity to methotrexate in BMT patients. Patients with lower MTHFR activity (homozygous for the polymorphism) had 36% higher mean oral mucositis than those with the wild-type genotype, as well as 34% slower recovery of platelet counts. Seemingly patients homozygous for C677T show higher methotrexate toxicity because of decreased nucleotide synthesis, resulting in a decreased DNA repair capacity and, consequently, delayed healing of the mucosa. Patients with polymorphisms in the *MTHFR* gene may be candidates for dose adjustment or use of alternative drugs [12].

14.4.4. β-Blocker Response in Heart Failure May Have Genetic Link

Activation of the renin-angiotensin system and elevation of circulating catecholamines both contribute to heart failure progression. Angiotensin converting enzyme (ACE) inhibitors and β-adrenergic receptor antagonists are mainstays of therapy. A considerable portion of the variability in ACE activity is genetically based and has been linked to a common polymorphism of the *ACE* gene. The mutation is known as the *ACE* deletion (*ACE D*) because it lacks a piece of DNA found in the normal gene. Although the mutation increases renin-angiotensin activation, its influence on patient outcome has been uncertain, and its pharmacogenomic interactions with β-blockers have not been studied.

A recent study suggests that patients with congestive heart failure who carry the *ACE* deletion are likely to respond especially well to β-blockers. The researchers followed 328 men and women with heart failure (mean left ventricular ejection fraction, 0.24) to assess the impact of the *ACE D* allele on transplant-free survival. Patients with two ACE deletion (short) versions of the *ACE* gene (*DD*) had the highest levels of ACE activity, and patients with two insertion (long) versions of the gene (*II*) had the lowest levels.

The adverse impact of the deletion on heart failure progression was dramatic in patients not treated with β-blockers. Two-year patient survival was 81% in *ACE* insertion, *II* genotype; 48% in *ACE* deletion, *DD* genotype; and 61% in the heterozygote, *ID* genotype. In contrast, patients who received β-blocker therapy showed no influence of the *ACE* genotype on transplant-free survival. Two-year patient survival was 70% in the *II*, 71% in the *ID*, and 77% in the *DD* genotypes.

These data clearly linked the *ACE D* allele with significantly poorer transplant-free survival. This effect was primarily evident in patients not treated with β-blockers and was not seen in patients receiving adrenergic therapy. These findings suggest a potential pharmacogenetic interaction between the *ACE D/I* polymorphism and therapy with β-blockers in the determination of heart failure survival [16].

14.4.5. Stent Restenosis, ACE Inhibitors, and Genetic Polymorphism

Some patients seem to have an inherent genetic susceptibility to restenosis following coronary stent implantation. Research has shown that patients homozygous for the deletion (*D*) allele of angiotensin converting enzyme (*ACE*) gene polymorphism lack functional enzyme and are at increased risk. These patients have elevated concentrations of ACE that could contribute to an accelerated growth response of smooth-muscle cells, leading to loss of patency. Thus one might suspect that ACE inhibition could modify the consequences of these increased ACE levels in *DD* patients. A recent report concerns the findings of a well-controlled pharmacogenetic trial to establish whether blockade of ACE with high doses of the ACE inhibitor quinapril (*Accupril*) could limit neointimal proliferation and reduce restenosis after stent implantation in patients carrying the *DD* genotype [17].

The *ACE* gene insertion/deletion (*I/D*) polymorphism was characterized in 345 patients who were undergoing coronary stenting. Among them, 115 had the *DD* genotype. The investigators assigned 91 of these patients to quinapril 40 mg daily or placebo. Treatment was initiated within 48 hours after stent implantation and continued for 6 months.

Surprisingly the primary end point of loss in minimum lumen diameter—a quantitative index of restenosis—was significantly higher in the quinapril group than in the placebo group. Secondary end points also showed consistent trends toward increased restenosis in the treatment group. The authors of the report concluded: "Contrary to our expectations, ACE inhibitor treatment did not reduce restenosis after coronary stent implantation in patients with *DD* genotype, but was associated with an exaggerated restenotic process when compared with administration of placebo [17]."

ACE inhibitors are widely prescribed for patients with atherosclerosis, hypertension, or diabetes, and after myocardial infarction, because of proven beneficial effects in each of these groups. In light of the results in the setting of stent restenosis, whether a patient with hypertension or some other target disorder who carries the *DD* genotype derives less benefit from ACE inhibitor therapy warrants consideration.

14.4.6. Prothrombotic Mutations Increase Risk of Acute MI in Women Taking HRT

Observational studies have suggested that the use of hormone replacement therapy (HRT) in postmenopausal women reduces the risk of coronary heart disease and have prompted the prescribing of HRT for this purpose. But estrogens are also prothrombotic, and high doses are associated with complications that include myocardial infarction, stroke, and venous thrombosis. The results of the Heart and Estrogen/Progestin Replacement Study (HERS), a secondary prevention trial that showed combined HRT was certainly no better than placebo at preventing coronary events in postmenopausal women, have renewed interest in the potential adverse effects of HRT. Indeed, in post hoc analyses of HERS data, treatment was associated with harm during the first year of follow-up and with benefits in years 4 and 5 of the study.

To explain this pattern, researchers hypothesized an immediate prothrombotic, proarrhythmic, or proischemic effect of treatment that is gradually outweighed by a beneficial effect on the underlying progression of atherosclerosis. According to this interpretation, a subgroup of patients, perhaps defined by a clinical characteristic, an environmental exposure, or a genetic trait, is susceptible to an early adverse effect of estrogen, while the rest of the population benefits from estrogen therapy.

In light of earlier evidence that genetic variations in prothrombin of human subjects are clearly associated with the risk of venous thrombosis, and before the results of HERS were known, investigators initiated a case-control study to assess whether prothrombotic mutations modify the association between HRT and the incidence of first MI. Cases were 232 postmenopausal women aged 50 to 79 years who had their first nonfatal MI between 1995 and 1998.

Controls were a sample of 723 postmenopausal women without MI who were matched to cases by age, calendar year, and hypertension status. The main outcome measure was risk of first nonfatal MI based on current use of HRT and the presence or absence of coagulation factor V Leiden and prothrombin 20210 G→A variants among cases and controls, stratified by hypertension.

In the study populations, 108 MI cases and 387 controls had hypertension. Among women with hypertension, the prothrombin variant was a risk factor for MI (odds ratio: 4.32). Compared with nonusers of HRT with wild-type genotype, women who were current users and who had the prothrombin variant had a nearly 11-fold increase in risk of a nonfatal MI. The interaction was absent among normotensive women. No interaction was found for factor V Leiden in either hypertensive or normotensive women.

The authors suggest that if their findings are confirmed, screening for the prothrombin 20210 G→A variant may permit a better assessment of the risks and benefits associated with HRT in postmenopausal women. It is important to note, however, that the prothrombin variant would account for only part of the pattern of early harm and late benefit seen in the HERS trials [18].

■ 14.5. INDIVIDUALIZED GENE-BASED MEDICINE: A MIXED BLESSING

Thanks to receptive media, scientific and medical discoveries have tantalized us with the prospect of individualized medicine in which drugs are prescribed based on each person's genetic makeup. A significant number of genes relating to human medical conditions has now been identified, and their location in chromosomes has been mapped. These details are now available as a part of the human genome project (Figure 14.1). This knowledge can now be put to use in developing genetic tests aimed at improving therapeutic management and preventing medical disorders and diseases. The technology for genetic testing is available and finding use, albeit limited. In some quarters this development is seen as a threat to the pharmaceutical industry, because it limits the target population for a drug and makes it more difficult to launch a drug destined to have blockbuster sales.

One company, Genaissance Pharmaceuticals, has invested heavily in genetic testing equipment in the hope of developing blood tests that, by providing a patient's genetic profile, would suggest which medicine has the best chance of working. One of its first targets is asthma medication. GlaxoSmithKline, however, has filed for patents on genetic tests for the effectiveness of its asthma medication fluticasone (*Flovent*). Observers surmise that GlaxoSmithKline has no intention of having the tests see the light of day nor allowing Genaissance or any other company to develop them.

The head of genetics at Glaxo said: "They can go screw with someone else's drugs" [*The Wall Street Journal Interactive Edition*, 18 June 2001]. GlaxoSmithKline is not opposed to all such tests. The company plans to use them to identify patients likely to suffer serious side effects from certain drugs and exclude them from clinical trials.

Despite the roadblocks, Genaissance is moving forward. Its asthma trial will involve 700 patients and test five drugs: *Serevent, Flovent, Pulmicort, Accolate*, and *Singulair*. Using complex algorithms, the investigators will attempt to correlate test results with how patients respond to each drug, taking into account variations in dozen of genes. A pilot study of albuterol in 121 asthma patients showed that the drug failed to produce meaningful bronchodilation in 11% of patients, who had one type of genetic variation, but significantly improved breathing in 28% of the group, who had other genetic variations [19].

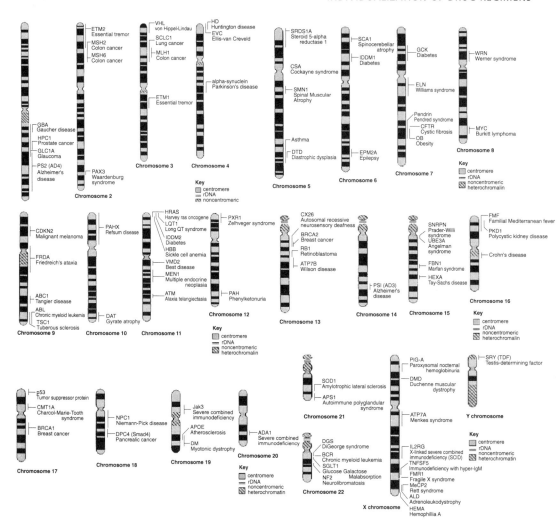

Figure 14.1. Human chromosomes and locations of genes linked to medical conditions. For some of the genes identified, recombinant DNA encoding the protein's RNA transcript and the amino-acid sequences, as well as the regulatory DNA sequences—including promoter binding regions, exons and introns, and splicing information—are reported and compiled in the National Center for Biotechnology Information (NCBI) database. (Figure reprinted from the noncopyrighted information in the database)

■ 14.6. CURRENT AND POTENTIAL APPLICATION OF PHARMACOGENETICS

The promise of pharmacogenetics has been realized for only a few drugs. One reason for the slow pace of development is that much of the science needed to practice gene-based medicine needs to be worked out. There is also caution because the concept is unconventional and widely perceived as adding to the cost of health care. Many clinicians view individualized drug therapy as impractical and question whether such an approach may ever be realized in light of the delay in making a decision and the cost of implementing genotyping or other tests to measure surrogate markers of gene function.

While no biopharmaceutical is approved with the requirement of genotyping as a part of its therapeutic indication, a recently approved monoclonal antibody therapy against breast cancer, trastuzumab (*Herceptin*), is indicated only for those tumors measurably expressing the protein expressed by the gene *erbB-2*. Because the oncogene product of *erbB-2* is elevated in no more than 30% of breast cancer patients, patients with no erbB-2 (HER2) protein in metastatic cancer cells derive no benefit from treatment with *Herceptin*. HER2 antigen testing was integral to the development of this product, and *Herceptin's* label directs that its use be based on a commercially available diagnostic test called Her2/neu (Box 14.1). Therefore

---

BOX 14.1. HOW *HERCEPTIN* WAS APPROVED AS A THERAPY THAT REQUIRES HER2 EXPRESSION TESTS ON PATIENT TUMOR TISSUES BEFORE USE

The search for genes that transform normal cells to tumor cells (oncogene research) in the 1980s led to the discovery of tumor-causing viral proteins capable of enhancing the growth rate of normal cells. Two viral oncogenes, *erbA* and *erbB*, identified from avian erythroblastosis virus, were shown to deregulate cell growth through receptor modifications (glycosylation and phosphorylation of the proteins) [20,21]. A cell homologue of viral erbB was later shown to be a proto-oncogene product that is a truncated form of epidermal growth factor (EGF). Cells that express erbB exhibit uncontrolled growth due to the lack of EGF receptor-regulated functions. This is proposed to be a mechanism of tumorogenesis [22]. About four years after discovery of cellular erbB expression, antibodies were developed to recognize erbB-2. By using these antibodies, researchers learned that about 10% of malignant tumor tissues collected from human organs overexpressed erbB-2.

By early 1990 Genentech had reported the development of murine monoclonal antibodies with therapeutic potential against tumor cells that overexpress erbB-2. One of these clones was humanized to produce a mouse/human chimeric form termed 4D5 and determined to be a suitable candidate for preclinical and clinical studies [23]. In 1998 the FDA approved the humanized monoclonal antibody, now known as trastuzumab (*Herceptin*) for the treatment of patients with metastatic breast cancer whose tumors overexpress erbB-2 (HER2) protein.

Only 20% to 33% of metastatic breast carcinomas robustly express HER2, and tumors that do not overexpress the protein are unlikely to benefit from treatment. Hence demonstration of clinical benefit in unselected breast cancer patients was a challenging task [24]. The need to keep patients on other chemotherapies added to the complexity of clinical trials. Nevertheless, the decision to first demonstrate overexpression of HER2 antigen levels as one of the key clinical trial inclusion criteria, reached in consultation with the FDA, allowed the investigators to demonstrate efficacy (Table 14.2).

The data on *Herceptin*—alone or in combination with current chemotherapies for metastatic breast cancer—clearly demonstrate that the degree of HER2 overexpression is related to treatment response. Inhibition of this tumor-transforming growth factor has proved to produce clinical effects but only in those patients with overexpressed phenotypes (2^+ and 3^+). These results formed the basis for the requirement to determine levels of HER2 on tumor tissues before deciding whether *Herceptin* is the best treatment strategy for an individual patient.

■TABLE 14.2. Levels of HER2 antigen expression and overall response rate to *Herceptin*

Patient HER2 Levels[a]	*Herceptin*	*Paclitaxel*	*Paclitaxel* and *Herceptin*	Anthracycline + Cyclophosphamide	Anthracycline + Cyclophosphamide and *Herceptin*
2^+	4%	16%	21%	43%	40%
3^+	17%	14%	44%	35%	53%

Data source: Biologic License Application to FDA.

[a]Patient tissues were evaluated using an immunohistochemical staining for HER2 proteins based on a 0–3 scale where 0 = none detected and 3^+ being strongly positive. Patient tissues with 2^+ scores are considered weakly positive for HER2 expression.

identifying the right breast cancer patient for *Herceptin* therapy is the first example of individualized drug therapy. *Herceptin* therapy, requiring prescreening of a patient's genetic makeup, is a significant step toward acceptance by clinicians of a new clinical practice paradigm.

Other examples illustrating that gene-based medicine can improve clinical outcomes are less widely used or less heralded but are important. Prescribing a smaller dose of azathioprine to those who lack thiopurine methyltransferase, the primary metabolic enzyme for this immunosuppressant, reduces the risk of toxicity in patients undergoing renal transplantation. As the toxicity of azathioprine is severe, and sometimes fatal, pharmacogenetic testing is becoming part of the standard of care in azathioprine therapy. Individuals who lack glucose-6-phosphate dehydrogenase show increased sensitivity toward dapsone-related hemolytic anemia. Some physicians routinely screen for this characteristic before prescribing dapsone to treat *Pneumocystis carinii* pneumonia in patients with AIDS. In each case, the benefit-to-risk ratio is poor or unacceptable without a test predictive for outcomes.

Despite its seeming benefit to individual patients, pharmacogenetics has not been widely embraced by the pharmaceutical industry. Certainly pharmacogenetic principles were applied in the development of the anticancer drugs *Herceptin* and *Gleevec*, both of which target a mutated gene or gene product. But these are exceptions. Pharmaceutical companies strive to develop drugs that are effective in as large a population as possible to reap maximum profits. Many currently available drugs are effective in no more than half of the target population. No pharmaceutical company desires to be told to develop a pharmacogenetic test to determine which patient will benefit from their product and which will not. This would be a time-consuming task and limit the market for their product. The same is true for the segment of the target population that displays adverse effects, unless they are serious ones.

Pharmaceutical companies will continue to develop targeted drugs like *Herceptin* as they accept the fact that, though the user population is limited, these drugs are likely to be profitable. Routine application of pharmacogenetics to identify patients who will safely benefit from a drug, however, is a long way off. The regulatory environment must change to accommodate this paradigm shift, the linkage between genetics and therapeutic response must be made manifest, and the cost of this strategy must be commensurate with its value.

■ 14.7. SUMMARY

Individualization of drug therapy to ensure optimal therapeutic response is a goal within reach. Application of pharmacogenetics in clinical practice is in its infancy, but there is every reason to think it will mature. Some unbridled enthusiasts believe that a day will come when each person carries a gene chip identification card, analogous to electronic IDs, or even an implanted chip, that contains complete genetic information, information that can be used for therapeutic decision-making. The technical feasibility, however, must be matched by a paradigm shift in clinical practice and an acute sensitivity to ethical concerns before that day comes. As the study and application of pharmacogenetics matures, associations of genetic variations linked to therapeutic response need to be validated, and genotyping assay must be improved to be more "user friendly" (i.e., rapid, sensitive, reproducible, readily available, simple to use, and cost effective). As exemplified by the use of *Herceptin*, new drugs will increasingly be targeted to only some patients, those who display a particular genetic marker that critically influences efficacy and safety.

REFERENCES

1. Brewer, T., and G.A. Colditz, *Postmarketing surveillance and adverse drug reactions: current perspectives and future needs.* Jama, 1999. **281**(9): 824–9.

2. Dormann, H., U. Muth-Selbach, S. Krebs, M. Criegee-Rieck, I. Tegeder, H.T. Schneider, E.G. Hahn, M. Levy, K. Brune, and G. Geisslinger, *Incidence and costs of adverse drug reactions during hospitalisation: computerised monitoring versus stimulated spontaneous reporting.* Drug Saf, 2000. **22**(2): 161–8.

3. Roses, A.D., *Pharmacogenetics and pharmacogenomics in the discovery and development of medicines.* Novartis Found Symp, 2000. **229**: 63–6; discussion 66–70.

4. Weber, W.W., *The history of pharmacogenetics.* Pharmaceutical News, 2000. **7**(6): 18.

5. Mahgoub, A., J.R. Idle, L.G. Dring, R. Lancaster, and R.L. Smith, *Polymorphic hydroxylation of Debrisoquine in man.* Lancet, 1977. **2**(8038): 584–6.

6. Scarlett, L.A., S. Madani, D.D. Shen, and R.J. Ho, *Development and characterization of a rapid and comprehensive genotyping assay to detect the most common variants in cytochrome P450 2D6.* Pharm Res, 2000. **17**(2): 242–6.

7. Katz, D., *The promise of pharmacogenetics.* Pharmaceutical News, 2001. **7**(6): 47–52.

8. Eichelbaum, M., and O. Burk, *CYP3A genetics in drug metabolism.* Nat Med, 2001. **7**(3): 285–7.

9. Kuehl, P., J. Zhang, Y. Lin, et al., *Sequence diversity in CYP3A promoters and characterization of the genetic basis of polymorphic CYP3A5 expression.* Nat Genet, 2001. **27**(4): 383–91.

10. Larkin, M., *Key to drug response may be in the SNP.* Lancet, 2001. **357**: 1020.

11. Klein, E., and R.J. Ho, *Challenges in the development of an effective HIV vaccine: current approaches and future directions.* Clin Ther, 2000. **22**(3): 295–314; discussion 265.

12. Hutchinson, E., *Working towards tailored therapy for cancer.* Lancet, 2001. **357**(9267): 1508.

13. Hines, L.M., M.J. Stampfer, J. Ma, J.M. Gaziano, P.M. Ridker, S.E. Hankinson, F. Sacks, E.B. Rimm, and D.J. Hunter, *Genetic variation in alcohol dehydrogenase and the beneficial effect of moderate alcohol consumption on myocardial infarction.* N Engl J Med, 2001. **344**(8): 549–55.

14. Hashida, T., S. Masuda, S. Uemoto, H. Saito, K. Tanaka, and K. Inui, *Pharmacokinetic and prognostic significance of intestinal MDR1 expression in recipients of living-donor liver transplantation.* Clin Pharmacol Ther, 2001. **69**(5): 308–16.

15. Stephenson, J., *As genes differ, so should interventions for cancer.* Jama, 2001. **285**(14): 1829–30.

16. McNamara, D.M., R. Holubkov, K. Janosko, A. Palmer, J.J. Wang, G.A. MacGowan, S. Murali, W.D. Rosenblum, B. London, and A.M. Feldman, *Pharmacogenetic interactions between beta-blocker therapy and the angiotensin-converting enzyme deletion*

polymorphism in patients with congestive heart failure. Circulation, 2001. **103**(12): 1644–8.

17. Meurice, T., C. Bauters, X. Hermant, V. Codron, E. VanBelle, E.P. Mc Fadden, J. Lablanche, M.E. Bertrand, and P. Amouyel, *Effect of ACE inhibitors on angiographic restenosis after coronary stenting (PARIS): a randomised, double-blind, placebo-controlled trial.* Lancet, 2001. **357**(9265): 1321–4.

18. Psaty, B.M., N.L. Smith, R.N. Lemaitre, H.L. Vos, S.R. Heckbert, A.Z. LaCroix, and F.R. Rosendaal, *Hormone replacement therapy, prothrombotic mutations, and the risk of incident nonfatal myocardial infarction in postmenopausal women.* Jama, 2001. **285**(7): 906–13.

19. Drysdale, C.M., D.W. McGraw, C.B. Stack, J.C. Stephens, R.S. Judson, K. Nandabalan, K. Arnold, G. Ruano, and S.B. Liggett, *Complex promoter and coding region beta 2-adrenergic receptor haplotypes alter receptor expression and predict in vivo responsiveness.* Proc Natl Acad Sci U S A, 2000. **97**(19): 10483–8.

20. Frykberg, L., S. Palmieri, H. Beug, T. Graf, M.J. Hayman, and B. Vennstrom, *Transforming capacities of avian erythroblastosis virus mutants deleted in the erbA or erbB oncogenes.* Cell, 1983. **32**(1): 227–38.

21. Hayman, M.J., G.M. Ramsay, K. Savin, G. Kitchener, T. Graf, and H. Beug, *Identification and characterization of the avian erythroblastosis virus erbB gene product*

as a membrane glycoprotein. Cell, 1983. **32**(2): 579–88.

22. Coussens, L., C. Van Beveren, D. Smith, E. Chen, R.L. Mitchell, C.M. Isacke, I.M. Verma, and A. Ullrich, *Structural alteration of viral homologue of receptor proto-oncogene fms at carboxyl terminus.* Nature, 1986. **320**(6059): 277–80.

23. Lewis, G.D., I. Figari, B. Fendly, W.L. Wong, P. Carter, C. Gorman, and H.M. Shepard, *Differential responses of human tumor cell lines to anti-p185HER2 monoclonal antibodies.* Cancer Immunol Immunother, 1993. **37**(4): 255–63.

24. Pegram, M.D., G. Pauletti, and D.J. Slamon, *HER-2/neu as a predictive marker of response to breast cancer therapy.* Breast Cancer Res Treat, 1998. **52**(1–3): 65–77.

25. Meyer, U.A., *Pharmacogenetics and adverse drug reactions.* Lancet, 2000. **356**(9242): 1667–71.

26. Phillips, K.A., D.L. Veenstra, E. Oren, J.K. Lee, and W. Sadee, *Potential role of pharmacogenomics in reducing adverse drug reactions: a systematic review.* Jama, 2001. **286**(18): 2270–9.

27. Ingelman-Sundberg, M., A.K. Daly, M. Oscarson, and D.W. Nebert, *Human cytochrome P450 (CYP) genes: recommendations for the nomenclature of alleles.* Pharmacogenetics, 2000. **10**(1): 91–3.

28. Ingelman-Sundberg, M., *Genetic susceptibility to adverse effects of drugs and environmental toxicants. The role of the CYP family of enzymes.* Mutat Res, 2001. **482**(1–2): 11–9.

15

GENE AND CELL THERAPY

■ SECTION ONE ■
15.1. OVERVIEW OF GENE
AND CELL THERAPEUTICS

■ SECTION TWO ■
15.2. DELIVERY AND EXPRESSION
OF GENES ENCODED FOR
FUNCTIONAL PROTEINS

■ SECTION ONE ■

■ 15.1. OVERVIEW OF GENE AND CELL THERAPEUTICS

Early examples of cell and gene therapy include bone marrow, organ, and tissue transplantation, where tissues and organs containing genetic information from the donor are transmitted to the recipient. While several definitions of cell and gene therapy have been advanced, a working definition must include embryonic cloning—generally known as animal and human cloning—via transfer of an engineered nucleus. Therefore we consider any strategy using nucleic acids—DNA or RNA derivatives—to treat, modify, or prevent a medical condition as cell and gene therapy.

Nucleic acids containing genetic information could act as inhibitors for gene expression. Anti-sense nucleic acid sequences designed to bind to target gene sequences and inhibit transcription, and ribozymes designed to degrade specific RNA transcripts, are also considered gene therapeutics.

The most common strategy is to use engineered nucleic acids that can be delivered to target cells and express functional protein. This strategy is discussed further in Section 15.2.

More than 5000 monogenic disorders have been linked to DNA alteration at a specific gene locus. Cell and gene therapy, although conceived primarily as a means to correct monogenic disorders, is now recog-

Biotechnology and Biopharmaceuticals, by Rodney J. Y. Ho and Milo Gibaldi
ISBN 0-471-20690-3   Copyright © 2003 by John Wiley & Sons, Inc.

■TABLE 15.1. Medical conditions that could benefit from cell and gene therapy

Monogenic Disorders	Acquired Disorders
Inborn errors of metabolism	*Infectious diseases*
Urea cycle	Human immunodeficiency virus (HIV)
Adenosine deaminase deficiency	Cytomegalovirus (CMV)
Phenylketourea	Viral hepatitis (i.e., hepatitis B and C)
Gaucher disease	
	Neurological disorders
Hemoglobinopathies	Parkinson's disease
Sickle cell disease	Neurodegenerative disease
Thalassemia	
	Cardiovascular diseases
Coagulation disorders	Atherosclerosis
Hemophilia A and B	Peripheral vascular disease
	Coronary restenosis
Pulmonary diseases	
Cystic fibrosis	*Cancer*
Alpha-1 antitrypsin deficiency	Many types using vaccine, toxin or endogenous regulatory protein expression strategies
Cardiovascular diseases	
Familial hypercholesterolemia	

nized also as a strategy with therapeutic potential for a wide range of acquired disorders including infections, neurological diseases, and cancers (Table 15.1).

While transplantation of bone marrow, liver, or kidney is a well-recognized strategy to correct medical conditions, the concept of using heterologous cells, with non-self genetic information, is relatively new, and it is an active and promising area of research. Recent advances in the development of human embryonic stem cells and the understanding of their cellular and molecular differentiation and maturation mechanisms have also enhanced progress in cell therapy. Some scientists consider the transplantation of engineered therapeutic cells that contain genetic information and are aimed at correcting medical conditions a form of gene therapy. Clearly, there is an interface between gene therapy and cell therapy.

In 1996 the delivery of Dolly, the sheep, developed from an embryonic cell in which the original nucleus was replaced with genetic information from an adult cell nucleus, introduced a new era of animal cloning (Box 15.1). Although replacing the nucleus of embryonic cells or cloning by nuclear transfer can be done with relative ease, the success rate of seeing the transformation of an embryo into an infant is relatively low. It took 277 attempts at cell fusion and 29 *in utero* implantations of embryos in 13 surrogate mothers to finally produce Dolly [1].

While the success in animal cloning is a major achievement, it also raises ethical concerns about attempts to use parallel technology to clone humans. This strategy has been shown to be a viable one for mice, pigs, goats, and cattle. The donor nucleus can be engineered and faithfully and efficiently produced as animal clones that

---

BOX 15.1. BASIC CONCEPT OF NUCLEAR TRANSFER TECHNOLOGY IN ANIMAL CLONING

When we think of gene therapy, we often imagine the replacement of a defective gene with a healthy one to correct genetically related medical disorders. In this context medical scientists identify and isolate the genetic defect, and subsequently engineer gene therapy strategies using the DNA sequences that are minimally required to correct the function.

A less likely idea is the use of a nucleus—the subcellular organelle containing all the genetic sequences—as a means to correct genetic defects. Cloning of an exact replica of a nucleus is of great interest to scientists, because the nucleus of all cells in each person contains sequences that determine unique traits. Cloning nuclei is of particular interest to biologists working to develop a group of identical transgenic animals so that, in theory, an exact copy of a recombinant protein produced in the transgenic founder animal can be produced by each genetically identical offspring. To these scientists, nuclear transfer technology is of paramount importance. After therapeutic genes are identified, it may take years or decades to develop a transgenic founder animal capable of producing therapeutic proteins, such as albumin, antibody, or blood-clotting factors. These proteins can now be produced in the milk of transgenic animals such as sheep or goats. This strategy appears to work for the founder animal but not necessarily in the offspring. Transgenes are often altered or lost in normal animal reproduction.

By transferring the nucleus of an adult cell into the nucleus-depleted embryo of another animal, all the transgenes in their original arrangement are propagated to genetically identical infants. This process, in theory, can provide a strategy to develop a herd of genetically identical transgenic animals for producing biopharmaceuticals. Because each transgenic cow can annually produce about 9000 liters of milk, production of recombinant proteins in a herd of genetically identical cows could substantially reduce costs with minimal concern about heterogeneity in the protein products.

In 1996, Ian Wilmut reported the successful cloning of a sheep, genetically identical to an adult sheep from which a nucleus of a mammary cell was obtained [1]. By replacing the nucleus of the egg with the nucleus from the donor, a sheep named Dolly was cloned. With proof of concept in place, biopharmaceutical companies have invested in animal cloning research to determine whether it is possible to raise a herd of genetically identical livestock to produce proteins; a concept that led to the coining of the word "pharming" [13]—farming of biopharmaceuticals.

---

express identical gene products. Organs harvested from engineered and cloned animals are of considerable scientific interest, because they can be made to express human histocompatability antigens [2]. Thus cell and gene therapy is any strategy using RNA and DNA derivatives, organized in subcellular, cellular, tissue, or other carrier forms, to treat, modify, or prevent a medical condition.

15.1.1. General Strategies in Gene and Cell Therapy

Gene Therapy

Most of the current gene therapies are designed to alter a disease state by delivering nucleic acids into a cell. These nucleic acids may be genes, portions of genes, oligonucleotides, or ribonucleic acids. In

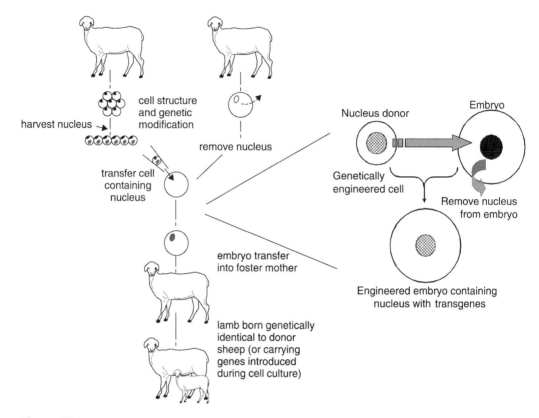

Figure 15.1. Schematic representation of nuclear transfer strategy to produce genetically identical offspring. (Adapted from PLL-Therapeutics)

theory, this strategy can be used to modify all cell types, including germ-line cells. There is strong opposition to the development of germ-line gene therapy because of potential transmission of genetic information to subsequent generations. Somatic gene therapy, designed to modify the biological processes of differentiated cells, constitutes the vast majority of gene therapy research and development.

The goal of gene therapy depends on the disease being treated. Treatment designed for inherited disorders such as the absence of factor VIII or IX demands life-long expression of gene products, while a vaccine designed to elicit immune responses against cancer cells requires only transient expression of gene products.

Some of the gene therapy strategies that require only transient expression of gene products include expression of (1) chemo-

protective genes in normal cells of cancer patients undergoing high-dose chemotherapy, (2) cancer antigens to elicit immune responses against cancer cells, (3) factors for improving coronary vascularization or preventing coronary restenosis, (4) antisense DNA sequences against viruses and bacteria to halt replication, (5) ribozymes—RNA enzymes—that recognize viral and bacterial target RNA sequences to interfere with their life cycle, and (6) antigens for developing vaccines against viral and other pathogens.

For gene therapy to be effective, it must be possible to deliver the therapeutic gene to the target cell. After delivery, the gene must be expressed at an appropriate level for the desired effect and for a sufficient time for this effect to realize benefits. Additionally, the delivery and expression of the therapeutic gene must do no harm.

All of these goals must be addressed to implement a gene therapy strategy. Clinical experience with gene therapy during the past decade shows that serious toxicity may occur when the gene and its package are delivered directly into the liver or lungs. Intravenous or intramuscular administration has had a clean safety profile up until now. Experience shows that gene transfer maneuvers can favorably alter the cellular phenotype *in vitro*, although a fundamental limitation exists *in vivo* because of insufficient gene delivery to and expression in target cells.

Implementation of gene therapy in humans has used two strategies. In some instances target cells have been removed from the body, genetically modified, and then reintroduced into the patient. This strategy is called the *ex vivo* approach. Alternatively, the *in vivo* approach involves directly delivering the therapeutic gene to the relevant target cells in an intact individual.

The advantages of the *ex vivo* approach are that it allows gene transfer to the target cells in a well-controlled setting in which delivery efficiency is optimized. This approach also allows modified cells to be characterized before they are reintroduced to the patient. Despite these advantages, this method may be limited to very select settings in which target cells can be propagated *ex vivo*; at present, this is a viable strategy for few tissue types. Furthermore the *ex vivo* approach requires individualized engineering of cells. Few patients are likely to benefit from this strategy because of the cost of the procedure and the small number of facilities capable of carrying out such procedures.

The *in vivo* approach overcomes some of these limitations because a single dosage form may be applicable to a large population of patients. Delivery *in vivo*, however, is fraught with considerably greater complexities and challenges than those encountered in the *ex vivo* approach. The *in vivo* approach must achieve delivery in the context of host barriers, including humoral, reticuloendothelial, and immunologic factors.

Early gene therapy protocols were principally of the *ex vivo* type and relied on recombinant retroviruses as gene transfer vehicles. Retroviruses can accomplish effective gene transfer to target cells despite being rendered replication-incompetent by genomic deletions. However, a variety of limitations have restricted their use to achieve *in vivo* gene transfer.

Retroviruses require proliferative target cells to mediate effective gene transfer. Lentiviruses, the class of retrovirus that includes human immunodeficiency virus, are an exception. They can also integrate in nondividing cells. An additional obstacle for retroviruses is the high susceptibility of the virus particle to humoral factors that ablate their gene transfer capacity. Perhaps most important, retroviruses, even when used *ex vivo*, pose a low but real risk of cancer.

To circumvent the limitations associated with recombinant retroviruses, alternative vector systems have been developed. They include both nonviral and viral approaches. In both approaches, the goal is to develop a system that can deliver genes *in vivo* after injection. The development of a successful vector system would have two very important consequences for potential gene therapy strategies: (1) it would allow the implementation of gene therapy strategies targeted to cell or tissue types that cannot currently be modified *ex vivo*, and (2) it would allow gene therapy to be carried out in a context other than highly specialized, tertiary care medical centers with the support facilities needed for the *ex vivo* strategy.

Nonviral systems have been developed and used to deliver genes *in vivo*. Delivery of genes by means of liposomes—artificial lipid bilayer vesicles—overcomes the potential safety hazards associated with viral gene sequences required for packaging. DNA-liposome complexes have been used in human clinical trials targeting pulmonary disorders and malignancies. Despite considerable systemic stability, delivery to target cells is nonspecific, because the liposomes used to date lack the mechanisms required to achieve targeting. Target-directed liposomes are being developed to exploit the endogenous cellular receptor-mediated pathway to transfer genes.

Nonretroviral viral vectors, based on recombinant adenoviruses and the parvovirus adeno-associated virus (AAV) are also under investigation. Adenoviruses have been evaluated for their ability to be directly delivered in vivo to a variety of organ systems, including the lung, liver, brain, and vasculature, but inflammation is a major concern. The AAV vector is less highly developed but shows promise; a practical problem is its limited DNA-carrying capacity. Additional details on DNA delivery vehicles, including their design and host expression requirement are discussed in Section 15.2.

Cell Therapy

Stem cell research is a major area of current efforts in cell therapy. As discussed in Chapter 6, Section 6.1, hematopoiesis is a tightly regulated system in which production of various blood cell types is responsive to specific functional demands. This system contains early multipotent stem cells with extensive proliferative potential but few differentiated features. They reside largely in the bone-encased marrow cavity. Stem cells have the potential, with appropriate inductive signals, to differentiate and give rise to populations of progenitor cells with progressively restricted renewal, proliferative, and lineage potential but with increasing functional characteristics defining a variety of specific lineages.

While the production of active blood cell types occurs predominantly in the bone marrow, the spleen, lymph nodes, and accessory lymphoid tissues are also ongoing sites of cell production. The earlier in ontogeny that stem cells are harvested, the greater their proliferative potential, as illustrated by the proliferative and growth potential of fetal liver and cord blood cells in clinical transplantation.

Over 70 cytokines maintain, stimulate, or inhibit hematopoiesis. Erythropoietin, macrophage-colony-stimulating factor (M-CSF), and granulocyte-CSF (G-CSF) are examples of cytokines with a relatively high degree of specificity. Most cytokines, however, have multiple actions, often referred to as pleiotropic effects. These cytokines are produced by a large variety of tissues and cell types. Most cells produce multiple cytokines, which can be differentially induced by various stimuli, including other cytokines, such as IL-1 (Chapter 6, Section 6.1). Cytokines act locally, and effective concentrations dissipate within a distance of only a few cells. The multipotent repopulating stem cell expresses receptors for most cytokines, but more mature cells have a more restricted distribution of receptors.

Stems cells are also the natural units of embryonic generation and tissue regeneration in adults. The list of tissues that use the model of differentiation from hematopoietic stem cells to progenitor cells to mature cells, now includes both the central and peripheral nervous system and skeletal muscle. Evidence is mounting that many, if not all, organs and tissues contain tissue-specific stem cells.

Currently stem cells are the focus of a broad and burgeoning area of science aimed at marshaling their generative powers into new treatments, especially for degenerative disorders such as stroke, Parkinson's disease, and spinal-cord injury. These disorders are so destructive that the affected tissue generally cannot be repaired by the body's normal healing mechanisms.

Stem cell therapy has captured the imagination of many researchers because of the potential of stem cells to differentiate into desired cell types and to have trophic qualities that direct them to the site of degeneration. For example, preliminary studies indicate that stem cells can differentiate into dopamine neurons, the brain cells that are lost in patients with Parkinson's disease. In another report, scientists proved, in principle, the concept known as therapeutic cloning. They took skin cells from rats' tails, converted the cells into embryonic stem cells, then transformed them into dopamine-producing brain cells. Many scientists believe that embryonic stem cells should be able to generate all medically desired new tissues, once biologists find the right natural signals to coax them into any desired cell type [3].

15.1.2. Cell and Gene Therapy for Select Medical Conditions

Gene Therapy to Augments Chemotherapy for Head and Neck Cancer

Head and neck cancers afflict about one half million patients each year. First-line treatment with chemotherapy induces a response in only 30% to 40% of patients, and tumors frequently recur. Alternative or complementary strategies have been based on oncolytic viruses that selectively attack tumor cells but not normal ones. One of these viruses, Onyx-015, is a modified adenovirus, a DNA virus that takes over the cell's protein synthesis machinery, replicates, then lyses the host cell to release its progeny.

In wild-type adenovirus, the regulatory protein E1B-55kDa, binds to and inactivates the host cell's p53 protein to promote its own replication. Without the regulatory protein, adenoviruses are not able to replicate. Researchers, however, have observed that an E1B-55kDa mutant adenovirus, Onyx-015, could replicate in and lyse p53-negative but not p53-positive human cells. Since p53 is mutated in 45% to 70% of all cases of head and neck cancers, Onyx-015 was developed as a tumor-cell-specific therapeutic agent [4].

Tumor-selective destruction has been observed in squamous cell carcinoma patients treated with chemotherapy plus Onyx-015. But, when the virus was given alone, it produced a clinical response in less than 15% of patients. A multinational team of researchers has shown in a phase II trial of Onyx-015 with chemotherapy that the combined treatment was more effective than either therapy alone in patients with recurrent head and neck cancers [4].

Gene Therapy for Refractory Angina

Reports presented at the American College of Cardiology 49th Annual Scientific Session, suggested that gene therapy for the last-resort treatment of refractory angina is safe and effective (*Reuters Health*, 14 March 2000). One report presented 12-week results in 30 patients with moderate to severe angina who received a piece of human naked DNA containing the gene for vascular endothelial growth factor-2 (VEGF-2) directly to the heart by means of a mini-thoracotomy. All previous treatments had failed in this patient cohort, even coronary bypass surgery in 83% of them.

The investigators found that 70% of the group had improvement in clinical status and that, in the group as a whole, there was a significant improvement in exercise tolerance time compared with baseline. Nine patients were entirely free of angina at the 12-week point. All adverse events were related to the administration procedure. The single death that occurred seemed unrelated to treatment.

Other investigators reported the results of a phase II trial in which 337 patients were randomized to receive recombinant fibroblast growth factor-2 (FGF-2) or placebo during cardiac catheterization. Although the primary end point—a significant improvement in exercise tolerance time at 90 days—was not achieved, exercise tolerance time was significantly better in a subset of patients over age 63. FGF-2 also significantly reduced the number of anginal episodes compared with the placebo group. Those patients with the most severe symptoms at baseline had the greatest benefit with treatment.

The success observed in these studies is thought to be the result of neoangiogenesis, promoted by the growth factors VEGF-2 and FGF-2. The researchers had hoped that these agents would stimulate formation of new blood vessels around occlusions in the coronary arteries and improve cardiac perfusion.

Gene Therapy to Preventing Late Graft Failure after Bypass Surgery

Autologous saphenous vein coronary artery bypass graft surgery is complicated by late graft failure due to neointima formation and subsequent atherosclerosis. Atherosclerosis occurs within 10 years in 30% to 50% of bypass patients. Matrix metalloproteinases (MMPs), as well as growth factors, play a role in neointima formation by promoting the growth of vascular smooth muscle cells. Tissue inhibitor of metalloproteinase-3 (TIMP-3), an extracellular matrix-associated MMP inhibitor, has been shown to significantly reduce cell proliferation at the luminal surface of human saphenous veins before organ culture and in pig saphenous veins before grafting into carotid arteries by promoting apoptosis (programmed cell death) of smooth muscle cells [5].

Adenoviruses carrying the gene that expresses TIMP-3 has been shown to be able to overexpress TIMP3 within the extracellular matrix. As a result MMP activity was decreased and apoptosis levels in the neointima and medial layer were significantly elevated by TIMP-3 overexpression. Consequently neointimal formation declined by 84% in human veins at 14 days and by 58% in pig vein grafts.

Gene Therapy for the Treatment of Hemophilia

Hemophilia, an X chromosome-linked disease that afflicts about 1 in every 5000 people, is characterized by the absence of at least one of a family of key enzymes involved in blood clotting. Two of the most prominent enzymes are factor VIII, the culprit in hemophilia A, and factor IX, which, when defective, causes hemophilia B, the less common of the two disease forms.

The sickest patients, those making less than 1% of the normal amount of either clotting factor, suffer uncontrollable bleeding episodes and debilitating joint damage. They must inject themselves with recombinant blood factors up to three times a week at a cost of nearly $100,000 a year. These injections promote clotting and relieve joint pain, but they carry the risk, if contaminated, of hepatitis C infection. Patients with a more moderate form of the disease, those having 1% to 5% of normal levels of the enzyme, live a significantly better life, requiring far fewer injections. Thus the goal for therapists is to boost the levels of active enzymes above the benchmark 1%.

Many believe that the ultimate strategy to control hemophilia is to administer a therapeutic gene to make up for the defective one. Their first challenge, however, was to find and characterize the human genes for factor VIII and factor IX, a feat accomplished in the mid-1980s. Researchers spent the next five years determining how defects in those genes influence disease severity. In 1989 a leading researcher in the field, Katherine High, identified, cloned, and characterized the factor IX gene defect that causes hemophilia B in dogs. To create a more tractable disease model, other investigators in the early 1990s knocked out the gene for factor IX in a strain of mice. Another research laboratory did the same for factor VIII.

In the early 1990s researchers found that retroviruses did not deliver enough genetic material, and the other widely used vector, adenovirus, was easily recognized by the immune system. Host cells harboring adenoviruses and their corrective genes are quickly eliminated from the body. As a result many scientists turned their attention to a less immunogenic viral vector, adeno-associated virus (AAV).

Unlike adenovirus, AAV does not cause disease in humans or other mammals, and researchers hoped it might be able to avoid detection and annihilation by the host immune system. AAV enters cells and homes in on chromosome 19. There the strand of viral DNA inserts itself and becomes a permanent part of the host cell's chromosome. Furthermore AAV also

seemed to be able to target nonreproducing cells. On the downside, AAV was difficult to grow in culture and could not accommodate large genes, such as factor VIII. Moreover, once loaded with smaller corrective genes, the virus no longer integrates into its spot on chromosome 19 but, rather, is inserted randomly throughout the genome.

By splicing factor IX to become shorter and without affecting protein function, researchers were able to package the engineered factor IX gene in AAV and express the protein in human cells. More details of this strategy, which has led to human testing is presented in Box 15.2.

BOX 15.2. ENGINEERING OF ADENO-ASSOCIATED VIRUS AS A GENE THERAPY FOR HEMOPHILIA

In the early 1990s scientists began to explore the use of adeno-associated virus (AAV) as a gene-delivery vehicle. Working in Katherine High's laboratory, they succeeded in splicing the gene for factor IX, which is smaller than that for factor VIII, into AAV. In 1997, High's group injected AAV-carrying human factor IX genes into the leg muscles of mice. Soon the rodents were steadily producing therapeutic levels of factor IX. The next year the group corrected hemophilia in genetically altered mice by injections into either leg muscles or livers. Then, in January 1999, High and colleagues reported that they had partially corrected hemophilia B in a dog colony.

Another gene therapy scientist, Mark Kay, has also advanced the treatment of hemophilia. His research team had engineered its own version of AAV, also carrying the gene for factor IX. The researchers injected the vector directly into the portal veins of hemophiliac mice and dogs. Delivery into the liver, although more invasive and riskier, proved to be more effective than injection into muscle. In both High's and Kay's procedures, the treated dogs produced at least 1% of normal blood

levels of factor IX. The liver protocol, however, required only one-tenth the number of viruses to accomplish this.

High and Kay soon decided to collaborate and move their work into humans, blending High's expertise in hemophilia with Kay's ability to manipulate vectors. For safety reasons they decided to start with muscle instead of liver injections. Although AAV takes 6 to 12 weeks to settle in a site within the cell nucleus, the gene rarely strays far from the site of injection. If the vector were injected into the muscle, it would stay there. If there is an unanticipated untoward effect, the clinical investigator can go in and resect the muscle.

Their proposal won FDA approval, the first gene therapy trial ever to inject AAV. In June 1999 the clinical team injected the first patient with genetically altered AAV. The procedure was uneventful. Two more patients followed. As the team reported in the March 2000 issue of *Nature Genetics*, at the suboptimal dose used in the study, meant only to detect manifest safety problems, the first three patients showed no untoward effects and no detectable antibodies [6]. Within 12 weeks after the injec-

(Continued on next page)

BOX 15.2. Continued

tion the researchers identified normal factor IX genes, and the protein itself, in biopsies taken from the patients' legs. Factor IX was also measured in the blood. Even at the low doses used in this safety study, one patient showed an increase in circulating levels of factor IX, exceeding the therapeutic threshold, and the other two reported needing fewer injections of recombinant factor IX to treat and prevent bleeding episodes.

On 17 September 1999, as the clinical team was preparing to inject the next three patients with a higher dose, a report was published detailing the death of a young man with ornithine transcarbamylase deficiency in a corrective gene therapy trial involving injection of the vector directly into the liver. Although the work in hemophiliacs was not directly affected, because the ill-fated study used adenovirus, not AAV, all gene therapy studies fell under intense scrutiny.

In light of the altered climate, High and Kay decided to lower the next dosing regimen to a half-log increment and to delay presentation of their proposed liver injection trial to the NIH's Recombinant DNA Advisory Committee. The delay proved wise, allowing issues to crystallize. On 9 March, RAC approved the proposal to use AAV, injected into the hepatic artery, to carry a corrective gene into hemophilia patients and the project moved to the offices of the FDA. In the interim, studies in three more patients with a slightly higher dose of genetically engineered AAV injected into leg muscles, did not yield much of an improvement over that seen in the first three patients [7]. As the researchers wait for final approval from the FDA, they are joined by hemophilia patients and their families in hoping that injection directly into the liver will provide a major increase in efficacy.

Recently another novel method has been described for factor VIII gene transfer in patients with hemophilia A that avoids the use of viral vectors. The transfer was safe and well tolerated, and provided a modicum of benefit [8]. Up until now the most promising strategy for factor VIII gene transfer had relied on adeno-associated virus (AAV).

The investigators evaluated the safety of their nonviral somatic-cell gene therapy system, which they call transkaryotic implantation, in six patients with severe hemophilia. The procedure involved isolation of dermal fibroblasts from the patients' upper arms. The fibroblasts were then transfected with a factor VIII gene-bearing plasmid. Cells that expressed factor VIII were cloned, propagated, and implanted into the patients' abdomens. This technique can be considered as a less invasive form of *ex vivo* gene therapy.

No patient experienced a significant adverse effect related to the procedure. None developed antibodies against factor VIII, and there was no cellular immune response to the fibroblasts. Four months later, four patients had higher levels of factor VIII than they had at baseline, although none was able to completely forgo replacement therapy. The findings are a first step toward clinical application, but many issues need to be clarified, especially why two patients failed to show any response.

Stem Cell Therapy to Repair Damage from Heart Attacks

Two research teams have reported the results of promising experiments that suggest stem cells harvested from a patient's own bone marrow could one day be used to repair the damage caused by an acute myocardial infarction. The studies showed that adult stem cells injected into rats and mice can promote angiogenesis, generate new myocardium in damaged cardiac tissue, and restore the heart's capacity to pump blood. If the technique proves effective, it would be a remarkable weapon against the ravages of heart disease, especially heart failure, which afflicts nearly five million Americans.

One research group identified stem cells in adult human bone marrow responsible for blood vessel development, expanded the cells *in vitro*, and injected them into rats that had suffered experimentally induced heart attacks two days earlier. The cells seemed to home in on the damaged heart tissue and quickly stimulate formation of new blood vessels. Rats given the treatment had sustained improvement in heart function of 30% to 40% greater than rats not treated with stem cells. The investigators concluded that "The use of cytokine-mobilized autologous human bone-marrow-derived angioblasts for revascularization of infarcted myocardium . . . has the potential to significantly reduce morbidity and mortality associated with left ventricular remodeling" [9].

Another group took a different approach. The investigators injected adult mouse stem cells directly into hearts of mice, near the area damaged by an infarction. The stem cells had been genetically engineered to include a gene that codes for a fluorescent marker, which allowed the researchers to determine whether the new heart muscle had come from donor stem cells. Within 7 to 11 days the stem cells multiplied and transformed into heart muscle cells and migrated into the damaged area. The marker identified the new myocardium as coming from donor stem cells. The researchers also found improved heart function and concluded, "Our studies indicate that locally delivered bone marrow cells can generate *de novo* myocardium ameliorating the outcome of coronary artery disease [10]."

Human embryonic stem cells can also differentiate into spontaneously contracting cells that have the structural and functional properties of human cardiomyocytes [11]. This technologic advance brings us closer to the possibility of increasing the number of functioning myocytes in the depressed region of the heart following an infarction by transplantation of myogenic cells.

Stem Cell Therapy to Produce Insulin

Researchers have developed a strategy to elaborate insulin-producing cells from mouse embryonic stem cells [12]. The breakthrough is a potential boon to hundreds of thousands of patients with type 1 diabetes. Clinical investigators have reported promising results in transplanting pancreatic cells from cadavers into type 1 diabetic patients, enabling them to stop insulin injections for a prolonged period and perhaps indefinitely. But the demand for pancreatic cells is much greater than the supply. An unlimited source of cells that can produce insulin in response to the body's signals is highly sought.

In their report the researchers describe a culturing technique that can turn mouse embryonic stem cells into cell clusters that resemble pancreatic islets. The clusters' inner cells produced insulin, while outer cells produced glucagon and somatostatin, two other proteins typically synthesized by pancreatic cells. Most important, the embryonic stem cell-derived pancreas cells produce insulin in response to glucose, the fundamental role of beta cells that regulate insulin secretion. The major shortcoming of the system at this time is the low levels of insulin production. Refinements in culture technique or drug manipulation may be needed to achieve therapeutic levels.

15.1.3. Summary

The advances in genetic and cellular research have allowed development of cell and gene therapeutic strategies for treating wide-ranging medical conditions. The field of gene therapy founded on molecular and cellular principles is still in its infancy and many technical and ethical issues are yet to be addressed. Although some cells and gene therapies have had limited success, efficiency and safety concerns must be addressed before they can be thought of as a mainstream therapeutic strategy. Some of the molecular designs of gene constructs and intricacies in gene delivery systems are addressed in Section 5.2.

REFERENCES

1. Campbell, K.H., J. McWhir, W.A. Ritchie, and I. Wilmut, *Sheep cloned by nuclear transfer from a cultured cell line.* Nature, 1996. **380**(6569): 64–6.

2. Vogel, G., *No moratorium on clinical trials.* Science, 1998. **279**(5351): 648.

3. Wakayama, T., V. Tabar, I. Rodriguez, A.C. Perry, L. Studer, and P. Mombaerts, *Differentiation of embryonic stem cell lines generated from adult somatic cells by nuclear transfer.* Science, 2001. **292**(5517): 740–3.

4. Anderson, W.F., *Gene therapy scores against cancer.* Nat Med, 2000. **6**(8): 862–3.

5. George, S.J., C.T. Lloyd, G.D. Angelini, A.C. Newby, and A.H. Baker, *Inhibition of late vein graft neointima formation in human and porcine models by adenovirus-mediated overexpression of tissue inhibitor of metalloproteinase-3.* Circulation, 2000. **101**(3): 296–304.

6. Kay, M.A., C.S. Manno, M.V. Ragni, et al., *Evidence for gene transfer and expression of factor IX in haemophilia B patients treated with an AAV vector.* Nat Genet, 2000. **24**(3): 257–61.

7. Gura, T., *Hemophilia: after a setback, gene therapy progresses . . . gingerly.* Science, 2001. **291**(5509): 1692–7.

8. Roth, D.A., N.E. Tawa, Jr., J.M. O'Brien, D.A. Treco, and R.F. Selden, *Nonviral transfer of the gene encoding coagulation factor VIII in patients with severe hemophilia A.* N Engl J Med, 2001. **344**(23): 1735–42.

9. Kocher, A.A., M.D. Schuster, M.J. Szabolcs, S. Takuma, D. Burkhoff, J. Wang, S. Homma, N.M. Edwards, and S. Itescu, *Neovascularization of ischemic myocardium by human bone-marrow-derived angioblasts prevents cardiomyocyte apoptosis, reduces remodeling afnd improves cardiac function.* Nat Med, 2001. **7**(4): 430–6.

10. Orlic, D., J. Kajstura, S. Chimenti, I. Jakoniuk, S.M. Anderson, B. Li, J. Pickel, R. McKay, B. Nadal-Ginard, D.M. Bodine, A. Leri, and P. Anversa, *Bone marrow cells regenerate infarcted myocardium.* Nature, 2001. **410**(6829): 701–5.

11. Kehat, I., D. Kenyagin-Karsenti, M. Snir, H. Segev, M. Amit, A. Gepstein, E. Livne, O. Binah, J. Itskovitz-Eldor, and L. Gepstein, *Human embryonic stem cells can differentiate into myocytes with structural and functional properties of cardiomyocytes.* J Clin Invest, 2001. **108**(3): 407–14.

12. Lumelsky, N., O. Blondel, P. Laeng, I. Velasco, R. Ravin, and R. McKay, *Differentiation of embryonic stem cells to insulin-secreting structures similar to pancreatic islets.* Science, 2001. **292**(5520): 1389–94.

13. Pennisi, E., *After Dolly, a pharming frenzy.* Science, 1998. 279(5351): 646–8.

■ **SECTION TWO** ■

SEAN SULLIVAN

■ 15.2. DELIVERY AND EXPRESSION OF GENES ENCODED FOR FUNCTIONAL PROTEINS

The development of cell and gene therapy was driven by the desire to treat the disease rather than the symptoms. For genetic diseases, gene delivery is the only therapy that promises cure. Delivery of a gene to replace a defective or deleted gene can now be achieved.

Two clinical applications have validated the field of gene therapy. The first involves *ex vivo* transfer of a gene that encodes for the γ_c chain of the cytokine receptor family. The gene is delivered to hematopoietic stem cells of infants lacking cytokine receptor functions. These infants do not have the ability to develop a competent immune system because they lack functional T and NK cells. Consequently any infections these infants incur are life threatening.

Ex vivo gene therapy can control gene delivery by limiting gene transfer to a specific cell type and controlling the number of cells that are transfected. However, *ex vivo* gene delivery is very complex with regard to the manufacture of gene-laden cells. Each patient becomes her or his own source of cells to be transfected, and therefore no more than one patient's cells can be processed at one time. A more practical therapeutic approach is to administer the same gene to multiple patients.

The potential utility of this strategy has been demonstrated using an adeno-associated virus (AAV) to deliver the genes encoding blood-clotting factors to hemophilia patients. AAV is nonpathogenic and can provide long-term expression. Intramuscular administration of an engineered AAV expressing factor IX, for example, has reduced bleeding episodes in these patients.

This section will describe the features and clinical applications for each gene delivery system. The description will start with a common feature of both viral and nonviral gene delivery systems, the expression cassette for the therapeutic protein.

15.2.1. Optimization of Therapeutic Gene Expression

In addition to the DNA sequence that encodes for the therapeutic protein, other DNA sequences flanking the coding region are required to control and ensure gene expression. The combination of these flanking sequences, plus the coding region, makes up the expression cassette depicted in Figure 15.2.

The expression cassette typically includes a promoter that contains 5′ untranslated DNA sequences that are recognized by RNA polymerase II. Next to the promoter is an initiation sequence signaling the start of a protein transcript. The next sequence is a G-methyl cap signal. The capping protein binds to this site and places a G-methyl cap at the 5′ end of the RNA transcript. The 5′G-methyl cap is also a signal for the RNA transcript to be exported out of the nucleus, thereby preventing the newly synthesized transcript from being degraded by RNA nucleases.

Most cellular genes are composed of intron and exon sequences. During maturation of the RNA transcript, the introns are excised and the exons are ligated together. This processing step also facilitates export of RNA from the nucleus into the cytoplasm for protein translation. At least one intron and one exon are almost always included in the therapeutic expression cassette to ensure that the engineered transcript is processed in the same manner as the natural cellular transcript.

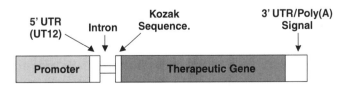

Figure 15.2. Gene expression cassette. The drawing shows the features of the expression cassette. These include the promoter, 5′untranslated sequence, intron, Kozak sequence, therapeutic gene, 3′ untranslated sequence, and the poly-adenosine signal sequence.

All protein transcripts start with a methionine, which is depicted as an AUG codon. Because proteins have multiple methionines, a Kozak sequence just upstream of the AUG serves as a signal for ribosomes to initiate protein synthesis. The coding region of the therapeutic gene also contains a stop codon that signals the ribosome to halt translation. The last feature is the poly-A signaling sequence. This sequence is recognized by a polymerase that polyadenylates the transcript. With each round of translation, an adenosine is removed. When a critical number of adenosines are removed, the transcript is degraded. Hence the length of the poly A tail determines the longevity of the mRNA and the amount of gene product.

Other factors that may affect RNA turnover are the untranslated regions (UTRs) of the transcript. They exist at the 5′ and 3′ ends of the gene expression cassette. The 5′UTR is defined as the DNA sequence from the cap to the AUG initiation codon. The 3′ UTR is defined as the region from the stop codon to the poly-A tail signal. Additional sequences, such as the recognition sequences for restriction endonucleases (designed to allow insertion of therapeutic proteins), are often introduced into these regions. These insertions may affect turnover and intracellular trafficking of RNA transcripts. To prevent false initiation of protein synthesis, scientists must ensure that there are no other AUGs prior to the initiation codon.

The intron/exon, 5′UTR, 3′UTR, and the poly-A tail all affect gene expression by modulating the turnover and intracellular trafficking of the transcript. The promoter, usually accompanied by an enhancer element, controls the number of transcripts that are synthesized. The enhancer, acting through the promoter, can dramatically increase gene expression.

The enhancer element contains binding sites for activator proteins that promote transcription. These can be located adjacent to the promoter sequence or can be distal to the promoter. Various viral promoters have been used for gene therapy. For example, the enhancer/promoter of cytomegalovirus (CMV) has been used routinely to achieve a high rate of transcription. The combination of the CMV enhancer sequence with a chick beta-actin promoter sequence has recently been shown to dramatically increase gene expression. Proper selection of the promoter and enhancer is important for successful gene expression.

Comparison of cellular promoters to viral promoters shows that the cellular promoters are weaker than the viral promoters, as reflected in the level of gene expression. However, addition of an enhancer element can increase promoter strength as noted above with the combination of the CMV enhancer and the cellular chick beta-actin promoter.

Another approach is to coexpress a synthetic transactivator. This is accomplished by fusing a DNA-binding protein with a transcription activator. A DNA sequence recognized by the DNA-binding protein replaces the promoter region of the expression cassette just upstream of the TATA box at the 5′ end. The DNA sequence is usually multimerized to increase the affinity of the synthetic activator. The synthetic transactivator can then be put under the control of a tissue-specific promoter. An example is the fusion of the yeast DNA-binding protein, GAL4, to the Nfκb transcription activator (p65). A 6X multimerized GAL4 recognition sequence is placed upstream of the TATA box of the therapeutic gene. By delivering transactivator and therapeutic gene using a two-plasmid system, researchers showed that this strategy increased gene expression 10-fold compared with delivering the therapeutic gene without transactivator.

The promoters described above are constitutively expressed, meaning that gene expression is always on. However, for certain applications, regulated rather than constitutive expression is desirable. Several regulated gene expression systems have been developed. The first is a tetracycline-dependent gene switch in which an *E. Coli* protein is designed to repress gene expression in the presence of doxycycline or other antibiotics [1]. This system can also employ an activator protein to turn on gene expression in the presence of the same antibiotic. A second system activates gene expression through the binding of the drug rapamycin to two fusion proteins, FRAP and FKBP. The drug-induced heterodimer binds to a DNA sequence upstream of the therapeutic gene and activates transcription [2].

Another system is derived from the baculovirus. It forms drug-dependent heterodimers from mammalian retinoid X receptors and modified Drosophila or Bombyx ecdysone receptors [3]. Yet another combines the components of the amplification system described above (i.e., the yeast GAL4 binding protein and the Nfκb p65 transcription activation subunit) with a mutated progesterone binding domain. This fusion protein, upon binding mifepristone, a progesterone analogue, results in a homodimer that then binds to DNA and activates expression of a gene downstream of the binding site [4–6].

15.2.2. Gene Delivery Systems

Gene delivery systems can be divided into viral and non-viral vectors. Viral systems are comprised of DNA and RNA viruses that contain the therapeutic gene encoded in the viral genome. Critical genes necessary for viral replication are deleted in all viral gene delivery vectors. The engineered virus retains its ability to infect cells but lacks the ability to produce viral progeny. When possible, pathogenic and immunogenic viral proteins are modified to further improve safety. Upon transfection, the viral genome enters the cell and is transported to the nucleus where it either integrates into the

cellular DNA or exists as an episome—DNA that exists outside of genomic DNA.

Nonviral gene delivery is the cellular delivery of plasmid DNA containing the gene itself. Plasmid DNA by itself has limited *in vivo* gene transfer capability. Local administration of the plasmid with external stimuli, such as electricity (electroporation) [7–10] or ultra-high-frequency sound waves (sonoporation) [11], can dramatically increase the level of gene expression. Cationic amphiphiles, peptides, and polymers can bind to plasmid DNA forming particles that serve two purposes. The first is to facilitate cellular binding and intracellular delivery of the plasmid. The second is to protect the DNA from degradative enzymes before cellular entry. The following sections describe each of the gene delivery systems in greater detail and list their preclinical and clinical applications. Key features of each gene delivery system are summarized in Table 15.2.

Retrovirus

Retrovirus Features. Retroviruses are double stranded RNA (dsRNA) viruses. The two strands of RNA are identical and partially hybridized. The viral RNA opens reading frames labeled *gag, pro, pol,* and *env.* The *gag* open reading frame encodes for three different proteins: the matrix, the capsid, and nucleocapsid. The *pro* region encodes for an aspartic protease involved in virion maturation. The *pol* region encodes for reverse transcriptase, which is responsible for converting the viral RNA to proviral DNA, and for integrase, which is responsible for integrating the proviral DNA into the host DNA. The RNA is packaged in a viral capsid that is surrounded by a viral envelope membrane composed of phospholipids and an integral membrane protein. The viral envelope protein binds to a cell surface receptor, resulting in the fusion of the viral envelope membrane with the cell's plasma mem-

■TABLE 15.2. Key features of viral and nonviral gene delivery systems

Features	Retrovirus	Lentivirus	Adenovirus	Adeno-Associated Virus	Plasmid
Gene size	7.5 kb	7.5 kb	30 kb	4.7 kb	Unlimited
Administration route	*Ex vivo*	*Ex vivo*	*Ex vivo/ in vivo*	*Ex vivo/ in vivo*	*Ex vivo/ in vivo*
Integration	Yes	Yes	No	Yes/no	Limited
Duration of expression	Short	Long	Short	Long	Short
Stability	Good	Not tested	Good	Good	Very good
Ease of preparation	Easy	Not known	Easy	Difficult	Easy
Immunogenicity	Slight	Slight	Extensive	Not reported	Not Reported
Pre-existing host immunity	Unlikely	Unlikely	Yes	Unlikely	No
Safety concerns	Insertional mutagenesis	Insertional mutagenesis	Inflammation	Insertional mutagenesis	None

Source: Verma and Somia [12].

Figure 15.3. Retroviral vector plasmid and its features. The "long terminal repeats" (LTRs) contain nucleotide sequences necessary for reverse transcription of dsRNA and, upon conversion to DNA, a retroviral promoter.

brane, and subsequently releases the viral capsid containing the RNA of the therapeutic gene into the cytoplasm.

Once inside the host cell, the viral capsid releases proviral DNA, which subsequently enters the nucleus and integrates into the host DNA. The essential genes for viral replication are polymerase (*pol*), and envelope glycoprotein (*env*). These viral genes are deleted from the genetically engineered retroviral vectors for gene therapy to avoid production of viral progeny. The replication-defective virus can be produced by transfecting a cell with three plasmids; one carries *gag* and *pol* to provide structure, another carries *env* to ensure viral entry for infection, and the final one carries the therapeutic gene. The replication-defective virus contains all of the elements necessary for entry and expression in host

cells. Alternatively, a viral packaging cell line such as 3T3 cells or A293 cells that constitutively expresses viral *gag, pol,* and *env* proteins can be transfected with the plasmid that encode for the therapeutic gene to generate replication defective virus for use in therapy.

A plasmid with a therapeutic gene inserted between two LTR sequences, as shown in Figure 15.3, can be transfected into a packaging cell line to yield replication-defective virions. This is accomplished by including the Ψ packaging sequence with the therapeutic gene. The viral LTR sequence serves as recognition sites for RNA polymerase II to synthesize the RNA. Two copies of the antisense sequence produced in the cells dimerize to provide a compact structure capable of being transported to the cytoplasm. The Ψ packaging sequence in the plasmid is recognized by the viral capsid proteins for assembling the dsRNA-containing therapeutic gene sequences into virions. These virions are now complete with all the necessary machinery for binding and delivery of the therapeutic gene.

Clinical Applications of Retroviral Vectors. A retrovirus was the first vector to be used for gene therapy aimed at ameliorating genetic diseases that required long-term gene expression, such as severe combined immunodeficiency syndrome (SCIDS), cystic fibrosis, and muscular dystrophy. The first successful gene therapy treatment was performed in 1990 for the treatment of SCIDS. The disease is caused by a lack of adenosine deaminase. T-lymphocytes from patients were transfected with a replication deficient retrovirus containing a healthy gene in vitro and then reintroduced back into the patients. One patient produced therapeutic levels of adenosine deaminase, whereas the other patient did not. It was later discovered that the difference in performance was due to a difference in viral transduction. The demonstration of a successful outcome in the one patient was proof of concept and sparked the next 12 years of research to apply this technology to reproducibly treat the disease [13].

Dr. Alain Fisher and his research team have successfully treated two infants with XSCIDS, a severe form of SCIDS that occurs only in boys. These patients lack functional T cells and natural killer cells (NK) due to mutations in the γ_c chain of the cytokine receptor family that recognizes interleukins (i.e., IL-2, -4, -7, -9, and -15). Ex vivo gene transfer was employed. The researchers delivered the wild-type sequence for the γc chain cytokine receptor subunit to hematopoietic stem cells isolated from these patients using a nonreplicating murine retrovirus [14,15].

Preclinical results in γ_c chain cytokine receptor subunit-deficient mice showed that gene expression could be restored resulting in normal T-cell and NK-cell development. Also gene transfer to canine bone marrow resulted in long-term expression. Since XSCIDS is a chronic disease, long-term gene expression is required.

In the two infants a 30% *ex vivo* transfection efficiency was achieved. Following the re-introduction of the stem cells, each patient's blood was analyzed on a regular basis for receptor expression in appropriate cell types, and for immunological function. After 10 months, the results showed that T- and NK-cell counts were comparable to matched normal infants. Moreover the T and NK cells were found to be completely functional.

Safety concerns for retroviruses include integration into an essential gene, resulting in the disruption of expression for that particular gene and causing permanent cellular malfunction. Another safety concern for retroviruses is the potential activation of silent genes in the host cell, such as oncogenes. These viruses contain repetitive DNA sequences at the 5′ and 3′ ends of the viral genome, termed long terminal repeats (LTRs), that exhibit promoter activity. Upon integration of the viral vector into host DNA, placement of the LTR next to a silent host gene may result in activation of the gene, potentially leading to cellular malfunction. In fact, one of the two XSCIDS infants who received gene therapy was reported recently to be exhibiting leukemia-like symptoms. The likelihood of these unwanted side effects is based on probabilities derived from the number of genomes integrated per cell, the number of integration events, and the number of infected cells.

Adenovirus

Adenovirus Features. Adenovirus is a nonintegrating double-stranded DNA virus that is packaged into an icosahedra structure consisting of seven different proteins. The vertices of the icosahedrons contain a penton base with a protruding trimeric fiber protein that contains an immunoglobulin fold. The fiber binds to a 46 kDa cell surface protein designated CAR for coxsackievirus and adenovirus receptor [16]. The hexon protein recognizes an $\alpha_V\beta_5$ integrin, a cell matrix protein. Both the integrin and CAR on target cell are required for adenoviral infection.

The virus enters a cell by endocytosis, and a package protease in the virion is activated by the acidic environment of the endosomes, resulting in proteolysis of the virion [17]. Ultimately the viral DNA is released into the cytoplasm where it is transported to the nucleus [18]. By deleting adenoviral genes (*E1A*, *E2*, and *E4*) required for the early stage of viral replication, the resultant adenoviral vector can infect the cells without eliciting viral replication. The deletion also creates a space for the insertion of an entire therapeutic gene. The engineered virus can be propagated easily using a packaging cell line that expresses the necessary gene to maintain viral replication. The virus can be produced in high titers on a pharmaceutical scale.

Adenovirus is able to infect dividing and nondividing cells, whereas a retrovirus is only capable of infecting dividing cells. This broadens the spectrum of cells that can be transfected by adenovirus and, in so doing, increases potential gene therapy applications. In addition adenovirus provides high gene transfer efficiency to cells expressing the correct receptor. One disadvantage of adenovirus is the expression of viral proteins. These proteins can cause an immune response and present serious problems to patients. The immune reactions can cause acute inflammatory and other cellular responses that can be life-threatening in exceptional cases. Immune responses appear to have contributed to hepatic failure and death in a subject enrolled in a gene therapy trial for evaluating an *adenoviral* vector that encodes for ornithine transcarbamylase gene. Also, because adenovirus is a human pathogen, a large percentage of the population can mount an immune response against it, thereby inhibiting gene transfer following repeated injections.

In theory, the immune responses against adenoviral proteins can be completely eliminated by deleting all adenoviral genes from the engineered vector. This is accomplished by using a CRE-LOX bacterial recombinase system and cell lines capable of packaging the replication-defective virus containing the therapeutic gene [19,20]. The packaging cell line is first transformed to constitutively express the CRE recombinase. The helper plasmid contains a DNA sequence recognized by CRE recombinase (termed LOX sites) that flanks the viral packaging signal. A second plasmid contains the therapeutic gene-expression cassette plus stuffer sequences with flanking LOX recombinase sites. Both helper and therapeutic plasmids are co-transfected into the packaging cells to produce virions capable of transfecting target cells but not replicating therein. Intracellularly CRE recombinase exchanges the packaging sequence from the helper plasmid to the therapeutic gene plasmid. This process results in the packaging of the therapeutic gene plus stuffer sequences into the final adenoviral vector to be used for gene therapy. This process also prevents adenoviral sequences in helper virus from incorporating into the viral progeny produced by the packaging cells.

The elimination of the adenoviral genes reduced the immune response to virally infected cells and resulted in prolonged gene expression. Because the recombinase-dependent sequence exchange process is not always error free, a small percentage of nonreplicating but immunogenic wild-type virus may find its way into the final population of virus particles. The small percentage of wild-type virus may continue to elicit immune side effects. Nevertheless, while not eliminated completely, the immune response to adenovirus is dramatically decreased.

Clinical Applications of Adenovirus. Adenoviral vectors have been developed successfully for the treatment of cancer and cardiovascular disease. For the treatment of cancer, a first generation, *E1A* gene-deleted adenovirus is used to encode a tumor suppressor gene for p53. Cellular expression of p53, a transcription factor that responds to DNA damage, has been shown to suppress tumor cell growth. Should DNA damage be detected during cell replication, p53 protein activates genes that prevent the cell from progressing further into the cell cycle, allowing the repair of the DNA. When DNA damage exceeds the capacity of the DNA repair machinery, p53 activates genes that trigger apoptosis. There are several tumor types, in which p53 mutations occur, that lack protection against proliferation.

Two research programs are based on the strategy of using an adenovirus encoding the *p53* gene and delivering it to tumor cells for the purpose of triggering apoptosis. In one, *p53*-containng adenovirus is injected directly into lung tumors [21]. This has produced some success, but antitumor activity is improved when combined with more conventional chemotherapies.

The other program involves systemic administration of the gene-laden adenovirus via a hepatic artery catheter to treat hepatocellular carcinoma (HCC). The hepatic artery supplies the normal liver with 25% of its blood supply. However, it is the sole blood supply for the carcinoma. Preclinical results demonstrated efficacy in a rat HCC model. In a small, open-label clinical trial of patients positive for HCC, but also having post-hepatitis cirrhosis, patient response was marginal [22].

A very different approach is to use wild type adenovirus to treat tumors that have a mutated p53 gene. This strategy is based on the premise that wild-type p53 inhibits adenoviral replication. Hence tumors with an inactive form of p53 will allow the virus to replicate and lyse cancerous cells [23].

Several clinical trials have been conducted [24], in patients with head and neck cancer [25], liver metastasis derived from colon cancer [26], and epithelial ovarian cancer [27].

Other successful applications of adenovirus vectors have been reported in the setting of cardiovascular disease. Expression of fibroblast growth factor (FGF) by gene transfer to cardiac muscle has resulted in the production of new blood vessels that allow blood to bypass clogged and blocked coronary arteries. The non-replicating adenovirus containing the FGF gene in its viral genome is administered by means of a catheter to the coronary heart muscle [28]. Phase I clinical trials have shown that the procedure is safe in the short term [29].

Following the limited success of this procedure, a multicenter phase II/III study is planned in European Union and the United States. The US component of the trial will study the effect of gene therapy in patients with stable angina. The EU component will evaluate patients with advanced coronary artery disease who are not considered candidates for interventions such as angioplasty or coronary artery bypass graft surgery.

There are also plans to deliver a gene therapy product nonsurgically for the treatment of patients with heart failure. In a controlled trial in a pig model of pacing-induced heart failure, intracoronary delivery of human FGF-4 showed significant improvement in regional cardiac function and a reduction in the size of the heart. If these results translate favorably to humans, FGF-4 gene therapy may be a therapeutic option for patients with cardiomyopathy.

Adeno-Associated Virus

Adeno-associated virus is the most innocuous of viral vectors. It is nonpathogenic and its genome can be engineered to only encode for the therapeutic gene. There are no immunological issues with regard to expression of viral proteins. Unlike adenovirus and retroviruses that yield maximal expression within a few days after transfection, it takes AAV a few weeks before reaching a maximum level of gene expression in cells. The initial low-level gene expression is followed by a slow and gradual increase that may make the immune system tolerant to foreign proteins. This would be particularly helpful for expression of therapeutic target proteins for which endogenous protein is absent. The absent protein generated by gene therapy may have a high tendency to induce immune responses even if it is a "natural" protein for normal individuals. In the case of hemophilia or muscular dystrophy, patients have never been exposed to the therapeutic proteins (delivered by gene therapy).

Adeno-Associated Virus Features. Adeno-associated virus is a single-stranded DNA virus that is a member of the *Parvoviridae* family. The viral DNA has two large open reading frames with a single polyadenylation signal. The open reading frame in the 5′ half of the AAV genome encodes for nonstructural regulatory proteins (Rep 78, Rep 68, Rep 57, and Rep 40). The 3′ half of the viral genome encodes for three structural proteins (Vp1, Vp2, and Vp3). The 5′ and 3′ ends of the viral genome are flanked by 145 nucleotide sequences, termed inverted terminal repeat sequences (ITRs). The first 125 nucleotides of these sequences are palindromic. This allows the sequence to fold back on itself, creating a hairpin structure at each end. This structure is recognized by the Rep 68 protein and facilitates viral genome integration into the host DNA. In human cells, integration occurs at a specific site on chromosome 19. Deletion of the Rep proteins from the viral genome results in random integration.

Viral production is accomplished by transfecting A293 cells with a plasmid expressing the Rep and capsid proteins, and another plasmid containing the therapeutic gene expression cassette flanked at the 5′ and 3′ ends with ITRs. The ITRs prompt a signal that results in packaging of the therapeutic gene expression cassette into the virion. The yield of AAV virions is significantly lower that of adenovirus virions. However, AAV virus titer can be dramatically enhanced by infecting the transfected cells with wild type adenovirus type 5. The viral genome is 4674 nucleotides long, relatively small for a viral vector. The combination of both ITRs and the expression cassette cannot exceed this number of nucleotides or virus titer will exponentially decrease.

The virus infects cells that express heparin sulfated proteoglycan and two coreceptors: fibroblast growth factor receptor and $\alpha_V\beta_5$ integrin [30–32]. The capsid proteins form the ligand-binding site for these cell surface receptors. Animal studies have shown that gene transfer efficiency varies with the type of virus. For example, in mice, following intramuscular administration, a several-fold increase in gene expression was observed with type 5 virus compared with type 2 virus [33]. This is the result of a difference in the primary sequence of the capsid proteins between type 2 and type 5 viuses. Specific amino-acid sequences have been inserted into the capsid proteins to create new ligands that recognize other cell surface receptors expressed on specific cell types [34–36].

On cellular entry, the AAV viral genome undergoes intermolecular circular concatamerization. This permits the introduction of two pieces of a single gene using two different AAV vectors, which then transplice to yield transcripts greater than 5 kbs. Concatamerization allows expression of therapeutic genes exceeding the AAV packaging restraint of 5 kbs.

Clinical Applications of Adeno-Associated Virus. There have been many preclinical demonstrations of successful production of therapeutic proteins by engineered AAV [37–39]. The major clinical application has been the treatment of hemophilia [40,41]. Studies have focused on type A hemophilia, a disorder in which patients fail to express blood-clotting factor IX. Following treatment, factor IX expression was sufficient to allow blood coagulation. There are also preclinical studies that show effective expression of factor VIII, which is deficient in patients with the more prevalent type B hemophilia [42]. This application requires the administration of two AAV vectors, one carrying the heavy chain and the other carrying the light chain of factor VIII.

Nonviral or Plasmid-Based Gene Delivery

Nonviral gene therapy delivers genes in the form of a circular DNA, termed a plasmid. These plasmids can be propagated in bacteria and purified to homogeneity in pharmaceutical quality. A typical plasmid contains the following features: a prokaryotic origin of replication sequence to amplify the number of plasmids per bacteria, a drug resistance gene to selectively grow only those bacteria containing the plasmid, and a eukaryotic expression cassette to express the therapeutic gene in eukaryotic cells. The manufacture of plasmids for pharmaceutical use is much easier than the manufacture of engineered viral vectors.

Although plasmid DNA is highly negatively charged, which portends difficulty in breaching the negatively charged cell membrane, and has a molecular weight greater than 10^6 daltons, which also hinders cellular penetration, plasmids can be administered into tumor or muscle tissues and express therapeutic proteins [43,44]. However, the level of expression is relatively low. Nevertheless, expression is sufficient for some clinical applications, such as the development of vaccines [45]. Increasing the level of expression may extend therapeutic application to allow conversion of transfected muscles into a bioreactor for

secretion of systemically active therapeutic proteins. Expression levels can be increased by several means: (1) administering the plasmid in a large volume, [46,47], (2) applying external stimuli, such as electroporation [48,49] or ultrasound, and (3) applying brief but high (100–600 psi) pressure (generated either by gunpowder or helium pulse) to promote penetration of the DNA plasmids that have been adsorbed on colloidal gold particles, the so-called gene gun approach [50].

Therapeutic applications for delivery of naked or unformulated (DNA) plasmids include expression of endothelial growth factors for promoting new blood vessel growth in ischemic tissue [51–53]. Large volume injection of plasmids through the hepatic portal vein has been evaluated as a means to enhance intracellular gene delivery through temporary induction of local osmotic shock to enhance plasmid entry into hepatic cells. The results showed that gene transfer to the liver occurs and that hepatocytes are the primary target [46].

Plasmid delivery can also be enhanced by inclusion of polymers or lipids in DNA formulations to facilitate cell uptake and gene transfer. The polymers can be a neutral, noninteractive molecule, such as polyvinylpyrolidone [54] or poloxamer, or positively charged, such as poly-L-lysine, chitosan or polyethyleneimine. The neutral polymers are effective for local administration to muscle or tumor mass, although the mechanism by which gene transfer occurs is unknown. These formulations have been used for treatment of head and neck cancer [55]. Positively charged polymers bind to DNA, creating a positively charged complex that then binds to the negatively charged plasma membrane of cells and triggers internalization. Preclinical studies have focused on poly-L-lysine. Derivatization with targeting ligands followed by plasmid DNA complexation has led to selective gene transfer *in vitro*, and some successes have been reported for *in vivo* gene transfer, although these applications are in a very early stage of development [56,57].

Positively charged lipids, first discovered in the late 1980s, have also been used to enhance plasmid delivery [58]. The lipids are composed of a positively charged head group attached to hydrophobic acyl chains. The head group either ion pairs with the phosphate backbone or hydrogen bonds to the DNA. The lipids are prepared in the form of liposomes and added to the plasmid. Upon binding of the cationic liposomes to the DNA, a transfection complex is formed. These complexes are effective for systemic delivery and yield gene transfer to the lungs, liver and spleen [59,60]. In tumor-bearing mice, with both subcutaneous and lung lesions, systemic administration of lipid-based transfection complexes has yielded gene transfer to the tumors and resulted in suppression of tumor growth [61–63]. Combinations of external stimuli, such as ultrasound, with systemic administration has yielded selective increases in gene transfer to the tumor [11].

A cationic lipid transfection complex encoding VEGF was compared to adenovirus encoding the same gene in a clinical trial that sought to reduce restenosis following coronary intervention. The gene delivery vectors were administered into a ballooned vessel using a catheter. New blood vessel growth was clearly observed in three of four patients for both gene delivery systems [24]. This was one of the first demonstrations that nonviral gene therapy yielded an equivalent biological response to that obtained by an adenoviral vector in a clinical setting. Nonviral gene delivery is still in an early stage of development. These initial studies promise to demonstrate some success but will also identify deficiencies that need to be addressed in developing new gene delivery systems. In so doing, the studies will also expand the diversity of medical disorders that can be treated with this technology.

15.2.3. Summary

The gene delivery systems presented in this section represent those with the most information regarding characterization and application. Each system has advantages and disadvantages. For this reason, there may be a disease application for one type of system that is not suitable for another. A few years ago, gene therapy was criticized for the number of clinical trials that had been completed with little or no success. Only in the past few years have demonstrations of benefit been realized to validate the concept. The question is no longer *will* this form of therapy be employed to treat disease, but *when* it will be employed.

REFERENCES

1. Gosen, M., and H. Bujard, *Tight control of gene expression in mammalian cells by tetracycline responsive promoters.* Proc Natl Acad Sci USA, 1992. **89**: 5547–51.
2. Rivera, V.M., T. Clackson, and S. Natesan, *A humanized system for pharmacological control of gene expression.* Nat Med, 1996. **2**: 1028–32.
3. No, D., T.P. Yao, and R.M. Evans, *Ecdysone-inducible gene expression in mammlian cells and transgenic mice.* Proc Natl Acad Sci USA, 1996. **93**: 3346–51.
4. Abruzzese, R.V., D. Godin, M. Burcin, V. Mehta, M. French, Y. Li, B.W. O'Malley, and J.L. Nordstrom, *Ligand-dependent regulation of plasmid-based transgene expression in vivo.* Hum Gene Ther, 1999. **10**(9): 1499–507.
5. Abruzzese, R.V., D. Godin, V. Mehta, J.L. Perrard, M. French, W. Nelson, G. Howell, M. Coleman, B.W. O'Malley, and J.L. Nordstrom, *Ligand-dependent regulation of vascular endothelial growth factor and erythropoietin expression by a plasmid-based autoinducible GeneSwitch system.* Mol Ther, 2000. **2**(3): 276–87.
6. Draghia-Akli, R., P.B. Malone, L.A. Hill, K.M. Ellis, R.J. Schwartz, and J.L. Nordstrom, *Enhanced animal growth*

via ligand-regulated GHRH myogenic-injectable vectors. Faseb J, 2002. **16**(3): 426–8.

7. Cupp, C.L., and D.C. Bloom, *Gene therapy, electroporation, and the future of wound-healing therapies.* Facial Plast Surg, 2002. **18**(1): 53–8.

8. Shibata, M.A., J. Morimoto, and Y. Otsuki, *Suppression of murine mammary carcinoma growth and metastasis by HSVtk/GCV gene therapy using in vivo electroporation.* Cancer Gene Ther, 2002. **9**(1): 16–27.

9. Yamashita, Y., M. Shimada, S. Tanaka, M. Okamoto, J. Miyazaki, and K. Sugimachi, *Electroporation-mediated tumor necrosis factor-related apoptosis-inducing ligand (TRAIL)/Apo2L gene therapy for hepatocellular carcinoma.* Hum Gene Ther, 2002. **13**(2): 275–86.

10. Nakano, A., A. Matsumori, S. Kawamoto, H. Tahara, E. Yamato, S. Sasayama, and J.I. Miyazaki, *Cytokine gene therapy for myocarditis by in vivo electroporation.* Hum Gene Ther, 2001. **12**(10): 1289–97.

11. Anwer, K., G. Kao, B. Proctor, I. Anscombe, V. Florack, R. Earls, E. Wilson, T. McCreery, E. Unger, A. Rolland, and S.M. Sullivan, *Ultrasound enhancement of cationic lipid-mediated gene transfer to primary tumors following systemic administration.* Gene Ther, 2000. **7**(21): 1833–9.

12. Verma, I.M., and N. Somia, *Gene therapy— promises, problems and prospects.* Nature, 1997. **389**(6648): 239–42.

13. Dunbar, C., L. Chang, C. Mullen, et al., *Amendment to Clinical Research Project. Project 90-C-195. April 1, 1993. Treatment of severe combined immunodeficiency disease (SCID) due to adenosine deaminase deficiency with autologous lymphocytes transduced with a human ADA gene.* Hum Gene Ther, 1999. **10**(3): 477–88.

14. Hacein-Bey, H., M. Cavazzana-Calvo, F. Le Deist, A. Dautry-Varsat, C. Hivroz, I. Riviere, O. Danos, J.M. Heard, K. Sugamura, A. Fischer, and G. De Saint Basile, *gamma-c gene transfer into SCID X1 patients' B-cell lines restores normal high-affinity interleukin-2 receptor expression and function.* Blood, 1996. **87**(8): 3108–16.

15. Hacein-Bey, S., F. Gross, P. Nusbaum, C. Hue, Y. Hamel, A. Fischer, and M. Cavazzana-Calvo, *Optimization of retrovi-*

ral gene transfer protocol to maintain the lymphoid potential of progenitor cells. Hum Gene Ther, 2001. **12**(3): 291–301.

16. Bergelson, J.M., J.A. Cunningham, G. Droguett, E.A. Kurt-Jones, A. Krithivas, J.S. Hong, M.S. Horwitz, R.L. Crowell, and R.W. Finberg, *Isolation of a common receptor for Coxsackie B viruses and adenoviruses 2 and 5.* Science, 1997. **275**(5304): 1320–3.

17. Greber, U.F., M. Willetts, P. Webster, and A. Helenius, *Stepwise dismantling of adenovirus 2 during entry into cells.* Cell, 1993. **75**(3): 477–86.

18. Greber, U.F., M. Suomalainen, R.P. Stidwill, K. Boucke, M.W. Ebersold, and A. Helenius, *The role of the nuclear pore complex in adenovirus DNA entry.* Embo J, 1997. **16**(19): 5998–6007.

19. Hardy, S., M. Kitamura, T. Harris-Stansil, Y. Dai, and M.L. Phipps, *Construction of adenovirus vectors through Cre-lox recombination.* J Virol, 1997. **71**(3): 1842–9.

20. Sandig, V., R. Youil, A.J. Bett, L.L. Franlin, M. Oshima, D. Maione, F. Wang, M.L. Metzker, R. Savino, and C.T. Caskey, *Optimization of the helper-dependent adenovirus system for production and potency in vivo.* Proc Natl Acad Sci USA, 2000. **97**(3): 1002–7.

21. Roth, J.A., S.F. Grammer, S.G. Swisher, R. Komaki, J. Nemunaitis, J. Merritt, and R.E. Meyn, *P53 gene replacement for cancer— interactions with DNA damaging agents.* Acta Oncol, 2001. **40**(6): 739–44.

22. Habib, N., H. Salama, A. Abd El Latif Abu Median, I. Isac Anis, R.A. Abd Al Aziz, C. Sarraf, R. Mitry, R. Havlik, P. Seth, J. Hartwigsen, R. Bhushan, J. Nicholls, and S. Jensen, *Clinical trial of E1B-deleted adenovirus (dl1520) gene therapy for hepatocellular carcinoma.* Cancer Gene Ther, 2002. **9**(3): 254–9.

23. Ries, S., and W.M. Korn, *ONYX-015: mechanisms of action and clinical potential of a replication-selective adenovirus.* Br J Cancer, 2002. **86**(1): 5–11.

24. Kirn, D., *Oncolytic virotherapy for cancer with the adenovirus dl1520 (Onyx-015): results of phase I and II trials.* Expert Opin Biol Ther, 2001. **1**(3): 525–38.

25. Kirn, D., *Replication-selective oncolytic adenoviruses: virotherapy aimed at genetic*

targets in cancer. Oncogene, 2000. **19**(56): 6660–9.

26. Reid, T., E. Galanis, J. Abbruzzese, D. Sze, J. Andrews, L. Romel, M. Hatfield, J. Rubin, and D. Kirn, *Intra-arterial administration of a replication-selective adenovirus (dl1520) in patients with colorectal carcinoma metastatic to the liver: a phase I trial.* Gene Ther, 2001. **8**(21): 1618–26.

27. Vasey, P.A., L.N. Shulman, S. Campos, J. Davis, M. Gore, S. Johnston, D.H. Kirn, V. O'Neill, N. Siddiqui, M.V. Seiden, and S.B. Kaye, *Phase trial of intraperitoneal injection of the E1B-55-kd-gene-deleted adenovirus ONYX-015 (dl1520) given on days 1 through 5 every 3 weeks in patients with recurrent/refractory epithelial ovarian cancer.* J Clin Oncol, 2002. **20**(6): 1562–9.

28. *FGF-4 gene therapy GENERX—collateral therapeutics.* BioDrugs, 2002. **16**(1): 75–6.

29. Grines, C.L., M.W. Watkins, G. Helmer, W. Penny, J. Brinker, J.D. Marmur, A. West, J.J. Rade, P. Marrott, H.K. Hammond, and R.L. Engler, *Angiogenic Gene Therapy (AGENT) trial in patients with stable angina pectoris.* Circulation, 2002. **105**(11): 1291–7.

30. Qing, K., C. Mah, J. Hansen, S. Zhou, V. Dwarki, and A. Srivastava, *Human fibroblast growth factor receptor 1 is a co-receptor for infection by adeno-associated virus 2.* Nat Med, 1999. **5**(1): 71–7.

31. Summerford, C., J.S. Bartlett, and R.J. Samulski, *AlphaVbeta5 integrin: a co-receptor for adeno-associated virus type 2 infection.* Nat Med, 1999. **5**(1): 78–82.

32. Summerford, C., and R.J. Samulski, *Membrane-associated heparan sulfate proteoglycan is a receptor for adeno-associated virus type 2 virions.* J Virol, 1998. **72**(2): 1438–45.

33. Chao, H., Y. Liu, J. Rabinowitz, C. Li, R.J. Samulski, and C.E. Walsh, *Several log increase in therapeutic transgene delivery by distinct adeno-associated viral serotype vectors.* Mol Ther, 2000. **2**(6): 619–23.

34. Wu, P., W. Xiao, T. Conlon, J. Hughes, M. Agbandje-McKenna, T. Ferkol, T. Flotte, and N. Muzyczka, *Mutational analysis of the adeno-associated virus type 2 (AAV2) capsid gene and construction of AAV2 vectors with altered tropism.* J Virol, 2000. **74**(18): 8635–47.

35. Ried, M.U., A. Girod, K. Leike, H. Buning, and M. Hallek, *Adeno-associated virus capsids displaying immunoglobulin-binding domains permit antibody-mediated vector retargeting to specific cell surface receptors.* J Virol, 2002. **76**(9): 4559–66.

36. Girod, A., M. Ried, C. Wobus, H. Lahm, K. Leike, J. Kleinschmidt, G. Deleage, and M. Hallek, *Genetic capsid modifications allow efficient re-targeting of adeno-associated virus type 2.* Nat Med, 1999. **5**(12): 1438.

37. Ma, H.I., S.Z. Lin, Y.H. Chiang, J. Li, S.L. Chen, Y.P. Tsao, and X. Xiao, *Intratumoral gene therapy of malignant brain tumor in a rat model with angiostatin delivered by adeno-associated viral (AAV) vector.* Gene Ther, 2002. **9**(1): 2–11.

38. Song, S., M. Scott-Jorgensen, J. Wang, A. Poirier, J. Crawford, M. Campbell-Thompson, and T. Flotte, *Intramuscular administration of recombinant adeno-associated virus 2 alpha-1 antitrypsin (rAAV-SERPINA1) vectors in a nonhuman primate model: safety and immunologic aspects.* Mol Ther, 2002. **6**(3): 329.

39. Goudy, K., S. Song, C. Wasserfall, Y.C. Zhang, M. Kapturczak, A. Muir, M. Powers, M. Scott-Jorgensen, M. Campbell-Thompson, J.M. Crawford, T.M. Ellis, T.R. Flotte, and M.A. Atkinson, *Adeno-associated virus vector-mediated IL-10 gene delivery prevents type 1 diabetes in NOD mice.* Proc Natl Acad Sci USA, 2001. **98**(24): 13913–18.

40. High, K.A., *AAV-mediated gene transfer for hemophilia.* Ann NY Acad Sci, 2001. **953**: 64–74.

41. Kay, M.A., C.S. Manno, M.V. Ragni, et al., *Evidence for gene transfer and expression of factor IX in haemophilia B patients treated with an AAV vector.* Nat Genet, 2000. **24**(3): 257–61.

42. Burton, M., H. Nakai, P. Colosi, J. Cunningham, R. Mitchell, and L. Couto, *Coexpression of factor VIII heavy and light chain adeno-associated viral vectors produces biologically active protein.* Proc Natl Acad Sci USA, 1999. **96**(22): 12725–30.

43. Lu, Y., Y. Tian, L. Wei, and X. Zhao, *[Advances in the researches of vaccines against tumors].* Chung Hua I Hsueh I Chuan Hsueh Tsa Chih, 2000. **17**: 288–90.

44. Yang, J.P., and L. Huang, *Direct gene transfer to mouse melanoma by intratumor injection of free DNA.* Gene Ther., 1996. **3**: 542–8.

45. Fynan, E.F., R.G. Webster, D.H. Fuller, J.R. Haynes, J.C. Santoro, and H.L. Robinson, *DNA vaccines: a novel approach to immunization.* Int J Immunopharmacol, 1995. **17**(2): 79–83.

46. Zhang, G., V. Budker, P. Williams, K. Hanson, and J.A. Wolff, *Surgical procedures for intravascular delivery of plasmid DNA to organs.* Methods Enzymol, 2002. **346**: 125–33.

47. Zhang, G., V. Budker, P. Williams, V. Subbotin, and J.A. Wolff, *Efficient expression of naked dna delivered intraarterially to limb muscles of nonhuman primates.* Hum Gene Ther., 2001. **12**(4): 427–38.

48. Lucas, M.L., L. Heller, D. Coppola, and R. Heller, *IL-12 plasmid delivery by in vivo electroporation for the successful treatment of established subcutaneous B16.F10 melanoma.* Mol Ther, 2002. **5**(6): 668–75.

49. Mir, L.M., M.F. Bureau, J. Gehl, R. Rangara, D. Rouy, J.M. Caillaud, P. Delaere, D. Branellec, B. Schwartz, and D. Scherman, *High-efficiency gene transfer into skeletal muscle mediated by electric pulses.* Proc Natl Acad Sci USA, 1999. **96**(8): 4262–7.

50. Boyle, C.M., and H.L. Robinson, *Basic mechanisms of DNA-raised antibody responses to intramuscular and gene gun immunizations.* DNA Cell Biol, 2000. **19**(3): 157–65.

51. Losordo, D.W., P.R. Vale, R.C. Hendel, C.E. Milliken, F.D. Fortuin, N. Cummings, R.A. Schatz, T. Asahara, J.M. Isner, and R.E. Kuntz, *Phase 1/2 placebo-controlled, double-blind, dose-escalating trial of myocardial vascular endothelial growth factor 2 gene transfer by catheter delivery in patients with chronic myocardial ischemia.* Circulation, 2002. **105**(17): 2012–18.

52. Bashir, R., P.R. Vale, J.M. Isner, and D.W. Losordo, *Angiogenic gene therapy: preclinical studies and phase I clinical data.* Kidney Int, 2002. **61**(Suppl 1): 110–14.

53. Baumgartner, I., *Therapeutic angiogenesis: theoretic problems using vascular endothelial growth factor.* Curr Cardiol Rep, 2000. **2**(1): 24–8.

54. Blezinger, P., J. Wang, M. Gondo, A. Quezada, D. Mehrens, M. French, A. Singhal, S. Sullivan, A. Rolland, R. Ralston, and W. Min, *Systemic inhibition of tumor growth and tumor metastases by intramuscular administration of the endostatin gene.* Nat Biotechnol, 1999. **17**(4): 343–8.

55. Mendiratta, S.K., A. Quezada, M. Matar, N.M. Thull, J.S. Bishop, J.L. Nordstrom, and F. Pericle, *Combination of interleukin 12 and interferon alpha gene therapy induces a synergistic antitumor response against colon and renal cell carcinoma.* Hum Gene Ther, 2000. **11**(13): 1851–62.

56. Wu, G.Y., J.M. Wilson, F. Shalaby, M. Grossman, D.A. Shafritz, and C.H. Wu, *Receptor-mediated gene delivery in vivo: partial correction of genetic analbuminemia in Nagase rats.* J Biol Chem, 1991. **266**(22): 14338–42.

57. Nishikawa, M., S. Takemura, F. Yamashita, Y. Takakura, D.K. Meijer, M. Hashida, and P.J. Swart, *Pharmacokinetics and in vivo gene transfer of plasmid DNA complexed with mannosylated poly(L-lysine) in mice.* J Drug Target, 2000. **8**(1): 29–38.

58. Felgner, P.L., T.R. Gadek, M. Holm, R. Roman, H.W. Chan, M. Wenz, J.P. Northrop, G.M. Ringold, and M. Danielsen, *Lipofection: a highly efficient, lipid-mediated DNA-transfection procedure.* Proc Natl Acad Sci USA, 1987. **84**(21): 7413–17.

59. Hofland, H.E., D. Nagy, J.J. Liu, K. Spratt, Y.L. Lee, O. Danos, and S.M. Sullivan, *In vivo gene transfer by intravenous administration of stable cationic lipid/DNA complex.* Pharm Res, 1997. **14**(6): 742–9.

60. Mahato, R.I., K. Anwer, F. Tagliaferri, C. Meaney, P. Leonard, M.S. Wadhwa, M. Logan, M. French, and A. Rolland, *Biodistribution and gene expression of lipid/plasmid complexes after systemic administration.* Hum Gene Ther, 1998. **9**(14): 2083–99.

61. Anwer, K., G. Kao, B. Proctor, A. Rolland, and S. Sullivan, *Optimization of cationic lipid/DNA complexes for systemic gene transfer to tumor lesions.* J Drug Target, 2000. **8**(2): 125–35.

62. Anwer, K., C. Meaney, G. Kao, N. Hussain, R. Shelvin, R.M. Earls, P. Leonard, A. Quezada, A.P. Rolland, and S.M. Sullivan, *Cationic lipid-based delivery system for systemic cancer gene therapy.* Cancer Gene Ther, 2000. **7**(8): 1156–64.

63. Liu, Y., A. Thor, E. Shtivelman, Y. Cao, G. Tu, T.D. Heath, and R.J. Debs, *Systemic gene delivery expands the repertoire of effective* *antiangiogenic agents.* J Biol Chem, 1999. **274**(19): 13338–44.

64. Freedman, S.B., P. Vale, C. Kalka, M. Kearney, A. Pieczek, J. Symes, D. Losordo, and J.M. Isner, *Plasma vascular endothelial growth factor (VEGF) levels after intramuscular and intramyocardial gene transfer of VEGF-1 plasmid DNA.* Hum Gene Ther, 2002. **13**(13): 1595–603.

INTEGRATION OF DISCOVERY AND DEVELOPMENT: THE ROLE OF GENOMICS AND PROTEOMICS

■ 16.1. OVERVIEW

The development of a new pharmaceutical product continues to be costly and time consuming. In 1997 the FDA approved 38 new molecular entities for marketing, and the pharmaceutical industry spent about 35 billion dollars in research and development. This amounts to about $1 billion for each new drug brought to market that year. A more conservative estimate, based only on the direct cost of launching a new drug, is still considerable, about $500 million.

In the 1980s advances in DNA cloning and expression facilitated discovery and development of novel biopharmaceuticals. These achievements expanded the ability to develop the therapeutic molecules that are the focus of this book. Those molecular cloning and expression techniques, however, were cumbersome and only provided a modest increase in the number of

Biotechnology and Biopharmaceuticals, by Rodney J. Y. Ho and Milo Gibaldi
ISBN 0-471-20690-3   Copyright © 2003 by John Wiley & Sons, Inc.

■TABLE 16.1. Human gene sequences with putative function verified to date

Functional Protein	Number of Genes	Percentage of Total Genes
Cell adhesion	11	0.6
Chaperone	16	0.9
Cytoskeletal structural protein	20	1.2
Extracellular matrix	12	0.7
Hydrolase	80	4.7
Intracellular transporter	51	3.0
Ion channel	7	0.4
Isomerase	21	1.2
Kinase	69	4.0
Ligase	9	0.5
Lyase	12	0.7
Miscellaneous	72	4.2
Motor	13	0.8
Nucleic acid enzyme	221	12.9
Oxidoreductase	64	3.7
Proto-oncogene	23	1.3
Receptor	23	1.3
Select regulatory molecule	88	5.1
Structural protein of muscle	8	0.5
Synthase and synthetase	64	3.7
Transcription factor	81	4.7
Transfer/carrier protein	11	0.6
Transferase	70	4.1
Transporter	44	2.6
Viral protein	4	0.2
Molecular function unknown	613	35.8
Total number of human genes verified	1707	99.4

Source: Adapted from Venter et al. [5].
Note: Data as of 2001.

targets available before the 1980s. It has been estimated that the total number of targets available for drug screening was limited to about 500 between 1940 and 1990 [1]. Drug targets are now becoming less of an issue.

The significant increase in the number of drug targets can be attributed to the exponential growth of data collected by the Human Genome Project, an international effort. The human genome sequencing effort has provided researchers with access to a large set of unique genes with assigned protein products. About 1700 of the genes within the human genome sequences are verified to produce functional proteins suitable for serving as drug targets (Table

16.1). The number of drug targets will continue to grow exponentially as more efficient gene product identification and expression techniques are developed to verify protein activities.

Capitalizing on human genomic information, scientists introduced a new area of research called *proteomics*. Proteomics is beginning to blossom and holds promise for rapid expansion of drug targets. Researchers believe that proteomics, though still in its infancy, will increase our ability to efficiently identify functionally important proteins encoded by the nucleotide sequences of putative genes. Currently the methodologies that link the protein expressed in the cell to either the disease state or the gene that directly controls the expression of the putative target are not fully developed.

While the definition continues to evolve, proteomics includes identification and quantification of proteins in cells, determination of molecular variations, protein interactions, and, ultimately, prediction of physiologic function. Early proteomic experiments focused mainly on identification strategies such as analyzing and cataloging proteins from cellular materials, separated on two-dimensional gel electrophoresis as unique spots. The introduction of mass spectrometry for protein identification allowed miniaturization of analysis and high-throughput screening. Mass spectrometry has also increased the ability to study protein interactions and molecular variations. Miniaturization has permitted the introduction of array technology. Array technology coupled with innovative computer algorithms for automated experimentation and data analysis have fueled the exponential growth in proteomics. In essence, proteomics is now considered as any procedure that is used to characterize large sets of proteins [2].

This chapter will first discuss how drug discovery and the development processes are being integrated to optimize the efficiency of bringing new drugs to the market. This will be followed by a discussion of the role of genomics and proteomics to increase the number of drug targets. A discussion of how these tools may improve the integration of discovery and development processes will ensue. Finally, we will conclude with a commentary on challenges and future prospects for using these tools in development of biopharmaceuticals.

■ 16.2. INTEGRATION OF DISCOVERY AND DEVELOPMENT OF THERAPEUTIC CANDIDATES

The current paradigm for developing a new biopharmaceutical divides the discovery and development phases. This division is arbitrary, and integration of the two phases may improve the productivity of the drug industry. Medicinal chemists, pharmacologists, and molecular scientists in discovery teams focus on identifying disease or molecular targets and potential drug candidates to achieve the highest possible potency. Increasingly, the efforts of discovery teams extend far beyond the identification of active compounds, and some members of the team contribute significantly through the development process up to phase I clinical testing.

Preclinical drug development involves understanding of dosage, formulation, and route of administration, typically in animal models, so that safe and effective doses for human administration can be estimated. The drug development phase extends from preclinical assessment through clinical testing.

The effort to bring a new drug to market requires multiple and sequential iteration of discovery and development. Time-consuming iterations can be reduced by using a project team approach. Nevertheless, fewer than 40% of potential drug candidates identified by discovery teams survive preclinical testing and graduate to human testing, and less than a few percent make it to market [3]. The primary reasons for failure are lack of efficacy, toxicity, and poor biopharmaceutical and pharmacokinetic properties. Biopharmaceutical and pharmacokinetic properties, including drug stability at absorption sites, permeation characteristics, and ADME (absorption, distribution, metabolism, and excretion) characteristics, determine how often the new drug needs to be administered and at what dose to produce efficacy and toxicity. Because these properties play a crucial role in the outcome of efficacy and toxicity studies, many research-driven pharmaceutical companies begin their efforts by characterizing these properties. Pharmacokinetics continues to influence development as one learns more about

■TABLE 16.2. Evolution of acceptable predictors of drug-like characteristics or desirable biopharmaceutic and pharmacokinetic properties

Timeline	Drug-like Characteristics
Before 1990	Acceptable oral bioavailability, based on *in vitro* drug dissolution studies— fraction of an oral dose expected to reach systemic circulation
After 1990	Reasonable permeation across mucosal (i.e., Caco-2) cells mimicking intestinal permeation for drugs whose bioavailability is limited by intestinal absorption
	Acceptable drug metabolism and interaction profiles based on microsomal and recombinantly expressed cytochrome P450 isoenzyme studies
Recent addition	Toxicogenomic data for early assessment of alteration in expression of RNA transcripts (and therefore proteins) related to toxic responses so that modification of safer and more potent variants of NME could be made and identified early

the molecular characteristics that may predict the disposition of a new drug (Table 16.2).

More recently high-throughput procedures for assessing toxicologic characteristics of new drugs are being developed and validated [4]. The basis of toxicology is further informed by pharmacogenomics. Using arrays of validated genes for detecting changes in RNA transcripts of key proteins with established pharmacologic and toxicologic links, pharmacogenomics can improve the odds of success.

While the properties to be emphasized when screening compounds to identify drug candidates are debated, there is a pressing need for the pharmaceutical industry to devise strategies to increase the number of viable drug candidates in the development pipeline. This must be accomplished by improving the success rate of drug candidates through preclinical and clinical testing. For example, a surrogate predictor of success is the absence of drug interactions. The most important group of drug-metabolizing oxidative enzymes, the cytochrome P450s, has now been cloned and is routinely used to identify drug interactions that are likely to lead to adverse effects.

The benefits of drug screening have now been merged with the power of combina-

torial chemistry. Combinatorial chemistry allows simultaneous synthesis of large numbers of structurally related compounds. Consequently large libraries of compounds structurally related to drug candidates have been developed and used to identify alternative drug candidates with more suitable properties. While drug candidates now abound, the rate-limiting step of the overall process is screening. More efficient screening procedures and identification of better predictors of drug safety and efficacy are urgently needed.

■ 16.3. GENOMICS: THE FIRST LINK BETWEEN SEQUENCES AND DRUG TARGETS

Genomics encompasses all genetic information integrated in a cell of a given species (i.e., DNA sequences organized in chromosomes and genes). Yeast and human cells are frequently used species for genomic analysis. Genomics often encompasses the technologies developed to acquire genetic information. Early work focused mainly on information collection, organization of gene structures, and mining the data for genetic associations with disease. It was soon realized, however, that genomics can only infer rather than directly demonstrate gene function. To directly demonstrate function of a putative gene, it must first be transcribed into RNA and subsequently translated to a protein with a defined function. Nevertheless, genomics provides the linkage between DNA sequences and putative proteins that might serve as drug targets. Some of the approaches used to study genomics are listed in Table 16.3.

The rapid advancement in molecular and computational biotechnologies has led us to these exciting times, with human genomic information on 2.91 billion nucleotide sequences available to scientists, physicians, and the general public for use in discovering the basis of incapacitating medical conditions and developing treatment strategies [5,6]. In 1990 only about 1000 human genes were recorded in the Genebank database. The development in 1991 of expressed sequence tags (ESTs), a computational analysis, and the effort organized for human genome mapping have led to an exponential increase in the number of unique human genes. In 1996, 16,000 human genes were identified. This increased to 30,261 by 1998. Based on genomic information, the total number of expressed human genes is estimated to be 60,000 to 70,000. Recent reports of human genome sequence analysis, however, have prompted scientists to re-assess that estimate [5,6].

The most recent assessment of the human genome (with more than 90% of human genome sequences in hand) has placed the number of expressed human genes at about 30,000 [5,7]. This estimate has been confirmed using serial analysis of gene expression (SAGE), a laboratory and computational analysis, which allows simultaneous assessment of a large number of transcripts in cells. This technique suggests that about 25,000 to 30,000 unique gene transcripts are expressed in human cells [5,7]. The new estimate is considerably less than the initial estimates, which were based

solely on computational (i.e., EST) analyses. Based on the estimated gene counts for fruit fly, roundworm, and mustard plant—14,000, 19,000, and 26,000, respectively—some investigators have noted, "Human [gene] complexity . . . is roughly a fly plus a worm or the equivalent of a plant" [8]. These comparative data challenge the statement in some biology textbooks that a higher genomic complexity is expected in higher organisms. Based on this assumption, one would expect the human genome to contain over 100,000 genes.

Additional analysis of DNA sequences in the human genome has revealed that large blocks of human genes are filled with repeated elements, including long interspersed repetitive elements (LINEs) and short interspersed repetitive elements (SINEs). Short interspersed repetitive elements such as *Alu* sequences are often used as target sequences for DNA fingerprinting.

Gene duplication at the DNA level is called *segmental duplication*, and at the RNA level it is known as *retrotransposition*. Duplication provides multiple copies of the same gene [5,6]. Sometimes a copy may not be perfect. The human genome exhibits gene duplication at the DNA level at about 2 to 3 orders of magnitude higher than the fruit fly and worm genomes. This difference could, in part, explain the additional biological process complexity of proteins expressed in human cells than in those of lower organisms. Additional protein complexity may derive from post-translational modifications, such as protein glycosylation, present in eukaryotes (humans) but not in prokaryotes (lower organisms). These complexities are not accurately predicted based on DNA sequence alone.

The limitations of genomic analysis are being addressed by proteomic analysis. DNA sequence analysis has helped to create protein product libraries and relational databases that hold promise to

■TABLE 16.3. Some approaches used to study genomic information

Approach	Description
DNA sequencing	Development of batch process in 1987, coupled with fluorescent dideoxy-terminator labeling on target DNA, has allowed determination of fluorescence-tagged DNA sequences, separated on high-resolution slab-gels and more recently separated by capillary electrophoresis. Both separation methods are capable of sequencing up to 700 bases for each reaction. The automated DNA sequencer can simultaneously process up to 100 samples at a time within 3–4 hours and generate data for 100 unique DNA sequences with about 600–700 bases each.
Bioinformatics in genomics	Human genome initiatives have generated voluminous gene sequence data that require development of computer algorithms to decipher. In practical terms, bioinformatics uses computing tools to identify the start, intervening sequences, and end of a coding region, as well as promoter motif, which is often found in the upstream of coding (RNA transcript producing) regions of DNA. In conjunction with databases developed to allow ready access to the large amount of data and sharing of information around the world through high-speed Internet links, bioinformatics is now taking central stage in capitalizing on the highly enriched DNA sequence data.
Chromosome and comparative gene map	The individual DNA sequences, detected as overlapping fragments, are assembled into genes, and the information is organized into chromosomes. This information is used to compare individuals and species. While gross chromosomal abnormality has been used for diagnosis of karyotype in medical genetics, the more recent comparative gene and chromosome mapping allows identification of genetic alteration related to human disease at the micro level. Such studies are called *pharmacogenomics*.
Pharmacogenomics	Genetic polymorphisms (e.g., single nucleotide changes or gene deletion, insertion, and duplication among individuals) related to drug response in population studies are identified. The genetic polymorphisms that influence how individuals with specific variations in gene loci related to drug response can then be used as a diagnostic screen to predict the outcome of a given drug regimen. For example, there is intense interest in using genetic information gathered from polymorphically expressed cytochrome P450 drug metabolizing enzymes to prevent adverse drug reactions. This area of research is called *pharmacogenetics*, and is a part of pharmacogenomics applications.
Functional genomics	To take a step closer to defining the function of DNA sequences, the sequences are fluorescence tagged to identify transcriptional activity of genes (but not translation of protein), which then can be related to the disease state or medication. This information allows researchers to hunt for genes that are affected in disease states (e.g., mutation in CFTR gene regulating a chloride channel in cystic fibrosis patients)
Structural genomics	For proteins that are members of gene families, such as cytochrome P450, available structural models derived from X-ray crystallography and NMR data allow inference of the structure of newly discovered family members. The inference is based on the sequence variations derived from comparing the reference and new DNA sequences. While a high degree of computational power is needed, predictions based on this approach provide preliminary estimates of protein structure without a huge investment in expression, purification, and crystallization. A full structural elucidation requires several years for each protein, even if it can be crystallized.

transform the drug discovery process. Proteomic data are mainly derived from large-scale sequencing of full-length cDNA, which is reverse transcribed from messenger RNA transcripts. Messenger RNA sequences are capable of being translated into unique proteins that exhibit specific functions (e.g., enzymatic or receptor-binding activity). These data are then confirmed with gene orthologs (evolutionary counterparts) expressed and characterized in other organisms using the worldwide data banked in public and fee-for-service databases [5,6]. The identified genes that are potentially useful as drug targets are categorized according to their functions. Functions are listed in Table 16.1 [5,6].

The cDNA sequence analysis allows scientists to express proteins directly rather than searching for blocks of DNA on the gene that assembles messenger RNA. Finding the gene, however, without a cDNA sequence, is cumbersome and time-consuming because it requires extensive mapping. cDNA sequences can be used to identify sequences (exons) on the gene encoding for a protein of interest. Analysis of exon organization can provide information about RNA reorganization processes that lead to additional complexity in mammalian cells.

While human genome sequencing efforts and innovations of the tools for sequence interpretation continue to advance, the information already annotated with markers for medical conditions is an invaluable resource (Figure 14.1, Chapter 14). It serves as a starting point for identifying disease targets, which in turn can facilitate the development of highly targeted drugs.

■ 16.4. PROTEOMICS: FROM SEQUENCES TO FUNCTIONS

Before the era of proteomics, structure and function were characterized, often in intricate detail, one protein at a time. The work of protein chemists has elucidated the underlying mechanisms of important protein-mediated processes and has led to the discovery of medical therapy. During this period many proteins central to metabolism, replication, transcription and translation, transport and secretion, and cytoskeletal and intracellular signaling processes were isolated and their functions characterized. Advances in protein chemistry were attributed to (1) scientists from different biomedical disciplines—including genetics, cell biology, biochemistry, biostructure, and pharmacology—attacking similar problems from different perspectives; (2) the highly conserved cellular mechanisms across species that allowed extrapolation of mammalian and human protein functions based on information of prokaryotic proteins, and (3) the development and standardization of highly effective and efficient tools for molecular analysis such as recombinant DNA techniques, DNA sequencing, and peptide synthesis and sequencing, just to name a few [2].

As we begin to appreciate that there is much information about drug targets that genomics cannot provide, the focus has shifted from genomics to proteomics. The shift in focus parallels the appreciation of the complexity of proteins, which exceeds the complexity of DNA sequences. While a DNA sequence may allow one to predict the amino-acid sequence of a protein—in cases where an open reading frame sequence (with a start and stop codon) is apparent—it can neither assure expression nor provide information about protein function.

Proteins could be glycosylated, phosphorylated, acylated, ubiquintinated, farnesylated, and sulfated. These processes are not easily predicted by DNA sequence analysis. In addition a single gene could encode multiple proteins through several mechanisms, including alternative splicing of messenger RNA and varying start and stop protein translation sites. Furthermore, protein synthesis and stability may also be regulated by constantly changing cellular processes and extracellular signals such as heat shock proteins, growth factors, and toxins. The end result of these processes could lead to changes in protein localization and interactions, generation of protein fragments, and alteration in protein function and turnover rates.

Protein fragmentation, on the other hand, may be needed for functional activity of some proteins, such as chymotrypsin and insulin, which assume active forms after removal of amino-acid sequences in chymotrypsinogen and proinsulin. Additional complexity in analytical methodologies to deduce protein function *in situ* could also arise from a single protein exhibiting more than one function. Conversely, a given function may require integration of multiple proteins, or that many other proteins can perform the same function.

Appreciating the complexity involved in protein analyses, many believe that more research in proteomics, particularly the development of a streamlined and standardized set of tools, is needed to effectively and efficiently study the large number of protein sequences derived from genomics. While many of the current tools developed for protein separation and structure and function analyses are effective on a small scale, they cannot be adapted to study simultaneously a large number of proteins. The need for additional capabilities for protein analysis in the cellular environment is highlighted by the large number of proteins essential to the performance of basic cellular functions. For example, about 15 proteins are involved in yeast mRNA processing [9], and 29 proteins are implicated in *Caenorhabditis elegans* development [10]. Highly efficient protein analysis by means of a yeast two-hybrid system (Box 16.1) has revealed that more than 2000 proteins are involved in more than 2700 interacting

cellular processes [2]. Standardizing automated high-throughput protein analysis could enhance the growth of proteomics.

Some high-throughput assays in development (yet to be validated) for processing of large numbers of samples include genome-wide messenger RNA expression profiling (using oligonucleotide chips [11,12], serial analysis of gene expression (SAGE) [13], green fluorescent protein (GFP) fusion proteins [14], DNA microarrays [15]), systematic analysis of protein–protein interactions (two-hybrid screens and mass spectrometry) [16]), and systematic gene knockouts in cells or animals (allowing confirmation of gene functions by deletions and correlating the physiological effects in a living system) [17]. Recently an integrated technique was developed employing the yeast two-hybrid system (Box 16.1), transposon tagging, and fluorescence-tagged antibody analysis (immunocytochemistry) to identify interactions and localization of more than 1300 proteins related to about 2000 genes using 11,000 recombinant yeast strains [18]. The information gathered using this integrated strategy is a significant advance, and is a starting point for further refinement. Certainly questions remain as to the most effective ways to validate these data and to narrow down gene products of interest. In addition it would be useful if such a strategy could be adapted for cells of higher eukaryotes such as mammals and humans.

Computer algorithms have been developed to analyze large sets of data to predict protein function. These computational algorithms provide more information than one could derive from amino-acid sequence alignment and domain predictions alone. At least three approaches—phylogenic profiling [19], fused genes or gene pair analysis for domain interaction [20,21], and gene cluster analysis [22]—have been developed to designate protein function. The descriptions of these three approaches are listed in Table 16.4.

Regardless of the strategy, protein analysis has revealed that many human proteins contain common functional domains associated with different receptor, ligand, or substrate-specific active sites. These functional domains can be considered as modular building block combinations. By insertion, shuffling, or modification of recognition domains, a protein such as one with a serine protease basic building block can become plasminogen, apolipoprotein, urokinase, protease specific antigen, coagulation factors IX and X, or complement C1r.

Other examples of basic binding blocks include (1) ATP binding cassette transporter with a characteristic nucleotide-binding domain signature, (2) cyclic AMP or cyclic GMP receptor proteins, and (3) BH domain signature found in Bcl-families. A more complete description of these proteins with common structural and functional domains is listed in Table 16.5. The development of protein databases and computational tools has allowed scientists unsurpassed access to invaluable information at their computer desktops. Many of these common domains could serve as drug targets for developing therapies, but these therapies will have a multiplicity of actions. Subsequent screening for new drug candidates within a family of related proteins provides the resolution needed to attack only the target proteins.

Collection of genomic and proteomic data and sharing of information across the globe through public and private databases has resulted in an extraordinary volume of data. Consequently, biomedical researchers and pharmaceutical scientists have an unprecedented menu. Nevertheless, selecting the gene and target protein for in-depth studies is daunting and akin to finding a needle in a haystack. The detailed study of a selected target protein is now the rate-limiting step in discovering new biopharmaceuticals. Hence, there is a pressing need to streamline or even automate structure-

---

BOX 16.1. YEAST TWO-HYBRID SYSTEMS PERMIT EFFICIENT STUDY OF PROTEIN–PROTEIN INTERACTIONS *IN SITU*

Traditionally protein–protein interactions studies have been performed *in vitro* after isolation and purification of individual proteins. While some *in vivo* or in situ protein-protein interaction studies can be performed by traditional methods using microinjection of purified proteins into oocytes, technical complexities limit the number of proteins that can be studied. Furthermore, many putative proteins of interest, predicted by genomic analysis, are not characterized and cannot be used in such studies. Some of the limitations posed by traditional methods have been overcome by use of yeast two-hybrid systems. These systems allow studies of many recombinant test proteins expressed in unique clones of yeast cells. These yeast cells can be induced to form diploid cells, as part of the mating process, thereby allowing study of interactions between two recombinantly expressed test proteins.

Yeast cells can exist as haploids of opposite mating types (either *a* or α). When an *a* and an α cell are allowed to mate, they form a diploid cell (a/α). To study interactions between two proteins, cDNA sequences of a protein of interest (PT1) are expressed as a fusion protein, linked to a DNA-binding domain (DBD) of a yeast gene-transcript activator in a haploid cell (e.g., *a*). cDNA sequences corresponding to another test protein (PT2) are linked to the

(*Continued on next page*)

BOX 16.1. Continued

gene of yeast gene-transcript activation domain (AD) and expressed in another strain of yeast (e.g., α cell). Interactions of DBD and AD, which ordinarily are not able to bind to each other, are essential for yeast cells to express a functional reporter gene (i.e., *HIS3*) that allows the diploid cells to grow on histidine selection media (Figure 16.1). Binding of the two test proteins, co-expressed as fusion products of DBD or AD, allows DAB-AD interactions to turn on *HIS3* in diploid yeast cells. As a result, these hybrid yeast cells can now grow on histidine selection media. Linking the *HIS3* gene

with an enzyme, such as β-galactosidase, that can be detected easily as a colored product permits verification of protein–protein interactions by diploid cells.

With DBD and AD plasmids already available for ready insertion of unique cDNA sequences from a large library of genes, a large number of unique yeast clones can be developed for evaluation of protein–protein interactions using such yeast two-hybrid systems. Thereby, protein-interaction studies can be automated in a high-throughput format for improved efficiency.

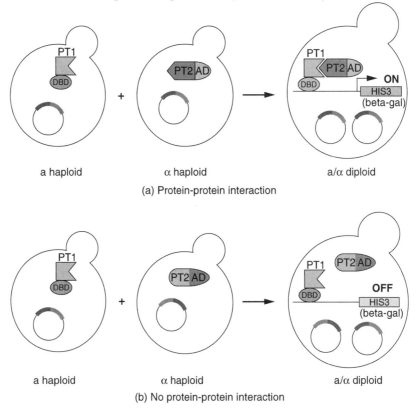

(a) Protein-protein interaction

(b) No protein-protein interaction

Figure 16.1. Schematic representation of the yeast two-hybrid system for evaluation of protein–protein interactions. Haploid yeast of *a* and α cells can mate to form (*a/α*) diploid cells. (A) If two test proteins, PT1 and PT2—expressed in (*a/α*) diploid cells as fusion proteins of DNA binding domains (DAB) and activation domains (AD) of yeast gene–transcript activator proteins—bind to each other, the binding interaction allows the diploid cells to grow in histidine selection media. Histidine selection media is permissive for diploid cells that express the *HIS3* reporter gene only if PT1 and PT2 interact. (B) If PT1 and PT2 do not interact, no HIS3 gene product is expressed and the hybrid cell cannot grow in histidine media.

■TABLE 16.4. Computational approaches used to predict functions of unknown proteins

Approach	Assumptions	What Is Needed	Functional Prediction
Phylogenic profiling	Proteins that function together in a pathway or structural complex are likely to evolve together.	Protein profile analysis for degree of match or similarity	Function and structure linkages to other proteins that participate in a common structural complex or metabolic pathway [19]
Fused gene or gene pair analysis to estimate domain interaction	Selection pressure promotes retention of gene fusion in ensuring protein domain interactions over the course of evolution.	Symmetry and sequence clustering of gene sequences to detect fusion genes	Functional associations of proteins [21,20]
Gene clusters analysis	Functional coupling exists for conserved gene clusters on genomes.	Ability to detect conserved gene clusters from a set of genomic sequences	Functional coupling of putative proteins derived from unique genes in the genomes of interest [22]

function studies of proteins. This will require contributions by scientists from different disciplines to reap the fruits of biotechnology.

■ 16.5. INTEGRATING PROTEOMIC AND GENOMIC TOOLS TO ACCELERATE DRUG DEVELOPMENT

The cost of drug development continues to spiral upward. Inflation and increased regulatory requirements, however, only account for a small portion of this increase. At this time, productivity is a major issue. A review of 198 new drug candidates that reached phase I clinical studies indicates a 60% failure rate due to poor pharmacokinetic properties or toxicity [23]. On the average, less than 2% of the drug failures could be attributed to drug interactions that resulted in adverse reactions [24]. In elderly patients, however, drug interactions could contribute up to 25% of the failure rate [25]. Given the cost and time required for clinical testing, preclinical identification of risk factors for failure could strikingly increase productivity. Most pharmaceutical companies now subscribe to the notion of "fail early—fail cheaply" [26]. The industry believes this can be achieved by developing lists of common risk factors, characterized *in vitro*, and then validating and applying them.

The quest to establish factors predictive of pharmacokinetic characteristics and drug toxicity that can be evaluated *in vitro* (microsomes) began as early as 1960 with the establishment of *in vitro* drug metabolism correlates of *in vivo* drug disposition [27]. The introduction of recombinant DNA technology and bioanalytical capabilities in the 1990s allowed expression and characterization of purified recombinant proteins that are key to drug metabolism. For many drugs these enzymes determine *in vivo* drug availability, and for most

■TABLE 16.5. Proteins with common structure and functional domains

Common Domain	Protein	Feature	Total Number of Genes Reported
Nucleotide binding domain or ATP-binding cassette (ABC transporters)	P-glycoprotein—a multidrug resistance efflux transporter Cystic fibrosis transmembrane conductance regulator Antigen peptide transporters (TAP1 and TAP2) Adrenoleukodystrophy protein Multidrug resistance-associated protein transporter 70kDa peroxisome membrane protein Sulfonylurea receptor Fungal elongation factor 3 Yeast proteins, STE6, ATM1, MDL 1 & 2, SNQ2 Yeast sporidesmin-resistance protein (*PDR5, STS1* or *YDR1* product)	The nucleotide binding site contains about 200 conserved amino-acid sequences.	5811
BH domain of apoptosis regulator protein Bcl-2 family	Bcl-2 Bcl-x (Bcl-x L and S isoforms) Mammalian Bax Mammalian Bak Mammalian Bcl-w Mammalian bad Human Bik Human myeloid leukemia cell differentiation protein Mcl-1 Mouse Bid Neuronal death protein Dp5	All these proteins contain a BH1, BH2, BH3, or BH4 domain, and some may contain more than one domain.	153
Serine protease	Blood coagulation factors VII, IX, X, XI, and XII Thrombin Plaminogen	These proteases contain signature sequences composed of well-conserved serine,	1694

	Protein C Cathepsin G Chymotrypsins Complement components—C1r, C1 and C2 Complement factors—B, D and I Cytotoxic cell proteases—granzymes A to H Mast cell proteases— Plasminogen activators (tPA and urokinase) Trypsin types I–IV Collagenase Aproprotein a Enterokinase (enteropeptidase) Elastases—1, 2, 3A, 3B (protease E), medullasin Hepatocyte growth factor activator Glandular kallikreins—EGF-binding protein A–C, NGF-γ, renin-γ, prostate-specific antigen (PSA), and tonin Mite fecal allergen Der pIII	aspartate, and histidine residues in close proximity to form active sites.	
Cyclic nucleotide-binding domain	cGMP-gated cation channel alpha 1 (rod photoreceptor cGMP-gated channel alpha subunit) cAMP-gated channel in olfactory signal transduction cAMP- and cGMP-dependent protein kinases Prokaryotic catabolic gene activator (CAP) (cAMP receptor protein)	A conserved cAMP- or cGMP-binding structure spans about 120 amino-acid residues assembled into 3 α-helices, and 8 strands assembled in antiparallel giving rise to a β barrel.	614

Data sources: Expasy and InterPro, and NCBI databases, accessed August 2002.

drugs they determine toxicity. These *in vitro* and *in vivo* correlation studies, first with cytochrome P450 isoenzymes, now include monoamine oxidase, peptide transporters (PEPT1 and PEPT2), and efflux transporters (Pgp, MRPs). The maturation of these *in vitro–in vivo* studies has influenced the FDA's "Guidance to the Industry" to endorse *in vitro* screening for potential drug interactions early in the course of drug development so that the most appropriate clinical drug interaction studies could be designed and carried out in a later phase of drug development.

Although few biopharmaceuticals are administered orally, they nevertheless are affected by drug metabolizing enzymes and transporter proteins. These proteins may play an important role in distribution and elimination. The development of a biopharmaceutical can also be expedited by identifying *in vitro* surrogates predictive of *in vivo* pharmacokinetics and toxicity. For example, cyclic peptides such as cyclosporine are a substrate of P-glycoprotein (Pgp), a transporter found in intestinal epithelium and lymphocyte membranes that regulates drug availability at both the systemic and cellular level. Peptide transporters such as PEPT1 and PEPT2 may influence the rate and extent of gastrointestinal absorption of short peptides as well as their distribution. Bile acid transporters may play an important role in reducing accumulation of potentially toxic protein and peptide metabolites. However, a systematic and concerted effort is needed to evaluate how and to what extent these processes influence the pharmacokinetics and toxicity of biotechnology products.

In developing a chemically based low-molecular-weight drug, it is common practice in the industry to synthesize about 40 structurally related analogues for identifying the drug candidate with an optimal pharmacokinetic and toxicologic profile. These chemically based analogues can be synthesized with relative ease by using common approaches such as combinatorial synthesis. Although this strategy cannot be directly implemented for protein-based analogues, one could use a set of closely related macromolecules such as IgG and its F_{ab} fragments as a starting point. We now have a collection of more than 20 IgGs and IgG derivatives with different specificity and indications for clinical use. These could serve as a set of analogues for identifying key *in vitro* correlates that are predictive of *in vivo* responses for biotechnology products.

Such a brute force effort to identify proteins and processes central to drug disposition early in the course of drug development, which seemed insurmountable just a few years ago, is now within reach. The congruence of technological advances such as high-throughput methodology, computation, and analytical methods (initially designed to facilitate genomic and proteomic studies) has made this effort a realistic one. The availability of these highly efficient automated technologies now allow us to increase efficiency by several thousandfold. Instead of performing drug assays one test tube at a time, we can perform 384 assays simultaneously in less than one-tenth of the time. By extrapolation, availability of such technologies in the 1960s could have allowed us to leapfrog by about 38 years.

One highly efficient high-throughput screening tool is microarray technology. DNA microarray technology allows immobilization of DNA templates in thousands of discrete dots that line up in rows and columns on glass slides or silicon chips. The presence of DNA sequences complementary to the DNA templates can be assessed quickly. A similar strategy could be used to screen and profile protein regulation by analyzing the cellular mRNA transcripts that regulate disease processes such as viral infection, inflammation, and cancer. Already such microarray technology, the product of genomics, is routinely used in proteomic studies to identify membrane-bound and secreted proteins, and to monitor the changes in their distribution in normal and abnormal cells [28,29]. These studies, done in the absence and presence of a drug candidate, could allow prediction of pharmacologic effects. Such studies are commonly called *pharmacogenomic studies*. As a result of these developments, the scope of pharmacogenomics now includes profiling drug efficacy and side effects of new and existing drug therapies.

Another area for potential integration in drug development is toxicologic studies done at genomic levels. Currently toxicologic screens of drug candidates are typically done in preclinical drug development with the hope of quickly learning potential safety pitfalls. The input of toxicologists to the project team during the discovery phase has been descriptive, limited to histopathological and gross toxicity analysis. However, this is rapidly changing as toxicologists begin to apply genomic tools.

Pharmaceutical toxicologists are exploring the utility of tools developed to analyze gene expression, in particular, microarry technology, for detecting untoward drug effects that could be correlated with modifications of the gene expression profile in response to drug exposure in cells and whole animals. If validated, this would allow quantitative measurements of tens of thousands of genes simultaneously on a slide no larger than a microscope slide within a few days, instead of months of preclinical toxicology studies in small and large animals (Box 16.2). To be practical, however, the overall cost of high-throughput screening methods and the lack of publicly available databases requires attention. The *in vitro* and *in vivo* data generated over the past few years, if made available, would allow predication of toxicologic profiles for many structurally related drug candidates, before initiating any laboratory experiments. The effort, however, will require tremendous computing power for data processing and retrieval [4].

BOX 16.2. HOW MICROARRAY TECHNOLOGY PROVIDES QUANTITATIVE MEASUREMENT OF CHANGES IN TENS OF THOUSANDS OF GENE EXPRESSIONS DUE TO DRUGS IN ANIMALS OR HUMANS

Microarray technology exploits the ability of mRNA transcripts isolated from a living system to bind specifically or hybridize to DNA templates immobilized on a solid support, typically in the format of rows of DNA strands lined up in dots on a microscope glass slide or silicon chip. By using an array of small dots containing complementary DNA sequences of genes involved in cellular processes, scientists can simultaneously determine hundreds or even thousands of genes within a cell or tissue by quantifying the amount of mRNA bound to each dot on the array. The sensitivity of the assay is greatly enhanced by use of high-yield (extinction coefficient) fluorescence tags.

To assess the effect of drugs in preclinical development, a rat is treated with a defined dose of drug given at a defined frequency. At the end of the experiment, mRNA transcripts from the tissues such as liver and kidney are isolated and hybridized on DNA arrays. The increase and decrease in expression of RNA due to drug treatment is analyzed with the help of a computer to determine which gene is affected and to what extent the drug influences expression. However, prior knowledge of genes that relate to toxicity and efficacy is needed to estimate potential toxicity and efficacy of the treatment.

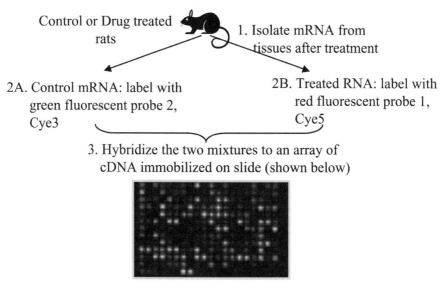

Control or Drug treated rats

1. Isolate mRNA from tissues after treatment

2A. Control mRNA: label with green fluorescent probe 2, Cye3

2B. Treated RNA: label with red fluorescent probe 1, Cye5

3. Hybridize the two mixtures to an array of cDNA immobilized on slide (shown below)

4. Analyze intensity differences reflective of the abundance of a given transcript (dot) corresponding to the same gene in the control and treated animal. Expression of upregulated gene transcripts emerge as high intensity (red) dots, down regulated as low intensity (green) dots and unaffected gene transcripts as medium intensity (blue).

Figure 16.2. Schematic representation of microarray technology used to measure changes in expression of RNA of cellular genes after drug exposure.

Recognizing the need for systematic toxicogenomic studies and developing databases to catalog the results, the National Institute of Environmental Health Sciences (NIEHS), in June 2000, established the National Center for Toxicogenomics to foster integrated studies of genetics, genome-wide mRNA expression, cell- and tissue-wide protein expression, and bioinformatics to understand the role of gene–environment interactions in disease. Substantial progress in the application of toxicogenomic tools to drug development, however, will require the support, input, and participation of the pharmaceutical industry. At this time many pharmaceutical companies view these technologies as experimental, requiring more research and validation.

Some of these sentiments are well founded. Many microarray-based assays still need to be standardized for drug dose and exposure of cells and animals, RNA transcript extraction, and profiling and data analysis. Establishing standardized procedures is extremely important for validation and lending credence to microarray results that are often variable. Corporate leaders are also very much aware that developing and employing genomic and proteomic tools will require substantial investment. New personnel with expertise in a variety of biological, chemical, physical, medical, and bioinformatic disciplines must be identified, hired, and integrated into the pharmaceutical development process. All this could take years of significant capital commitment before any benefits are reaped [1].

The introduction of toxicogenomics in the pharmaceutical industry will also add ongoing costs to drug development, especially if the FDA makes such information mandatory for NDA or BLA submissions.

While toxicogenomics seems to offer an exciting prospect, its adoption by the pharmaceutical industry will not come quickly. Until toxicogenomic information is demonstrated to predict clinical drug response, it is unlikely to have a major impact on drug development.

■ 16.6. SUMMARY

It is only a matter of time until technological advances, the continuous accumulation of genomic data, and innovations in proteomics have significant impact beyond drug discovery. The human genome sequencing project is in the final stages, and now we need to begin putting the name with the face, by identifying the functions of proteins that newly discovered genes encode. Many of these newly identified targets still lack information on how and when they act in cells and whole living systems. It is clear that a crossdiscipline, team approach will be needed to anticipate, early in the drug development process, how newly identified molecular targets and molecular entities may behave in living systems. Maturation of genomics and proteomics along with integration of sciences involved in drug discovery and development will allow weeding out of drug candidates with poor drug-like properties and unacceptable safety profiles early in the research and development process. A successful integration of discovery and development with the help of effective and efficient tools developed to study genomics and proteomics is likely to reduce costs and increase the number of novel drugs produced each year by research-based pharmaceutical companies.

REFERENCES

1. Goodfellow, P.N., *The impact of genomics on drug discovery.* Novartis Found Symp, 2000. **229**: 131–2; discussion 132–5.
2. Fields, S., *Proteomics. Proteomics in genomeland.* Science, 2001. **291**(5507): 1221–4.
3. Borchardt, R.T., *Integration of pharmaceutical discovery and development: case histories.* Pharmaceutical biotechnology; v. 11. 1998, New York: Plenum Press. xxix, 607.
4. Ulrich, R., and S.H. Friend, *Toxicogenomics and drug discovery: will new technologies help us produce better drugs?* Nat Rev Drug Discov, 2002. **1**(1): 84–8.
5. Venter, J.C., M.D. Adams, E.W. Myers, et al., *The sequence of the human genome.* Science, 2001. **291**(5507): 1304–51.
6. Lander, E.S., L.M. Linton, B. Birren, et al., *Initial sequencing and analysis of the human genome.* Nature, 2001. **409**(6822): 860–921.
7. Zhang, L., W. Zhou, V.E. Velculescu, S.E. Kern, R.H. Hruban, S.R. Hamilton, B. Vogelstein, and K.W. Kinzler, *Gene expression profiles in normal and cancer cells.* Science, 1997. **276**(5316): 1268–72.
8. Subramanian, G., M.D. Adams, J.C. Venter, and S. Broder, *Implications of the human genome for understanding human biology and medicine.* Jama, 2001. **286**(18): 2296–307.
9. Fromont-Racine, M., J.C. Rain, and P. Legrain, *Toward a functional analysis of the yeast genome through exhaustive two-hybrid screens.* Nat Genet, 1997. **16**(3): 277–82.
10. Uetz, P., and R.E. Hughes, *Systematic and large-scale two-hybrid screens.* Curr Opin Microbiol, 2000. **3**(3): 303–8.
11. Cho, R.J., M. Fromont-Racine, L. Wodicka, B. Feierbach, T. Stearns, P. Legrain, D.J. Lockhart, and R.W. Davis, *Parallel analysis of genetic selections using whole genome oligonucleotide arrays.* Proc Natl Acad Sci USA, 1998. **95**(7): 3752–7.
12. Cho, R.J., M.J. Campbell, E.A. Winzeler, L. Steinmetz, A. Conway, L. Wodicka, T.G. Wolfsberg, A.E. Gabrielian, D. Landsman, D.J. Lockhart, and R.W. Davis, *A genome-wide transcriptional analysis of the mitotic cell cycle.* Mol Cell, 1998. **2**(1): 65–73.

13. Velculescu, V.E., L. Zhang, W. Zhou, J. Vogelstein, M.A. Basrai, D.E. Bassett, Jr., P. Hieter, B. Vogelstein, and K.W. Kinzler, *Characterization of the yeast transcriptome.* Cell, 1997. **88**(2): 243–51.

14. Niedenthal, R.K., L. Riles, M. Johnston, and J.H. Hegemann, *Green fluorescent protein as a marker for gene expression and subcellular localization in budding yeast.* Yeast, 1996. **12**(8): 773–86.

15. DeRisi, J.L., V.R. Iyer, and P.O. Brown, *Exploring the metabolic and genetic control of gene expression on a genomic scale.* Science, 1997. **278**(5338): 680–6.

16. Fields, S., *The future is function.* Nat Genet, 1997. **15**(4): 325–7.

17. Winzeler, E.A., D.D. Shoemaker, A. Astromoff, et al., *Functional characterization of the S. cerevisiae genome by gene deletion and parallel analysis.* Science, 1999. **285**(5429): 901–6.

18. Ross-Macdonald, P., P.S. Coelho, T. Roemer, et al., *Large-scale analysis of the yeast genome by transposon tagging and gene disruption.* Nature, 1999. **402**(6760): 413–18.

19. Pellegrini, M., E.M. Marcotte, M.J. Thompson, D. Eisenberg, and T.O. Yeates, *Assigning protein functions by comparative genome analysis: protein phylogenetic profiles.* Proc Natl Acad Sci USA, 1999. **96**(8): 4285–8.

20. Dandekar, T., B. Snel, M. Huynen, and P. Bork, *Conservation of gene order: a fingerprint of proteins that physically interact.* Trends Biochem Sci, 1998. **23**(9): 324–8.

21. Enright, A.J., I. Iliopoulos, N.C. Kyrpides, and C.A. Ouzounis, *Protein interaction maps for complete genomes based on gene fusion events.* Nature, 1999. **402**(6757): 86–90.

22. Overbeek, R., M. Fonstein, M. D'Souza, G.D. Pusch, and N. Maltsev, *The use of gene clusters to infer functional coupling.* Proc Natl Acad Sci USA, 1999. **96**(6): 2896–901.

23. Kennedy, T., *Managing the dicovery/development interface.* Drug Discovery Today, 1997. **2**: 436–444.

24. Fuhr, U., M. Weiss, H.K. Kroemer, G. Neugebauer, H. Rameis, W. Weber, and B.G. Woodcock, *Systematic screening for pharmacokinetic interactions during drug development.* Int J Clin Pharmacol Ther, 1996. **34**(4): 139–51.

25. Doucet, J., P. Chassagne, C. Trivalle, I. Landrin, M.D. Pauty, N. Kadri, J.F. Menard, and E. Bercoff, *Drug–drug interactions related to hospital admissions in older adults: a prospective study of 1000 patients.* J Am Geriatr Soc, 1996. **44**(8): 944–8.

26. Ekins, S., B.J. Ring, J. Grace, D.J. McRobie-Belle, and S.A. Wrighton, *Present and future in vitro approaches for drug metabolism.* J Pharmacol Toxicol Methods, 2000. **44**(1): 313–24.

27. Lien, E.J., and C. Hansch, *Correlation of ratios of drug metabolism by microsomal subfractions with partition coefficients.* J Pharm Sci, 1968. **57**(6): 1027–8.

28. Diehn, M., M.B. Eisen, D. Botstein, and P.O. Brown, *Large-scale identification of secreted and membrane-associated gene products using DNA microarrays.* Nat Genet, 2000. **25**(1): 58–62.

29. Diehn, M., A.A. Alizadeh, and P.O. Brown, *Examining the living genome in health and disease with DNA microarrays.* JAMA, 2000. **283**(17): 2298–9.

Appendix I

DOSAGE FORM, PHARMACOKINETICS, AND DISPOSITION DATA

Biotechnology and Biopharmaceuticals, by Rodney J. Y. Ho and Milo Gibaldi
ISBN 0-471-20690-3   Copyright © 2003 by John Wiley & Sons, Inc.

TABLE I. Dosage form, pharmacokinetics, and disposition data

Category	Drug Name	Trade Name	Dosage Form and Dose	Route of Administration	Clearance	Volume of Distribution	Half-life	Effective Dose	Toxic Dose	Major Route of Elimination
Antibodies and derivatives	Abciximab, F_{ab} fragment of chimeric human-murine monoclonal antibody 7E3 directed against the glycoprotein (GP) IIb/IIIa ((alpha) IIb beta 3) receptor of human platelets	ReoPro	Sterile solution 10 mg in 5 ml vials (2 mg/ml)	IV	NA	NA	$\alpha = \sim 10$ min; $\beta = 30$ min	0.25 mg/kg intravenous bolus administered 10–60 min; followed by 0.125 µg/kg/min (to a maximum of 10 µg/min) for 12 h	NA	NA
Antibodies and derivatives	Alemtuzumab, anti-CD52 humanized monoclonal antibody, CHO product	Campath	30 mg in 3 ml solution vial	IV	NA	NA	~12 d	3 mg over 2 h IV infusion, individualized as maintenance dose after*	>80 mg dose	Liver and renal catabolism and elimination
Antibodies and derivatives	Basiliximab, chimeric murine-human monoclonal IgG₁ against IL-2 receptor alpha (CD25), produced by rDNA technology	Simulect	Lyophilized 20 ml vial	IV	41 ± 19 ml/h	8.6 ± 4.1 liters	7.2 ± 5.1 d	Two 20 mg doses at 2 h before transplantation and 4 d after transplantation surgery	No serious effects in renal transplant patients with up to 60 mg dose	Receptor-mediated processes; renal and hepatic (predicted)
Antibodies and derivatives	Capromab Pendetide, murine monoclonal 7E11-C5.3 IgG, Indium 111 labeled	ProstaScint, ^{111}In	0.5 mg in 1 ml vial	IV	42 ± 22 ml/h	4 ± 2 liters	67 ± 11 h	0.5 mg (5 mCi)	Up to 10 mg dose without additional adverse response	Renal
Antibodies and derivatives	Cytomegalovirus Immune globulin, human hyperimmuneglobulin, purified	CytoGram	50 ml vial (50 mg/ml)	IV	NA	NA	$\alpha = 15$–30 h; $\beta = 21$–26 d	150 mg/kg (range 50–150 mg/kg) q. 2 w for 16 weeks.	Limited data available	Receptor-mediated processes; hepatobilliary and renal (predicted)
Antibodies and derivatives	Daclizumab, anti-CD25 humanized IgG₁, recombinant product	Zenapax	25 mg in 5 ml solution vial	IV	15 ml/h	6 liters	20 (11–38)	1 mg/kg in over 15 min; every 14 d for a total of 5 doses	Under the age of 18	Receptor-mediated processes; liver and kidney (predicted)
Antibodies and derivatives	Etanercept, recombinant fusion protein containing TNF receptor linked to F_c domain of IgG₁, produced in CHO cells	Enbrel	25 mg/vial, preservative-free, lyophilized powder	SC	89 ml/h	0.4 liter	115 h (range: 98–300 h)	25 mg given two times a week	No dose limiting toxicities recorded up to 60 mg/m² IV	Receptor-mediated processes (evident by increase in plasma levels

									(N), 32 mg/m² (RA)	upon repeated dosing); hepatobiliary and renal (predicted metabolic processes)
Antibodies and derivatives	Gemtuzumab-oxogamici (anti CD33-caliceamicin conjugate); recombinant humanized IgG₄	Mylotarg	5 mg in 20 ml vial	IV	NA	NA	45 h; 60 h (second dose)	9 mg/m²; can be given a 2nd dose 14 d later for a total of two doses	Exhibit mortality 2 mg/kg (~1.3× higher than recommended dose) in rats; 4.5 mg/kg (~6× higher than recommended dose) in male monkeys	Liver; oxidative metabolism of oxogamicin and antibody catabolism
Antibodies and derivatives	Hepatitis B immune globulin (human), purified by chromatography procedure	Nabi-HB	312 U in 1 ml or 1560 in 5 ml single dose vial	IM	0.433 ± 0.144 liter/d	15.3 ± 6.2 liters	24.8 ± 5.6 d	0.06 ml/kg IM ASAP; preferably within 24 hours	NA	Liver catabolism (predicted)
Antibodies and derivatives	Human immune globulin, purified by cold ethanol fractionation procedure	BayGam	15–18% protein solution in 2 or 10 ml single dose vial	IM	NA	NA	23 d	0.02 ml/kg prophylaxis to 1.2 ml/kg as a varicella-zoster immue globulin substitute	NA	NA
Antibodies and derivatives	Ibritumomab tiuxetan, monoclonal antibody directed against the CD20 antigen, produced in CHO cells	Zevalin	6 mg in 0.6 ml saline	IV	NA	1–2.5 mg/kg	30 h	Individualized dosage*	NA	NA
Antibodies and derivatives	Imciromab pentetate, Fab fragment of anti-myosin IgG₂ monoclonal antibody bound to DTPA for labeling with In 111	Myoscint (In 111)	0.5 mg of Imciromab in 1 ml for In 111 chloride labeling	NA	NA	NA	α = 1.5 h, β = 20.2	2 mCi (74 MBq) or 1.8–2.2 mCi (67–81 MBq)	NA	NA
Antibodies and derivatives	Immune globulin (human)	Gammar-PIV	Lyophilized, 1, 2.5, 5, 10 g vial, 2–400 mg/kg 3–4 times/week	IV	NA	NA	NA	200–400 mg/kg q. 21–28 d	NA	Liver catabolism

■ TABLE I. *Continued*

Category	Drug Name	Trade Name	Dosage Form and Dose	Route of Administration	Clearance	Volume of Distribution	Half-life	Effective Dose	Toxic Dose	Major Route of Elimination
Antibodies and derivatives	Immune globulin (human) against Rho	WinRho SDF	600, 1500, 5000 IU vial	IV, IM	NA	NA	NA	Initial dose of 50 mg/kg; followed by reduced maintenance dose according to Hgb levels	NA	Receptor-mediated processes; liver and kidney (predicted)
Antibodies and derivatives	Immune globulin (human) against Rho (D), purified IgG Cohn fraction II	BayRho-D	820 (300 µg) (full) or 137 (mini) IU vial or syringe	IM	NA	NA	NA	300 µg within 72h of delivery, 28-week gestation, following miscarriage or termination of ectopic pregnancy where appropriate	NA	Hepatobilliary and renal (predicted)
Antibodies and derivatives	Immune globulin (human), (IGIV)	Sandoglobulin	Lyophilized, 1, 3, 6, 12 g vial, 2–400 mg/kg 3–4 times/week	IV; not to exceed 2 mg Ig/kg/min	NA	NA	NA	0.4 g/kg of body weight on 2-5 consecutive d	NA	Receptor-mediated processes; liver and kidney (predicted)
Antibodies and derivatives	Immune globulin D (RH1) (human), purified by cold alcohol fractionation	RhoGam Rho (D), MICRhoGAM Rho (D)	820 IU (300 µg/ml) in 1 ml prefilled syringe, smaller prefilled doses are available as MICRhoGAM	IM	NA	NA	NA	300 µg equivalent to 820 IU	NA	NA
Antibodies and derivatives	Immune globulin against hepatitis B (human)	BayHep B	217 IU/ml in 1 or 5 ml vial, 0.06 ml/kg	IM	NA	NA	17.5–25 d	0.06 ml/kg, IM within 24 postexposure	NA	Receptor-mediated processes; liver and kidney (predicted)
Antibodies and derivatives	Immune globulin Intravenous (human) N-10% & 5%	Gamimmune	0.1 g/ml in 10, 50, 100 or 200 ml	IV, not to exceed 0.08 ml/kg without saline	NA	NA	21 d	0.01 to 0.02 ml/kg body weight per minute for 30 minutes; if	NA	NA

tolerated, can be increased to 0.08 ml/kg; 400–1000 mg/kg q. 28 d

Category	Description	Trade name	Formulation	Route	Clearance	Volume of distribution	Half-life	Dose	Toxicity	Elimination
Antibodies and derivatives	Infliximab, anti tumor necrosis factor alpha TNF(alpha), chimeric human IgG1	Remicade	Lyophilized 100 mg vial; reconstitute to 20 ml	IV	13–22 ml/h	4.3–5.7 liters	3.2–8 d dose-dependent*	3 mg/kg (Rheumatoid arthritis), 5 mg (Crohn's disease), IV at 0, 2, 6 weeks; every 8 weeks thereafter, usually given in combination with methotrexate	>20 mg/kg dose	Receptor-mediated processes; liver and kidney (predicted)
Antibodies and derivatives	Muromonab-CD3, anti-CD3 antibody	Orthoclone OKT3	5 mg in 5 ml	IV	NA	6.5 liters	18 h	2.5 mg ($<$30 kg) or 5 mg ($>$30 kg) per d for 10–14 d	10 mg/d may produce clinical symptoms including hyperthermia, severe chills, myalgia, vomiting, diarrhea, edema, oliguria, pulmonary edema, and acute renal failure	Receptor-mediated processes; hepatobiliary and renal (predicted)
Antibodies and derivatives	Nofetumomab, F_{ab} fragment of a monoclonal NR-LU-10 IgG_{2b}(murine) against small lung cancer associated antigen, T_c 99 labeled	Verluma, technetium Tc 99m	10 mg in 1 ml vial	IV	NA	NA	1.46 h	5–10 mg (30 mCi = 1.110 MBq Tc 99) in 15–20 ml saline, IV over 3–5 min	No major toxicity observed with 19.9 mg dose	Renal (68% within 22 h); hepatobiliary (secondary)
Antibodies and derivatives	Palivizumab, humanized monoclonal IgG_1, against F protein of respiratory syncytial virus (RSV), recombinant DNA product	Synagis	Lyophilized, sterile powder in 50 or 100 mg vial	IM, 15 mg/kg	NA	NA	18 d	15 mg/kg, IM, q. 30 d	No apparent toxicity at 50 mg/kg (rabbit) or up to 5 monthly recommended dose	Receptor-mediated processes; hepatobiliary and renal (predicted)

■ TABLE I. *Continued*

Category	Drug Name	Trade Name	Dosage Form and Dose	Route of Administration	Clearance	Volume of Distribution	Half-life	Effective Dose	Toxic Dose	Major Route of Elimination
Antibodies and derivatives	Rabies immune globulin (human)	BayRab	2 or 10ml vials, 150 U/ml 20 U/kg	Infiltrated area; IM (gluteal)	NA	NA	NA	NA	NA	NA
Antibodies and derivatives	Rabies immune globulin (human)	Imogam Rabies-HT	3 or 10ml vials, 150 U/ml	Infiltrated area; IM (gluteal)	NA	NA	NA	20IU/kg	NA	NA
Antibodies and derivatives	Respiratory syncytial virus immune globulin intravenous (human), RSV IVIG	Respigam	50ml single dose vial (SDV)	IV	NA	NA	NA	NA	NA	Similar to other IVIG
Antibodies and derivatives	Rituximab, chimeric murine/human IgG$_1$ monoclonal against CD20 (B-cell suface antigen), produced in CHO cells	Rituxan	10 or 50ml vial (10mg/ml)	IV	0.046liter/h (1st dose); 0.015liter/h (4th dose)	NA	average = 60h (variable depending on tumor burden); 68h (1st dose), 190h (4th dose)	375mg/m 2 IV infusion (50mg/h) on d 1, 8, 15, & 22 (4 doses)	Tolerated up to 50mg/m²	Receptor-mediated processes, including antibody-mediated cytotoxic (ADCC) pathway; hepatobiliary and renal elemination of catabolized antibody molecules (predicted)
Antibodies and derivatives	Tetnus immune globulin (human) isolated from solubilized Cohn Fraction II	BayTet	250U vial or syringe	IM (deep)	NA	NA	23d	250IU for (>7yr old) or 5U/kg (<7yr old)	NA	Liver catabolism
Antibodies and derivatives	Thymoglobulin, anti-thymocyte globulin (rabbit), purified rabbit IgG against human thymocytes	Sangstat	Lyophilized 25mg vial, reconstitute to 5ml	IV	NA	7ml vials	2.3d	1.25-1.5mg/kg/d 4-hour IV infusion for 7-14d	Overdose may cause elukopenia and/or throbocytopeni	Receptor-mediated processes; hepatobiliary and renal
Antibodies and derivatives	Trastuzumab, humanized monoclonal IgG1 against HER2	Herceptin	Lyophilized, sterile powder nominally containing 440mg	IV	Depends on HER-2 concentration (~5.1)	44ml/kg (P)	1.7 and 12d at the 10 and 500mg dose	4mg/kg over 90 min initially, followed by 2 mg/kg over 30 min every week	None reported up to 500mg single dose	NA
Enzymes	Adenosine deaminase (ADA), PEG modified (bovine), [11–17 × PEG5000]	Adagen	250IU/ml 1.5ml vial, 15IU/kg weekly dose	IM	NA	NA	3–6d	10IU/kg,15IU/ kg, 20IU/kg; 1st, 2nd, 3rd dose, respectively	NA	Hepatic (predicted)
Enzymes	Alteplase, tPA human recombinant, produced	Activase	Lyophilized 50 or 100mg,	IV	380–570 ml/min	~ (P) volume, 4	~5min	Front-loaded double IV	Under 18 years of age	Liver

	in *E. coli*			reconstitute to 1 mg/ml			liters/kg		infusion of 100 mg for patients >67 kg	NA	NA
Enzymes	Ceradase, placental-derived	Ceradase	Solution for injection in 10 IU/ml or 80 IU/ml vial	IV	~6–25 ml/min/kg	0.05–0.28 liters/kg	3.6–20 min	2.5 IU/kg 3 times weekly to 60 IU/kg 2 times weekly; depending on disease severity	NA	NA
Enzymes	DNAse, dornase alfa, human recombinant deosyribonuclease I (rhDNAse), produced in CHO cells	Pulmozyme	2.5 mg in 2.5 ml solution vial for use with approved nebulizer	Inhalation of aerosol generated by one of the recommended nebulizers	NA	NA	Limited systemic levels	2.5 mg single-use ampule inhaled once daily using one of the recommended nebulizers	Up to 180 times normal dose was tolerated in rats and monkeys	Pulmonary (catabolism in lung) and renal
Enzymes	Drotrecogin alfa (acitvated), human recominant of activated protein C	Xigris	Lyophilized power vial 5, 20mg drotrecogin alfa	IV	42 liters/h	12–30 µg/kg/h	13 min	24 µ/kg/h infusion over 96 h	NA	Systemic circulation
Enzymes	Imiglucerase-recombinant (beta)-glucocerebrosidas	Cerezyme	Lyophilized 200 or 400 IU vial	IV infusion over 1–2 h	14.5 ± 4.0 ml/min/kg	0.12 ± 0.02 liter/kg	3.6–10.4 min	2.5 IU/kg 3 times weekly to 60 IU/kg, IV infusion over 1–2 h once every 2 weeks	Up to 240 IU/kg two times weekly without apparent toxicity	Not well understood; liver metabolism (predicted)
Enzymes	Pegaspargase, L-asparaginase conjugated to PEG (5000), *E. coli*-derived	Oncaspar	5 ml vial (750 IU/ml)	IM (preferred), IV	NA	2.1 liters/m², ~plasma volume	5.7 ± 3.24 d	2500 IU/m² every 14 d	10,000 IU/m² may produce symptomatic toxicity. IM, route may reduce incidence of hepatotoxicity, coagulopathy and GUI and renal disorder	Hepatobiliary (no detectable levels in urine)
Enzymes	Reteplase, nonglycosylated genetically modified	Retavase	Lyophilized, 18.1 mg vial containing 10.4 IU, reconstitute to 10 IU/ml	IV bolus	250–450 ml/min	NA	13–16 min	10 + 10 IU double bolus IV over 2 min at time 0, and 30 min after	NA	Liver and kidney
Enzymes	Streptokinase, purified bacterial protein	Streptase	Lyophilized 0.25, 0.75, or	IV, intracoronary	NA	NA	23 min	1,500,000 IU within 60 min	NA	NA

■ TABLE I. *Continued*

Category	Drug Name	Trade Name	Dosage Form and Dose	Route of Administration	Clearance	Volume of Distribution	Half-life	Effective Dose	Toxic Dose	Major Route of Elimination
	product, plasminogen activator		1.5 MU vial reconstitute to 6.5 ml	(IC)				IV or 20,000 IU by bolus followed by 2000 IU/min. for 60 min IC (acute); 100,000 IU/h, for 24–72 h for thrombosis or embolism with 30 min bolus 250,000 IU loading dose		
Enzymes	Tenecteplase, human recombinant tissue plasminogen activator (tPA) produced in CHO cells	TNKase	Lyophilized 50 mg vial reconstituted in 10 ml	IV	99–104 ml/min	~Plasma volume	α = 20–24 min β = 90–13 min	Single IV bolus doses of TNKase 30–50 mg, based on body weight	No unexpected toxicity observed with rabbits and dogs up to 30–50 mg/kg	Liver metabolism
Gene therapeutics	Fomivirsens sodium, antisense nucleotide polymer against CMV	Vitravene	6.6 mg/ml in 0.25 ml vial	Intravitreal injection	NA	NA	NA	330 µg/0.05 ml two doses, every other week	NA	Exonucleases found in eyes
Hematologic and coagulation factor	Darbepoetin alfa, human recombinant produced in the CHO cells	Aranesp	25,40,60,100, 200 µg in 1 ml solution	SC, IV	NA	50 U/kg	21 h	1 ml weekly	Hemoglobin levels of patient over 12 g/dl	Hepatic and renal
Hematologic and coagulation factor	Pegfilgrastim	Neulasta	0.6 ml solution vials ~6 mg pegfilgrastim	SC	NA	NA	15–80 h	5 mcg/kg/d	300	NA
Hematologic and coagulation factors	Antihemophilic factor (human), purified factor VIII	Koate-DVI, also available as Koate-HP	Lyophilized 250, 500, or 100 U vials (Koate-DVI includes a second viral inactivation step in the manufacturing process)	IV	2.88–4.44 ml/kg/h	0.054–0.80 liter/kg	12–17 h	Depends on the type and severity of bleeding episode; ~14–25 U/kg	NA	NA
Hematologic and coagulation	Antihemophilic factor, human recombinant	Kogenate; Kogenate-FS	Lyophilized 250, 500, or	IV	2.88–4.44 ml/kg/h	0.054–0.80 liter/kg	~13 h	According to the needs of the	NA	NA

factors	factor VIII produced in BHK cells	(Helixate FS)	1000 IU, reconstituted to 2.5 ml; Kogenate contains albumin; revised purification of rFactor VIII with FS formulation					patient, the severity of the deficiency, the severity of the hemorrhage, the presence of inhibitors, and the FVIII level desired*		Hepatic (predicted)
Hematologic and coagulation factors	Antihemophilic factor/von Willebrand factor complex (human)-factor VIII	Humate-P	Lyophilized; reconstitute to 20 or 40 IU/ml (contains albumin and other proteins)	IV	2.88–4.44 ml/kg/h	0.054–0.80 liter/kg	8.4–17.4 h	15–50 IU/kg loading dose; 15–25 IU/kg every 8 h		
Hematologic and coagulation factors	Antihemophilic factor-human recombinant heavy and light chains combination (factor VIII) produced in CHO cells	Recombinate	Lyophilized 250, 500, or 1000 IU vials	IV	2.9–4.4 ml/kg/h	0.054–0.8 liter/kg	14.4 ± 4.9 h	Calculated based on plasma concentration and dosing interval as described on product label*	NWD	NA
Hematologic and coagulation factors	Antithrombin III, human, purified AT-III, alpha2-glycoprotein	Thrombate III	Lyophilized, 500 or 1000 U vial, reconstitute to 50 U/ml	IV	NA	NA	22 h	Target concentration 12.5 mg/dL; 1.4% increase in plasma level for each 1 mg/kg dose administered	NA	NA
Hematologic and coagulation factors	Bivalirudin, reversible thrombin inhibitor, recombinant hirudin	Angiomax	250 mg vial	IV	3.4 ml/min/kg (renal function-dependent)	(P) min protein binding	25 min	1 mg/kg @ 2.5 mg/kg/h for 4 h; followed by 0.2 mg/kg/h up to 20 h as needed	NA	Hepatic and renal elimination
Hematologic and coagulation factors	Coagulation factor IX (Human recombinant), produced in CHO cells	BeneFix	Lyophilized, 250, 500, or 1000 IU vial, reconstitute and use within 3 h	IV over several minutes	NA	250, 500, or 1000 IU vial	19.4 (11–36) h	Individualized dosage*	NA	Hepatic (predicted)

■ TABLE I. *Continued*

Category	Drug Name	Trade Name	Dosage Form and Dose	Route of Administration	Clearance	Volume of Distribution	Half-life	Effective Dose	Toxic Dose	Major Route of Elimination
Hematologic and coagulation factors	Coagulation factor VIIa (human recombinant), produced in BHK cells	NovoSeven	Lyophilized 60 or 240 IU equivalent to 1.2 or 4.8 mg vial (does not contain albumin)	Intravenous bolus	33 ml/kg/h (range 27–49 ml/kg/h)	103 ml/kg (range; 78–139 ml/kg)	2.3 h (range: 1.7–2.7 h)	90 µg/kg every 2h (range 35–120) until hemostasis is achieved.	No complication reported up to 986 µg/kg for five consecutive d	Hepatic (predicted)
Hematologic and coagulation factors	Epoetin alfa, human recombinant, produced in CHO cells	Epogen, Procrit	2, 3, 4, 10, or 40,000 IU/ml	IV or self-administered SC	0.032–0.055 ml/min/kg	0.021–0.063 liter/kg	6–13 h	50–300 IU/kg	Under 1 year of age	Receptor-mediated processes; hepatobiliary
Hematologic and coagulation factors	Eptifibatide, reversible platelet aggregation inhibitor	Integrilin	20 mg in 10 ml vial, 75 mg or 200 mg in 100 ml vial	IV	55–58 ml/kg/h (P)	185 ml/kg	2.5 h (α = 5 min)	Condition-specific*	NA	Renal elimination
Hematologic and coagulation factors	Factor VIII human recombinant glycosylated protein, produced in CHO cells	ReFacto	Lyophilized 250, 500, or 1000 IU/vial	IV	127 ml/kg-min	159.6 liters/kg	14.5 ± 5.3 h (range: 7.6–27.7 h)	Based on individual needs* (range 20–100 IU/dl)	>1250 IU/kg	Renal
Hematologic and coagulation factors	Filgrastim, G-CSF, human recombinant produced in *E. coli*, nonglycosylated	Neupogen	300 or 480 µg in each single-use vial or prefilled syringe	IV or subcutaneous bolus	0.5–0.7 ml/min/kg	150 ml/kg	α = 30 min β = 3–4 h	5 µg/kg/d, administered as a single daily injection by SC bolus injection, by short IV infusion (15–30 minutes), or by continuous SC or continuous IV infusion (up to 2 weeks)	1150 µg/kg/d sub-chronic or chronic administration in monkey may lead to cardiopulmonary insufficiency and 15–28 fold increases in PBL and death.	Renal and hepatic
Hematologic and coagulation factors	Lepirudin-thrombin inhibitor, Hirudin-human recombinant produced in yeast	Refludan	Lyophilized 50 mg vial	IV infusion after bolus dose	195 ml/min	12.2–18 liters (increase with age and declining renal function)	1.3 h	0.4 mg/kg IV bolus over 15–20 s; followed by 0.15 mg/kg/h over 2–10 d (up to 110 kg body weight)	NA	Renal excretion

Category	Description	Trade name	Formulation	Route	Clearance	Volume of distribution	Half-life	Dose	Toxicity	Elimination/metabolism
Hematologic and coagulation factors	Oprelvekin, human recombinant IL-11, a thrombopoietic growth factor that stimulates stem cells and megakaryocyte progenitor, produced in E. coli.	Neumega	Lyophilized 5 mg vial, reconstitute in 1 ml, use within 3 h after reconstitution	SC	2.2–2.7 ml/min/kg	112–152 ml/kg	7.0 h	50 µg/kg q. d up to 21 d	>50µg/kg may produce cardiovascular events	Receptor-mediated processes; kidney
Hematologic and coagulation factors	Sargramostatin, human recombinant GM-CSF, produced in a yeast (S. cerevisiae) expression system	Leukine	Lyophilized 250 µg (2.4 MIU); 500 µg (2.8 MIU) in 1 ml	SC, IV	420–529 ml/min/m^2	NA	60 min	250 µg/m^2	>100 µg/kg/d for 7–18 days may produce reversible adverse events, including dyspnea, malaise, nausea, fever, rash, sinus tachycardia, headache and chills	Receptor-mediated processes; hepatobiliary
Hormones	Calcitonin, salmon	Miacalcin	2 ml vial, 2200 IU/ml, 200 IU/0.09 ml dose	Nasal spral (~0.09 ml/actuation) and IM/SC injection	NA	NA	43 min	100 IU (0.5 ml) per d, SC or IM	NA	Kidney metabolism
Hormones	Cosyntropin: alpha 1–24 corticotropin (a ACTH subunit), synthetic	Cortrosyn	Lyophilized 0.25 mg, reconstitute in 1 ml	IM or IV	NA	NA	3 h	NA	NA	Renal elimination and adrenal metabolism
Hormones	Follitropin alpha, human follicle-stimulating hormone (hFSH), recombinant	Gonal-F	Lyophilized 37.5, 75, or 150 IU vial; reconstitute in 1 ml	SC	0.6 liter/h (F)	10 ± 3 liters	~2–2.5 h	75 IU/d	NA	Renal elimination
Hormones	Follitropin beta, human follicle-stimulating hormone (hFSH), recombinant	Follistim	75 IU in 5 ml	SC, IM	NA	~8 liters	Dose-dependent; 43.9 h (300 IU), 30.1 (150 IU), 26.9 (75 IU)	75–150 IU/d IM or SC	NA	NA
Hormones	Ganirelix acetate, synthetic decapeptide	Antagon	250 µg in 0.5 ml	SC	2.4–3.3 liter/h (single	43.7–76.5 liters (single	12.8–16.2 h (single vs.	250 µg/d	NA	Hepatic (biliary 75%); renal (22%)

■ TABLE I. *Continued*

Category	Drug Name	Trade Name	Dosage Form and Dose	Route of Administration	Clearance	Volume of Distribution	Half-life	Effective Dose	Toxic Dose	Major Route of Elimination
	with antagonistic activity against gonadotropin releasing hormone (GnRH)				vs. multiple dose)	vs. multiple dose)	multiple dose)			
Hormones	Glucagon, recombinant	Glucagon for injection/ GlucaGen	Lyophilized 1 mg vial (Glucagon), lyophilized powder 2 ml vial (Glucagen)	IM, SC, IV	13.5 ml/ min/kg	0.25 l/kg	8–18 min	1 mg (>25 kg); 0.5 mg (<25 kg)	NA	NA
Hormones	Goserelin: luteinizing hormone-releasing hormone (LHRH) decapeptide analogue	Zoladex	3.6 mg	SC	110.5 ± 47.5 ml/min	44.1 ± 13.6 liters	4.2 ± 1.1 h	3.6 mg implant administered every 28 d (prostate cancer, endometriosis, breast cancer, endometrial thinning)	NA	Hepatic metabolism and urinary excretion (predominant)
Hormones	Human chorionic gonadotropin (HCG), placental extracts	Novarel	Lyophilized 10,000 U vial, 10 ml	NA	NA	NA	NA	4000 U 3×/wk for 3 wk	NA	Renal
Hormones	Human chorionic gonadotropin (HCG), placental extracts	Pregnyl	Lyophilized 10,000 U vial, 10 ml; 4–5000 U, t.i.w	IM	NA	NA	~2–3 h	Cryptorchidism 4000 U IM 3×/wk × 3 wks, hypogonadism 500–1000 U IM 3×/wk × 3 wks, then 500–1000 units 2×/wk × 3 wks, ovulation induction 5000–10,000 U one d after last dose of menotropins	NA	Renal elimination
Hormones	Human chorionic gonadotropin (HCG), placental extracts	Profasi	Lyophilized 2, 5, or 10,000 U vial; 4–5000 U, t.i.w	IM	NA	NA	5.6 h	4000 U 3×/wk, 3 wk	NA	Renal
Hormones	Insulin, human recombinant (*E. coli*)	Humulin	100 IU/ml available in R,	SC	NA	0.26–0.36 liter/kg	25 min (0.1 IU/kg);	Variable, according to	NA	Renal and hepatic

Class	Drug	Trade name	Formulation	Route	Clearance	Vd	Half-life	Dosage		Metabolism
			N, L, or U formulations				52 min (0.2 IU/kg)	patient's need*	NA	Renal and hepatic
Hormones	Insulin, human recombinant (yeast, S. cerevisiae)	Novolin	Available in N, R, or L formulations, 10ml (100IU/ml) vial or prefilled syringe or cartriges	SC	NA	0.15 liter/kg	141 min	Individualized dosage*	NA	Renal and hepatic
Hormones	Insulin, human recombinant, yeast (S. cerevisiae), infusion pump or syringe refills	Velosulin BR	10ml (100IU/ml) vial, formulated for insulin infusion pump and syringe	Infusion pump and SC	NA	0.15 liter/kg	NA	Individualized dosage*	NA	Renal and hepatic
Hormones	Insulin, Lispro variant, human, recombinant, rDNA origin	Humalog	100IU/ml in 10ml vial	SC	NA	0.26–0.36 liter/kg	Dose-dependent 26 min (0.1 IU/kg) and 52 min (0.2 IU/kg)	Variable, according to patient's need*	NA	Renal and hepatic
Hormones	Insulin, pork-derived	Ilectin II	100 U/ml, avialbe in R, N, or L formulation	SC, not for IM	NA	0.15 liter/kg	5–15 min (IV); SC (dose-dependent) @ 90 U ~198 min	Variable, according to patient's need, ~0.2 U/kg	NA	Renal and hepatic
Hormones	Insulin-asp (B28), human recombinant (yeast S. cerevisiae)	NovoLog	100IU in 10ml vial, 3ml PenFill cartridges	SC	NA	NA	81 min	Variable, according to patient's need, 0.5–1.0 IU/kg per	NA	Renal and hepatic
Hormones	Insulin-Glargine (recombinant); Asn (A21) > Gly; addition of two Arg to B20 and B21, more rapid onset	Lantus	100IU/ml in 5 or 10ml vial	SC	NA	NA	NA	Variable, depends on patient's need, ~2–100IU*	NA	Renal and hepatic
Hormones	Leuprolide acetate, GnRH or LH-RH analogue	Lupron, Lupron depot	5 mg/ml in 2.8ml vial; 3.75mg in sustained	SC injection or various depot formulation for IM injection	7.6 liters/h	27 liters	2.9–3.6 h	3.75 mg IM depot injection (endometriosis), 1 mg SC daily	NA	NA

■ TABLE I. Continued

Category	Drug Name	Trade Name	Dosage Form and Dose	Route of Administration	Clearance	Volume of Distribution	Half-life	Effective Dose	Toxic Dose	Major Route of Elimination
			release microsphere					(prostate cancer)		
Hormones	Menotropins, human chorionic gonadotropins (hCG) (extracted)	Humegon	Lyophilized 75 U ampule to be reconstituted to 2 ml	IM	NA	NA	31–37 h	75 U/d for 7–10d	NA	Renal elimination
Hormones	Menotropins, human chorionic gonadotropins (hCG) (extracted)	Pergonal	Lyophilized 75 U in 2 ml ampule	NA	NA	NA	2.2–2.9 h	75 IU/d for 10 d	NA	Renal excretion
Hormones	Menotropins, human chorionic gonadotropins (hCG) (extracted)	Repronex	NA	IM, SC	0.8–1.44 liter/h	39–57 liters	0.06–0.12 h	150–450 IU/d	NA	NA
Hormones	Nafarelin acetate, gonadotropin-releasing hormone (GnRH) agonist	Synarel	200 μg per spray; Nasal spray for central precocious puberty, or endometriosis; packaged in 0.5 oz 8 ml Nasal spray 2 mg/ml solution	Intranasal	NA	NA	~3 h	400 μg daily total*	NA	Hepatic metabolism and renal excretion (predominant)
Hormones	Octreotide, somatostatin, octapeptide	Sandostatin	10, 20, 30 mg in PLGA depot formulation; 50, 100, or 500 μg in 1 ml ampule	SC, IV	10 liters/h	13.6 liters	$\alpha = 0.2$ h $\beta = 1.7$ h	50–100 μg, 2–3 times daily, upward titration in doses	Mortality occurred in mice and rats given 72 mg/kg and 18 mg/kg IV, respectively	Liver and renal
Hormones	Oxytocin	Pitocin	1 ml ampoule, 1 ml syringe, 1 ml vial, 10 ml multiple dose vial; each with 10 IU/ml	IV, IM	NA	NA	3–5 min	Depending on uterine response; 0.5–2 IU/min infusion, adjunctive therapy for	NA	Kidney and liver metabolism

Class	Drug	Trade Name	Formulation	Route	Clearance	Volume of Distribution	Half-life	Dosing	Dose	Elimination
Hormones	Proirelin, throtropin-releasing hormone (TRH), synthetic tri-peptide, 5-oxo-(pro-his-pro)-ami	Thyrel TRH	0.5 mg in 1 ml ampule	IV over 15–30 sec	NA	NA	NA	incomplete or inevitable abortion NA	200–500 mcg (adult) or 7 mcg/kg (6 < children < 16); maximum dose 500 mcg	NA
Hormones	Somatotropin, human recombinant	Genotropin	Lyopilized 1.5 mg; reconstitute with 1.13 ml H$_2$O	SC	0.3 liter/kg/h	1.3 ± 0.8 liters/kg	0.4 h (N), 3 h (GHD)	Individualized, dosing dependent on indication*	NA	NA
Hormones	Somatotropin, human recombinant (E. coli)	Humatrope	Lyophilized 5 mg vial, reconstitute in 5 ml	SC, IM	0.14 liter/h/kg	0.07 liter/kg	0.36 h (IV) 3.8–4.9 (apparent with SC & IM)	Individualized, dosing varies with indications*	NA	Liver and kidney
Hormones	Somatrem-recombinant human pituitary growth hormone produced in E.coli	Protropin	Lyophilized 5 or 10 mg vial	SC, IM	NA	NA	NA	0.1 mg/kg, 3 times a week	NA	NA
Hormones	Somatropin	Nutropin AQ	10 mg in 2 ml	SC	116–174 ml/h/kg	150 ml/h-kg	19.5 ± 3.1 min	Individualized, dosing dependent on indication*	5.3–8 mg/d	Liver and kidney (predominant)
Hormones	Somatropin	Saizen	5 mg (15 IU)	SC, IM	14.6 ± 2.8	12.0 ± 1.08 liters	0.6 h	Individualized, dosing dependent on indication*	5.3–8 mg/d	Liver metabolism and Renal elimination (predominant)
Hormones	Somatropin	Serostim	4 mg, 5 mg, and 6 mg vials	SC	0.0015 ± 0.0037 liter/h	12.0 ± 1.08 liters	4.28 ± 2.15 h	Individualized, dosing dependent on Indication*	NA	Liver metabolism and kidney clearance (predominant)
Hormones	Thyrotropin alfa, thyroid stimulating horme (TSH), human recombinant, produced in CHO cells	Thyrogen	Lyophilized 1.1 mg vial; reconstitute to 1.2 ml	IM	NA	NA	25 ± 10 h	0.9 mg IM q. d × 2 or q 72 h × 3	2.7–3 mg dose produces symptomatic untoward events; IV instead of IM administration	Liver and kidney (predicted)

■ TABLE I. *Continued*

Category	Drug Name	Trade Name	Dosage Form and Dose	Route of Administration	Clearance	Volume of Distribution	Half-life	Effective Dose	Toxic Dose	Major Route of Elimination
Interferons and cytokines	Aldesleukin, human recombinant interleukin-2, produced	Proleukin	Lyophilized, 22×10^6 IU vial, reconstitute and use within 48h	IM	268 ml/min	1.2 ml Sterile water for injection	$\alpha = 6$–10 min; $\beta = 60$–90 min	600,000 IU/kg	Dose limiting toxicity; life-threatening toxicities may be ameliorated by the intravenous administration of dexamethasone, which may also result in loss of the therapeutic effects may also produce adverse symptoms	Renal and hepatic
Interferons and cytokines	Anakinra, human IL-1Ra	Kineret	1 ml in solution vial ~ 100 mg anakinra	SC	NA	NA	4–6h	100 mg/day	875 mg/72h	Renal and hepatic
Interferons and cytokines	Denileukin diftitox, IL-2 Diphtheria conjugate	Ontak	300 μg in 2 ml	IV infusion	1–2 ml/min/kg	0.06–0.08 liter/kg	$\alpha = 2$–5 min; $\beta = 70$–80 min	9 or 18 μg/kg/d × 5 d every 3 weeks	Dose limiting toxicity at >31 μg/kg/d	Liver and kidney (predicted)
Interferons and cytokines	Interferon alfa-2a	Roferon-A	Sterile solution, SDV and prefilled syringes containing 3, 6, 9, 36 MIU	IV, SC	2.14–3.62 ml/min/kg	0.223–0.748 liter/kg	5 h	3 MIU 3 times weekly for 52 weeks	Avoid repeated high dose administration	Renal filtration
Interferons and cytokines	Interferon alfa-2b, human recombinant product produced in E. coli	Intron A	Lyophilized 5, 10, 18, 25, or 50 MIU vial	IM, SC, IV, intralesional	2.14–3.62 ml/min/kg	25–31 liters	2–3h	2 MIU 3 times weekly for 6 months (hairy cell leukemia)	Avoid repeated high dose administration	Kidney metabolism, renal elimination
Interferons and cytokines	Interferon alfacon-1, human recombinant, produced in E. coli	Infergen	Sterile solution, SDV and prefilled syringes containing 9 μg in 0.3 ml or 15 μg in 0.5 ml	SC	2.0 ml/min/kg	~ P volume	5.1 h ($\alpha = 5$–9 min)	9 μg 3 times weekly for 24 weeks	Avoid repeated high dose administration	Renal glomeruli elimination

Category	Description	Trade name	Formulation	Route				Dose		Elimination
Interferons and cytokines	Interferon alfa-n1, Lymphoblastoid (natural product)	Wellferon	3 MU in 1 ml	IM, SC	85 ml/min	NA	7–10 h	3 MIU q.3.w.	15 MU/m², q.3.w.	Renal filtration
Interferons and cytokines	Interferon alfa-n3 (human leukocyte derived)	Alferon-N	5 MU in 1 ml vial (~200 MU/mg protein)	Local wart injection	NA	NA	NA	0.05 ml (250,000 IU) per condylomata acuminata wart, q.2.w. up to 8 weeks	NA	NA
Interferons and cytokines	Interferon beta-1a	Rebif	22, 44 μg in 0.5 ml solution	SC	33–55 liters/h	NA	69 h	44 μg 3×/wk and increased according to individual*	>44 μg 3×/wk	NA
Interferons and cytokines	Interferon beta-1a, human recombinant product produced in CHO cells	Avonex	Lyophilized 200 μg, 30 μg, reconstitute in 1 ml	IM	334 ml/h/kg	62 liters	10 h (IM); 4 h (IV); 8.6 h (SC)	30 μg weekly	Not known	Renal elimination (predicted)
Interferons and cytokines	Interferon beta-1b, human recombinant produced in E. coli	Betaseron	Lyophilized 32 MIU equivalent to 0.3 mg per vial (albumin added as a stabilizer)	SC	9.4–28.9 ml/min/kg	0.25–2.88 liters/kg	variable 8 min to 4.3 h	0.25 mg, every other d	NA	NA
Interferons and cytokines	Interferon gamma-1b, human recombinant produced in E. coli	Actimmune	100 μg 0.5 ml vial; (30 MIU/mg)	SC	1.4 liters/min	NA	5.9 h (SC)	50 μg/m 2 (1 MIU/m 2) per dose 3 times weekly	NA	Receptor-mediated processes, hepatobiliary
Interferons and cytokines	Peginterferon alfa-2b, human recombinant produced in E. coli	PEG-Intron	2 ml lyophilized powder–100, 160, 240, 300 μg/ml	SC	22 ml/kg*h	1.0 μg/kg	40 h	NA	1.0 μg/kg/h	Renal elimination (30%)
Interferons and cytokines	Peginterferon alfa-2b, PEG-modified human recombinant inteferon alfa-2b (Intron A, produced in E. coli)	PEG-Intron	Lyophilized 74, 118, 177, or 222 μg in 2 ml vial	SC	NA	1.2–1.4 liters/kg	33–37 h	40–50 μg/kg SC every week for 52 weeks	NA	Receptor-mediated processes (saturable)
Other	Becaplermin, PDGF, human recombinant produced in yeast, S. cerevisiae	Regranex Gel	Topical gel preparation, 100 μg becaplermin/g gel	Topical	NA	NA	NA	10 μg used daily over the ulcer	NA	NA

■TABLE I. *Continued*

Category	Drug Name	Trade Name	Dosage Form and Dose	Route of Administration	Clearance	Volume of Distribution	Half-life	Effective Dose	Toxic Dose	Major Route of Elimination
Other	Botulinum toxin A, neutrotoxin complex	Botox	Vacuum dried 100 U vial	IM	NA	NA	NA	NA	NA	NA
Other	Fibrin sealant	TISSEEL VH Kit	Four vials in different sizes (0.5, 1.0, 2.0, 5.0): lyophilized fibrin sealant, fibrinolysis inhibitor (bovine), thrombin (human), calcium chloride	IV	NA	1 ml, 2 ml, 4 ml, 10 ml vials	NA	NA	NA	NA
Other	Proteinase inhibitor (human), (alpha1-PI), alpha-1 antitrypsin	Prolastin	Lyophilized 500 or 1000 mg vial, reconstituted to 25 mg/ml	IV	NA	NA	~4 d	60 mg/kg once a week	NA	Hepatic (predicted)
Vaccine	Haemophilus b conjugate vaccine (meningococcal protein conjugate)	PedvaxHIB	0.5 ml solution contains 7.5 µg haemophilus b capsular polysaccharide (PRP), 125 mcg Neisseria meningitidis outer membrane protein complex (OMPC) and 225 mcg alum	IM	NA	NA	NA	1–2 0.5 ml dose (depending on age); q.8w	NOT RECOMMENDED FOR USE IN INFANTS YOUNGER THAN 6 WEEKS OF AGE	NA
Vaccine	Hepatitis A vaccine, inactivated	Havrix	1 ml adult dose contains ~1440 U viral antigen, 0.5 mg alum	IM	NA	NA	NA	2 doses given at 0 and 6–12 months	NA	NA
Vaccine	Rabies vaccine	Imovax	Lyophilized, reconstituted 1 ml dose contains 2.5	IM	NA	NA	NA	Three 1 ml doses given at 0, 7, and 21–28 d	NA	NA

IU rabies antigen

Vaccines	BCG live	TICE BCG (TICE strain), TheraCys (Cannaught strain)	10–80 million CFU: 50 mg BCG, lyophilized: reconstitute to 50 ml; use within 2 h	Intravesicular instillation	NA	50 mg BCG suspended in saline	NA	50 ml intravesicular instillation, q.w. 6–10 months	NA	Renal
Vaccines	BCG live	PACIS	10–80 million CFU: 50 mg BCG, lyophilized: reconstitute to 50 ml; use within 2 h	Intravesicular instillation, SC, IV	NA	120 mg semisolid dry	NA	50 ml intravesicular instillation, q.w. 6–10 months	NA	Renal
Vaccines	Diphtheria and tetanus toxoids and acellular pertussis vaccine	Infanrix	0.5 ml dose containing 25U DT, 10U TT, 25 μg PT, 25 μg FHA, and 8 μg pertactin absorbed onto 0.625 mg alum in 0.5 ml vials	IM	NA	NA	NA	4 doses (q.8w)	Under 15 months of age	NA
Vaccines	Diphtheria and tetanus toxoids and acellular pertussis vaccine adsorbed	Acel-Imune	0.5 ml dose containing 9U DT, 5U TT, 3 μg PT, 34 μg FHA, and 2 μg pertactin absorbed onto 0.23 mg alum	IM	NA	NA	NA	4 doses (q.8w)	NA	NA
Vaccines	Diphtheria and tetanus toxoids and acellular pertussis vaccine adsorbed	Tripedia	0.5 ml dose containing 6.9U DT, 5U TT, 23 μg PT, 23 μg FHA absorbed	IM	NA	NA	NA	3 doses, q.8w	NA	NA

TABLE I. *Continued*

Category	Drug Name	Trade Name	Dosage Form and Dose	Route of Administration	Clearance	Volume of Distribution	Half-life	Effective Dose	Toxic Dose	Major Route of Elimination
Vaccines	Haemophilus b conjugate (meningococcal protein conjugate) and hepatitis b (recombinant)	COMVAX	0.5 ml dose containing 7.5 µg haemophilus B (Ross strain)-capsular polysaccharide (PRP), 125 µg of Neisseria meningitidis outer membrane protein complex (OMPC), 5 µg of HBsAg, 0.225 mg alum onto 0.17 mg alum	IM	NA	NA	NA	0.5 ml dose at 2, 4, and 12 months of age	NA	NA
Vaccines	Haemophilus b conjugate vaccine (diphtheria CRM 197 protein conjugate)	HibTITER	0.5 ml contains 10 µg haemophilus b saccharide conjugated to 25 µg of CRM 197 DT protein (DT CRM 197 is a nontoxic variant of diphtheria toxin)	IM	NA	NA	NA	1–3 0.5 ml doses depending on age; q.8w; adult single dose without booster	NA	NA
Vaccines	Haemophilus b conjugate vaccine (tetanus toxoid conjugate)	ActHIB, OmniHIB	Reconstituted 0.5 ml contains 10 µg capsular polysaccharide conjugated to 24 µg of inactivated tetanus toxoid	IM	NA	NA	NA	2 doses at 0 and 2 months	NA	NA
Vaccines	Hepatitis A vaccine, inactivated	Vaqta	0.5 ml dose contains ~ 25U hepatitis A virus antigen	IM	NA	NA	NA	2 doses at 0 and 6–18 months	NA	NA

Type	Description	Trade name	Composition/Dose	Route			Schedule		
Vaccines	Hepatitis A, inactivated and hepatitis B (recombinant) vaccine	TWINRIX	1.0 ml dose contains 720 U inactivated hepatitis A virus, 20 µg of recombinant HBsAg, 0.45 mg alum & 0.225 mg alum (PED); 1 ml dose contains ~50 U hepatitis A virus antigen & 0.45 mg alum (Adult)	IM	NA	NA	3 doses, on 0-, 1-, and 6-month schedule	NA	NA
Vaccines	Hepatitis B vaccine (recombinant), produced in yeast *S. cerevisiae*	Recombivax	1 ml dose contains 10 µg (adult); 40 µg (dialysis) of hepatitis B surface antigen, 0.5 mg alum, 50 µg thimerosal (preservative); 0.5 ml dose contains 5 µg antigen without preservative (pediatric)	IM	NA	NA	3 doses as designed for appropriate population; at 0, 1, and 6 months	NA	NA
Vaccines	Hepatitis B vaccine (recombinant); recombinant hepatitis B surface antigen produced in *S. cerevisiae* (yeast)	Engerix-B	0.5 ml suspension contains 10 µg hepatitis B surface antigen adsorbed on 0.25 mg alum	IM	NA	NA	10 µg in 0.5 ml (children) 20 µg in 1 ml (adult) at 0, 1, 6 months; 40 µg in 2 ml (hemodialysis) given at 0, 1, 2, 6 months	NA	NA
Vaccines	Lyme disease vaccine (recombinant OspA), acylated	LYMErix	5 ml dose contains 30 µg lipoprotein OspA, 0.5 mg alum	IM	NA	NA	30 µg given at 0, 1, 12 months	NA	NA

■ TABLE I. *Continued*

Category	Drug Name	Trade Name	Dosage Form and Dose	Route of Administration	Clearance	Volume of Distribution	Half-life	Effective Dose	Toxic Dose	Major Route of Elimination
Vaccines	Pneumococcal 7-valent conjugate vaccine (diphtheria CRM197 protein)	Prevnar	0.5 ml dose contains 2 μg each antigen for serotype 4, 9V, 14, 18C, 19F, and 23F, 4 mcg each serotype 6B, 20 mcg CRM197, 0.124 mg alum	IM	NA	NA	NA	4 doses at 2, 4, 6, and 12–15 months of age	NA	NA
Vaccines	Pneumococcal vaccine, polyvalent, 23 types of *S. pneumoniae*	PNU-Immune-23	0.5 ml dose contains 26 μg polysaccharide antigen purified from 23 serotypes of *S. pneumoniae*	IM, SC	NA	NA	NA	Single 0.5 ml dose	NA	NA
Vaccines	Pneumococcal vaccine, polyvalent-23 pneumococcal isolates	Pneumovax 23	0.5 ml dose contains 22 μg polysaccharide antigens purified from 23 serotypes of *S. pneumoniae*	IM, SC	NA	NA	NA	Single 0.5 ml dose	NA	NA
Vaccines	Rabies vaccine	RabAvert	Lyophilized, 1 ml dose contains 2.5 IU rabies antigen	IM	NA	NA	NA	Three 1 ml doses given on d 0, 7, and 18–21	NA	NA
Vaccines	Varicella virus vaccine, live, Oka/Merck strain of live, attenuated varicella virus	Varivax	Lyophilized, 0.5 ml reconstituted dose contains 1350 plaque forming units (PFU)	SC	NA	NA	NA	A single 0.5 ml dose (<12 yrs old); two 0.5 ml doses (>13 yrs old) at 0 and 4–8 weeks	NA	NA

Note: Standards and practices in medicine continue to change as new data become available. Therefore each medical professional should consult additional sources, as needed. Before prescribing medications, the user is advised to check the product information sheet accompanying each drug to verify conditions of use and identify any changes in dosage schedule or contraindications, particularly if the agent to be administered is new, infrequently used, has a narrow therapeutic range or suspected drug interactions. See "User Agreement (page iv)" for additional information.

Abbreviations: IU, international units; MIU, million international units; MU, million units; P volume, plasma volume—about 5 liters for a 80 kg person.

*Consult product label and other sources.

Appendix II

MOLECULAR CHARACTERISTIC AND THERAPEUTIC USE

Biotechnology and Biopharmaceuticals, by Rodney J. Y. Ho and Milo Gibaldi
ISBN 0-471-20690-3   Copyright © 2003 by John Wiley & Sons, Inc.

■ **TABLE II. Molecular characteristic and therapeutic use**

Category	Drug Name	Trade Name	MW (kDa)	Therapeutic Use	Nucleotide Sequence[a]	Amino acid Sequence[a]	Protein Structure[b]
Antibodies and derivatives	Abciximab, F$_{ab}$ fragment of chimeric human-murine monoclonal antibody 7E3 directed against the glycoprotein (GP) IIb/IIIa ((alpha) IIb beta 3) receptor of human platelets	ReoPro	48	Adjunct to percutaneous coronary intervention for the prevention of cardiac ischemic complications	Human clone 7E3 immunoglobulin Light chain variable region (VkJ) GI: 882313	AAA69059.1	NA[c]
Antibodies and derivatives	Alemtuzumab, anti CD52 humanized monoclonal antibody, CHO product	Campath	150	Treatment of patients with B-cell chronic lymphocytic leukemia who have been treated with alkylating agents and who have failed fludarabine therapy	F$_{ab}$ fragment Heavy chain GI: 5542161 Light chain GI: 5542160	NA	Antibody to Campath-1H humanized F$_{ab}$ PDB ID: 1BEY
Antibodies and derivatives	Basiliximab, chimeric muring/human monoclonal IgG$_1$ against IL-2 receptor alpha (CD25), produced by rDNA technology	Simulect	144	Prophylaxis of acute organ rejection in patients receiving renal transplantation when used as part of an immunosuppressive regimen that includes cyclosporine and corticosteroids	Light chain variable region GI: 16508644	NA	PDB ID: 1MIM
Antibodies and derivatives	Capromab Pendetide, murine monoclonal 7E11-C5.3 IgG, [111]Indium labeled	ProstaScint, [111]In	~150	For the preparation of Indium 111 Capromab Pendetide to be used as a diagnosing imaging agent in newly-diagnosed patients with biopsy-proven prostate cancer, thought to be clinically localized after standard diagnostic evaluation, who are at high risk for pelvic lymph	F$_{ab}$ fragment Heavy chain GI: 5542161 Light chain GI: 5542160	NA	PDB ID: 1CE1

Category	Name	MW	Description	Indication	GI / Sequence	Accession	Structure
Antibodies and derivatives	CytoGram	~150	Cytomegalovirus Immune globulin, human hyper immuneglobulin, purified.	node metasteses, and in postprostatectomy patients with a rising PSA and a negative or equivocal standard metastatic evaluation in whom there is a high clinical suspicion of occult metastatic disease / Prophylaxis of cytomegalovirus disease associated with transplantation of kidney, lung, liver, pancreas and heart. In transplants of these organs other than kidney from CMV seropositive donors into seronegative recipients, prophylactic CMV-IGIV should be considered in combination with ganciclovir	IgG light chain GI: 1355190 Heavy chain GI: 32013	CAA4225.1	SM00407.5
Antibodies and derivatives	Zenapax	144	Daclizumab, anti-CD25 humanized IgG1, recombinant product	Prophylaxis of acute organ rejection in patients receiving renal transplants, to be used as part of an immunosuppressive regimen that includes cyclosporine and corticosteroids	Chain H, IgG F_{ab} fragment (CD25-Binding): GI: 2194043. Chain L GI: 2194044	XP_172016.1 GI: 2204476 (similar to IgG1/ kappa antibody)	IgG F_{ab} fragment (CD25-Binding) PDB ID: 1MIM
Antibodies and derivatives	Enbrel	~93	Etanercept, recombinant fusion protein containing TNF receptor linked to Fc domain of IgG$_1$, produced in CHO cells	Reduction in signs and symptoms of moderately to severely active rheumatoid arthritis in patients who have had an inadequate response to one or more disease-modifying antirheumatic drugs (DMARDS)	Etanercept GI: 339752	P20333	Structure of TNF receptor- associated factor 2 (Traf2) PDB ID: 1CA4
Antibodies and derivatives	Mylotarg TM	151–153	Gemtuzumab-oxoga micin (anti CD33-calicheamicin conjugate); recombinant	Treatment of patients with CD33-positive acute myeloid leukemia in first relapse who are 60 years of age or older and who are not considered	Recombinant IgG$_4$ heavy chain GI: 9857758. IgG$_4$ light chain variable region (CDR	H: AAG00912.1 L: AAB35404.1	SM00406.5

■ TABLE II. *Continued*

Category	Drug Name	Trade Name	MW (kDa)	Therapeutic Use	Nucleotide Sequence[a]	Amino acid Sequence[a]	Protein Structure[b]
	humanized IgG$_4$			candidates for cytotoxic chemotherapy	region, acceptor Ab "Eu") [human, peptide, 108 aa]: GI: 1195515		
Antibodies and derivatives	Hepatitis B immune globulin (human), purified by chromatograpy procedure	Nabi-HB	~150	For the treatment of acute exposure to HBsAg, perinatal exposure of infants born to HBsAg-positive mothers, sexual exposure to HBsAg-positive persons and household exposure of infants to persons with acute HBV	NA	NA	NA
Antibodies and derivatives	Human immune globulin, purified by cold ethanol fractionation procedure	BayGam	~150	Given prophylactically or soon after exposure to hepatitis A; given to prevent or modify measles (Rubeola), Varicella (Zoster), Rubella, and immunoglobulin deficiency	NA	NA	NA
Antibodies and derivatives	Ibritumomab tiuxetan, monoclonal antibody directed against the CD20 antigen, produced in CHO cells	Zevalin	~150	Treatment of patients with relapsed or refractory low-grade, follicular, or transformed B-cell non-Hodgkin's lymphoma, including patients with Rituximab (Rituxan) refractory follicular non-Hodgkin's lymphoma. The therapeutic regimen includes Rituximab, Indium-111 Ibritumomab Tiuxetan, and Yttrium-90 Ibritumomab Tiuxetan	GI: 16902038 GI: 16902040	AAL27649 AAL27650	SM00406.5
Antibodies and derivatives	Imciromab pentetate, Fab fragment of anti-myosin IgG$_2$ monoclonal antibody	Myoscint (In 111)	~50	Preparation of In 111 Imciromab pentetate to be used as a cardiac imaging agent for detecting the presence and/or identifying	Heavy chain immunoglobulin G2 variable region [synthetic construct]:	*H:* AAK39434.1 *L:* AAK39435.1	Structure of IgG$_{2A}$ F$_{ab}$ fragment PDB ID: 1YEH

	bound to DTPA for labeling with Indium 111			the location of myocardial injury in patients with suspected myocardial infarction	GI: 13785651 Light chain immunoglobulin G2 variable region [synthetic construct]: GI: 13785653	NA	NA	NA
Antibodies and derivatives	Immune globulin (human)	Gammar-PIV	~150	Adults, children and adolescents with primary defective antibody synthesis such as agammaglobulinemia or hypogammaglobulinemia, who are at increased risk of infection	NA	NA	NA	NA
Antibodies and derivatives	Immune globulin (human) against Rho (D)	WinRho SDF	150	Treatment of nonsplenectomized, Rho (D)-positive children with chronic or acute ITP, adults with chronic ITP, or children and adults with ITP secondary to HIV infection	NA	NA	NA	NA
Antibodies and derivatives	Immune globulin (human) against Rho (D), purified IgG cohn fraction II	BayRho-D full and mini dose	~150	Prevention of Rh hemolytic disease of the newborn by its administration to the Rho (D)-negative mother within 72 hours after birth of a Rho (D)-positive infant	NA	NA	NA	NA
Antibodies and derivatives	Immune globulin (human), (IGIV)	Sandoglobulin	~150	Treatment of patients with primary immunodeficiencies, childern and adults with immune thrombocytopenic purpura (ITP)	NA	NA	NA	Average structure SM00407.5
Antibodies and derivatives	Immune globulin against D (Rh₁) (human), purified by cold alchol fractionation	RhoGam Rho (D), MICRhoGAM Rho (D)	~150	Known or suspected that fetal red blood cells have entered the circulation of an Rh-negative mother unless the fetus or the father can be shown conclusively to be Rh-negative; it is also used to	NA	NA	NA	NA

■ TABLE II. *Continued*

Category	Drug Name	Trade Name	MW (kDa)	Therapeutic Use	Nucleotide Sequence[a]	Amino acid Sequence[a]	Protein Structure[b]
				neutralize Rh-positive red blood cells or component such as platelets or granulocytes prepared from Rh-positive blood given to Rh-negative female of childbearing age			
Antibodies and derivatives	Immune globulin against hepatitis B (human)	BayHep B	~150	A postexposure immunotherapy regimen used in combination with hepatitis B immune globulin (human) with hepatitis B vaccine	NA	NA	NA
Antibodies and derivatives	Immune globulin Intravenous	Gamimmune N-10% & 5%	~150	Treatment of primary immunodeficiency states in which severe impairment of antibody forming capacity; idiopathic thrombocytopenic purpura (IPT); bone marrow transplantation; pediatric HIV infection	NA	NA	NA
Antibodies and derivatives	Infliximab, anti-tumor necrosis factor alpha TNF(alpha), chmeric human	Remicade	149	Treatment of moderately to severely active Crohn's disease for the reduction of the signs and symptoms in patients who have an inadequate response to conventional therapies; and treatment of patients with fistulizing Crohn's disease for the reduction in the number of draining enterocutaneous fistula(s)	Anti-TNF immunoglobulin GI: 11275300 GI: 11275314	H: BAB18259 L: BAB18257	NA
Antibodies and derivatives	Muromonab-CD3, anti-CD3 antibody IgG$_{2a}$	Orthoclone OKT3	~150	Treatment of acute allograft rejection in renal transplant patients, and steroid-resistant acute allograft rejection in cardiac and hepatic transplant patients	OKT3 heavy chain variable region [Mus musculus]: GI: 1565186 OKT3 light chain [Mus musculus]:	H: BAA11539.1 L: CAA01594.1	Average IgG$_{2a}$ PDB ID: 1AIF

Antibodies and derivatives	Nofetumomab, Fab fragmet of a monoclonal NR-LU-IO IgG$_{2b}$(murine) against small lung cancer associated antigen, Tc 99 labeled	Verluma, technetium Tc 99m	~40	Detection of extensive stage disease in patients with biopsy-confirmed, previously untreated, small cell lung cancer	Immunoglobulin heavy chain variable region [Mus musculus]: GI: 14029680 Mus spatrazine-specific IgG$_{2b}$ light chain variable region (Ig VL) mRNA, complete cds: GI: 545340	H: AAK52786.1 L: AAB29872.2	Dithiol melanotropin peptide cyclized via rhenium metal coordination PDB ID: 1B0Q
Antibodies and derivatives	Palivizumab, humanized monoclonal IgG$_1$ gaginst F protein of respiratory syncytial virus (RSV), recombinant DNA product	Synagis	148	Prophylaxis of serious lower respiratory tract disease, caused by respiratory syncytial virus, in pediatric patients at high risk of RSV disease	Homo sapiens recombinant IgG$_1$ heavy chain gene, partial cds: GI: 9857752 Mus musculus anti-human transferrin IgG$_1$ light chain variable region (Ig VL) mRNA, partial cds: GI: 786586	H: AAG00909.1 L: AAB33233.2	Average F$_{ab}$ of human IgG1; fragment (Cbr96) complexed with Lewis Y nonoate methyl ester PDB ID: 1CLY
Antibodies and derivatives	Rabies immune globulin (human)	BayRab	~150	All persons, who are not immunized, but suspected of exposure to rabies	NA	NA	NA
Antibodies and derivatives	Rabies immune globulin (human)	Imogam Rabies-HT	~150	Individuals who have not been previously immunized, but suspected of exposure to rabies	NA	NA	NA
Antibodies and derivatives	Respiratory syncytial virus immune globulin intravenous (human), RSV IVIG	Respigam	150	Prevention of serious lower respiratory tract infection caused by respiratory syncytial virus in children less than 24 months of age with bronchopulmonary dysplasia or a history of prematurity (less than or equal to 35 weeks gestation)	NA	NA	NA

■ TABLE II. *Continued*

Category	Drug Name	Trade Name	MW (kDa)	Therapeutic Use	Nucleotide Sequence[a]	Amino acid Sequence[a]	Protein Structure[b]
Antibodies and derivatives	Rituximab, chimeric murine/human IgG$_1$ monoclonal against CD20 (B-cell suface antigen), produced in CHO cells	Rituxan	145	Treatment of patients with relapsed or refractory low-grade or follicular, B-cell non-Hodgkin's lymphoma	Homo sapiens recombinant IgG$_1$ heavy chain gene, partial cds: GI: 9857752 Mus musculus anti-human transferrin IgG$_1$ light chain variable region (Ig VL) mRNA, partial cds: GI: 786586	H: AAG00909.1 L: AAB33233.2 PDB ID: 1CLY	IgG F$_{ab}$ (human IgG$_1$) chimeric fragment (Cbr96) complexed with Lewis Y nonoate methyl ester
Antibodies and derivatives	Tetnus immune globulin (human), isolated from solubilized Cohn fraction II	BayTet	~150	Prophylaxis against tetanus following injury in patients whose immunization is incomplete or uncertain	NA	NA	NA
Antibodies and derivatives	Thymoglobulin, anti-thymocyte globulin (rabbit), purified rabbit IgG against human thymocytes	Sangstat	~150	Treatment of acute rejection in renal transplant patients	NA	NA	NA
Antibodies and derivatives	Trastuzumab, humanized monoclonal IgG$_1$ against HER2	Herceptin	~150	For treatment of patients with metastatic breast cancer whose tumors overexpress the HER2 protein and who have received one or more chemotherapy regimens for their metastatic disease; trastuzumab in combination with paclitaxel is indicated for treatment of patients with metastatic breast cancer whose tumors overexpress HER2 protein and who have not received	Anti-DNA immunoglobulin heavy chain IgG [Mus musculus]: GI: 1872464 Anti-DNA immunoglobulin light chain IgG [Mus musculus]: GI: 1870507	H: AAB49171.1 L: AAB48814.1	Average strutture of F$_{ab}$ (human IgG$_1$ chimeric fragment (Cbr96) complexed with Lewis Y methyl ester nonoate PDB ID: 1CLY

chemotherapy for their metastatic disease

Enzymes	Adagen	Adenosine deaminase (ADA), PEG modified (bovine), [11-17xPEG5000]	90+ (170 contributed by PEG conjugate)	Enzyme replacement therapy for adenosine deaminase (ADA) deficiency in patients with severe combined immunodeficiency disease (SCID) who are not suitable candidates or who have failed bone marrow transplantation	Adenosine deaminase (Adenosine Aminohydrolase): GI: 5902736	PID: g5902736	Murine adenosine deaminase (D295E) PDB ID: 1FKW
Enzymes	Activase	Alteplase, tPA human recombinant, produced in E. coli	~52	Management of acute ischemic stroke in adults, and acute massive pulmonary embolism, for improving neurological recovery, and reducing the incidence of disability	NM 000930; GI: 14702165	NP 001794.1	NA
Enzymes	Ceradase	Ceradase (placental-derived)	59.3	Long-term enzyme replacement therapy for patients with a confirmed diagnosis of type 1 Gaucher disease	GI: 121283	PID: g121283	NA
Enzymes	Pulmozyme	DNAse, dornase alfa, human recombinant deosyribonuclease I (rhDNAse), produced in CHO cells	37	Management of cystic fibrosis patients to improve pulmonary function	Deoxyribonuclease 1 precursor (DNAse I) (Dornase alfa): GI: 118919	PID: g118919	Atomic structure of the actin: DNAse I complex. PDB ID: 1ATN
Enzymes	Xigris	Drotrecogin alfa (acitvated), human recominant of activated protein C	55	Reduction of mortality in adult patients with severe sepsis (sepsis associated with acute organ dysfunction) who have a high risk of death (e.g., as determined by APACHE II)	K02059.1 GI: 190322	XP_118651	1AUT L GI: 2392175
Enzymes	Cerezyme	Imiglucerase-recombinant (beta)-glucocerebrose	60.4	Long-term enzyme replacement therapy for patients with a confirmed diagnosis of type 1 Gaucher	Glucosylceramidase precursor (imiglucerase): GI: 121283	PID: P04062	NA

■ TABLE II. Continued

Category	Drug Name	Trade Name	MW (kDa)	Therapeutic Use	Nucleotide Sequence[a]	Amino acid Sequence[a]	Protein Structure[b]
Enzymes	Pegaspargase, L-asparaginase conjugated to PEG (5000), *E. coli*-derived enzyme	Oncaspar	NA	In combination chemotherapy for the treatment of patients with acute lymphoblastic leukemia who are hypersensitive to native forms of L-asparaginase	L-asparaginase II precursor (L-asparagine amidohydrolase II) (L-asnase II) (Colaspase): GI: 114252	PID: g114252	PDB ID: 3ECA *E. coli* L-asparaginase, an enzyme used in cancer therapy
Enzymes	Reteplase, a nonglycosylated genetically modified variant of tPA	Retavase	39.6	In the management of acute myocardial infarction (AMI) in adults for the improvement of ventricular function following AMI, the reduction of the incidence of congestive heart failure and the reduction of mortality associated with AMI	(RETEPLASE): GI: 137119P	PID: g137119	NA
Enzymes	Streptokinase, purified bacterial protein product, plasminogen activator	Streptase	NA	Acute myocardial infarction (AMI) in adults; lysis of intracoronary thrombi, improvement of ventricular function, and the reduction of mortality associated with AMI; reduction of infarct size and congestive heart failure associated with AMI	Streptokinase A precursor: GI: 14195682	PID: g14195682	Complex of the catalytic domain of human plasmin and streptokinase PDB ID: 1BML
Enzymes	Tenecteplase, human recombinant tissue plasminogen activator (tPA) produced in CHO cells	TNKase	~53	Reduction of mortality associated with acute myocardial infarction (AMI)	GI: 14777515	PID: g14777515	NA
Gene therapeutics	Fomivirsens sodium, antisense nucleotide polymer against CMV	Vitravene	7.1	Use to treat CMV retinitis in patients with AIDS	5'-GCG TTT GCT CTT CTT CTT GCG-3'	NA	NA

Category	Description	Brand name		Indication			Structure
Hematologic and coagulation factor	Darbepoetin alfa, human recombinant produced in the CHO cells	Aranesp	30–37	Treatment of anemia associated with chronic renal failure of patients on and not on dialysis	NA	NA	PDB ID: 1BUY for Erythropoietin A
Hematologic andcoagulation factor	Pegfilgrastim	Neulasta	39 (19 without glycosyl gp)	To decrease the incidence of infection, as manifested by febrile neutropenia, in patients with nonmyeloid malignancies receiving myelosuppressive anticancer drugs associated with a clinically significant incidence of febrile neutropenia	NA	P09919 (Filgrastim only)	Granulocyte colony-stimulating factor precursor (G-CSF) (Filgrastim only): P09919 GI: 117564
Hematologic and coagulation factors	Antihemophilic factor (human), purified factor VIII	Koate-DVI, also available as Koate-HP	NA	Treatment of classic hemophilia (hemophilia A)	NM_000123; GI: 10518504	NP_000123.1	Coagulation factor VIII, NMR structure PDB ID: 1FAC
Hematologic and coagulation factors	Antihemophilic factor, human recombinant factor VIII produced in BHK cells	Kogenate; Kogenate-FS	80–90	Treatment of classic hemophilia (hemophilia A) in which there is a demonstrated deficiency of activity of the plasma clotting factor FVIII (Compared to Kogenate, some patients who developed inhibitors on study with Kogenate-FS continued to manifest a clinical response when inhibitor titers were less than 10 Bethesda units (BU) per ml).	Factor VIII precursor	PID: g119767	Coagulation factor VIII, NMR structure PDB ID: 1FAC
Hematologic and coagulation factors	Antihemophilic factor/von Willebrand factor complex (human)	Humate-P	NA	Indication for use: For use in adult patients for treatment and prevention of bleeding in hemophilia A (classic hemophilia) and in adult and pediatric patients for treatment of spontaneous and trauma-induced bleeding	Noncovalent complex of antihemophilic factor VIII: NM_000123; GI: 10518504 and von Willebrand	Noncovalent complex of NP_000123 (antihemophilic factor) and NP_000543 (von Willebrand facor)	A1 domain of von Willebrand factor PDB ID: 1AUQ

■ TABLE II. *Continued*

Category	Drug Name	Trade Name	MW (kDa)	Therapeutic Use	Nucleotide Sequence[a]	Amino acid Sequence[a]	Protein Structure[b]
				episodes in severe von Willebrand disease and in mild and moderate von Willebrand disease where use of desmopressin is known or suspected to be inadequate	factor: NM_000552; GI: 9257255		
Hematologic and coagulation factors	Antihemophilic factor-human recombinant heavy and light chain combination (factor VIII) produced in CHO cells	Recombinate	NA	For the prevention and control of hemorrhagic episodes and perioperative management in hemophilia A patients	GI: 2829726, factor VIII precursor; GI: 119767, antihemophilic factor	PID: g119767	Coagulation factor VIII, NMR structure PDB ID: 1FAC
Hematologic and coagulation factors	Antithrombin III, human, purified AT-III, alpha2-glycoprotein	Thrombate III	58	Hereditary antithrombin III deficiency	Antithrombin-III (ATIII):	PID: g2829726	The intact and cleaved human antithrombin III complex as a model for serpin-proteinase interactions PDB ID: 1ATH
Hematologic and coagulation factors	Bivalirudin, reversible thrombin inhibitor, recombinant hirudin	Angiomax	2.2	Use as an anticoagulant in patients with unstable angina undergoing percutaneous transluminal coronary angioplasty (PCTA); intended for use with asprin	(Hirudin P18): GI: 124998	PID: g124998	Human thrombin complex with Hirudin variant PDB ID: 1HXF
Hematologic and coagulation factors	Coagulation factor IX (human recombinant), produced in CHO cells	BeneFix	55	For use in the control and prevention of hemorrhagic episodes in patients with hemophilia B, including the perioperative management of hemophilia B patients undergoing surgery	NM_000133; GI: 10518507	NP_000124.1	Human coagulation factor IXa in complex with p-amino benzamidine PDB ID: 1RFN

Class	Description	Trade name	Mass	Indication	Accession	PID/ID	Structure
Hematologic and coagulation factors	Coagulation factor VIIa (human recombinant), produced in BHK cells	NovoSeven	50	For use in the treatment of bleeding episodes in hemophilia A or B patients with inhibitors to factor VIII or factor IX. NovoSeven should be administered to patients only under the direct supervision of a physician experienced in the treatment of hemophilia	Coagulation factor VII precursor (serum prothrombin conversion accelerator) (Eptacog-alfa): GI: 119766	PID: g119766	Structure of human factor VIIa and its implications for the triggering of blood coagulation PDB ID: 1QFK
Hematologic and coagulation factors	Epoetin alfa, human recombinant, produced in CHO cells	Epogen, Procrit	30.4	Treatment of anemia associated with chronic renal failure (CRF)	Erythropoietin precursor: GI: 119526	PID: g119526	P01588 GI: 119526 PDB ID: 1BUY
Hematologic and coagulation factors	Eptifibatide, reversible platelet aggregation inhibitor	Integrilin	0.8; heptapeptide, synthetic	Acute coronary syndrome (UA/NQMI), including patients who are to be managed medically and those undergoing percutaneous coronary intervention (PCI)	NA	*N6*-(amino-iminomethyl)-*N2*-(3-mercapto-1-oxo propyl-L-lysylglycyl-L-(alpha)-aspartyl-L-tryptophyl-L-prolyl-L-cysteinamide, cyclic (1–6)-disulfide	NA
Hematologic and coagulation factors	Factor VIII, human recombinant glycosylated protein, produced in CHO cells	ReFacto	170	Control and prevention of hemorrhagic episodes and for surgical prophylaxis in patients with hemophilia A	E00527; GI: 2168806	NP_000122	Crystal structure of active site-inhibited human coagulation factor VIIa (Des-Gla) PDB ID: 1CVW
Hematologic and coagulation factors	Filgrastim, G-CSF, human recombinant produced in *E. coli*, nonglycosylated	Neupogen	18.8	To decrease the incidence of infection, as manifested by febrile neutropenia, in patients with nonmyeloid malignancies receiving myelosuppressive anticancer drugs associated with a significant incidence of severe neutropenia with fever	G-CSF precursor GI: 117564 (Pluripoietin) (Filgrastim) (Lenograstim)	PID: g117564	Crystal structure of canine and bovine granulocyte-colony stimulating factor (G-CSF). PDB ID: 1BGC

▮ TABLE II. *Continued*

Category	Drug Name	Trade Name	MW (kDa)	Therapeutic Use	Nucleotide Sequence[a]	Amino acid Sequence[a]	Protein Structure[b]
				(Glycosylated rG-CSF product: Lenograstim produced by Phone-Poulenc Rorer—may produce more potent initial peak response—is currently not approved by FDA in the US)			
Hematologic and coagulation factors	Lepirudin-thrombin inhibitor, Hirudin-human recombinant	Refludan	7.0	Anticoagulation in patients with heparin-induced thrombocytopenia and associated thromboembolic disease	Hirudin variant-1 (Lepirudin): GI: 124981	PID: g124981	Nuclear magnetic resonance solution structure of Hirudin(1–51) and comparison with corresponding three-dimensional structures determined using the complete 65-residue Hirudin polypeptide chain PDB ID: 1HIC
Hematologic and coagulation factors	Oprelvekin, human recombinant IL-11, a thrombopoietic growth factor that stimulates stem cells and megakaryocyte progenitor, produced in *E. coli*	Neumega	19 nonglycosylated	Prevention of severe thrombocytopenia and the reduction of the need for platelet transfusions following myelosuppressive chemotherapy in patients with nonmyeloid malignancies who are at high risk of severe thrombocytopenia	IL-11precursor GI: 15341755 Oprelvekin GI: 124294	P30809	NA
Hematologic and coagulation factors	Sargramostatin, human recombinant GM-CSF, produced in a yeast (*S. cerevisiae*) expression system	Leukine	19.5, 16.8, 15.5 (mixture)	Mobilization of hematopoietic progenitor cells into peripheral blood	GM-CSF precursor GI: 117561	P04141	Cytokyne-binding region of gp130 PDB ID: 1CSG

Hormones	Calcitonin, salmon	Miacalcin	3.4	Treatment of symptomatic Paget's disease of bone	CAS Registry: 47931-85-1 Calcitonin 1 precursor: GI: 115483	PID: g115483	Glycosylated Eel Calcitonin PDB ID: 1BZB
Hormones	Cosyntropin; alpha 1–24 corticotropin (a ACTH subunit), synthetic	Cortrosyn	2.9	Diagnostic agent to detect adrenocortical insufficiency in suspected individual	CAS Registry: 16960-16-0	(1–24)-tetracosapeptide; Ser-Tyr-Ser-Met-Glu-His-Phe-Arg-Trp-Gly-Lys-Pro-Val-Gly-Lys-Lys-Arg-Arg-Pro-Val-Lys-Val-Tyr-Pro	NA
Hormones	Follitropin alpha, human follicle-stimulating hormone (hFSH), recombinant	Gonal-F	~9 (alpha), ~11 (beta)	Induction of ovulation, induction of spermatogenesis in men	GI: 120552 Follitropin-beta precursor (follicle-stimulating hormone-beta) (FSH-beta)	PID: g120552	Human follicle stimulating hormone PDB ID: 1FL7
Hormones	Follitropin beta, human follicle-stimulating hormone (hFSH), recombinant	Follistim	~9 (alpha), ~11 (beta)	Development of multiple follicles, and induction of ovulation and pregnancy	GI: 12644092 follitropin beta chain precursor (FSH-beta) (FSH-B)	PID: g12644092	Human follicle stimulating hormone PDB ID: 1FL7
Hormones	Ganirelix acetate, synthetic decapeptide with antagonistic activity against gonadotropin releasing hormone (GnRH)	Antagon	1.6	Inhibition of premature LH surges in women undergoing controlled ovarian hyperstimulation	NA	N-acetyl-3-(2-naphthyl)-d-alanyl-4-chloro-d-phenylalanyl-3-(3-pyridyl)-d-alanyl-L-seryl-L-tyrosyl-9 N 9, N10-diethyl-d-homoarginyl-L-leucyl-N 9, N10-diethyl-L-homoarginyl-L-prolyl-d-10 alanylamide acetate	NA

■ TABLE II. Continued

Category	Drug Name	Trade Name	MW (kDa)	Therapeutic Use	Nucleotide Sequence[a]	Amino acid Sequence[a]	Protein Structure[b]
Hormones	Glucagon, recombinant	Glucagon for injection/ GlucaGen	3.5	Emergency kit for controlling severe hypoglycemia	GI: 121484 glucagon precursor	PID: g121484	Structure of a glucagon analogue PDB ID: 1BH0
Hormones	Goserelin: luteinizing hormone-releasing hormone (LHRH) decapeptide analogue	Zoladex	1.3	Palliative treatment of advanced carcinoma of the prostate, advanced breast cancer, and management of endometriosis	Gonadotropin-releasing hormone 1 (leutinizing-releasing hormone) GI: 14782453	PID: g14782453	NA
Hormones	Human chorionic gonadotropin (HCG), placental extracts	Novarel	NA	Prepubertal cryptorchidism, hypogonadism, induction of ovulation	GI: 116184 HCG-beta	PID: g116184	Crystal structure of human chorionic gonadotropin PDB ID: 1HRP
Hormones	Human chorionic gonadotropin (HCG), placental extracts	Pregnyl	NA	Prepubertal cryptorchidism, hypogonadism, ovulation induction	HCG family [Homo sapiens]: GI: 15277242	PID: g15277242	Crystal structure of human chorionic gonadotropin PDB ID: 1HRP
Hormones	Human chorionic gonadotropin (HCG), placental extracts	Profasi	NA	Prepubertal cryptorchidism, hypogonadism, ovulation induction	GI: 116184 HCG-bet	PID: g116184	Crystal structure of human chorionic gonadotropin PDB ID: 1HRP
Hormones	Insulin, human recombinant (E. coli)	Humulin	5.8	Treatment of patients with diabetes mellitus for the control of hyperglycemia	GI: 124617 insulin precursor	PID: g124617	Lys(B28)Pro(B29)-human insulin PDB ID: 1LPH
Hormones	Insulin, human recombinant (yeast, S. cerevisiae)	Novolin	5.8	Treatment of patients with diabetes mellitus for the control of hyperglycemia	GI: 124617 insulin precursor	PID: g124617	Lys(B28)Pro(B29)-human insulin PDB ID: 1LPH
Hormones	Insulin, human recombinant, yeast (S. cerevisiae), infusion pump or syringe refills	Velosulin BR	5.8	Treatment of patients with diabetes mellitus for the control of hyperglycemia	GI: 124617 insulin precursor	PID: g124617	Lys(B28)Pro(B29)-human insulin PDB ID: 1LPH

Hormones	Insulin, lispro variant, human, recombinant, rDNA origin	Humalog	5.8	Treatment of patients with diabetes mellitus for the control of hyperglycemia; often used to produce a more rapid onset and shorter duration of action than human regular insulin	GI: 124617 insulin precursor	PID: g124617	Lys(B28)Pro(B29)-human insulin PDB ID: 1LPH
Hormones	Insulin, pork derived	I lectin II	5.8	Treatment of patients with diabetes mellitus for the control of hyperglycemia	GI: 124617 insulin precursor	PID: g124617	Lys(B28)Pro(B29)-human insulin PDB ID: 1LPH
Hormones	Insulin-asp (B28), human recombinant (yeast *S. cerevisiae*)	NovoLog	5.8	Treatment of patients with diabetes mellitus for the control of hyperglycemia	GI: 124617 insulin precursor	PID: g124617	Lys(B28)Pro(B29)-human insulin PDB ID: 1LPH
Hormones	Insulin-glargine (recombinant); Asn (A21) > Gly; addition of two Arg to B20 and B21, more rapid onset	Lantus	6.1	Control of hyperglycemia for type 1 and type 2 diabetes mellitus, for a more rapid response	GI: 124617 insulin precursor	PID: g124617	Lys(B28)Pro(B29)-human insulin PDB ID: 1LPH
Hormones	Leuprolide acetate, GnRH or LH-RH analogue	Lupron, Lupron-depot	1.3	Management of endometriosis, uterine leiomyomata (Fibroids)	CAS Registry: 53714-56-0 (leuprorelin); 74381-53-6 (leuprorelin acetate)	5-oxo-L-prolyl-L-his tidyl-L-tryptophyl-L-seryl-L-tyrosyl-D-le ucyl-L-leucyl-L-argi nyl-N-ethyl-L-prolin amide acetate	NA
Hormones	Menotropins, human chorionic gonadotropins (hCG) (extracted)	Humegon	NA	Induction of ovulation, stimulation of spermatogenesis in men	HCG family [Homo sapiens]; GI: 15277242	PID: g15277242	Crystal structure of human chorionic gonadotropin PDB ID: 1HRP
Hormones	Menotropins, human chorionic gonadotropins (hCG) (extracted)	Pergonal	NA	Induction of ovulation; with concomitant hCG for stimulating spermatogenesis	HCG family [Homo sapiens]; GI: 15277242	PID: g15277242	Crystal structure of human chorionic gonadotropin PDB ID: 1HRP
Hormones	Menotropins, human chorionic	Repronex	NA	Multiple follicular development (controlled ovarian stimulation)	HCG family [Homo sapiens];	PID: g15277242	Crystal structure of human chorionic

■ TABLE II. *Continued*

Category	Drug Name	Trade Name	MW (kDa)	Therapeutic Use	Nucleotide Sequence[a]	Amino acid Sequence[a]	Protein Structure[b]
	gonadotropins (hCG) (extracted)			and ovulation induction in patients who have previously received pituitary suppression	GI: 15277242		gonadotropin PDB ID: 1HRP
Hormones	Nafarelin acetate, gonadotropin-releasing hormone (GnRH) agonist	Synarel	NA	Management of endometriosis or treatment of central precocious puberty (CPP) (gonadotropin-dependent precocious puberty) in children of both sexes	Gonadotropin-releasing hormone-1 (leutinizing-releasing hormone) GI: 14782453	5-oxo-L-prolyl-L-histidyl-L-tryptophyl-L seryl L-tyrosyl-3-(2-naphthyl)-D alanyl-L-leucyl-L-arginyl-L-prolyl-glycinamide acetate PID: g14782453	NA
Hormones	Octreotide, somatostatin, octatpeptide	Sandostatin	1.0	Treatment of carcinoid and vasoactive intestinal peptide; acromegaly (reduce blood levels of growth hormone), symptomatic treatment of metastatic carcinoid tumors, and profuse water diarrhea associated with intestinal tumors	Somatostatin precursor [contains: somatostatin-28 and somatostatin -14]: GI: 134557	L-Cysteinamide, D-phenylalanyl-L-cysteinyl-L-phenyl-alalyl-D-tryptophyl-L-lysyl-L-threonyl-N-[2-hydroxy-1-(hydroxy methyl) propyl]-, cyclic (2 → 7)-disulfide; [R-(R* · R*)] PID: g134557	NMR results of the backbone conformational equilibria PDB ID: 1SOC (sheet structure)
Hormones	Oxytocin	Pitocin	1.0	Induction of labor at term, control uterine bleeding	Oxytocin-neurophysin-1 precursor (OT-NPI) [contains: oxytocin (ocytocin); neurophysin-1]: GI: 128071	Cys-Tyr-Ile-Gln-As n-Cys-Pro-Leu-Gly-NH₂ cyclic (1→6) disulphide; [2-Leucine, 7-isoleucine] vasopressin PID: g128071	Crystal structure analysis of deamino-oxytocin: conformational flexibility and receptor binding PDB ID: 1XY1
Hormones	Proirelin, throtropin-releasing	Thyrel TRH	NA	Adjustment of thyroid hormone dosage given to patients with	Thyrotropin-releasing hormone	5-oxo-L-prolyl-L-his tidyl-L-proline	NA

		hormone (TRH), synthetic tri-peptide, 5-oxo-(pro-his-pro)		primary hypothyroidism; can also be used as an diagnostic assessment of thyroid function	[Homo sapiens]: GI: 15294438	amide; CAS# 24305-27-9 PID: g15294438	
Hormones	Genotropin	Somatotropin, human recombinant	22	Growth failure due to an inadequate secretion of endogenous growth hormone, Prader-Willi syndrome (PWS), adults with growth hormone deficiency (GHD) of either childhood- or adult-onset etiology	Somatotropin precursor (growth hormone) GI: 134703	PID: g134703	A heuristic approach to predicting the tertiary structure of bovine somatotropin PDB ID: 1BST
Hormones	Humatrope	Somatotropin, human recombinant (E. coli)	22	Pediatric patients who have growth failure due to an inadequate secretion of normal endogenous growth hormone	Somatotropin precursor (growth hormone) GI: 134703	PID: g134703	A heuristic approach to predicting the tertiary structure of bovine somatotropin PDB ID: 1BST
Hormones	Protropin	Somatrem recombinant human pituitary growth hormone produced in E. coli	22	Children with documented lack of human growth hormone	Somatotropin precursor (growth hormone) GI: 134703	PID: g134703	NA
Hormones	Nutropin AQ	Somatotropin	22	Use as a growth factor to increase growth rate stunted due to a lack of adequate endogenous GH secretion	NA	NA	NA
Hormones	Saizen	Somatotropin	22	Long-term treatment of children with growth failure due to inadequate secretion of endogenous growth hormone	NA	NA	NA
Hormones	Serostim	Somatotropin	22	Treatment of AIDS wasting or cachexia	NA	NA	NA
Hormones	Thyrogen	Thyrotropin alfa, thyroid-stimulating	~9.2	NA	Thyrothropin beta chain	PID: g136443	NA

■ **TABLE II.** *Continued*

Category	Drug Name	Trade Name	MW (kDa)	Therapeutic Use	Nucleotide Sequence[a]	Amino acid Sequence[a]	Protein Structure[b]
	hormore (TSH), human recombinant, produced in CHO cells				precursor (thyroid-stimulating hormone; TSH-B) GI: 136443		
Interferons and cytokines	Aldesleukin, human recombinant interleukin-2, produced in *E. coli*	Proleukin	15.5	For the treatment of adults 18 years of age or older with metastatic renal cell carcinoma	NM_000586; GI: 10835148 GI: 22219369	NP_063916.1	PID: 1M4C PID: 1IRL (a variant structure)
Interferons and cytokines	Anakinra, human IL-1Ra	Kineret	17.3	Reduction in signs and symptoms of moderately to severely active rheumatoid arthritis, in patients 18 years of age or older who have failed one or more disease-modifying antirheumatic drugs (DMARDS)	AJ271338.1 GI: 6729586 GI: 124308	P01589 P14778	PID: 1ILT
Interferons and cytokines	Denileukin diftitox, IL-2 diphtheria conjugate	Ontak	58	Treatment of patients with persistent or recurrent cutaneous T-cell lymphoma whose malignant cells express the CD25 component of the IL-2 receptor	GI: 10835148 for IL-2	NP_063916.1	PID: 1M4C
Interferons and cytokines	Interferon alfa-2a	Roferon-A	19	Treatment of chronic hepatitis C	Interferon-alpha2 precursor GI: 124449	P01563	PID: 1ITF NMR structure
Interferons and cytokines	Interferon alfa-2b, human recombinant product produced in *E. coli*	Intron A	19.3	Use in conjunction with chemotherapy in patients with follicular lymphoma, hairy-cell leukemia, malignant melanoma, AIDS-related Kaposi's sarcoma, and chronic hepatitis C	Interferon-alpha-2 precursor A chain GI: 2624437 B chain GI: 2624438	1RH2A 1RH2B	PDB ID: 1RH2

Category	Name	Brand	MW	Indication	Precursor/GI	Accession	Structure
Interferons and cytokines	Interferon alfacon-1, human recombinant,	Infergen	19.5	Treatment of chronic hepatitis C virus (HCV) infection in patients 18 years of age or older with compensated liver disease who have anti-HCV serum antibodies and/or the presence of HCV RNA	GI: 2118860	I51970	NA
Interferons and cytokines	Interferon alfa-n1, Lymphoblastoid (natural product)	Wellferon	~19 (a mixture of natural interferons)	For the treatment of chronic hepatitis C in patients 18 years of age or older without decompensated liver disease	Interferon-alpha-2 precursor GI: 124449	P01563	Interferon-2a, NMR structures PDB ID: 1ITF
Interferons and cytokines	Interferon alfa-n3 (human leukocyte-derived)	Alferon-N	~16–27	Intralesional treatment of refractory or recurring external condylomata acuminata in patients 18 years of age or older	NA	NA	Interferon-2A, NMR, 24 structures PDB ID: 1ITF
Interferons and cytokines	Interferon beta-1a	Rebif	22.5	Treatment of patients with relapsing forms of multiple sclerosis to decrease the frequency of clinical exacerbations and delay the accumulation of physical disability	GI: 4504603	NP_002167	Interferon-beta PID: 1RMI
Interferons and cytokines	Interferon beta-1a, human recombinant product produced in CHO cells	Avonex	22.5	Treatment of relapsing forms of multiple sclerosis to slow the accumulation of physical disability and decrease the frequency of clinical exacerbations	Interferon-beta precursor GI: 124469 (fibroblast interferon)	P01574	Human interferon-beta crystal structure PDB ID: 1AU1
Interferons and cytokines	Interferon beta-1b, human recombinant	Betaseron	18.5	Treatment of patients with relapsing- remitting multiple	Interferon-beta precursor	P01574	Human interferon-beta crystal structure

■ TABLE II. *Continued*

Category	Drug Name	Trade Name	MW (kDa)	Therapeutic Use	Nucleotide Sequence[a]	Amino acid Sequence[a]	Protein Structure[b]
	produced in *E. coli*			sclerosis to reduce the frequency of clinical exacerbation	GI: 124469 (fibroblast interferon)		PDB ID: 1AU1
Interferons and cytokines	Interferon gamma-1b, human recombinant produced in *E. coli*	Actimmune	16.5	Delaying time to disease progression in patients with severe, malignant osteopetrosis; reducing the frequency and severity of serious infections associated with chronic granulomatous disease	Interferon-gamma precursor GI: 124479 (immune interferon)	P01574	Crystal structure of IFN-beta PDB ID: 1AU1
Interferons and cytokines	Peginterferon alfa-2b, human recombinant produced in *E. coli*	PEG-Intron	31 kD	Treatment of chronic hepatitis C in patients not previously treated with interferon alfa who have compensated liver disease and are at least 18 years of age	IFN-alfa-2 precursor GI: 124449	P01563	PID: 1ITF NMR-based IFN-alfa structure
Interferons and cytokines	Peginterferon alfa-2b, PEG modified human recombinant inteferon alfa-2b (intron A, produced	PEG-Intron	31 (19.2 intron A + 12 contributed by PEG modification)	Treatment of chronic hepatitis C in patients not previously treated with interferon alfa who have compensated liver disease and are at least 18 years of age; use in combination with ribavirin	NA	NA	Interferon back bone PDB ID: 1RH2
Other	Becaplermin, PDGF, human recombinant produced in yeast, *S. cerevisiae*	Regranex Gel	25	Treatment of lower extremity diabetic neuropathic ulcers that extend into the subcutaneous tissue or beyond, and have adequate blood supply	GI: 129724 Platelet-derived growth factor, B chain precursor adequate blood supply	P01127	PDB ID: 1PDG
Other	Botulinum toxin A, neutrotoxin complex	Botox	NA	NA	Botulinum neurtotoxin type A precursor (BONT/A) GI: 399133	P10845	Crystal structure PDB ID: 1F31; 1I1E

Other	Fibrin sealant	Tisseel VH Kit	NA	Adjunct to hemostasis in surgeries involving cardiopulmonary bypass and treatment of splenic injuries due to blunt or penetrating trauma to the abdomen, when control of bleeding by conventional surgical techniques, including suture, ligature, and cautery, is ineffective or impractical; also indicated as an adjunct for the closure of colostomies	Fibrinogen-beta precursor GI: 399492	P02675	Crystal structure of modified bovine fibrinogen PDB ID: 1DEQ; 1FZA
Other	Proteinase inhibitor (human), (alpha1-PI), alpha-1 antitrypsin	Prolastin	52	NA	Serine (or cysteine) proteinase inhibitor, clade A (alpha-1 antiproteinase, antitrypsin), member 3 [Homo sapiens]: GI: 13097704	BC003559	PDB ID: 1QLP
Vaccine	Haemophilus b conjugate vaccine (meningococcal protein conjugate)	PedvaxHIB	Irrev	Vaccination against invasive disease caused by Haemophilus influenzae type b in infants and children 2 to 71 months of age	NA	NA	NA
Vaccine	Hepatitis A vaccine, inactivated	Havrix	NA	Immunization of persons ≥ 2 years of age against disease caused by hepatitis A virus	Polyprotein [Hepatitis A virus]: GI: 12018152	AF31428	Crystal complex of the 3C proteinase from hepatitis A virus with its inhibitor and implications for the polyprotein processing in HAV PDB ID: 1QA7

TABLE II. Continued

Category	Drug Name	Trade Name	MW (kDa)	Therapeutic Use	Nucleotide Sequence[a]	Amino acid Sequence[a]	Protein Structure[b]
Vaccine	Rabies vaccine	Imovax	NA	Pre- and postexposure immunization and treatment against rabies	NA	NA	NA
Vaccines	BCG, Live	TICE BCG (TICE strain), TheraCys (Cannaught strain)	Irrev[c]	Expand the indication for intravesical instillation, to include adjunct treatment of stage Ta or TI papillary tumors of the bladder	GI: 15147222	NM_022154	NA
Vaccines	BCG, Live	PACIS	Irrev	Treatment of carcinoma in situ (CIS) in the absence of associated invasive cancer of the bladder	GI: 1154722	NM_022154	NA
Vaccines	Diphtheria and tetanus toxoids and acellular pertussis vaccine	Infanrix	Irrev	Primary and booster immunization of infants and children except as a fifth dose in children who have previously received four doses of DTaP	NA	NA	NA
Vaccines	Diphtheria and tetanus toxoids and acellular pertussis vaccine, adsorbed	Acel-Imune	69 (pertactin antigen)	New indication for use as a three-dose primary series in children at least six weeks of age and for the fourth and fifth dose in children who have received three doses of DTaP or diphtheria and tetanus toxoids and pertussis vaccine	NA	NA	NA
Vaccines	Diphtheria and tetanus toxoids and acellular pertussis vaccine adsorbed	Tripedia	Irrev	Active immunization against diphtheria, tetanus and pertussis (whooping cough) simultaneously in infants and children 6 weeks to 7 years of age (prior to seventh birthday)	NA	NA	NA

Vaccines	Haemophilus b conjugate (meningococcal protein conjugate) and hepatitis B (recombinant)	COMVAX	Irrev	Immunization of persons 6 weeks to 15 months of age born to hepatitis B surface antigen (HBsAg)-negative mothers	NA	NA	NA
Vaccines	Haemophilus b conjugate vaccine (diphtheria CRM 197 protein conjugate)	HibTITER	Irrev	Immunization of children 2 months to 71 months of age against invasive diseases caused by haemophilus influenzae type b	NA	NA	NA
Vaccines	Haemophilus b conjugate vaccine (tetanus toxoid conjugate)	ActHIB, OmniHIB	Irrev	Combined with AvP DTP vaccine by reconstitution is indicated for the active immunization of infants and children 2 through 18 months of age for the prevention of invasive disease caused by H influenzae type b and/or diphtheria, tetanus, and pertussis	NA	NA	NA
Vaccines	Hepatitis A vaccine, inactivated	Vaqta	NA	Active pre-exposure prophylaxis immunization against disease caused by hepatitis A virus in persons 2 years of age and older; primary immunization should be given at least 2 weeks prior to expected exposure to HAV	Polyprotein [hepatitis A virus]: GI: 12018153	AAG45423	Hepatitis A virus 3C proteinase PDB ID: 1HAV
Vaccines	Hepatitis A, inactivated and hepatitis B (recombinant) vaccine	TWINRIX	NA	Active immunization of persons 18 years of age or older against disease caused by hepatitis A virus and infection by all known subtypes of hepatitis B virus	Sequence 56 from US patent 6072049: GI: 12807112	AAE43009.1	Crystal complex of the 3c proteinase from hepatitis A virus with its inhibitor and implications for the polyprotein processing in HAV PDB ID: 1QA7

■ TABLE II. Continued

Category	Drug Name	Trade Name	MW (kDa)	Therapeutic Use	Nucleotide Sequence[a]	Amino acid Sequence[a]	Protein Structure[b]
Vaccines	Hepatitis B vaccine (recombinant), produce in yeast S. cerevisiae	Recombivax	NA	Vaccination against infection caused by all known subtypes of hepatitis B virus Recombivax HB dialysis formulation is indicated for vaccination of adult predialysis and dialysis patients against infection caused by all known subtypes of hepatitis B virus	Sequence 56 from US patent 6072049: GI: 12807112	AAE43009	NA
Vaccines	Hepatitis B vaccine (Recombinant); recombinant hepatitis B surface antigen produced in S. cerevisiae (yeast).	Engerix-B	NA	Immunization against infection caused by all known subtypes of hepatitis B virus. As hepatitis D (caused by the delta virus) does not occur in the absence of hepatitis B infection, because risk factors for hepatitis C are similar to those for hepatitis B, immunization with hepatitis B vaccine is recommended for individuals with chronic hepatitis C.	Sequence 56 from US patent 6072049: GI: 12807112	AAE43009	NA
Vaccines	Lyme disease vaccine (recombinant OspA), acylated	LYMErix	31	For active immunization against Lyme disease in individuals 15–70 years of age	Sequence 2 from US patent 6221363: GI: 15111607	AAE68480	NA
Vaccines	Pneumococcal 7-valent conjugate vaccine (diphtheria CRM197 protein)	Prevnar	NA	Immunization of infants and toddlers against invasive disease caused by S. pneumoniae due to capsular serotypes included in the vaccine (4, 6B, 9V, 14, 18C,	Diphtheria toxin CRM45/197: GI: 224021	1007216A	NA

19F, and 23F). 2, 4, 6, and 12–15 months of age to prevent invasive pneumococcal disease

Vaccines	Pneumococcal vaccine, polyvalent, 23 types of *S. pneumoniae*	PNU-Immune-23	NA	Immunization against pneumococcal disease caused by those pneumococcal types included in the vaccine	Streptococcus pneumoniae pneumococcal vaccine antigen A (pvaA) gene, complete cds: GI: 13345018	AF291698	NA
Vaccines	Pneumococcal vaccine, polyvalent-23 pneumococcal	Pneumovax 23	NA	Vaccination against pneumococcal disease caused by those pneumococcal types included in the vaccine	Streptococcus pneumoniae pneumococcal vaccine antigen A (pvaA) gene, complete cds: GI: 13345018	AF291698	NA
Vaccines	Rabies vaccine	RabAvert	NA	Pre-exposure and postexposure immunization of children and adults against rabies	Sequence 2 from US patent 5698202: GI: 3205564	AAC19417	NA
Vaccines	Varicella virus vaccine, live, Oka/Merck strain of live, attenuated varicella virus	Varivax	NA	For the active immunization of persons 12 months of age and older against varicella	GI: 593786	AAA55393	NA

[a]Nucelotide and amino-acid sequence data can be accessed through Gene Bank at URL: *http://www.ncbi.nlm.nih.gov:80/entrez/query.fcgi*

[b]Protein structure data can be accessed through Protein Data Bank at URL: *http://www.rcsb.org/pdb*; some structural data, designated as "SM" can be assessed through the Simple Architecture Research Tool, SMART at URL: *http://smart.embl-heidelgerg.de.*

[c]NA, not available; irrev, irrelevant.

Appendix III

NOMENCLATURE OF BIOTECHNOLOGY PRODUCTS

The Unite States Adopted Names (USAN) council is the designated drug nomenclature agency responsible to develop rules designed for coining informative, nonproprietary names of drugs and related chemical and biological substances. In collaboration with the World Health Organization (WHO), the USAN council coins nonproprietary names for drug and biotechnology products based on the standardized and established rules governing the classification of new substances. [Note: Adapted from USAN council: New Names. J. Clin Pharmacol Therap, publish monthly].

In collaboration with the FDA's CBER and the WHO's, International Nonproprietary Names Committee, the USAN created specific guidelines for coining names for biological and recombinant products, including interferons, interleukins, growth hormones, colony-stimulating factors, cytokines, and monoclonal antibodies. Some of these identifiers are included as prefixes, suffixes, and infixes (internally located) within the name of a new molecular entity to signify different class of biologic molecules.

Biotechnology and Biopharmaceuticals, by Rodney J. Y. Ho and Milo Gibaldi
ISBN 0-471-20690-3   Copyright © 2003 by John Wiley & Sons, Inc.

■TABLE III.1. Nomenclature of biological products

Biological Products	Guiding Principle	Prefix	Suffix	Most Commonly Used Forms	Conventions	Examples
Interferons	A class name for a family of species-specific proteins (or glycoproteins) that are produced according to information encoded by species of interferon genes, and exert complex antineoplastic, antiviral, and immunomodulating effects			Interferon alfa (formerly known as leukocyte or lymphoblatoid interferon)	Appropriate Greek letter (spelled out) is the second word of the name: alfa, beta, gamma	Interferon alfa-2a Interferon alfa-2b Interferon beta-1a Interferon beta-1b Interferon gamma-1a
				Interferon beta (formerly fibroblast interferon)	Arabic numerals and letters are appended to the Greek letter by a hyphen to delineate subcategories; the lowercase letter is assigned by the drug nomenclature agencies to differentiate one manufacturer's interferon from another's	
				Interferon gamma (formerly immune) interferon	For mixtures of naturally occurring interferons, the lowercase letter *n* precedes the number, (i.e., interferon alfa-*n*1 or interferon alfa-*n*2)	

Category	Rule	Examples
Interleukins	Suffix –leukin is used in naming interleukin 2 (IL-2) type substances	Aldesleukin Celmoleukin Teceleukin
Somatotropins	Prefix *som-* is used for growth hormone derivative Suffix –*bov* (and the *som*-prefix) are required for bovine somatotropin Suffix –*por* (and the *som*-prefix) are required for porcine somatotropin derivatives	*Human* 1. Somatropin for human growth hormone 2. Somatrem for methionyl HGH *Bovine* 1. Somidobove 2. Sometribove 3. Somagrebove *Porcine* 1. Somalapor 2. Somenopor 3. Sometripor 4. Somfasepor
Colony-stimulating factors	Suffix –*grastim* is used for granulocyte colony-stimulating factors (G-CSF) Suffix-*gramostim* is used for granulocyte macrophage colony stimulating factors (GM-CSF) Suffix –*mostim* is used for macrophage colony-stimulating foactors (M-CSF) Suffix –*plestim* is used for interleukin 3 (IL 3 factors classifed as pleiotropic colony-stimulating factors	*Grastim* 1. Lenograstim 2. Filgrastim *Gramostim* 1. Molgramostim 2. Regramostim 3. Sargramostim *Mostim* 1. Mirimostim 2. Muplestim 3. Daniplestim

■ **TABLE III.1.** *Continued*

Biological Products	Guiding Principle	Prefix	Suffix	Most Commonly Used Forms	Conventions	Examples
Erythropoietins	The word *epoetin* is used for recombinant human erythropoietin Epoetin describes erythropoietin preparations that have an amino-acid sequence identical to the endogenous cytokine				Erythropoietin, is followed by the appropriate Greek letter (spelled out) Alfa, beta, and gamma are added to designate the preparations that differ in composition and the nature of the carbohydrate moieties	Epoetin alfa Epoetin beta Epoetin gamma
Monoclonal antibodies[a]	Identification of the animal source of the product is an important safety factor based on the number of products that may cause source-specific antibodies to develop in patients		Suffix –*mab* is used for monoclonal antibodies and fragments		Identifiers used as infixes preceding the –*mab* suffix stem u = human e = hamster o = mouse i = primate a = rat xi = chimera zu = humanized General disease state subclass must be incorporated into the name by use of a code syllable	Umab (human) Omab (mouse) Ximab (chimera) Zumab (humanized)

[a]For additional details on guiding principles involved in coining USAN for antibodies, please see World Health Assembly (WHA) resolution 4619.

Appendix IV

SYNONYM OF TRADE, COMMON, AND SCIENTIFIC NAMES

Drug Name	Trade Name	Synonyms
2-166-Interferon beta 1 (human fibroblast reduced), 17-L-serine-; interferon betaser, recombinant	Betaseron	*Interferon Beta-1b, rDNA [BIO]* Interferon betaser, rDNA [BIO] Betaseron [TR reg. to Schering AG] Interferon beta-1b [FDA USAN INN BAN] 2-166-Interferon beta1 (human fibroblast reduced), 17-L-serine-[CAS; beta is Greek letter] Interferon beta1, 17-L-serine-, (2S-(2R*,5R*))-[SY] Human Recombinant Betaser17 Interferon [SY] Betaferon [TR used in Europe; assigned to Schering AG] 145155-23-3; 96778-78-8 [CAS RN]
Abciximab, F_{ab} fragment of chimeric human-murine monoclonal antibody 7E3 directed against the glycoprotein (GP) IIb/IIIa ((alpha)IIb beta 3) receptors of human platelets	ReoPro	*Platelet Mab, rDNA [BIO]* ReoPro [TR (assigned to Eli Lilly & Co.)] CenteoRx [TR former in US] Abciximab [FDA USAN INN BAN] Immunoglobulin G, (human-mouse monoclonal c7E3 clone p7E3VGHhhCgamma4) F_{ab} fragment anti-human glycoprotein IIb/IIIa receptor), disulfide with human-mouse monoclonal c7E3 clone p7E3VkappahCkappa light chain [CAS] F_{ab} fragment of the chimeric human-murine monoclonal antibody 7E3 [SY] Ati-GPIIb/IIIa monoclonal antibody [SY] GPIIb/IIIa monoclonal antibody [SY]

Biotechnology and Biopharmaceuticals, by Rodney J. Y. Ho and Milo Gibaldi
ISBN 0-471-20690-3   Copyright © 2003 by John Wiley & Sons, Inc.

Drug Name	Trade Name	Synonyms
		c7E3 [SY]
		human-murine monoclonal antibody 7E3 [SY]
		c7E3 F$_{ab}$ fragment [SY]
		7E3, human-murine monoclonal antibody [SY]
		143653-53-6 [CAS RN]
Activated protein C, human recombinant	Xigris	*Activated Protein C Products [BIO]* Drotrecogin alfa (activated) NDC 0000-2285-73; -76; 0002-7561-20; 0002-7559-5 [NDC]
Adenosine deaminase (ADA), PEG modified (bovine), [11-17xPEG5000]	Adagen	*Adenosine Deaminase, PEG-[BIO]* Pegademase Bovine [FDA USAN] Deaminase, adenosine, cattle, reaction product with succinic anhydride, esters with polyethylene glycol mono-methyl ether [CAS] PEG-Adenosine Deaminase [SY] EC 3.5.4.4 [Enzyme Commission EC] 9026-93-1 [CAS RN]
Aldesleukin, human recombinant interleukin-2, produced in *E. coli*	Proleukin	*IL-2, rDNA [NAM BIOPHARMA]* Aldesleukin [NAM USAN INN BAN] 2-133-Interleukin 2 (human reduced), 125-L- serine- [NAM CAS] 125-L-Serine-2-133-interleukin 2 (human reduced) [NAM CAS] IL-2 [NAM SY] Des-alanyl-1, serine-125 human interleukin-2 [NAM SY] Interleukin-2 (recombinant human) [NAM SY] T-cell growth factor [NAM SY] TCGF [NAM SY] Leuferon-2 [NAM SY] Thymocyte stimulating factor [NAM SY] Lymphocyte mitogenic factor [NAM SY] Macrolin [NAM TR used in France, perhaps other European countries] NSC 373364; 600664 [NAM SY NCI] 110942-02-4 [CAS RN]
Alemtuzumab, anti CD52 humanized monoclonal antibody, CHO product	Campath	*CD52 Mab, rDNA [BIO]* CAMPATH-1H [TR reg. to Millennium] Alemtuzumab [FDA] CD52 monoclonal antibody, humanized [SY] CAMPATH-1 [TR former; reg. to Welcome Foundation]
Anakinra, recombinant form of the human IL-1 IL-1Ra, expressed through *E. coli*	Kinerret	*IL-1Ra, rDNA* FIL1-ETA [CAS] Interleukin 1 Superfamily E [CAS] IL1H2 [CAS] IL1 Alpha [CAS] IL1 Beta [CAS]

Drug Name	Trade Name	Synonyms
Antihemophilic factor, human recombinant factor VIII produced in BHK cells	Kogenate; Kogenate-FS (Helixate FS)	*Factor VIII products* Antihemophilic factor, human [FDA for blood-derived product] Antihemophilic factor (recombinant) [FDA] Factor VIII [SY]
Antihemophilic factor, human recombinant factor VIII produced in CHO cells	Recombinate; ReFacto	*Factor VIII, rDNA, Mab purif./Baxter [BIO]* Antihemophilic factor (recombinant) [FDA] Antihemophilic factor [USAN] Antihemophilic factor, human [former USAN] Factor VIII (rDNA) [BAN] Antihemophilic factor USP [USP] Blood-coagulation factor VIII, complex [CAS] Factor VIII, recombinant [SY] Factor VIII: von Willebrand's factor complex [SY for an intermediate] Factor VIII: C [SY for an intermediate] NDC 0944-2938-01; -02; -03 [NDC]
Antithrombin III, human, purified AT-III, alpha2-glycoprotein	Thrombate III	*Antithrombin III/Bayer [BIO]* Antithrombin III (Human) [FDA] Antithrombin III [CAS INN BAN] AT-III [SY] Heparin cofactor B [SY] Heparin cofactor I [SY] EINECS 232-568-2; EINECS 290-516-4 [EINECS] 52014-67-2; 90170-80-2 [CAS RN]
Antithrombin III, human, purified AT-III, alpha2-glycoprotein	Thrombate III	*Antithrombin III/P&U [BIO]* Atnativ [TR] Antithrombin III (Human) [FDA] Antithrombin III [CAS INN BAN] AT-III [SY] Heparin cofactor B [SY] Atenotiv [SY] heparin cofactor I [SY] EINECS 232-568-2; EINECS 290-516-4 [EINECS] 52014-67-2; 90170-80-2 [CAS RN]
Antithrombin III, human, purified AT-III, alpha2-glycoprotein	Thrombate III	*Thrombin/GenTrac [BIO]* Thrombin-JMI [TR current, assigned to Jones Pharma] ThromboGen [TR former, when marketed by GenTrac] Thrombin [FDA] Thrombin, topical (bovine origin) [FDA former; as originally approved] Thrombin, Topical USP [USP] NDC 05604-7100-1; -7102-1; -7104-3; -7105-3; -7106-1; -7102-2; -7104-2; -7105-2; -7354-2; -7355-2; [NDC]

Drug Name	Trade Name	Synonyms
Antithrombin III, human, purified AT-III, alpha2-glycoprotein	Thrombostat	*Thrombin/W-L [BIO]* Thrombin [SY]
Basiliximab, a "chimeric" recombinant monoclonal antibody	Simulect	*IL-2R Mab, conc. rDNA/Novartis [BIO]* Simulect Lyophilisate for Injection [TR] Basiliximab [FDA] Interleukin-2 (IL-2) receptor monoclonal antibody [SY] Recombinant chimeric (murine/human) monoclonal antibody (IgGl K) anti-IL-2Ra (CD25) [SY] CD25 monoclonal antibodies [SY] Chimeric anti-Tac monoclonal antibody [SY] Simulect lyophilizate for injection [SY TR used in some FDA doc.] SDZ CHI 621 [Sandoz code]
BCG, live bacterium *Mycobacterium bovis—* Bacillus Calmette-Guerin	PACIS	*BCG live [BIO]* TICE BCG (TICE strain) [SY] Theracys (cannaught strain) [SY] PACIS [TR in US and other countries; assigned to BioChem Pharma] BCG, Live [FDA] Bacillus Calmette-Guerin (BCG), substrain Montreal [SY] Mycobacterium bovis (bacillus of Calmette and Guerin strain), live [SY] Armand-Frappier strain of BCG [SY]
Becapiermin, PDGF, human recombinant produced in yeast, *S. cervisiae*	Regranex Gel	*PDGF rDNA/J&J [BIO]* Regranex [TR] Becaplermin [FDA USAN INN] Platelet-derived growth factor B, recombinant [CAS] 165101-51-9 [CAS RN] RhPDGF [SY] RhPDGF-BB [SY] PDGF [SY] CTAP-III [SY] RWJ 60235 [SY] rPDGF [SY]
Beta-Glucocerebrosidase, recombinant, glycoprotein enzyme purified from human placenta tissues	Ceredase	*Glucocerebrosidase, rDNA [BIO]* Alglucerase [FDA USAN INN BAN] Glucosylceramidase (human placenta isoenzyme protein moiety reduced) [CAS] Ceramidase, glucosyl- [CAS] Beta-D-glucosyl-*N*-acylsphingosine glucohydrolase [SY] Beta-Glucocerebrosidase [SY] Glucocerebrosidase, beta- [SY]

Drug Name	Trade Name	Synonyms
		Macrophage-targeted beta-glucocerebrosidase [SY] 143003-46-7; 37228-64-1 [CAS RN]
Bivalirudin, reversible thrombin inhibitor, recombinant hirudin	Angiomax	*Bivalirudin [BIO FDA]* Hirulog [TR former] Efludan [TR former] BG 8967 [SY Biogen code name] Huridin-derived synthetic peptide [SY]
Calcitonin	Miacalcin	*Calcitonin [BIO]* Calcimar [TR outside US] Calcitonin-salmon [SY] Miacalcin Nasal [SY] Calcitonin Gene-Related Peptide [SY] Katacalcin [CAS] Cibacalcin [CAS]
Capromab Pendetide, murine monclonal 7E11-C5.3 IgG, Indium 111	ProstaScint	*Prostrate Mab radioconj. [BIO]* In111 [SY] ProstaScint [TR for Capromab Pendetide immune conjugate] Indium In 111 ProstaScint [TR for final radioimmune conjugate] Capromab Pendetide [FDA USAN for ProstaScint immune conjugate] Kit for the Preparation of Indium In 111 Capromab Pendetide [FDA for radioimmune conjugate preparation kit] Indium In 111 Capromab Pendetide [FDA for final radioimmune conjugate] Capromab [INN for Mab] Immunoglobulin G 1 (mouse monoclonal 7E11-C5.3 antihuman prostatic carcinoma cell), disulfide with mouse monoclonal 7E11-C5.3 light chain, dimer, $N6$-(N-[2-[[2-(bis (carboxymethyl)amino]ethyl] (-carboxymethyl)-amino)ethyl]-N-(carboxymethyl)glycyl]-$N2$-(N-glycyl-L-tyrosyl)-L-lysine conjugate [CAS for ProstaScint] 145464-28-4 [CAS RN for final radioimmune conjugate] CYT-356 [SY for Mab with linker] 111In-CYT-356 [SY for Mab with linker] 7E11-C5.3-GYK-DTPA [SY for Mab with linker] 7E11-C5 [SY for Mab] GYK-DTPA.HCl [SY for tripeptide linker-chelator] GYK-DTPA-HCl.CYT-356 [SY for Mab with linker-chelator] CYT-351 [SY Mab]

Drug Name	Trade Name	Synonyms
		In-111-Capromab Pendetide [SY for final radioimmune conjugate] CYT-063 [SY for tripeptide linker-chelator] Glycyl-tyrosyl-(N-epsilon-diethylenetriaminopentaacetic acid-lysine hydrochloride [SY for tripeptide linker-chelator] NDC 57902-817-01 [NDC for Capromab Pendetide kit] Myoscint (Imicro pentetate, F_{ab} fragment of anti-myosin IgG2 monoclonal antibody bound to DTPA for labeling with Indium 111)
CMV antisense	Vitravene	*CMV Antisense Drug [BIO]* Fomivirsen sodium [FDA NAM] ISIS 2922 [IC]
Coagulation factor IIa (human recombinant), produced in BHK cells	NovoSeven	*Factors IIa rDNA [BIO]* NovoSeven [TR assigned to Novo Nordisk A/S] Coagulation factor VIIa (recombinant) [FDA] rFVIIa [SY] FVIIa, recombinant [SY] Activated factor VII [SY] NDC 0169-7060; -01; -7062-01; -7064-01; [NDC]
Coagulation factor IX (human recombinant), produced in CHO cells	Benefix	*Factor IX, rDNA/GI [BIO]* Coagulation factor IX (recombinant) [FDA] Antihemophilic factor IX [SY] Factor IX [SY] Factor IX and factor IXa [MESH] rFIX [SY] NDC 58394-001; -002; -003-01 [NDC]
Collagenase, enzyme derived from fermentation of the bacterium Clostridium hisolyticum	Santyl ointment	*Collagenase [BIO]* Santyl [TR reg. by Knoll] Iruxol [TR outside U.S.] Novuxol [TR outside U.S.] Clostridopeptidase A [SY] Biozyme-C [TR former] Iruxol [TR foreign] 9001-12-1; 37288-86-1; 39433-96-0 [CAS RN] NDC 0044-5270-02; -03 [NDC]
Cosyntropin; alpha 1–24 corticotropin (a ACTH subunit); synthetic	Cortrosyn	*ACTH, rDNA [BIO]* Synacthen [CAS] Tetracosactide [CAS] NDC 0052-0731-10 [NDC]
Daclizumab, anti-CD25 humanized IgG1, recombinant product	Zenapax	*IL-2 Mab, rDNA/Roche [BIO]* Zenapax [TR reg. to Roche] Daclizumab [FDA current] Dacliximab [FDA former] Immunoglobulin G1, anti-(human interleukin 2 receptor) (human-mouse monoclonal clone

Drug Name	Trade Name	Synonyms
		1H4 *gamma*1-chain), disulfide with human-mouse monoclonal clone 1H4 light chain, dimer [CAS]
		Immunoglobulin G$_1$(human-mouse monoclonal clone 1H4 *gamma*-chain anti-human antigen Tac), disulfide with human-mouse monoclonal clone 1H4 light chain, dimer [SY]
		Immunoglobulin G$_1$(human-mouse monoclonal clone 1H4 *gamma*-chain anti-human interleukin 2 receptor), disulfide with human-mouse monoclonal clone 1H4 light chain, dimer [SY]
		SMART anti-TAC antibody [SY]
		Daclizumab [SY]
		IL-2R monoclonal antibody [SY]
		Ro 24-7375 [SY]
		Humanized anti-Tac monoclonal antibody [SY used in PDL press releases]
		152923-56-3 [CAS RN]
		NDC 0004-0501-09 [NDC]
Denileukin difitox-diphtheria conjugate	Ontak	*IL-2, rDNA/Seragen [BIO]*
		N-L-Methionyl-387-L-histidine-388-L-alanine-1-388-toxin (Corynebacterium diphtheriae strain C7) (388-2¥) protein with 2-133-interleukin 2 (human clone pTIL2-21a) [CAS]
		Interleukin-2 Fusion Protein [SY]
		Interleukin-2/diphtheria toxin fusion protein, recombinant [SY]
		LY335348 [SY]
		DAB389IL-2 [SY; used by Ligand]
		173146-27-5 [CAS RN]
		NDC 64365-503-01 [NDC]
Diphtheria and tetanus toxoids and acellular pertussis vaccine	Infanrix	*Haemophilus b Vaccine (PRP-T) [BIO]*
		ActHIB [TR]
		Haemophilus b conjugate vaccine (tetanus toxoid conjugate) [FDA]
		NDC 49281-545-05; NDC 49281-549-10 [NDC]
Diphtheria and tetanus toxoids and acellular pertussis vaccine	Infanrix	*Tetanus & diphtheria toxoids/PMC [BIO]*
		Tetanus and diphtheria toxoids adsorbed for adult use [FDA]
		Tetanus and diphtheria toxoids combined aluminum
		Phosphate precipiatated [FDA former]
		NDC 49821-800-10; -83 [NDC]
Diphtheria and tetanus toxoids and acellular pertussis vaccine	Infanrix	*Tetanus & diphtheria toxoids/Lederle [BIO]*
		Tetanus and diphtheria toxoids adsorbed purogenated for adult use [TR Purogenated is the reg. trademark]

Drug Name	Trade Name	Synonyms
		Tetanus and diphtheria toxoids adsorbed for adult use [FDA] Tetanus and diphtheria toxoids combined aluminum phosphate adsorbed [or alum precipitated] [FDA former] TD vaccine [SY] NDC 0005-1875-31; -47 [NDC]
Diphtheria and tetanus toxoids and acellular pertussis vaccine	Infanrix	*Tetanus & diphtheria toxoids/Mass. [BIO]* Tetanus and diphtheria toxoids adsorbed for adult use [FDA] Tetanus and diphtheria toxoids combined aluminum phosphate precipitated [FDA former] TD vaccine [SY]
Diphtheria and tetanus toxoids and acellular pertussis vaccine	Infanrix	*Tetanus & Diphtheria Toxoids/Wyeth [BIO]* Tetanus and diphtheria toxoids adsorbed for adult use [FDA] Tetanus and diphtheria toxoids combined alum precipitated (for adult use) [FDA former] TD Vaccine [SY]
Diphtheria and tetanus toxoids and acellular pertussis vaccine	Infanrix	*Tetanus Toxoid Adsorbed/PMC [BIO]* Tetanus toxoid adsorbed [FDA USAN] Tetanus toxoid adsorbed USP [USP] NDC 49281-800-83; -84 [NDC]
DNAse, Dornase alfa human recombinant deosyribonuclease I (rhDNAse), produced in CHO cells	Pulmozyme	*DNase, rDNA [BIO]* Dornase alfa [USAN BAN INN] Deoxyribonuclease (human clone 18-1 protein moiety) [CAS] Deoxyribonuclease I [SY] E.C. 3.1.4.5 [NUM EC] Alkaline deoxyribonuclease [SY] 143831-71-4; 9003-98-9 [CAS RN] NDC 50242-100-40 [NDC]
Epoetin alfa, human recombinant, produced in CHO cells	Epogen	*EPO, rDNA/Amgen [BIO]* Darbepoetin alfa [FDA NAM] Aranesp [FDA TR] Erythropoietin [CAS] EPO, rDNA/Amgen [BIO] Epoetin alfa [FDA CAS USAN INN BAN] 1-165-Erythropoietin (human clone lambda HEPOFL13 protein moiety) [CAS] 1-165-Erythropoietin (human clone lambdaHEPOFL13 protein moiety), glycoform alpha [CAS] Erythropoietin [SY] Erythropoietin, recombinant [SY] 113427-24-0 [CAS RN]

Drug Name	Trade Name	Synonyms
Epoetin alfa, human recombinant, produced in CHO cells	Procrit	*EPO, rDNA/Ortho [BIO]* Epoetin Alfa [FDA NAM CAS USAN INN BAN] 1-165-Erythropoietin (human clone lambda HEPOFL13 protein moiety) [NAM CAS 9CI] 1-165-Erythropoietin (human clone lambdaHEPOFL13 protein moiety), glycoform alpha [NAM CAS] Erythropoietin [NAM SY] Erythropoietin, recombinant [NAM SY] EPO [NAM SY] rHuEPO [NAM SY] Eprex [NAM TR used in most European and other foreign countries; formerly used in US] Espo [NAM TR in Japan] Erypo [NAM TR in Germany and Austria] 113427-24-0 [CAS RN] NDC 55513-126; -267; -148; -144; -823; -283; -478-10 [NDC]
Eptifibatide, reversible platelet aggregation inhibitor	Intergrilin	*Eptifibatide [BIO]* B Integrilin [SY] Interegrelin [SY] NDA 20-718; -718/S-002; -718/S-010; -718/S-013 [CAS RN] NDC 0085-1777-01 [NDC]
Etanercept, recombinant fusion protein containing TNF receptor linked to F_c domain of IgG1, produced in CHO cells	Enbrel	*TNF receptor-IgG Fc, rDNA [BIO]* Etanercept [FDA INN USAN] TNF-alpha receptor-immune globlulin G1 Fc fusion protein, Recombinant [SY] Cachectin receptor [SY for TNF receptor] p75-IgG Fc, rDNA [SY] Tumor necrosis factor receptor2-immune globlulin G1 Fc Fusion protein, recombinant [SY] rhuTNFR/Fc [SY] TNFR/Fc, rDNA [SY] NDC 58406-425-34 [NDC]
Fibrin sealant	TISSEEL VH Kit	*Fibrin sealant/Baxter [BIO]* Fibrin sealant, two-component fibrin sealant, vapor-heated kit [FDA] Fibrin sealant kit [SY] Thrombin Solution [FDA for a component] Sealer protein concentrate (human) [FDA for a component] Fibrinogen [FDA for a component] Fibrinolysis inhibitor solution (bovine) [FDA for a component]

Drug Name	Trade Name	Synonyms
		Thrombin (human) [FDA for a component] Duploject [TR for application device] Aprotinin [FDA for a component]
Fligrastim, G-CSF, human recombinant produced in *E. coli*, nonglycosylated	Neupogen	*G-CSF, rDNA/Amgen [BIO]* Filgrastim [USAN INN BAN MESH] Colony-stimulating factor (human clone 1034), *N*-L-methionyl-[CAS] 121181-53-1 [NUM CAS RN] none [NUM CAS according to Amgen MSDS] r-metHuG-CSF [SY] NSC 614629 [SY NCI] Granulocyte Colony Stimulating Factor [SY] Gran [TR in Japan] Granulokine [TR in Italy] Neulasta [FDA TR] PEG-filgrastim [FDA] NDC 55513-530; NDC 55513-546-10 [NDC]
Follitropin alpha, human follicle-stimulating hormone (hFSH), recombinant	Gonal F	*Follicle-stimulating hormone (FSH) products* Follitropin alfa [FDA NAM] follicle-stimulating hormone [SY] FSH [SY] FSH rDNA/Serono [BIO] Follitropin alfa [FDA INN] Follicle-stimulating hormone, glycoform alpha [CAS for alpha subunit] Follicle-stimulating hormone (human clone lambda 15B beta-subunit protein moiety reduced [CAS for beta subunit] Follicle-stimulating hormone, recombinant [SY] rhFSH [SY] FSH alpha [SY] Follitropin alpha [SY] 56832-30-5; 110909-60-9; 9002-68-0; 9002-67-9 [CAS RN]
Follitropin alpha, human follicle-stimulating hormone (hFSH), recombinant	Follistim	*Follicle-stimulating hormone (FSH) products/ Organon* Follitropin beta Follistim [TR in U.S.] Follitropin beta [FDA USAN] Puregon [TR in Europe] NDC 0052-0306-18; NDC 0052-0306-31 [NDC]
Ganirelix acetate, synthetic decapeptide with antagonistic activity against gonadotroplin releasing hormone (GnRH)	Antagon	*Ganirelix [BIO]* X Antagon [SY] Antagon Injection [SY] NDA 21-057 [CAS RN] NDC 0052-0301-01; -71[NDC]
Gemtuzumab-oxogamicin, anti CD33-calicheamincin	Mylotarg	*CD33 Immunotoxin, rDNA [BIO]* Gemtuzumab-ozogamicin [FDA]

Drug Name	Trade Name	Synonyms

conjugate, recombinant humanized IgG4		CMA-676 [SY]
		hP67.6 monoclonal antibody/*N*-acetyl-gamma
		    calicheamicin conjugate [SY]
		N-acetyl-gamma calicheamicin/hP67.6
		    monoclonal antibody conjugate [SY]
		Calicheamicin-CD33 monoclonal antibody
		    conjugate [SY]
		NDC 0008-4510-01 [NDC]
		CD33 monoclonal antibody/*N*-acetyl-gamma
		    calicheamicin conjugate [SY]
		Humanized mP67.6 Mab/*N*-acetyl-gamma
		    calicheamicin conjugate [SY]

Glucagon, recombinant	Glucagon for injection/ GlucaGen	*Glucagon, rDNA* Products, none
Glucagon, recombinant	Glucagon for injection/ GlucaGen	*Glucagon, rDNA/Lilly* NA
Glucagon, recombinant	Glucagon for injection/ GlucaGen	*Glucagon, rDNA/Novo [BIO]* GlucaGen [TR] Glucagon, recombinant for injection [FDA] Glucagon [CAS USAN INN BAN JAN] Glucagon USP [USP] 16941-32-5; 9007-92-5 [CAS RN] NDC 55390-004-1 [NDC]
Gonadotropin-releasing hormones	Zoladex	*GnRH/Zeneca [BIO]* Buserelin [SY] FSH/LH-RH [CAS] GnRH [BIO] Gonadorelin [CAS] Goserelin [FDA] Histrelin[SY] Nafarelin [SY] Goserelin acetate [CAS] Nafarelin acetate [BIO] Synarel [BIO]

Granulocyte/macrophage colony-stimulating factor, recombinant	Leukine	*GM-CSF, rDNA/Immunex [BIO]*
		Sargramostim [FDA USAN]
		Colony-stimulating factor 2 (human clone
		    pHG25 protein moiety), 23-L-leucine [CAS]
		123774-72-1 [CAS RN]
		83869-56-1 [CAS RN]
		GM-CSF [SY]
		Granulocyte-macrophage colony-stimulating
		    factor [SY]
		NSC 617589 [SY NCI]
		NSC 613795 [SY NCI]
		rhGM-CSF [SY]
		NDC 58406-002-01; -33; -050-30; -001-01 [NDC]

Drug Name	Trade Name	Synonyms
Haemophilus b conjugate vaccine (diphtheria CRM 197 protein conjugate)	HibTITER	*Haemophilus b vaccine (PRP-D) [BIO]* ProHIBiT [TR] Haemophilus b conjugate vaccine (diphtheria toxoid conjugate) [FDA] Hib-Vaccinol [NAM SY] PRP-D [SY] NDC 49281-0541-10; -61 [NDC]
Haemophilus b conjugate vaccine (diphtheria CRM 197 protein conjugate)	HibTITER	*Haemophilus b vaccine/Lederle [BIO]* Haemophilus b conjugate vaccine (diphtheria CRM197 protein conjugate) [FDA] 126161-67-9 [CAS RN] NDC 0005-0104-41; -0201-10 [NDC current] NDC 53124-0201-05; -10; -0104-41 [NDC former]
Haemophilus b conjugate vaccine (diphtheria CRM 197 protein conjugate)	HibTITER	*Haemophilus b vaccine/SKB [BIO]* ActHIB [TR] Haemophilus influenzae type b conjugate vaccine [FDA] Haemophilus b conjugate vaccine (tetanus toxoid conjugate) [SY] PRP-T [SY]
Haemophilus b conjugate vaccine (meningococcal protein conjugate)	PedvaxHIB	*Haemophilus b vaccine/Merck [BIO]* Haemophilus b conjugate vaccine (meningococcal protein conjugate) [FDA] PRP-OMP [SY] NDC 00006-4792-00; NDC 00006-4797-00; NDC 00006-4877-00; NDC 00006-4897-00 [NDC]
Hepatitis A vaccine, inactivated	Havrix	*Heptatitis A virus vaccine/SKB [BIO]* Heptatitis A virus vaccine, inactivated [SY] HAVsorbat SSW [TR foreign]
Hepatitis A, inactivated and hepatitis B (recombinant) vaccine	TWINRIX	*Hepatitis A virus vaccine/Merck [BIO]* VAQTA [TR] Hepatitis A virus vaccine inactivated [FDA]
Hepatitis B immune globulin (human), purified by chromatography procedure	Nabi-HB	*Hepatitis B immune globulin* BayHep B [FDA] Hep. B imm. glob./Nabi [BIO] Hepatitis B immune globulin (human) [FDA] Hepatitis B immune globulin (human) Solvent/ Detergent Treated and filtered [FDA full name on insert] H-BIG 5% SDF [TR used prior to launch] NDC 59730-4402-1; -4403-1; NDC 0026-0636-00; -01; -05 [NDC]
Hepatitis B vaccine (recombinant), produce in yeast *S. cerevisiae*	Recombivax	*Hepatitis B vaccine, rDNA/Merck [BIO]* Recombivax HB [TR] Hepatitis B vaccine (recombinant) [FDA]

Drug Name	Trade Name	Synonyms
		Hepatitis B virus (HBV) subtype *adw* surface antigen (rHBsAg)
		Nonglycosylated lipoprotein complex [SY]
		H-B-Vax [TR]
		Gen H-B-Vax [TR in Germany]
		Heptavax-II [TR in Japan]
		R-HV Vaccine [TR in Japan]
Hepatitis B vaccine (recombinant), produce in yeast *S. cerevisiae*	Engerix-B	*Hepatitis B vaccine, rDNA/SKB [BIO]*
		Hepatitis B vaccine (recombinant) [FDA]
		Tecnoquim [TR non-US]
		Hepatitis B virus surface antigen, recombinant [SY]
		HBsAg, rDNA [SY]
Hepatitis B vaccine inactivated	Hepatavax-B	*Hepatitis B vaccine, blood-derived [BIO]*
		Hepatitis B vaccine inactivated [FDA]
Herceptin; HER2 receptor monoclonal antibody, recombinant	Trastuzumab	*HER2 receptor Mab, rDNA [BIO]*
		Trastuzumab [FDA USAN]
		HER2 receptor monoclonal antibody, recombinant [SY]
		C-erbB2 monoclonal antibody [SY]
		MDX-210 [SY]
		520C9x22 [SY]
		Metavert [SY]
		Epidermal growth factor receptor-2 protein (HER2) monoclonal antibody [SY]
		NDC 50242-0134-60 [NDC]
Imiglucerase-recombinant (beta)-glucocerebrosidase	Cerezyme	*Imiglucerase [BIO]*
		Alglucerase [FDA]
		C Cerezyme [CAS RN]
		C Ceredase [CAN RN]
Immune globulin (human)	Venoglobulin-S	*Immune Globulin (IGIV)/Alpha [BIO]*
		Immune globulin intravenous (human), solvent detergent treated [FDA]
		Venoglobulin-I [TR upon original approval]
		Gamma globulin, intravenous [SY]
		Immune globulin, intravenous [SY]
		IVIG [SY]
		Intravenous immune globulin (IVIG) [SY]
		NDC 49669-1622-1; -1623-1; -1624-1 [NDC]
Immune globulin (human)	Polygam S/D	*Immune globulin (IGIV)/ARC [BIO]*
		Immune Globulin Intravenous (Human) [FDA]
		Gamma globulin, intravenous [SY]
		IGIV [SY]
		Intravenous immune globulin [SY]
		Immune globulin, intravenous [SY]
		Intravenous immune globulin (IVIG) [SY]

Drug Name	Trade Name	Synonyms
Immune globulin (human)	Gammar-PIV	*Immune globulin (IGIV)/Aventis [BIO]* Gammar-P I.V. [TR] Immune Globulin Intravenous (Human) [FDA] Intravenous immune globulin (IVIG) [SY] Gamma globulin [SY] IVIG [SY] Immune globulin, intravenous [SY] NDC 0053-7486-01; -02; -05; -06; -10 [NDC]
Immune globulin (human)	Gammagard S/D	*Immune globulin (IGIV)/Baxter [BIO]* Immune Globulin Intravenous (Human) [FDA] Gamma globulin, intravenous [SY] IVIG [SY] Immune globulin, intravenous [SY] Intravenous immune globulin (IVIG) [SY] Iveegam [TR former, for OIH/Immuno product]
Immune globulin (human)	Gamimune N	*Immune globulin (IGIV)/Bayer [BIO]* Immune Globulin Intravenous (Human) [FDA] Gamma globulin, intravenous [SY] Immune globulin, intravenous [SY] IVIG [SY] Intravenous immune globulin (IVIG) [SY] NDC 0026-0648-12; -20; -24; -71 [NDC]
Immune globulin (human) against Rho (D), purified IgG Cohn fraction II	BayRho-D full and mini dose	*Rho(D) Imm. Glob./Bayer [BIO]* BayRho-D full dose [TR] Rho(D) immune globulin (human) [FDA] Rho(D) immune globulin [USAN] Rho(D) immune human globulin [former USAN] Rho(D) immune globulin USP [USP] NDC 0026-0631-01; -10; -15; -22 [NDC]
Immune globulin (human), (IGIV)	Sandoglobulin	*Immune globulin (IGIV)/Novartis [BIO]* Immune globulin intravenous (human) [FDA] Gamma globulin [SY] Intravenous immune globulin (IVIG) [SY] Immune globulin, intravenous [SY] IGIV [SY] NDC 0078-0120-94; -0122-95; -0124-96; -01244-93 [NDC]
Immune globulin (human), (IGIV)	Panglobulin	*Immune Globulin (IGIV)/Swiss [BIO]* Immune globulin intravenous (human) [FDA] Gamma globulin [SY] Intravenous immune globulin (IVIG) [SY]
Immune globulin against D (RH1) (human), purified by cold alcohol fractionation	RhoGam Rho(D), MICRhoGAM Rho (D)	*Rho(D) Imm. Glob./Ortho [BIO]* RhoGAM ultra-filtered [TR] MICRhoGAM ultra-filtered [TR] Rho(D) immune globulin (human) [FDA]

Drug Name	Trade Name	Synonyms
		Rho(D) immune globulin [USAN]
		Rho(D) immune human globulin [former USAN]
		Rho(D) Immune Globulin USP [USP]
		NDC 0562-7807-05; -10; -25 [NDC]
Infliximab, anti tumor necrosis factor alpha TNF (alpha), chimeric human IgG1	Remicade	*TNF Mab, rDNA [BIO]*
		Infliximab [FDA]
		Tumor necrosis factor-alpha monoclonal antibody [SY]
		Chimeric IgG1 anti-human TNF MAb cA2 [SY]
		cA2 [SY]
		TNF-alpha MAb cA2 [SY]
		Avakine [TR former; dropped because suggested the product was a cytokine]
		CentTNF [TR former or non-US]
		57894-0030-01 [NDC]
Insulin, human recombinant (*E. coli*)	Humulin	*Insulin, rDNA/Lilly [BIO]*
		Humulin R [TR]
		Regular insulin human injection (rDNA origin) [FDA for Humulin R]
		Human insulin (rDNA origin) isophane suspension [FDA for Humulin N]
		Human insulin (rDNA origin) zinc suspension [FDA for Humulin L]
		Human insulin (rDNA origin) extended zinc suspension [FDA for Humulin U]
		70% Human insulin isophane suspension, 30% human insulin injection (rDNA origin) [FDA for Humulin 70/30]
		50% Human insuin isophane suspension, 50% human insulin injection (rDNA origin) [FDA for Humulin 50/50]
		Insulin biosynthetic human [FDA]
		9004-10-8 [CAS RN]
		Human insulin (rDNA origin) isophane suspension [SY for Humulin N]
		NPH [SY used by Lilly for Humulin N]
		Ultralente [SY used by Lilly for Humulin U (Ultralente is reg. to Novo Nordisk)]
		Lente [SY used by Lilly for Humulin L (Lente is reg. to Novo Nordisk)]
		Zinc-insulin crystals [SY]
		Huminsulin [TR in Germany, Switzerland]
		Humuline [TR in the Netherlands]
		Umuline [TR in France]
		Humulina [TR in Spain, Austria, Belgium]
Insulin, human recombinant (yeast, *S. cerevisiae*)	Novolin	*Insulin, rDNA/Novo [BIO]*
		Novolin R [TR in US]

Drug Name	Trade Name	Synonyms
		Novolin L [TR in US]
		Novolin Lente [TR in US]
		Velosulin BR [TR for buffered solution]
		Novolin N [TR in US]
		Novolin 70/30 [TR in US for 70%/30% mixture of Novolin N and Novolin L]
		Regular, human insulin injection (recombinant DNA origin) [FDA]
		Lente, human insulin zinc suspension (recombinant DNA origin) [FDA]
		Buffered regular human insulin injection (rDNA origin) [FDA]
		70% NPH, Lente, human insulin zinc suspension and 30% regular, human insulin injection (recombinant DNA origin) [FDA]
		NPH, Human insulin isophane suspension (recombinant DNA origin) [FDA]
		Insulin human [USAN INN BAN for uncomplexed protein]
		Human insulin (neutral protamine hagedorn) [SY for NPH]
		Human insulin (NPH) [SY]
		9004-10-8; 11061-68-0 [CAS RN]
Insulin, Lispro variant, human, recombinant, rDNA origin	Humalog	*Insulin Lispro rDNA [BIO]*
		Insulin lispro injection (rDNA origin) [FDA]
		Humalog Mix75/25 [TR]
		Insulin lispro protamine [FDA Orange Book for Humalog mix
		Insulin (human), 28b-L-lysine-29b-L-proline- [CAS; "b" is "b" is superscript]
		28b-L-Lysine-29b-L-proline insulin (human) [CAS; "b" is superscript]
		LY 275585 [SY]
		Lys(B28), Pro(B29) human insulin [SY]
		Insulin lispro protamine suspension [SY for NPL-type (NPH-like) component of Humalog Mix75/30]
		133107-64-9 [NUM CAS]
		NDC 0002-7510-01; -7515-59 [NDC]
Insulin-Glargine (recombinant); Asn (A21) > Gly; addition of 2 Arg to B20 and B21, more rapid onset	Lantus	*Insulin-glargine, rDNA [BIO]*
		Insulin glargine [rDNA origin] injection [FDA]
		Insuman [SY]
		HOE1 [SY]
		21.sup.A-Gly-30B.sup.a-L-13-Arg-30.sup.Bb-L-Arg-human
		Insulin [SY [SY (.sup. refers to following letter only)]
		Insulin, recombinant long-acting [SY]

Drug Name	Trade Name	Synonyms
		Gly(A21)-human insulin Arg(B31)-Arg(B32)-OH [SY MF] NDC 0088-2220-32; -33 [NDC]
Interferon alfa-2a	Roferon-A	*Interferon, alfa-2a rDNA [BIO]* Interferon alfa-2 [FDA] Interferon Alfa-2 [USAN INN BAN] Interferon Alfa-2 (genetical recombination) [JAN] Interferon alpha A (human leukocyte protein moiety reduced) [CAS; alpha is Greek letter] Interferon alpha-2a [SY] alpha-2a interferon [SY] Interferon alpha A [SY] NSC 377523 [NCI] rHuIFN-2a [SY] Ro 22-8181 [SY] Canferon [TR in Japan] alfa-2a interferon [SY] 76543-88-9 [NUM CAS] NDC 0004-2007-09; -2010-09; -2011-09; -2012-09; -2015-07; -2015-09; -2016-07; -2016-09; -2017-07; -2017-09 [NDC]
Interferon alfa-2a	Roferon-A	*Interferon alpha-2a, rDNA, PEG- [BIO]* Interferon alfa-2a, rDNA, PEG-PEGylated-40K interferon [SY used by Roche] Pegasys [TR] Interferon alpha-2a, recombinant, pegylated [SY]
Interferon alfacon-1, human recombinant, produced in *E. coli.*	Infergen	*Interferon alfacon-1, rDNA [BIO]* Interferon alfacon-1 [FDA USAN] Inferax [TR foreign] Consensus interferon [SY] R-metHuIFN-Con1 [SY] Wellferon (lymphoblastoid) [SY]
Interferon alfa-n3	Alferon N Injection	*Interferon alfa-n3 [BIO]* Interferon alfa-n3 (human leukocyte derived) [FDA] Interferon alfa-n3 [USAN] Leukocyte Interferon [former USAN] Interferons, alpha- [CAS; alpha is Greek letter] Interferon, leukocyte-derived [SY] Alpha-n3 interferon [SY] Cytopharm [TR in Germany] Altemol [TR in Mexico] 74899-72-2 [CAS RN]
Interferon beta-1a, human recombinant product produced in CHO cells	Avonex; Rebif	*Interferon beta-1a, rDNA [BIO]* Interferon beta-1a [FDA] Interferon beta-1a [USAN INN BAN]

Drug Name	Trade Name	Synonyms
		Interferon beta1 (human fibroblast protein reduced) [CAS] Interferon beta-1a [SY] Recombinant beta interferon [SY] rhIFN-beta1a [SY] BG-9015 [SY] Neoferon [TR foreign] 145258-61-3 [NUM CAS]
Interferon gamma-1b, human recombinant produced in *E. coli*	Actimmune	*Interferon Gamma-1b, rDNA [BIO]* Actimmune [TR formerly Genentech, now reg. to InterMune] Interferon gamma-1b [FDA] Interferon gamma-1b [USAN BAN INN] 1-139-Interferon gamma (human lymphocyte protein moiety reduced), N (sup2)-L-methionyl-[CAS; gamma is Greek letter] Immukine [TR in UK] Imuforgamma [TR in Germany] Imukin [TR in Spain, France, Denmark, Sweden, Austria] Antigen-induced interferon [SY] Type 2 interferon [SY] Mitogen-induced interferon [SY] pH2-labile interferon [SY] Acid-labile interferon [SY] 98059-61-1 [CAS RN]
Lepirudin-thrombin inhibitor, Hirudin-human recombinant produced in yeast	Refludan	*Hirudin, rDNA [BIO]* Lepirudin (rDNA) [FDA INN] Hirudin, 1-L-leucine-2-L-threonine-63-desulfo-(Hirudo medicinalis HV1) [CAS] HBW 023 [IC] Hoe-023 [IC] [Leu.sup.1-Thr.sup.2]-63-desulfohirudin [SY] 8001-27-2 [CAS RN] NDC 0088-2150-57 [NDC]
Leuprolide acetate, GnRH or LHRH analog	Lupron	*Leuprolide acetate [BIO]* Lupron depot [CAS] Lupron Depot-Ped [CAS] Lupron Depot-3 Month 11.25 mg [SY] Lupron Depot-3 Month 22.5 mg [SY] Lupron-3 Month SR Depot 22.5 mg [SY] Lupron Depot-4 Month 30 mg [SY] Viadur [CAS] Leuprorelin [CAS]
Lyme disease vaccine (Recombinant OspA), acylated	LYMErix	*Lyme Vaccine, rDNA/SKB [BIO]* Lyme disease vaccine (recombinant OspA) [FDA] Lyme vaccine, recombinant [SY]

Drug Name	Trade Name	Synonyms
		Borrelia burgdorfi outer surface protein A (OspA), recombinant [SY] L-OspA with aluminum hydroxide adjuvant [SY]
Menotropins, human chorionic gonadotropin (HCG)	Pergonal	*HCG, rDNA [BIO]* Repronex [SY] Humegon [SY] Ovidrel [TR] Choriogonadotropin [FDA] Ovidrelle [TR in Europe] Chorionic gonadotropin, recombinant [SY] Novarel (placental extracts) [BIO] Profasi (placental extracts) [BIO] Pregnyl (placental extracts) [BIO]
Muromonab-CD3, anti-CD3 antibody *IgG$_{2a}$*	Orthoclone OKT3	*CD3 Mab [BIO]* Muromonab-CD3 [FDA USAN INN JAN] Muromonab-CD3 Sterile Solution [FDA full name on insert/labeling] Murine monoclonal antibody for cluster of differentiation 3 antigen [SY] OKT-3 [SY] ant-CD3 [SY] T3 monoclonal antibody [SY] NDC 59676-101-01 [NDC]
Nofetumab, F$_{ab}$ fragment of a monoclonal NR-LU-10 IgG2b (murine) against small lung cancer associated antigen, Tc99 labeled	Verluma	*NR-LU-10 Mab Fab radioconj. [BIO]* Technetium [CAS] TC 99m [CAS] Verluma [TR (for kit)] Nofetumomab [FDA (for monoclonal antibody)] Kit for the preparation of Technicium Tc 99m Nofetumomab Merpentan [FDA] Technicium Tc 99m Nofetumomab Merpentan [FDA for final Radioimmune conjugate] Immunoglobulin G2b, anti-(human tumor) F$_{ab}$ fragment (mouse monoclonal NR-LU-10 gamma2b-chain), disulfide with mouse monoclonal NR-LU-10 kappa chain, oxo(N,N'-(1-(3-oxopropyl)-1, 2-ethandiyl)bis(2-mercaptoacetamido)) (4-*N,N′,S,S′*)technetate (1–), 99mTc conjugate [CAS for final radioimmune conjugate] Mofetumomab merpentan [SY for final radioimmune conjugate] NR-LU-10—99mTc [SY or final radioimmune conjugate] CD20 monoclonal antibody Fab fragment—Tc 99m [SY for final radioimmune conjugate]

Drug Name	Trade Name	Synonyms
NovoRapid; B28-Asp insulin; B28 asp regular human insulin, recombinant; Insulin Aspart [rDNA origin]	NovoLog	*Insulin-aspart, rDNA [BIO]* Insulin aspart [rDNA origin] [FDA] Insulin aspart [USAN INN] NovoRapid [TR in outside U.S.] Insulin aspart [rDNA origin] [FDA] IAsp [SY] Insulin X14 [SY] INA-X14 [SY] B28-Asp-insulin, [SY] B28 asp regular human insulin analog [SY] NDC 0169-7501-11; -3303-12 [NDC]
Octreotide	Sandostatin	*Octreotide/Novartis [BIO]* Sandostatin LAR Despot [SY] Octreotide [FDA] Octatpeptide [FDA] B Sandostatin [CAS] Sandostatin LAR Depot [CAS] NDC 0078-0180; -0181; -0182-03 [NDC]
Oprelvekin, human recombinant IL-11, a thrombopoietic growth factor that stimulate stem cells and megakaryocyte progenitor, produced in *E. coli*	Neumega	*IL-11, rDNA [BIO]* Oprelvekin [USAN FDA] 2-178 Interleukin-11 (human clone pXM/IL-11) [CAS] Interleukin-11 [MESH] Adipogenesis inhibitory factor [SY] AGIF [SY] IL-11 [SY] Recombinant human interleukin-11 [SY] rHuIL-11 [SY] rhIL-11 [SY] 145941-26-0 [CAS RN]
Oxytocin	Pitocin	*Oxytocin [BIO]* Syntocinon [Outside US] NDA 18-261/S-021; -261/S-020; -261/S-019 [CAS RN] NDC 00000-1157-310 [NDC]
Pegaspargase, L-asparaginase conjugated to PEG (5000), *E. coli* -derived enzyme	Oncaspar	*Asparaginase, PEG- [BIO]* Pegaspargase [FDA USAN INN] Asparaginase, reaction product with succinic anhydride, esters with polyethylene glycol, monomethyl ether [CAS] Monomethoxypolyethylene glycol succinimidyl L-asparaginase [CAS] Polyethyleneglycol-L-glutaminase-L-asparaginase [SY] PEG-L-glutaminase-L-asparaginase [SY] PEG-Asparaginase [SY] PEG-ASNase [SY] pegylated L-asparaginase [SY]

Drug Name	Trade Name	Synonyms
		PEG-L-asparaginase [SY] 130167-69-0 [CAS RN] NDC 0075-0640-05 [NDC]
Pneumococcal 7-valent conjugate vaccine (diphtheria CRM197 protein)	Prevnar	*Pneumococcal Vaccine-CRM197 [BIO]* Prevenar [TR former; used prior to approval] Pneumococcal 7-valent conjugate vaccine (diphtheria CRM197 protein) [FDA] PncCRM [SY] PCV7 [SY] NDC 0005-1970-67 [NDC]
Pneumococcal vaccine, polyvalent, 23 types of *S. pneumoniae*	PNU-IMUNE 23	*Pneumococcal Vaccine/Lederle [BIO]* PNU-IMUME 23 [TR (PNU-IMUME is reg. trademark)] Pneumococcal vaccine polyvalent [FDA] Pneumococcal vaccine, polyvalent [FDA former] Streptococcus pneumonia capsular polysaccharide vaccine [SY] NDC 0005-2309-31; -33; NDC 0006-4739; -4743; -4894-00 [NDC]
Pneumococcal vaccine, polyvalent, 23 types of *S. pneumoniae*	PNEUMOVAX 23	*Pneumococcal vaccine/Merck [BIO]* Pneumococcal vaccine polyvalent [FDA] Pneumococcal vaccine, polyvalent [FDA former] Streptococcus pneumonia capsular polysaccharide vaccine [SY] PNEUMOVAX II [TR foreign] PULMOVAX [TR foreign] NDC 0006-4739-00; NDC 0006-4894-00; NDC 0006-4943-00 [NDC]
Proiterlin, thrutropin-releasing hormone (TRH), synthetic tri-peptide, 5-oxo- (pro-his-pro)-amide	Thyrel TRH	*TRH [BIO]* Thyrotrophin-releasing peptide [CAS] Thyrotrophin-releasing hormone [CAS] Relefact TRH [CAS] NDC 55566-0081-5 [NDC]
Proteinase inhibitor (human), (alpha1-PI), alpha-1 antitrypsin	Prolastin	*Proteinase inhibitor [BIO]* Alpha-1-proteinase inhibitor (human) [FDA] Apha1-antitrypsin [CAS] 9041-92-3 [CAS RN] alpha1-trypsin inhibitor [SY] AAT [SY] A1AT [SY] A1PI [SY] alpha1-PI [SY] Prolastina [TR in Spain] CuttActive [TR non-US] NDC 0026-0601-30; -35 [NDC]
Rabies immune globulin (human)	BayRab	*Rabies immune globulin/Aventis [BIO]* IMOGAM RABIES—HT [TR]

Drug Name	Trade Name	Synonyms
		Rabies immune globulin (Human) [FDA] Rabies immune globulin [USAN] Antirabies serum [SY] RIG [SY] Human diploid cell rabies vaccine [SY] 50361-0180-02; NDC 50361-0180-10 [NDC]
Rabies immune globulin (human)	BayRab	*Rabies immune globulin/Bayer [BIO]* Rabies immune globulin (human) [FDA] Rabies immune globulin [USAN] Antirabies serum [SY] RIG [TR] Hyperab [TR for product prior to BayRab] Rabies vaccine/PMC Canada NDC 00026-0618-02; -10 [NDC]
Rabies immune globulin (human)	RABIE-VAX	*Rabies vaccine/PMC France [BIO]* Rabies vaccine [FDA USAN] Rabies vaccine, inactivated [FDA used on some approval documents] Rabies vaccine USP [USP], rabies vaccine [FDA USAN] Rabies vaccine, inactivated [FDA used on some approval documents] Rabies vaccine USP [USP]
Rabies vaccine	Imovax	*Rabies vaccine/PMC Canada [BIO]* Rabies vaccine/PMC France IMOVAX rabies [TR] Rabies vaccine [FDA] Rabies vaccine, human diploid cell (HDCV) [FDA former] Wistar rabies virus strain PM-1503-3m grown in human diploid cell cultures [SY] Human diploid cell rabies vaccine (HDCV) [SY] HDCV vaccine [SY] NDC 49281-250-10; -251-20 [NDC]
Rabies vaccine	RVA	*Rabies vaccine/BioPort [BIO]* Rabies vaccine adsorbed [FDA]
Rabies vaccine	Rab Avert	*Rabies vaccine/Chiron [BIO]* Rabies vaccine [SY] Purified chick embryo cell vaccine (PCEC) [SY] PCEC [SY] RabiPur [TR in Europe]
Respiratory syncytial virus (RSV) monoclonal antibody, recombinant	Synagis	*RSV Mab, rDNA [BIO]* Palivizumab [FDA USAN INN] Respiratory syncytial virus (RSV) monoclonal antibody [SY]

Drug Name	Trade Name	Synonyms
		MEDI-493 [IC] NDC 60574-4111-01; -2101-02 [NDC]
Reteplase, a nonglycosylated genetically modified variant of tPA	Retavase	*tPA, rDNA/Centocor [BIO]* Reteplase [FDA INN] 173-527-plasminogen activator (human tissue-type protein moiety reduced), 173-L-serine-174-L-tyrosine-175-L-glutamine- [CAS] Tissue plasminogen activator, recombinant [SY] BM-06022 [SY] Fibrinokinase [SY] Recormon [TR used by Boehringer Mannheim in Europe] Repilysin [TR used by Roche in Europe] 133652-38-7 [CAS RN] NDC 57894-040-01; -02 [NDC]
Rituximab, chimeric murine/human IgG1 monoclonal against CD20 (B cell surface antigen), produced in CHO cells	Rituxan	*CD20 Mab, rDNA [BIO]* Rituximab [FDA USAN INN] Anti-CD20 monoclonal antibody [SY] IDEC-C2B8 [SY] Pan-B antibodies [SY] CD20 monoclonal antibody, recombinant [SY] MabThera [TR used by Roche outside of US and Japan]
Satumomab pendetide, monoclonal antibody against TAG-72	OncoScrint	*TAG-72 Mab, conc. [BIO]* Satumomab concentrate (for further manufacturing use) [FDA] Satumomab [USAN INN BAN] B72.3 [SY] CYT-099 [SY] TAG-72 monoclonal antibody [SY] Anti-TAG-72 monoclonal antibody [SY] Tumor associated glycoprotein-72 Mab, conc. [SY] 144058-40-2 [CAS RN]
Somatotropin, human recombinant	Genotropin	*Somatropin, rDNA/P&U [BIO]* Somatropin [rDNA origin] for injection [FDA] Human growth hormone, recombinant [SY] hGH [SY] Somatomammotropin [SY] Crescormon [TR foreign] rhGH [SY] 12629-01-5 [CAS RN] NDC 0013-2649-02; -2654-02; -2650-02; -2655-02; -2651-02;-2656-02; -2652-02; -2657-02; -2653-02; -2658-02; -2626-94; -2626-81; -2616-94; -2616-81; -2646-94; -2646-81 [NDC]
Somatotropin, human recombinant (*E. coli*)	Humatrope	*Somatropin, rDNA/Lilly [BIO]* HumatroPen [TR for cartridge-loaded pen injector]

Drug Name	Trade Name	Synonyms
		Somatropin (rDNA origin) [FDA for active ingredient]
		Somatropin (rDNA origin) for injection vials [FDA for vials]
		Somatropin (rDNA origin) for injection cartridges for use with the Humatropen injection device [FDA for cartridges]
		Somatropin [USAN INN]
		Growth hormone (human) [SY]
		Human growth hormone, recombinant [SY]
		hGH [SY]
		LY137998 [SY]
		Crescormon [SY]
		Somatropina [SY Spanish]
		Somatropine [SY French]
		Somatropinum [SY Latin]
		somatomammotropin [SY]
		rhGH [SY]
		Umatrope [TR in France]
		EINECS 235-735-8 [EINECS]
		12629-01-5 [CAS RN]
		NDC 0002-7335-16; -8089-01; -8090-01; -8091-01 [NDC]
Somatotropin, human recombinant (*E. coli*)	Norditropin	*Somatropin, rDNA/Novo [BIO]*
		Norditropin SimpleXx [TR (for liquid formulation)]
		Dial-a-dose somatropin delivery devices [TR]
		Nordipen [TR for delivey device]
		Somatropin (rDNA origin) for injection [FDA]
		Human growth hormone, recombinant [SY]
		Somatotropin [SY]
		hGH [SY]
		Somatomammotropin [SY]
		Norditropine [TR in France]
		Norditropin S-chu [TR in Japan]
		Nordipen [TR for injector]
		NDC 0169-7774-11; -12 [NDC]
Somatrem-recombinant human pituitary growth hormone produced in *E. coli*	Protropin	*Somatropin, rDNA met- [BIO]*
		Somatrem [FDA USAN INN BAN JAN]
		Somatropin (human), N-L-methionyl- [CAS]
		Growth hormone, N-L-methionyl- [CAS]
		Growth hormone (human), N-L-methionyl- [CAS]
		Somatotropin, recombinant, *E. coli* expressed [SY]
		human growth hormone, recombinant [SY]
		hGH [SY]
		met-hGH [SY]
		Somatomammotropin [SY]

Drug Name	Trade Name	Synonyms
		82030-87-3 [CAS RN] NDC 50242-016-20 [NDC]
Somatropin	Nutropin AQ	*Somatropin (human) [FDA CAS]* Somatropin [USAN INN BAN] Human growth Hormone [JAN] Growth hormone (human) [CAS] Somatotropin, recombinant [SY] hGH, rDNA [SY] rhGH [SY] Somatomammotropin [SY]
Somatropin	Nutropin AQ	*Somatropin, rDNA, depot [BIO]* Nutropin Depot [TR] Somatropin (rDNA origin) for injectable suspension [FDA] hGH [SY] Human growth hormone [SY] NDC 50242-032-3; -034-41; -036-54 [NDC]
Somatropin	Nutropin AQ	*Somatropin, rDNA/Genentech [BIO]* Nutropin [TR] Somatropin (rDNA origin) for injection [FDA] Somatrem [SY] Human growth hormone, recombinant [SY] Somatomammotropin [SY] Somatonorm [TR in UK, Sweden, Switzerland, Japan, and other countries] hGH [SY] 12629-01-5 [CAS RN]
Somatropin	Saizen	*Somatropin, rDNA/Serono [BIO]* Saizen [TR for hGH deficiency] Serostim [for AIDS-related cachexia] Cool.click [TR for injector] Somatropin (rDNA origin) for injection [FDA] Somatropin (human) [CAS] Human growth hormone, recombinant [SY] rhGH [SY] hGH [SY] 12629-01-5 [CAS RN] NDC 44087-1005-2; [NDC for Saizen] NDC 44087-0004-7; -0005-7; -0006-7 [NDC for Serostim]
Streptokinase, purified bacterial protein product, plasminogen activator	Streptase	*Streptokinase/Pharmacia [BIO]* Streptokinase [NAM FDA] Anistreplase [NAM SY] Kabikinase [NAM TR] 81669-57-0; 9002-01-1 [CAS RN] NDC 00016-8025-2; -8026-2; -8027-2 [NDC]

Drug Name	Trade Name	Synonyms
Tenecteplase, human recombinant tissue plasminogen activator	TNKase; Activase	*tPA, rDNA/Genentech [NAM BIOPHARMA]* Alteplase, recombinant [NAM FDA] Alteplase USP [NAM USAN] Plasminogen activator (human tissue type 1 chain form protein moiety) [CAS] Tissue plasminogen activator type I (recombinant human) [SY] tPA [SY] rt-PA [NAM SY] t-PA [SY] Fibrinokinase [SY] Actilyse [TR in Europe, various Asian/Pacific Rim countries] Actiplas [NAM TR in Italy] Tenecteplase [FDA] Tissue plasminogen activator, TNK-, recombinant [SY] 105857-23-6 [CAS RN] NDC 50242-038-61; -004-13; -040-02 [NDC]
Tetanus immune globulin isolated from solublized Cohn faction II	BayTet	*Tetanus immune globulin/Bayer [BIO]* Tetanus immune globulin (human) solvent/ detergent treated [FDA full name on product insert] Tetanus immune globulin (human) [USAN] Tetanus immune human globulin [former USAN] Hyper-Tet [SY] NDC 0026-0634-01; -70; -86 [NDC]
Thymoglobulin. antithymocyte globulin (rabbit), purified rabbit IgG against human thymocytes	Sangstat	*Lymphocyte immune globulin [BIO]* Thymoglobulin [TR former PMC trademark; retained by SangStat] Lymphocyte immune globulin, anti-thymocyte globulin (rabbit) [FDA used in CBER Prod. and Est. listings] Anti-thymocyte globulin (rabbit) [FDA used in approval letter] Thymoglobuline [SY] Polyclonal rabbit anti-thymocyte globulin [SY] Rabbit thymocyte immune globulin [SY] NDC 62053-534-25 [NDC]
Thyrotopin alfa, thyroid stimulating hormone (TSH), human recombinant, produced in CHO cells	Thyrogen	*TSH, rDNA [BIO]* Thymopentin alfa for injection [FDA] Thyroid-stimulating hormone, recombinant [SY] rTSH [SY] TSH [SY] NDC 58468-1849-4 [NDC]
Thyrotopin alfa, thyroid stimulating hormone (TSH),	TYPHIM Vi	*Typhoid vaccine/PMC [BIO]* Typhoid Vi polysaccharide vaccine [FDA]

Drug Name	Trade Name	Synonyms
human recombinant, produced in CHO cells		ViCPS [SY] Salmonella typhi vaccine [SY] NDC 49281-0790-01; -20; -50 [NDC]
Thyrotopin alfa, thyroid stimulating hormone (TSH), human recombinant, produced in CHO cells	Vivotif Berna	*Typhoid vaccine/Swiss [BIO]* Typhoid vaccine live oral Ty21a [FDA] Salmonella typhi vaccine [SY] Vaccinum Febris Typhoidis Vivum Perorale (Stirpe Ty 21a) [SY Latin] Ty21a typhoid vaccine [SY] Typhoral [TR in Germany] NDC 58337-0003-01 [NDC]
Thyrotopin alfa, thyroid stimulating hormone (TSH), human recombinant, produced in CHO cells		*Typhoid vaccine/Wyeth [BIO]* Typhoid vaccine [FDA USAN] Typhoid vaccine USP [USP] Salmonella typhi vaccine [SY] NDC 0008-0343-01; -02 [NDC]
Varicella virus vaccine, Oka/ Merck strain of live, attenuated varicella virus	Varivax	*Varicella virus vaccine [BIO]* Varicella-Zoster virus (VZV) VARIVAX [TR] Varicella virus vaccine live (Oka/Merck) [FDA] Chickenpox vaccine [SY] NDC 0006-4826-00; -4827-00 [NDC]

Abbreviations: [BIO], biological based name commonly known to scientist; [TR], trade names that often are also trademarks; [FDA], other names used by FDA for tracking the BLA/PLA application of the unique product; [USAN], the nonproprietary name coined by the United States Adopted Names (USAN) council (see Appendix III for details on USAN naming convention); [CAS], the names assigned by Chemical Abstracts Services (CAS); [CAS RN], CAS registry numbers; [SY], synonyms and other common names; [USP], the names described in the United States Pharmacopia; [EC], EC numbers are the identity assigned by Enzyme Commission based on the specific chemical reactions catalyzed by unique enzymes, [INN], International Nonproprietary Names; [NCI], National Cancer Institute; [EINECS], the names listed in the European Inventory of Existing Commercial Chemical Substances; [BAN], the identification listed in the British Approved Name; [IC], the code name assigned by the respective pharmaceutical companies, often used in drug development; [NDC], National Drug Codes.

Appendix V

OTHER INFORMATION TABLES

Biotechnology and Biopharmaceuticals, by Rodney J. Y. Ho and Milo Gibaldi
ISBN 0-471-20690-3 Copyright © 2003 by John Wiley & Sons, Inc.

■TABLE V.1. Degenerative base triplet on ribonucleotides that codes for specific amino acids

		U	C	A	G	
U		UUU Phe	UCU Ser	UAU Tyr	UGU Cys	H
		UUC Phe	UCC Ser	UAC Tyr	UGC Cys	C
		UUA Leu	UCA Ser	UAA Stop (*och*)	UGA Stop (*opal*)	A
		UUG Leu	UCG Ser	UAG Stop (*amb*)	UGG Trp	G
C		CUU Leu	CCU Pro	CAU His	CGU Arg	U
		CUC Leu	CCC Pro	CAC His	CGC Arg	C
		CUA Leu	CCA Pro	CAA Gln	CGA Arg	A
		CUG Leu	CCG Pro	CAG Gln	CGG Arg	G
A		AUU Ile	ACU Thr	AAU Asn	AGU Ser	U
		AUC Ile	ACC Thr	AAC Asn	AGC Ser	C
		AUA Ile	ACA Thr	AAA Lys	AGA Arg	A
		AUG* Met	ACG Thr	AAG Lys	AGG Arg	G
G		GUU Val	GCU Ala	GAU Asp	GGU Gly	U
		GUC Val	GCC Ala	GAC Asp	GGC Gly	C
		GUA Val	GCA Ala	GAA Glu	GGA Gly	A
		GUG* Val (Met)	GCG Ala	GAG Glu	GGG Gly	G

Note: Bases are given as ribonucleotides. *GUG usually codes for valine, but it can code for methionine to initiate an mRNA chain translation.

Abbreviations used for purine and pyrimidine bases on ribonucleotides: U, uridine; C, cytosine; A, adenine; G, guanine. Stop (*och*) refers to the ochre termination triplet, Stop (*amb*) refers to the amber termination triplet, and Stop (*opal*) refers to the opal termination triplet.

■TABLE V.2. Common amino acids and their abbreviations

Name	3-Letter Code	1-Letter Code	Linear Formula	Structure
Alanine	**ala**	**A**	$CH_3-CH(NH_2)-COOH$	
Arginine	**arg**	**R**	$NH=C(NH_2)-NH-(CH_2)_3-CH(NH_2)-COOH$	
Asparagine	**asn**	**N**	$H_2N-CO-CH_2-CH(NH_2)-COOH$	
Aspartic acid	**asp**	**D**	$HOOC-CH_2-CH(NH_2)-COOH$	
Cysteine	**cys**	**C**	$HS-CH_2-CH(NH_2)-COOH$	
Glutamine	**gln**	**Q**	$H_2N-CO-(CH_2)_2-CH(NH_2)-COOH$	
Glutamic acid	**glu**	**E**	$HOOC-(CH_2)_2-CH(NH_2)-COOH$	
Glycine	**gly**	**G**	NH_2-CH_2-COOH	

■TABLE V.2. *Continued*

Name	3-Letter Code	1-Letter Code	Linear Formula	Structure
Histidine	**his**	**H**	$HN-CH=N-CH-CH-CH_2-CH(NH_2)-COOH$	
Isoleucine	**ile**	**I**	$CH_3-CH_2-CH(CH_3)-CH(NH_2)-COOH$	
Leucine	**leu**	**L**	$(CH_3)_2-CH-CH_2-CH(NH_2)-COOH$	
Lysine	**lys**	**K**	$H_2N-(CH_2)_4-CH(NH_2)-COOH$	
Methionine	**met**	**M**	$CH_3-S-(CH_2)_2-CH(NH_2)-COOH$	
Phenylalanine	**phe**	**F**	$Ph-CH_2-CH(NH_2)-COOH$	
Proline	**pro**	**P**	$NH-(CH_2)_3-CH-COOH$	

■**TABLE V.2.** *Continued*

Name	3-Letter Code	1-Letter Code	Linear Formula	Structure
Serine	**ser**	**S**	$HO-CH_2-CH(NH_2)-COOH$	
Threonine	**thr**	**T**	$CH_3-CH(OH)-CH(NH_2)-COOH$	
Tryptophan	**trp**	**W**	$Ph-NH-CH-C-CH_2-CH(NH_2)-COOH$	
Tyrosine	**tyr**	**Y**	$HO-p-Ph-CH_2-CH(NH_2)-COOH$	
Valine	**val**	**V**	$(CH_3)_2-CH-CH(NH_2)-COOH$	

■TABLE V.3. Abbreviations

Abbreviation	Explanation
α_2PI	α_2 plasminogen inhibitor
AAV	Adeno-associated virus
ACE	Angiotensin converting enzyme
ACE D	ACE deletion
AcP	Muscle acylphosphatase
ACR	American College of Rheumatology
ACTH	Adrenocortitropic hormone
ADA	Adenosine deaminase
ADCC	Antibody-dependent cell-mediated cytotoxicity
ADH	Antidiuretic hormone
ADH	Alcohol dehydrogenase
ADME	Absorption, distribution, metabolism, and elimination
aFGF	Acidic fibroblast growth factor
ALL	Acute lymphoblastic leukemia
ALS	Amyotropic lateral sclerosis
ALT	Alanine aminotransferase
alum	Aluminum salts
AMI	Acute myocardial infarction
AML	Acute myeloid leukemia
Anti-Hbe	Antibody to HBeAg
APACHE	Acute physiology and chronic health evaluation
Apo-E	Apolipoprotein E
APS	Antigen-presenting cells
AUC	Area under the curve
AUC_{po}	Area under the curve, oral dosing
BCG	Bacille Calmette-Guérin
B-CLL	B-cell chronic lymphocytic leukemia
Bcr	Breakpoint cluster region
bFGF	Basic fibroblast growth factor
BGG	Bovine g-globulin
BLA	Biologic License Application
BLE	Biologic License Establishment
BMT	Bone marrow transplantation
BO	Bilirubin oxidase
BPD	Bronchopulmonary dysplasia
BPH	Benign prostatic hypertrophy
BPTI	Bovine pancreatic trypsin inhibitor
BSA	Bovine serum albumin
BsAb	Bispecific antibodies
CAR	Coxsakievirus and adenovirus receptor
CBER	Center for Biologics Evaluation and Research
CDER	Center for Drug Evaluation and Research
cDNA	Complementary DNA
CDS	Chemical-based delivery system
CEA	Carcino-embryonic-antigen
cGMP	Current good manufacturing process (practice)
CHAPS	3-[(3-cholamidepropyl)-dimethylammonio]-1-propanesulfate
CHO	Chinese hamster ovarian cells

■**TABLE V.3.** *Continued*

Abbreviation	Explanation
CK	Creatine kinase
CL_H	Clearance, hepatic
CL_R	Clearance, renal
CLL	Chronic lymphocytic leukemia
Cmax	Maximum serum concentration
cmc	Carboxymethyl cellulose
CMC	Chemistry, manufacturing, and controls
CML	Chronic myeloid leukemia
CMV	Cytomegalovirus
CNS	Central nervous system
CPP	Central precocious puberty
CR	Complete remission
CSF	Cerebrospinal fluid
CSFs	Colony-stimulating factors
C_{SS}	Steady state plasma concentration, IV dosing
CT	Computerized tomography
CTCL	Cutaneous T-cell lymphoma
CTL	Cytotoxic T-lymphocytes
CYP	Cytochrome P450
CYP2C19	Cytochrome P450 2C19
CYP3A4	Cytochrome P450 3A4
Cyt c	Cytochrome c
dATP	Deoxyadenosine triphosphate
DDAVP	Desmopressin acetate
DHAP	Alkyl-dihydroxyacetone phosphate syntheseis
DHT	Dihydrotestosterone
DMARDs	Disease-modifying antirheumatic drugs
DMF	Dimethylformamide
DMSO	Dimethy sulfoxide
DNA	Deoxyribonucleic acid
DNase	Deoxyribonuclease I
DNAse	Dornase
DSC	Differential scanning calorimetry
DsRNA	Double stranded RNA
EC_{50}	Median inhibitory concentration against target cells
ED_{50}	Median effective dose
EDSS	Expanded Disability Status Scale
EGF	Epidermal growth factor
ELA	Establishment License Application
ELISA	Enzyme linked immunosorbent assays
env	Envelope glycoprotein
EPO	Erythropoietin
Epoetin	Erythropoietin
ESTs	Expressed sequence tags
EU	Endotoxin units
F_{ab}	Binding fragment of IgG antibody
FALDH	Fatty aldehyde dehydrogenase and fatty alcohol NAD-oxidoreducease phosphate oxidoreductase

■TABLE V.3. *Continued*

Abbreviation	Explanation
F_c	IgG antibody fragment that is easily crystallized
F_cRN	Brambell receptor
FDA	Food and Drug Administration
f*e*	Fraction excreted unchanged
FEV1	Forced expiratory volume in 1 second
FGF	Fibroblast growth factor
FIF	Fibroblast-derived IFN
FSH	Follicle stimulating hormone
G6PDH	Glucose-6-phosphate dehydrogenase
GAS	Gamma-activated sites
G-CSF	Granulocyte colony-stimulating factor
GDH	Glutamate dehydrogenase
GFP	Green fluorescent protein
GFR	Glomerular filtration rate
GH	Somatotropin or growth hormone
GHD	Growth hormone-deficiency
GHRH	Growth hormone release hormone
GI	Gastrointestinal
GM-CSF	Granulocyte-macrophage colony stimulating factor
GMPs	Good Manufacturing Procedures
GnRH	Gonadotropin-releasing hormone
GPCRs	G-protein coupled receptors
GUSTO	Global Utilization of Streptokinase and Tissue plasminogen activator for Occluded coronary arteries
GVHD	Graft-versus-host disease
H_2O_2	Hydrogen peroxide
HAMA	Human against mouse antibody (anti-murine antibody) reactions
HbA1c	Glycosylated hemoglobin
HbeAg	Hepatitis B *e* antigen
HbsAg	Hepatitis B surface antigen
HBV	Hepatitis B virus
HCC	Hepatocellular carcinoma
HCG	Human chorionic gonadotropin
hMG	Human menopausal gonadotropin
hCG	Human chorionic gonadotropin
HCL	Hairy cell leukemia
HCV	Hepatitis C virus
HEC	Hydroxyethyl cellulose
HERS	Heart and Estrogen/Progestin Replacement Study
hFSH	Human follicle-stimulating hormone
hGH	Human growth hormone
HGS	Human genome sciences
HHS	Health and human services
hIGF-I	Human insulin-like growth factor I
HIV	Human immunodeficiency virus
HLL	*Humicola lanuginosa* lipase
HP-β-CD	Hydroxypropyl-β-cyclodextrin
hPAH	Human phenylalanine hydroxylase
HPLC	High-pressure liquid chromatography

■TABLE V.3. *Continued*

Abbreviation	Explanation
HPMC	Hydroxypropyl methylcellulose
HRT	Hormone replacement therapy
HSA	Human serum albumin
HSV	Herpes simplex virus
HSV-2	Genital herpes
HuIFN-γ	Human interferon-gamma
I/D	Insertion/deletion
IC_{50}	Median inhibitory concentration against normal cells
ICH	International Conference on Harmonization
IFN-α	Interferon-alpha (or alfa)
IFN-β	Interferon-beta
IFN-γ	Interferon-gamma
IgA	Immunoglobulin A
IGF1	Insulin-like growth factor-1
IgG	Immunoglobulin G
IL	Interleukin
IL-1α	Interleukin-1 alpha
IL-1β	Interleukin-1 beta
IL-1Ra	Interleukin-1 receptor antagonist
IL-2	Interleukin-2
IND	Investigational new drug
IPV	Inactivated polio vaccine
IR	Infrared spectroscopy
IRA	Infarct-related artery
IRB	Institutional review board
ISGs	IFN-stimulated genes
ISRE	IFN-stimulated response element
ITRs	Inverted terminal repeat sequences
IU	International units
IV	Intravenous
k_{α}	Absorption rate
k_e	Elimination rate constant
KLH	Keyhole limpet hemocyanin
LAK	Lymphokine activated killer
LBM	Lean body mass
LD_{50}	Median lethal dose
LDH	Lactate dehydrogenase
LH	Luteinizing hormone
LINEs	Long interspersed repetitive elements
LTRs	Long terminal repeats
LPS	Lipopolysaccharides
LMW-UK	Low molecular weight urokinase
MAP	Mitogen activated protein
Mb	Myoglobin
M-CSF	Macrophage colony-stimulating factor
MDH	Maleate dehydrogenase
MDR1	Multidrug resistance protein
Met-hGH	Methionyl human growth hormone
MGDF	Megakarocyte growth and development factor

■TABLE V.3. *Continued*

Abbreviation	Explanation
mg/kg	Milligram per killogram
MHC	Major histocompatibility complex
MI	Myocardial infarction
MMPs	Matrix metalloproteinases
MIU	Million international units
MMR	Measles/mumps/rubella
MPIF-1	Myeloid progenitor inhibitor factor-1
mRNA	Messenger RNA
mRNA	Messenger ribonucleic acid
MRI	Magnetic resonance imaging
MRPs	Multidrug resistant-associated protein isoforms
MRT	Mean residence time
MS	Multiple sclerosis
MS	Mass spectroscopy
MTHFR	MethylenetetACErahydrofolate reductase
MTX	Methotrexate
MU	Million units
MW	Molecular weight
NADPH	Nicotinamide adenine dinucleotide (reduced form)
NAT	N-acetyltransferase
NCE	New chemical entitiy
NDA	New drug application
NDC	New drug candidate
NESP	Novel erythropoiesis stimulating protein
NHL	Non-Hodgkin's lymphoma
NIEHS	National Institute of Environmental Health Sciences
NIH	National Institutes of Health
NK	Natural killer
NME	New molecular entity
NMR	Nuclear magnetic resonance spectroscopy
NPH Insulin	Neutral Protamine Hagedorn Insulin equivalent to isophane insulin
NSAIDs	Nonsteroidal anti-inflammatory drugs
ORR	Overall response rate
OTA	Office of Technology Assessment
PAI-1	Plasminogen activator inhibitor-1
PAL	Phenylalanine ammonia-lyase
PCR	Polymerase chain reactions
PDUFA	Prescription Drug User Fee Act
PE	Pseudomonas exotoxin
PEG-Intron	Powder for injection
PEG	Polyethylene glycol
PEG-modified ADA	Pegylated adenosine deaminase
PHS	Public Health Service
PFK	Phosphofructokinase
Pgp	P-glycoprotein
PKR	Phosphokinase receptor
PKU	Phenylketonuria
PLA	Product License Application
PLG	Polylactide-co-glycolide

■**TABLE V.3.** *Continued*

Abbreviation	Explanation
Pol	Polymerase glycoprotein
PPIs	Proton pump inhibitors
PTCA	Percutaneous transluminal coronary angioplasty
PTH resistance	Pseudohypoparathyroidism
PTH	Parathyroid hormone
PVA	Polyvinylalcohol
PVP	Polyvinylpyrrolidone
PWS	Prader-Willi syndrome
QA/QC	Quality assurance/quality control
R	Regular
Rb	Retinoblastoma
R&D	Research and development
RA	Rheumatoid arthritis
rbSt	Recombinant bovine somatotropin
rconIFN	Recombinant consensus α-interferon
Rebif	Interferon beta-1a
RES	Reticuloendothelial system cells
rFIX	Recombinant factor IX
rFSH	Recombinant follicle-stimulating hormone
rFXIII	Recombinant factor XIII
rHA	Recombinant human albumin
rhCNTF	Recombinant human ciliary neuro-trophic factor
rhDNase	Recombinant human deoxyribonuclease
RhGM-CSF	Recombinant human GM-CSF
rhIL-1ra	Recombinant human interleukin-1 receptor antagonist
rhKGF	Recombinant human keratinocyte growth factor
rhMCSF	Recombinant human macrophage colony-stimulating factor
rhPDGF	Recombinant human platelet derived growth factor
rhPTH	Recombinant human parathyroid hormone
RhTNF-α	Human recombinant tumor necrosis factor-α
rIL-2	Recombinant interleukin-2
RNase A	Ribonuclease A
RRMS	Relapsing-remitting type of MS
RSV	Respiratory syncytial virus
RT-PCR	Reverse transcriptase polymerase chain reaction
SAGE	Serial analysis of gene expression
SAH	*S*-adenosylhomocystein hydrolase
SAR	Structure-activity relationship
SC	Subcutaneous (route of administration)
SCIDS	Severe combined immunodeficiency syndrome
SDS	Sodium dodecyl sulfate
SDS-PAGE	Sodium dodecyl sulfate polyacrylamide gel electrophoresis
SINEs	Short interspersed repetitive elements
SNPs	Single nucleotide polymorphisms
SOD	Superoxide dismutase
SST	Somatostatin
STAT	Signal transducers and activators of transcription
STDHF	Sodium tauro-24,25 dihydrofusidate
sTNFRs	Soluble TNF receptors

■**TABLE V.3.** *Continued*

Abbreviation	Explanation
$t_{1/2\beta}$	Half-life elimination
TAA	Tumor-associated antigens
TGF	transforming growth factor
TIGR	Institute for Genomic Research
TIMP-3	Tissue inhibitor of metalloproteinase-3
TNF	Tumor necrosis factor
TNFbp	Tumor necrosis factor binding protein
TNFR	Tumor necrosis factor receptor
TNKase	Tenecteplase
t-PA	Tissue plasminogen activator
TPO	Thrombopoietin
TRAIL/Apo2L	Tumor necrosis factor related apoptosis inducing ligand
TRH	Thyrotropin-releasing hormone
TSH	Thyrotropin stimulating hormone
TT	Tetanus toxoid
U-insulin	Ultralente insulin
uFSH	Urinary follicle-stimulating hormone
USP	United States Pharmacopoeia
UTG	UDP-glucuronosyl-teansferase
UTR	Untranslated regions
V_d	Volume of distribution
VEGF	Vesicular endothelial cell growth factors
VZV	Varicella zoster vaccine

INDEX

MONOGRAPHS INDEX

Biopharmaceutical	Trade Name	Page
INTERFERON ALFA-2A	*Roferon-A*	190
INTERFERON ALFA-2B	*Intron-A*	192
INTERFERON ALFACON-1	*Infergen*	188
INTERFERON BETA-1A	*Avonex*	193
INTERFERON BETA-1A	*Rebif*	206
INTERFERON BETA-1B	*Betaseron*	196
INTERFERON GAMMA-1B	*Actimmune*	198
LEPIRUDIN	*Refludan*	151
LEUPROLIDE	*Lupron*	234
MUROMONAB-CD3	*Orthoclone OKT3*	289
NAFARELIN	*Synarel*	232
OCTREOTIDE	*Sandostatin*	241
OPRELVEKIN	*Neumega*	142
OXYTOCIN	*Pitocin*	240
PALIVIZUMAB	*Synagis*	306
PEGADEMASE BOVINE	*Adagen*	258
PEGASPARGASE	*Oncaspar*	261
PEGFILGRASTIM	*Neulasta*	157
PEGINTERFERON ALFA-2B	*PEG-Intron*	205
RETEPLASE	*Retavase*	264
RITUXIMAB	*Rituxan*	302
SARGRAMOSTIM	*Leukine*	140
SOMATREM	*Protropin*	224
SOMATROPIN	*Humatrope, Genotropin, Norditropin, Nutropin, Saizen, Serostim*	226
TENECTEPLASE	*TNKase*	266
TRASTUZUMAB	*Herceptin*	304